[SQ]UARE

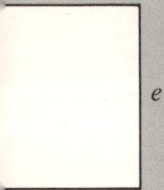

edge e
(special case of rectangle, $l = w = e$)

Area $= e^2$
Perimeter $= 4e$

RIGHT [TRIANGLE]

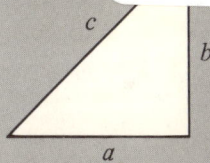

legs a, b; hypotenuse c

$$c^2 = a^2 + b^2$$

[CI]RCLE

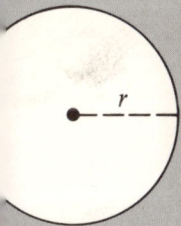

radius r

Area $= \pi r^2$
Circumference $= 2\pi r$

RECTANGULAR BOX (PARALLELEPIPED)

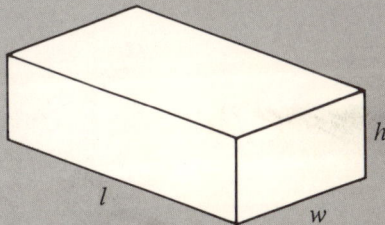

dimensions l, w, h

Volume $= lwh$
Surface area $= 2(lh + lw + wh)$

[CY]LINDER (RIGHT CIRCULAR)

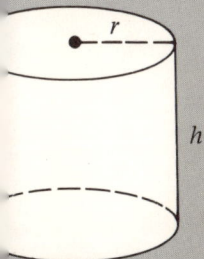

altitude h, radius of base r

Volume $= \pi r^2 n$
Lateral surface area* $= 2\pi rh$
Total surface area $= 2\pi rh + 2\pi r^2$

CONE (RIGHT CIRCULAR)

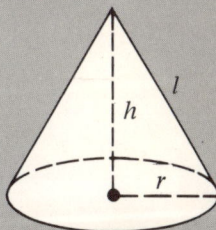

altitude h, radius of base r, slant height l

Volume $= \frac{1}{3}\pi r^2 h$
Lateral surface area* $= \pi rl$
Total surface area $= \pi rl + \pi r^2$

*Surface area of "side," not including circular base or top.

Functions and graphs

Functions
and graphs

calculus preparatory
mathematics

MALCOLM W. POWNALL
Professor of Mathematics
COLGATE UNIVERSITY, HAMILTON, NEW YORK

Prentice-Hall, Inc., Englewood Cliffs, New Jersey 07632

Library of Congress Cataloging in Publication Data

Pownall, Malcolm W.
   Functions and graphs.

   Includes Index.
   1. Functions. 2. Algebra—Graphic methods. I. Title.
QA331.P87 1983   515   82-11304
ISBN 0-13-332304-8

©1983 by Prentice-Hall, Inc., Englewood Cliffs, N.J. 07632

Editorial and production supervision by Maria McKinnon
Interior design by Maureen Eide and Maria McKinnon
Cover design by Maureen Eide
Manufacturing buyer: John Hall

ISBN 0-13-332304-8

Printed in the United States of America

10 9 8 7 6 5 4 3 2 1

Prentice-Hall International, Inc., *London*
Prentice-Hall of Australia Pty. Limited, *Sydney*
Editora Prentice-Hall do Brasil, Ltda., *Rio de Janeiro*
Prentice-Hall of Canada, Ltd., *Toronto*
Prentice-Hall of India Private Limited, *New Delhi*
Prentice-Hall of Japan, Inc., *Tokyo*
Prentice-Hall of Southeast Asia Pte. Ltd., *Singapore*
Whitehall Books Limited, *Wellington, New Zealand*

For Joe, Betsy, Kathy, and Tom

Contents

6 Complex zeros of polynomial functions 213

7 Composition and inverses 240

8 Exponential and logarithmic functions 256

9 The trigonometric functions 299

Preface

This text is intended to prepare students for a college level course in calculus. It has developed from notes prepared for a summer course at Colgate for prefreshmen enrolled in the University Scholars Program. This program includes students with diverse secondary school backgrounds, and widely differing educational objectives; a common purpose, however, has been their intention to pursue college mathematics at the calculus level.

It seems to me that good preparation for calculus demands attention on two fronts: (1) developing specific mathematical skills and knowledge that many students do not acquire in secondary school; and (2) gaining some insight and familiarity with the kinds of questions that arise in calculus. I have tried to address both these needs.

Topics from algebra, trigonometry, and coordinate geometry form the syllabus of the text, which is centered about the unifying concept of function. I have been selective in the choice of topics because most students (and instructors) are anxious to get on with calculus. I believe the material contained in this book provides very solid background for that purpose and have attempted to present these topics clearly at a level suitable for college students.

I have tried to set the stage for calculus by emphasizing functions, graphs, and word problems of the sort encountered in a traditional calculus course. Limits and continuity are done informally. I have treated rate of change, linearity, and linear interpolation in more detail than is common in a course of this type and have used interpolation repeatedly throughout the remainder of the book. Change of variables, substitution, composition and decomposition of functions are valuable tools and I have given considerable attention to them.

Inverses of functions are discussed in general and then applied in several special cases. The approach to trigonometry is analytic, as I believe it should be for calculus-bound students; those seeing trigonometry for the first time will not find this approach any more difficult, while students seeing it again are more likely to find this approach new.

The book is divided into two parts. Part I consists of material that most college students have studied in high school: the laws of algebra, integral exponents, the number line, inequalities, coordinate geometry, and the elements of graphing equations. However, for may students the initial exposure to these topics has not "taken", and so Part I offers an opportunity for review at the college level.

The main part of the text is contained in Part II, which deals with elementary functions and graphs. The instructor will find most of the standard topics included, with particular emphasis on those topics that are actually needed in calculus. Chapter 4 on functions, and Chapter 7 on composition and inverses, are especially detailed.

There is enough material in Parts I and II for 50 to 60 class periods (excluding tests and review periods). Accordingly, some omissions will have to be made in a brief course. Here are some possible shortcuts:

1. For well-prepared students, Part I may be covered quickly. This would save up to 6 class periods.

2. In Chapter 5, Sections 5-8 and 5-9 on asymptotes of rational functions could be deleted, since these topics are often included in the calculus syllabus. (The concept of an asymptote does recur in later chapters, and so the descriptive "definition" would have to be added at that time.) This omission would save about 3 class periods.

3. Chapter Six on complex zeros of polynomials is entirely optional. Its omission would save 3 or 4 class periods.

4. Many calculus courses include conic sections and polar coordinates. Omission of Section 10-5 and Chapter 11 would save 6 or 7 class periods.

I would like to thank Gertrude Pownall and Thomas Tucker, who made valuable suggestions after teaching from an early version of the manuscript. I am also indebted to the editorial staff of Prentice-Hall, Inc., and their reviewers for excellent advice and constructive criticism. I have gratefully accepted many of their helpful recommendations.

Finally I wish to express my heartfelt thanks to Lorraine Aveni who patiently and expertly typed many versions of the manuscript.

Hamilton, New York   Malcolm W. Pownall

A word to the student

The purpose of this book is to prepare you for a college level course in calculus. Calculus is one of the most important branches of mathematics, because it helps us to answer fundamental questions that arise in fields as diverse as mechanical engineering, health sciences, theoretical economics, and business adminsitration. Calculus has traditionally been a required course for many branches of science, and is often required today for entry into professional programs such as medicine and business administration.

Calculus is exciting and fun, but it is also demanding. In this book we will concentrate on reviewing and developing the knowledge and skills you will need if you are to succeed and get the most pleasure out of your calculus experience. Now and then we will glimpse ahead, to get some idea of what calculus is about, but we will not deal with calculus itself.

The main concept in our work is that of function. We will make a careful study of the linear and quadratic functions; the polynomial and rational functions; exponential, logarithmic, and trigonometric functions. You have undoubtedly encountered some of these functions in previous courses. In addition to their algebraic properties we will be strongly interested in discussing their graphs, because skill in graphing is one of the most important tools a student can bring to a calculus course. The functions we study have important applications, many of which we will study here; other applications will come later when you have the additional tools of calculus to work with.

Before we get underway with the study of functions and their graphs in Part II, we present some background material on the algebra and geometry of real numbers, and on the basics of graphing. Some of the topics covered in Part I may be familiar from high school, but a review of these items will help you to sharpen your skills. Other topics such as solving inequalities, and using symmetry to draw graphs, are less likely to be familiar but will prove to be valuable tools in this course and in calculus. I have written this book with the hope that you, the student, will be able to make good use of it in learning mathematics. I have tried to state ideas clearly and to follow them with examples that illustrate these ideas. When reading the text and studying examples you should have pencil and paper at hand so that you can follow through any details not completely carried out. Some details are left out in order to make you think actively while you read.

The book is not intended to replace your instructor—so when you have questions about a reading assignment that you cannot answer after a reasonable amount of thought and effort, be sure to ask.

Mathematics is not a spectator sport—participate! Your instructor will encourage your participation by assigning homework exercises. You should do them as carefully and completely as you can since your ability to do the exercises is the best test of whether you are learning the material. To help you know how well you are doing, answers to most odd numbered problems are in the back of of the book. Of course, you should not look at an answer until you have completed your own solution. Many students will not be able to do all of the problems assigned, and your instructor will be glad to have your ask questions after you have made a reasonable effort. Chances are that some others in the class will have the same questions.

Some of the exercises are labeled "FOR THE ENTHUSIASTS." Do not be scared by this heading. These are not necessarily harder problems. I have put them there for students who enjoy doing mathematics and have time for more than the standard fare, and for instructors who want to enrich their courses. Frequently these problems deal with more theoretical aspects of a topic, other times they simply offer interesting digressions, or perhaps they just give more detail on the material of the section. If you or your instructor are enthusiasts, please consider these problems an invitation.

Each chapter is followed by a summary and a set of review exercises. These are intended to assist you in reviewing the chapter or in preparing for tests. Warning: You should not use these as your *only* means of preparation, for no two-page summary can adequately cover the material of many pages! I suggest you *first* thoroughly review the sections of the chapter and *then* look at the summary, which lists some things you should know, and some things you should be able to do. If your initial review was thorough you should feel comfortable about most of these things and you should be able to do the review problems easily without referring back to the sections. If you find things you do not remember clearly, or problems you cannot do, then the corresponding section or sections should be thoroughly restudied.

Students in the Colgate University Program of University Scholars have studied from this text as it developed over a number of years. I have found the most conspicuous characteristic of these students to be a high degree of motivation, and I have been pleased at their diligence, enthusiasm, and success. I hope that you too will be diligent, that your experiences in this course will whet your appetite for more mathematics, and that you will be very successful!

Fundamentals

ONE

The algebra and geometry of real numbers

1

1-1 THE ALGEBRA OF REAL NUMBERS (A BRIEF REVIEW)

In this book we will do most of our work within the system of real numbers. This system includes the integers (whole numbers) 0, ± 1, ± 2, ± 3, and so on. It also includes numbers like $-8/3$, 3.651, π, $\sqrt{5}$, and $(-1 + \sqrt{5})/3$. The system of real numbers does *not* include complex or "imaginary" numbers, such as $i = \sqrt{-1}$ and $2 - 3i$.

Here we have listed for review and reference some of the algebraic laws governing the system of real numbers. You should be familiar from elementary algebra with the way in which these laws are used in working with algebraic expressions and equations.

Basic Properties of the Real Number System

The fundamental operations of addition and multiplication can always be carried out within the system of real numbers. These operations obey the following laws.

Closure Laws.
1. For addition. If a and b are any real numbers, then their **sum** $a + b$ is a real number; a and b are called **terms** in this sum.
2. For multiplication. If a and b are any real numbers, then their **product** ab is a real number; a and b are called **factors** in this product. The product ab is sometimes written $a \times b$ or $a \cdot b$.

3

The closure laws merely say that the addition of two real numbers results in a real number and the multiplication of two real numbers results in a real number.

Associative Laws.
1. For addition. If a, b, and c are any real numbers, then
$$(a + b) + c = a + (b + c).$$
2. For multiplication. If a, b, and c are any real numbers, then
$$(ab)c = a(bc).$$

As an illustration of the associative law for addition, the sum $2 + 3 + 4$ can be calculated as $(2 + 3) + 4$ or $2 + (3 + 4)$. The first method gives $5 + 4 = 9$; the second gives $2 + 7 = 9$.

Since the equation $(a + b) + c = a + (b + c)$ holds for all real numbers a, b, and c, we drop the parentheses and write $a + b + c$. In the same manner, the associative law for multiplication permits us to write abc instead of $(ab)c$ or $a(bc)$ because either method of calculating the product gives the same result.

Commutative Laws.
1. For addition. If a and b are any real numbers, then
$$a + b = b + a.$$
2. For multiplication. If a and b are any real numbers, then
$$ab = ba.$$

The commutative laws say that the order of the factors in a product, or of the terms in a sum, does not matter.

Distributive Laws.
1. If a, b, and c are any real numbers, then
$$a(b + c) = ab + ac.$$
2. If a, b, and c are any real numbers, then
$$(a + b)c = ac + bc.$$

The distributive laws enable us to simplify expressions that involve both sums and products. For example, $2(x + y)$ is the product of 2 and $x + y$; this product is the same as the sum of $2x$ and $2y$:
$$2(x + y) = 2x + 2y.$$
The distributive laws are important in deriving the equations in the section on special products and factoring that follows later in this section.

The identity and inverse laws are used (perhaps unconsciously) when solving equations. Let us look in detail at the solution of
$$2x + 3 = 9.$$
First, we add -3 to both sides.[1]
$$(2x + 3) + (-3) = 9 + (-3) = 6$$
Next, we apply the associative law for addition.
$$2x + [3 + (-3)] = 6$$
Next, we use the property of the inverse for addition, $3 + (-3) = 0$.
$$2x + 0 = 6$$
Then we use the property of the additive identity 0, $2x + 0 = 2x$.
$$2x = 6$$
To complete the solution, we multiply both sides by $\frac{1}{2}$, the reciprocal of 2,
$$\frac{1}{2}(2x) = \frac{1}{2} \cdot 6 = 3$$
and apply the associative, inverse, and identity laws for multiplication.

---

[1] It is a property of equality that if equals are added to equals, the results are equal; similarly, if equals are multiplied by equals, the results are equal.

$$\left(\frac{1}{2} \cdot 2\right)x = 3$$

$$1 \cdot x = 3$$

$$x = 3$$

So the solution is $x = 3$.

Classification of Real Numbers

The real numbers are classified as

1. positive.
2. negative.
3. zero.

Every real number belongs to one and only one of these three classes. The number 0 is neither positive nor negative but is in a class by itself.

The sum of two positive numbers is positive and so is the product. The sum of two negative numbers is negative, but the product of two negative numbers is positive. The product of a negative number with a positive number is negative. If x is positive, then $-x$ is negative; but if x is negative, then $-x$ is positive.

Properties of Zero

Zero is a special number in several respects. First, it serves as an identity for addition, as stated in the list of basic properties given earlier:

$$a + 0 = 0 + a = a$$

for every real number a. Secondly, 0 is the only real number that is neither positive nor negative. Thirdly, it can be proved from the basic properties that

$$a \cdot 0 = 0 \cdot a = 0$$

for every real number a; thus multiplication by 0 always results in 0. Finally, it can be proved from the basic properties that if a product ab is equal to 0, then at least one of the two factors must be 0.

$$\text{If } ab = 0, \text{ then } a = 0 \quad \text{or} \quad b = 0.$$

Subtraction and Division

The operations of subtraction and division are defined in terms of addition and multiplication.

Definition
Subtraction. The **difference** $a - b$ is defined by $a + (-b)$.
Division. If $b \neq 0$, then the **quotient** $a \div b$ is defined by $a \cdot (1/b)$.[2]

---

[2]The symbol $\neq$ means "not equal to."

Note that because zero has no reciprocal, the expression $a \div 0$ is meaningless—that is, *division by zero is not defined.*

We often write $a \div b$ as a fraction: a/b or $\frac{a}{b}$; a is called the **numerator** and b the **denominator** of this fraction. Zero is not permitted as a denominator.

Further Properties of Real Numbers

The following rules hold for all real numbers a, b, c, and d (except where a denominator of 0 is involved). They can be proved from the properties and definitions stated earlier.

$$-a = (-1)a$$
$$-(-a) = a$$
$$-(a + b) = (-a) + (-b)$$
$$-(a - b) = b - a$$
$$a - (-b) = a + b$$
$$(-a)(b) = -ab$$
$$(-a)(-b) = ab$$
$$\frac{a}{c} + \frac{b}{c} = \frac{a + b}{c}; \quad \frac{a}{c} + \frac{b}{d} = \frac{ad + bc}{cd}$$
$$\frac{a}{c} - \frac{b}{c} = \frac{a - b}{c}; \quad \frac{a}{c} - \frac{b}{d} = \frac{ad - bc}{cd}$$
$$\frac{a}{c} \cdot \frac{b}{d} = \frac{ab}{cd}$$
$$\frac{a}{c} \bigg/ \frac{b}{d} = \frac{a}{c} \cdot \frac{d}{b} = \frac{ad}{bc}$$

Powers

When a real number a is multiplied by itself, it is customary to write a^2, read "a-squared" or "a to the second power." Thus $a^2 = a \cdot a$.

Example 1

$$3^2 = 3 \cdot 3 = 9$$

This shorthand notation is even handier for products with more factors; a^3 denotes a product consisting of three a's:

$$a^3 = a \cdot a \cdot a \qquad \text{("a-cubed" or "a to the third power")}$$
$$a^4 = a \cdot a \cdot a \cdot a \qquad \text{("a to the fourth power")}$$

and so on. If n is a positive integer, then

$$a^n = a \cdot a \cdot \cdots \cdot a \qquad \text{(product of n a's)}.$$

The number n is called an **exponent** and a^n is read "a to the nth power." It is easy to see that, for any real number a,

$$a^m \cdot a^n = a^{m+n}$$
$$(a^m)^n = a^{mn}$$

whenever m and n are positive integers.

Example 2
$$7^3 \cdot 7^2 = (7 \cdot 7 \cdot 7) \cdot (7 \cdot 7) = 7 \cdot 7 \cdot 7 \cdot 7 \cdot 7$$
$$= 7^5$$
$$= 7^{3+2}$$

Example 3
$$(7^3)^2 = (7 \cdot 7 \cdot 7)^2 = (7 \cdot 7 \cdot 7) \cdot (7 \cdot 7 \cdot 7)$$
$$= 7^6 = 7^{3 \cdot 2}$$

If a and b are real numbers, then

$$(ab)^n = a^n b^n$$

for any positive integer n.

Example 4
$$(7 \cdot 5)^2 = (7 \cdot 5) \cdot (7 \cdot 5) = 7 \cdot 7 \cdot 5 \cdot 5$$
$$= 7^2 \cdot 5^2$$

Example 5
If x is any real number, then
$$(2x)^4 = 2^4 x^4 = 16x^4.$$

In elementary algebra you used the basic properties of real numbers and the notation of exponents to multiply expressions and combine terms as in the following examples.

Example 6
$$2x(x + 7y) = (2x) \cdot x + (2x)(7y)$$
$$= 2x^2 + 14xy$$

Example 7
$$(3x + 5y)(2x + 3y) = (3x + 5y)(2x) + (3x + 5y)(3y)$$
$$= (3x)(2x) + (5y)(2x) + (3x)(3y) + (5y)(3y)$$
$$= 6x^2 + 10xy + 9xy + 15y^2$$
$$= 6x^2 + 19xy + 15y^2$$

Example 8

$$(6 + y)(6 - y) = (6 + y)(6) + (6 + y)(-y)$$
$$= (36 + 6y) + (-6y - y^2)$$
$$= 36 + (6y - 6y) - y^2$$
$$= 36 + 0 - y^2$$
$$= 36 - y^2$$

Example 9

$$4x[(x + 2)(x - 3)] = 4x[(x + 2)x + (x + 2)(-3)]$$
$$= 4x[(x^2 + 2x) + (-3x - 6)]$$
$$= 4x[x^2 + (2x - 3x) - 6]$$
$$= 4x[x^2 - x - 6]$$
$$= 4x^3 - 4x^2 - 24x$$

It is important to realize that in applying the basic properties of real numbers, we can substitute for $a, b, c, \ldots$ any expressions that represent real numbers. Example 6, for instance, is an application of distributive law (1) with $a = 2x$, $b = x$, and $c = 7y$. The first step of Example 7 applies the same law with $a = 3x + 5y$, $b = 2x$, and $c = 3y$.

Special Products and Factoring

The following equations exhibit some important special products.

1. $(u + v)(u - v) = u^2 - v^2$
2. $(u + v)^2 = u^2 + 2uv + v^2$
3. $(u + a)(u + b) = u^2 + (a + b)u + ab$
4. $(au + b)(cu + d) = acu^2 + (bc + ad)u + bd$
5. $(u - v)(u^2 + uv + v^2) = u^3 - v^3$
6. $(u + v)(u^2 - uv + v^2) = u^3 + v^3$

The validity of each equation can be verified by multiplying out the left side and simplifying, using the basic properties. Because each equation is true for all real values of u, v, a, b, c, d, we can substitute for these variables any expressions that stand for real numbers.

The process of expressing a combination of terms as a product of several factors is known as factoring. This process is the reverse of "multiplying out" and often requires some trial and error.

Example 10

Factor $t^2 - 9$.

Solution. This expression is a difference of squares, as in Eq. (1), where $u = t$ and $v = 3$. Consequently,

$$t^2 - 9 = (t + 3)(t - 3).$$

Example 11

Factor $t^2 + 4st + 4s^2$.

Solution. We notice that the first and last terms are the squares of t and $2s$ and that the middle term is twice the product of t and $2s$. Thus by Eq. (2), with $u = t$ and $v = 2s$,

$$t^2 + 4st + 4s^2 = (t + 2s)^2.$$

Example 12

Factor $x^2 + 5x + 6$.

Solution. We let $u = x$ in (3) and try to find a and b such that $a + b = 5$ and $ab = 6$. The selection $a = 3$, $b = 2$ works and so

$$x^2 + 5x + 6 = (x + 3)(x + 2).$$

Example 13

Factor $2x^2 - 5x - 3$.

Solution. We will apply Eq. (4) to this problem. Letting $u = x$, we want the product ac of the coefficients of u to be 2 and the product of b and d to be -3. We try various combinations with the objective of making the middle coefficient on the right equal to -5. A suitable choice is $a = 2$, $b = 1$, $c = 1$, $d = -3$; so $2x^2 - 5x - 3 = (2x + 1)(x - 3)$.

Example 14

Factor $8t^3 - 27$.

Solution. Let $u = 2t$ and $v = 3$ in Eq. (5).

$$8t^3 - 27 = (2t)^3 - 3^3 = (2t - 3)[(2t)^2 + 2t \cdot 3 + 3^2]$$
$$= (2t - 3)(4t^2 + 6t + 9)$$

Example 15

Factor $\dfrac{x^3}{64} + \dfrac{y^3}{8}$.

Solution. Let $u = \dfrac{x}{4}$ and $v = \dfrac{y}{2}$. Then by Eq. (6)

$$\frac{x^3}{64} + \frac{y^3}{8} = \left(\frac{x}{4}\right)^3 + \left(\frac{y}{2}\right)^3 = \left(\frac{x}{4} + \frac{y}{2}\right)\left(\frac{x^2}{16} - \frac{xy}{8} + \frac{y^2}{4}\right).$$

Square Roots, Cube Roots

A **square root** of a real number a is a real number x such that $x^2 = a$. Every positive real number has *two* square roots, one positive and one negative. The two square roots of 36, for instance, are 6 and -6. We will denote the *positive* square root by the symbol $\sqrt{}$. Thus $\sqrt{36} = 6$; and if we wish to refer to the negative square root of 36, we simply prefix a minus sign: $-\sqrt{36} = -6$. It is important to remember that the $\sqrt{}$ sign is reserved for the nonnegative square root.

Zero has just one square root, $\sqrt{0} = 0$. Negative numbers do not have square roots within the real number system, because the square of any real number is non-negative.

For many numbers the square root can only be approximated in our decimal number system—for example, $\sqrt{2} \approx 1.414$.[3]

For nonnegative numbers a we always have

7. $(\sqrt{a})^2 = a$,

just by definition of square root.

A **cube root** of a real number a is a number x such that $x^3 = a$. *Every* real number has exactly one real cube root denoted by $\sqrt[3]{a}$. For instance, $\sqrt[3]{64} = 4$ because $4^3 = 64$; and $\sqrt[3]{-8} = -2$ because $(-2)^3 = -8$. By definition,

8. $(\sqrt[3]{a})^3 = a$

for every real number a.

The symbols $\sqrt{}$ and $\sqrt[3]{}$ are called radicals.

Example 16

Find $\sqrt{441}$.

Solution. Because $20^2 = 400$, the answer is near 20 but larger. We try 21: $21^2 = 441$. Thus $\sqrt{441} = 21$.

Example 17

Find $\sqrt{7}$.

Solution. You probably have a calculator with which you can easily find $\sqrt{7}$ by pressing two buttons. How could you find $\sqrt{7}$ without using a calculator? Trial and error is one way of estimating this number roughly. Clearly $\sqrt{7}$ is less than 3 (because $3^2 = 9$) but greater than 2 (because $2^2 = 4$). Let us try $2.5 = 5/2$. We find $(5/2)^2 = 25/4 = 6.25$; so 2.5 is too small. Next, we try 2.6; we get $(2.6)^2 = 6.76$; so 2.6 is still too small. But 2.7 is too large, for $(2.7)^2 = 7.29$. Thus $\sqrt{7}$ is between 2.6 and 2.7. Continuing this way, we find $(2.65)^2 = 7.0225$ and $(2.64)^2 = 6.9696$; so $\sqrt{7}$ is between 2.64 and 2.65. To three decimal places, $\sqrt{7}$ is given by 2.646. (A calculator gives 2.64575.)

[3]We use the symbol $\approx$ to mean "approximately equal to."

Example 18

Find $\sqrt[3]{7}$.

Solution. Using the same approach as in Example 17, we first observe that $\sqrt[3]{7}$ is between 1 and 2 and probably closer to 2 than to 1. We try 1.75: $(1.75)^3$ is approximately 5.359; so 1.75 is still too small. Next, try 1.9: $(1.9)^3$ is 6.859, again too small. Trying 1.91, we find that $(1.91)^3$ is approximately 6.968; so we are close. Because 1.92 is too big (its cube is approximately 7.078), we know that $\sqrt[3]{7}$ is between 1.91 and 1.92. (A calculator gives 1.91293.)

You learned in elementary algebra that, for nonnegative real numbers a and b,

$$9. \quad \sqrt{ab} = \sqrt{a}\sqrt{b} \quad \text{and} \quad \sqrt{\frac{a}{b}} = \frac{\sqrt{a}}{\sqrt{b}} \quad (b \neq 0)$$

whereas for all real numbers a and b

$$10. \quad \sqrt[3]{ab} = \sqrt[3]{a}\sqrt[3]{b} \quad \text{and} \quad \sqrt[3]{\frac{a}{b}} = \frac{\sqrt[3]{a}}{\sqrt[3]{b}} \quad (b \neq 0)$$

These laws help us to simplify expressions involving radicals, as the following examples show.

Example 19

Simplify $\sqrt{288}$ and evaluate.

Solution. A preliminary estimate shows that this number is between 15 and 20, probably closer to 15. By factoring the number inside the radical, we get

$$\sqrt{288} = \sqrt{144 \cdot 2} = \sqrt{144}\sqrt{2} = 12\sqrt{2}.$$

The approximate value is $12 \cdot (1.414)$ or 16.968.

Example 20

Simplify $\sqrt{72/7}$ and evaluate.

Solution.

$$\sqrt{\frac{72}{7}} = \frac{\sqrt{72}}{\sqrt{7}} = \frac{\sqrt{36 \cdot 2}}{\sqrt{7}} = \frac{6\sqrt{2}}{\sqrt{7}}$$

Using 1.414 for $\sqrt{2}$ and 2.646 for $\sqrt{7}$, we get 3.206. Another way to handle this is to multiply the numerator and denominator of $6\sqrt{2}/\sqrt{7}$ by $\sqrt{7}$, thus clearing the denominator of radicals. The result is

$$\frac{6\sqrt{2}}{\sqrt{7}} = \frac{6\sqrt{2}\sqrt{7}}{7}$$

and using our approximations as before, we get 3.207. (Do you see why this result is larger than the first one?)

Example 21

Simplify $\sqrt[3]{392}$ and evaluate.

Solution.

$$\sqrt[3]{392} = \sqrt[3]{8 \cdot 49} = \sqrt[3]{8 \cdot 7 \cdot 7}$$
$$= 2\sqrt[3]{7}\sqrt[3]{7}$$

Using the approximate value 1.913 for $\sqrt[3]{7}$, we get 7.319.

Implication

One statement **implies** another if the second statement can be deduced from the first. For instance, the statement

$$a + b = a + c$$

implies

$$b = c$$

because, given $a + b = a + c$, we can subtract a from both sides of this equation, thereby deducing $b = c$.[4]

The implication just established is an important law concerning the real number system.

Cancellation Law (Addition). If a, b, and c are real numbers, then

$$a + b = a + c \quad \text{implies} \quad b = c.$$

There is also a cancellation law for multiplication.

Cancellation Law (Multiplication). If a, b, and c are real numbers, and $a \neq 0$, then

$$ab = ac \quad \text{implies} \quad b = c.$$

Equivalent Statements

The word **equivalent** occurs frequently in mathematics. When we say that two statements are equivalent, we mean that the first statement holds **if and only if** the second statement holds; in other words, each statement implies the other. For example, consider again the equations

$$a + b = a + c$$

and

$$b = c.$$

[4]It is a property of equality that if equals are subtracted from equals, the results are equal. Similarly for the operations of addition, multiplication, and division (as long as division by zero is not involved).

We saw that $a + b = a + c$ implies $b = c$. Then given $b = c$, we can deduce that $a + b = a + c$ by adding a to both sides, so $b = c$ implies $a + b = a + c$. Each equation implies the other and hence they are equivalent.

Example 22

Show that the equations $2x = 10$ and $x = 5$ are equivalent.

Solution. Given $2x = 10$, we can divide both sides by 2 and thereby deduce $x = 5$; this proves that $2x = 10$ implies $x = 5$. Given $x = 5$, we can multiply both sides by 2 and thereby deduce $2x = 10$, proving that $x = 5$ implies $2x = 10$. This shows that the equations $2x = 10$ and $x = 5$ are equivalent.

Example 23

Are the equations $x^2 = 36$ and $x = 6$ equivalent?

Solution. From $x = 6$ we can deduce $x^2 = 36$ by squaring both sides. This proves that $x = 6$ implies $x^2 = 36$. We cannot, however, deduce $x = 6$ from $x^2 = 36$. Starting with $x^2 = 36$, we cannot conclude $x = 6$ because it might be that $x = -6$. Thus $x^2 = 36$ does not imply $x = 6$ and so the two equations are not equivalent.

EXERCISES 1-1

In Problems 1–14 evaluate the given number (e.g., $2^2 + 2^3 = 4 + 8 = 12$).

1. $\dfrac{2^3 + 2^4}{2^5}$
2. $\dfrac{2^5}{2^3 + 2^4}$
3. $2^1 + 2^2 + 2^3$

4. $1 + 7 \cdot 10^1 + 3 \cdot 10^2$
5. $4 + 8 \cdot 10^1 + 5 \cdot 10^2 + 10^3$

6. $(.1)^2$
7. $(.1)^3$
8. $(.02)^2$
9. $(.02)^3$

10. $(.02)^5$
11. $(.9)^2$
12. $(.9)^3$
13. $(.9)^4$

14. $(.9)^5$

In Problems 15–24 multiply out the given expression and simplify.

15. $(3a + b)(2a - b)$
16. $(5x + y)(x - 3y)$
17. $(\tfrac{1}{6}u + v)(3u - 2v)$

18. $4x(x + 2)(x - 3)$
19. $(1 - x)(1 - 2x)(1 - 3x)$

20. $(t^2 + 1)(t^2 + 3)$
21. $(1 - x^2)(2 + x^2)$
22. $(\tfrac{2}{3}x + y)(6x - 9y)$

23. $(1 + \sqrt{2}\,x)(1 - \sqrt{2}\,x)$
24. $(4 + \sqrt{3}\,t)(-5 - 2\sqrt{3}\,t)$

In Problems 25–42 factor the given expression.

25. $x^2 - 16$
26. $t^2 - 25$
27. $u^2 - 3uv + 2v^2$

28. $x^2 + 8xy + 7y^2$
29. $u^2 + uv - 2v^2$
30. $4x^2 - y^2$

31. $9x^2 + 9x + 2$
32. $a^2 + ab - 12b^2$
33. $6x^2 - 23x + 20$

34. $8x^2 - 22x - 6$
35. $x^2 + \dfrac{3}{2}x - 1$
36. $x^2 + \dfrac{19}{3}x + 2$

37. $\dfrac{4}{3}t^2 + \dfrac{2}{3}t - 10$
38. $t^3 - 8u^3$
39. $t^3 + 27u^3$

40. $\dfrac{u^3}{8} + v^3$
41. $\dfrac{x^3}{8} - \dfrac{y^3}{64}$
42. $\dfrac{8}{27}u^3 - \dfrac{1}{64}v^3$

43. (a) Fill in the blanks in the square root table shown without using a calculator. Estimate $\sqrt{5}$ by trial and error. To obtain $\sqrt{6}$ and $\sqrt{8}$, use equations $\sqrt{6} = \sqrt{3}\sqrt{2}$ and $\sqrt{8} = 2\sqrt{2}$.

Number	Square root
1	1.000
2	1.414
3	1.732
4	
5	
6	
7	2.646
8	
9	

†**(b)** Check your results with a calculator.

44. (a) Estimate $\sqrt[3]{9}$ to two decimal places by trial and error.
   (b) Estimate $\sqrt[3]{72}$, using your result in part (a).
†**(c)** Check your results with a calculator.

In Problems 45–68 simplify the given number, leaving your answer in terms of $\sqrt{2}$, $\sqrt{3}$, or $\sqrt{5}$, with no radicals in the denominator.

45. $\sqrt{72}$ **46.** $\sqrt{75}$ **47.** $\sqrt{16 + 4}$ **48.** $\sqrt{5^2 + 10^2}$

49. $\sqrt{243}$ **50.** $\sqrt{80}$ **51.** $\sqrt{48}$ **52.** $(2\sqrt{2})^5$

53. $(1 - 2\sqrt{2})^2$ **54.** $(10 + 2\sqrt{2})(10 - 2\sqrt{2})$ **55.** $(3 + 2\sqrt{2})(4 - \sqrt{2})$

56. $\dfrac{5}{\sqrt{2}}$ **57.** $\dfrac{7}{\sqrt{5}}$ **58.** $1 - \dfrac{1}{\sqrt{5}}$ **59.** $\sqrt{\dfrac{4}{5}}$

60. $\sqrt{\dfrac{1}{12}}$ **61.** $\sqrt{\dfrac{9}{20}}$ **62.** $\sqrt{.36}$ **63.** $\sqrt{1.8}$

64. $\sqrt{.8}$ **65.** $\sqrt{7.2}$ **66.** $\sqrt{.09}$ **67.** $\sqrt{.05}$

68. $\sqrt{.0025}$

In Problems 69–78 express the given number in terms of $\sqrt[3]{2}$ or $\sqrt[3]{3}$ with no radicals in the denominator.

69. $\sqrt[3]{81}$ **70.** $\sqrt[3]{24}$ **71.** $\sqrt[3]{-.125}$ **72.** $\sqrt[3]{250}$

73. $\sqrt[3]{\dfrac{2}{27}}$ **74.** $\sqrt[3]{-375}$ **75.** $\sqrt[3]{\dfrac{-8}{125}}$ **76.** $\sqrt[3]{.25}$

77. $\sqrt[3]{192}$ **78.** $\sqrt[3]{\dfrac{-27}{4}}$

In each of Problems 79–84 you are given a pair of equations (a) and (b). State whether they are equivalent or not. If not, determine whether either one implies the other.

79. (a) $x + 3 = 5$ **(b)** $x = 2$
80. (a) $x^2 + 3 = 5$ **(b)** $x^2 = 2$
81. (a) $x^2 + 3 = 5$ **(b)** $x = 2$

†A dagger preceding an exercise means that a calculator may be useful.

82. (a) $x^2 + 3 = 5$ (b) $x = -2$

83. (a) $3(x - 5) = 0$ (b) $x = 5$

84. (a) $3(x - 5) = 6$ (b) $x = 7$

85. Verify the cancellation law for multiplication.

1-2 INTEGRAL EXPONENTS AND THEIR APPLICATIONS

In Section 1-1 we reviewed the meaning and use of positive integral exponents. Here we look at some applications and discuss the use of zero and negative integers as exponents.

Interest

Interest is money paid by one individual or organization (such as a bank) for the use of another's money. The amount of money borrowed is called the **principal** and the ratio of the interest paid over a period of time to the principal is called the interest **rate**. The time period ordinarily used to express interest rates is one year.

> **Example 1**
>
> A college student borrows $2000 (principal) from a bank at a rate of 9%. At the end of one year the student owes the bank interest in the amount of .09 times $2000 or $180. To pay off his debt at that time, the student must pay the bank $2180.

In the case of a savings account, the bank is the borrower and pays interest to the depositor. The principal is the amount of money deposited into the account initially.

> **Example 2**
>
> You deposit $2000 in a savings account on January 1. The bank pays interest annually at a rate of 6%. One year later the bank owes you $2000 (.06) = $120 interest. So your account is then worth $2000 + $120 = $2120. Notice that the amount in your account can be thought of as
>
> $$\$2120 = \$2000(1 + .06).$$
>
> If you decide to leave this money in the account another year, you can regard $2120 as your new principal. After one more year you will have $2120 + $2120(.06) = $2120(1 + .06). This can be thought of as
>
> $$[(\$2000(1 + .06)](1 + .06)$$
>
> or $\$2000(1 + .06)^2 = \$2247.20.$
>
> At the end of 3 years the amount will be
>
> $$\$2000(1 + .06)^3 = \$2382.03$$
>
> and so on. For any positive integer n the amount in the account after n years will be
>
> $$\$2000(1 + .06)^n.$$

In Example 2 we assumed that after the first year the bank pays interest on interest previously earned. Such interest is called **compound interest**. Example 2 is a case of interest **compounded annually**.

Interest Compounded Annually. If n is a positive integer, the formula
$$A = P(1 + r)^n$$
gives the amount A in a savings account after n full years, where

$$P = \text{principal (initial investment)}$$

$$r = \text{annual interest rate}$$

assuming no withdrawals are made and that interest is compounded annually.

Example 3

A savings bank offers long-term savings accounts that earn annual interest at 8.45%, provided that you leave your principal in the account for at least 7 years. If you deposit $1000 and leave all interest on deposit, what will the account be worth at the end of 7 years?

Solution. Because $P = \$1000$, $r = .0845$, and $n = 7$, we want to calculate
$$A = 1000(1.0845)^7.$$

Using a calculator, we find $(1.0845)^7 = 1.76444$ approximately and so $A = \$1764.44$.

Example 4

Ms. Johnson invests $2500. After one year (with no withdrawals) her account is worth $2750. What is the annual interest rate on her account?

Solution. The interest after one year is $250; hence the interest rate is 10% per year.

Example 5

Mr. Johnson invests $2000. After 2 years (with no withdrawals) his account is worth $2205. What is the annual interest rate on his account?

Solution. Let r be the interest rate. Then $2205 = 2000(1 + r)^2$ and so
$$(1 + r)^2 = \frac{2205}{2000} = \frac{441}{400}.$$

This equation can be solved by taking square roots:
$$1 + r = \pm\sqrt{\frac{441}{400}} = \pm\frac{21}{20}$$
$$r = -1 \pm \frac{21}{20}.$$

Of the two solutions, one is negative and must be rejected; the solution we want is $r = 1/20$ or .05.

Example 6

Professor Jones has had a savings account for exactly 3 years at 8% per year and made no withdrawals. If his account is now worth $6613.49, how much did he invest 3 years ago?

Solution. Let P be the amount of Professor Jones' initial investment. Then $P(1.08)^3 = \$6613.49$ and so

$$P = \frac{6613.49}{(1.08)^3}.$$

Using a calculator, we find that $P = \$5250$.

Savings accounts differ widely not only in the interest rate paid but also in such details as the frequency with which interest is paid, the date on which an amount deposited begins to earn interest, and the minimum length of time that the principal must stay on deposit to earn full interest.

We will return to a discussion of interest in Chapter 8.

Population Growth

The growth of certain populations resembles the growth of money in a bank account.

Example 7

The number of cells in a population of bacteria increases because of cell division. Suppose that each cell divides into two cells after one minute. If there were 1000 cells at the start of an experiment, then after one minute there would be $1000 \times 2 = 2000$ cells, after 2 minutes there would be $(1000 \times 2) \times 2 = 1000 \times 2^2 = 4000$ cells, and after n minutes there would be 1000×2^n cells. You can verify that after 10 minutes the population will slightly exceed one million cells.

Decay

A radioactive substance decays with time. There is a fixed length of time H, called the **half-life** of the substance, such that in any time interval of length H the amount of the substance remaining at the end of the time interval is one-half the amount at the beginning of that time interval.

Example 8

Suppose that the half-life of a certain radioactive substance is 2 days.

(a) If on a certain day one gram of the substance is present, how much will be present 4 days later? $2n$ days later?

(b) How many grams decomposed between the end of the fourth day and the end of the sixth day?

Solution. (a) The amount remaining 2 days after the start is $\frac{1}{2}$ gram. After 2 more days the amount remaining is $\frac{1}{2} \cdot \frac{1}{2} = \frac{1}{4}$ gram. Thus at the end of the fourth day $\frac{1}{4}$ gram remains. At the end of the $2n$th day $(\frac{1}{2})^n$ of one gram remains.

(b) At the end of the sixth day $\frac{1}{8}$ gram remains and so $\frac{1}{4} - \frac{1}{8} = \frac{1}{8}$ gram decomposed between the fourth day and the sixth day.

Zero and Negative Integral Exponents

The following basic properties of positive integral exponents are easy to verify from the definition given in the last section.

If a and b are nonzero real numbers and if m and n are positive integers, then

1. $a^{m+n} = a^m a^n$.

2. $a^{m-n} = \dfrac{a^m}{a^n}$ if m is greater than n.

3. $a^{mn} = (a^m)^n$.

4. $(ab)^n = a^n b^n$.

5. $\left(\dfrac{a}{b}\right)^n = \dfrac{a^n}{b^n}$.

Example 9

The following equations are numerical examples of properties (1) to (5).

1. $3^{2+1} = 3^2 3^1 = 9 \cdot 3 = 27$

2. $3^{4-2} = \dfrac{3^4}{3^2} = \dfrac{81}{9} = 9$

3. $3^6 = (3^2)^3 = 9^3 = 729$

4. $(3 \cdot 2)^3 = 3^3 \cdot 2^3 = 27 \cdot 8 = 216$

5. $\left(\dfrac{2}{3}\right)^4 = \dfrac{2^4}{3^4} = \dfrac{16}{81}$

The definition of a^n can be extended in a natural way to include the case where n is zero or a negative integer, provided that $a \neq 0$. Consider $n = 0$. It is desirable to make basic property (1) hold when $n = 0$. But this means

$$a^{m+0} = a^m a^0$$
$$a^m = a^m a^0$$

and dividing both sides by a^m, we see that we should have

$$1 = a^0.$$

For this reason, a^0 is defined to be 1. It is easy to verify that (1), (2), (3), (4), and (5) now hold for all nonnegative integers. In fact, the equation in (2) even holds for such integers when $m = n$.

Now suppose that k is a negative integer, say $k = -n$, where n is positive. It is desirable to make (2) hold, and choosing $m = 0$, we see that we should have

$$a^{0-n} = \frac{a^0}{a^n} = \frac{1}{a^n}.$$

Thus a^{-n} is defined to be $1/a^n$.

Definition.
$$a^0 = 1$$
$$a^{-n} = \frac{1}{a^n}$$

With these definitions it can be proved that for $a, b \neq 0$, the equations in (1), (2), (3), (4), and (5) hold for all integers m, n and that the restriction that m be greater than n in (ii) can be eliminated entirely.

Theorem 1-1. If a and b are any real numbers different from zero and if m and n are any integers, then

1. $a^{m+n} = a^m a^n$.
2. $a^{m-n} = \dfrac{a^m}{a^n}$.
3. $(a^m)^n = a^{mn}$.
4. $(ab)^n = a^n b^n$.
5. $\left(\dfrac{a}{b}\right)^n = \dfrac{a^n}{b^n}$.

The proof of each part of Theorem 1-1 involves many cases. As an illustration, we give the proof of (3) for the case in which m is a positive integer and n a negative integer, say $n = -p$. Then

$$(a^m)^n = (a^m)^{-p} = \frac{1}{(a^m)^p}$$

by definition of negative exponents. But p is a positive integer so $(a^m)^p = a^{mp}$ and

$$(a^m)^n = \frac{1}{(a^m)^p} = \frac{1}{a^{mp}} = a^{-mp}.$$

Now $-mp = mn$ and, finally,

$$(a^m)^n = a^{-mp} = a^{mn}.$$

Example 10

$2^0 = 1$	(by definition)
$2^{-3} = \dfrac{1}{2^3} = \dfrac{1}{8}$	(by definition)
$2^{-3}2^6 = 2^{-3+6} = 2^3 = 8$	[by Theorem 1-1, part (1)]
$2^{10} = (2^5)^2 = 32^2 = 1024$	[by Theorem 1-1, part (3)]
$(.5)^{-2} = \left(\dfrac{1}{2}\right)^{-2} = (2^{-1})^{-2} = 2^2 = 4$	[by definition and Theorem 1-1, part (3)]

Example 11

If a is a nonzero real number,

$$(a^2 - 2a)(3a)^{-1} = (a^2 - 2a)\frac{1}{3a} = a^2 \cdot \frac{1}{3a} - (2a) \cdot \frac{1}{3a}$$

$$= \frac{1}{3}a - \frac{2}{3}.$$

Example 12

If a is a nonzero real number,

$$(a + a^{-1})^{-2} = \frac{1}{(a + a^{-1})^2}$$

$$= \frac{1}{a^2 + 2aa^{-1} + (a^{-1})^2}$$

$$= \frac{1}{a^2 + 2 + a^{-2}}$$

$$= \frac{a^2}{a^2} \cdot \frac{1}{a^2 + 2 + a^{-2}}$$

$$= \frac{a^2}{a^4 + 2a^2 + 1}.$$

Example 13

If x is a nonzero real number, then

$$(x^4 + 2x^{-4})^2 = (x^4)^2 + 2x^4 \cdot 2x^{-4} + (2x^{-4})^2$$

$$= x^{4 \cdot 2} + 4x^{4-4} + 4x^{-4 \cdot 2}$$

$$= x^8 + 4x^0 + 4x^{-8}$$

$$= x^8 + 4 + 4x^{-8}.$$

Scientific Notation

In our decimal system of numeration "365" stands for $300 + 60 + 5$; that is,

$$365 = 3 \times 10^2 + 6 \times 10^1 + 5 \times 10^0.$$

The digits 3, 6, and 5 are coefficients of powers of 10. Digits occurring after the decimal point are associated with negative powers of 10; for example,

$$.362 = 3 \times 10^{-1} + 6 \times 10^{-2} + 2 \times 10^{-3}$$

$$= \frac{3}{10} + \frac{6}{100} + \frac{2}{1000},$$

$$475.86 = 4 \times 10^2 + 7 \times 10^1 + 5 \times 10^0 + 8 \times 10^{-1} + 6 \times 10^{-2}$$

$$= 400 + 70 + 5 + \frac{8}{10} + \frac{6}{100}.$$

Multiplication by 10 merely shifts the decimal point one position to the right.

$$10 \times (1.362) = 10(1 \times 10^0 + 3 \times 10^{-1} + 6 \times 10^{-2} + 2 \times 10^{-3})$$

$$= 1 \times 10^1 + 3 \times 10^0 + 6 \times 10^{-1} + 2 \times 10^{-2}$$

$$= 13.62$$

Similarly,

$$10^2(1.362) = 136.2.$$

In general, multiplication by a power of 10 shifts the decimal point to the right if the exponent is positive, to the left if the exponent is negative.

Example 14

(a) $10^3 \times 5.81624 = 5816.24$

(b) $10^3 \times 481.6 = 481600$

(c) $10^{-3} \times 481.6 = .4816$

(d) $10^{-3} \times 5.81624 = .00581624$

Scientists deal with some numbers that are very large, others that are very small. An astronomer, for instance, might determine that at a certain point in its orbit the planet Pluto is 3,700,000,000 miles from the sun. A physicist might need to make a calculation that involves the diameter of a molecule of water: .000000038 centimeter. In order to deal with these cumbersome numbers, it is convenient to use **scientific notation**.

$$3,700,000,000 = 3.7 \times 10^9$$

$$.000000038 = 3.8 \times 10^{-8}$$

Each positive real number can be expressed in scientific notation as a number between 1 and 10 times a suitable power of 10. Calculations with these numbers are frequently simplified by using properties of exponents.

Example 15

The distance of the earth from the sun is approximately 93 million miles. Express this number in scientific notation.

Solution. 9.3×10^7 miles.

Example 16

The velocity of light is approximately 3×10^8 meters per second. How many seconds does it require for a ray of light to travel 6 miles? Express the answer in scientific notation and as an ordinary decimal.

Solution. There are about 1610 meters in a mile; so 6 miles is about 9660 meters. The time in seconds required for a ray of light to travel this distance is about

$$\frac{9660}{3 \times 10^8} = \frac{966 \times 10}{3 \times 10^8} = \frac{322}{10^7} = 3.22 \times 10^{-5}.$$

As an ordinary decimal, this is .0000322.

Example 17

The mass of the sun is 1.991×10^{30} kilograms whereas the mass of the earth is 5.979×10^{24} kilograms. How many times greater is the mass of the sun than the mass of the earth?

Solution. The ratio is

$$\frac{1.991 \times 10^{30}}{5.979 \times 10^{24}} = \frac{1.991}{5.979} \times \frac{10^{30}}{10^{24}} = \frac{1.991}{5.979} \times 10^6.$$

Using a calculator, the coefficient of 10^6 is approximately .333; so the ratio is $.333 \times 10^6 = 3.33 \times 10^5$ and the sun has mass approximately 333,000 times the mass of the earth.

EXERCISES 1-2

In Problems 1–13 evaluate the given expression.

1. $(2^3 + 2^4)2^{-5}$　　　　　　　　　**2.** $2^{-5}(2^{-3} + 2^{-4})$

3. $2^3(2^{-1} + 2^2)$　　　　　　　　　**4.** $(2^{-1} + 2^{-2})^{-2}$

5. $\dfrac{3^2 + 3^{-2}}{2}$　　　　　　　　　**6.** $2^{-1} + 2^{-2} + 2^{-3}$

7. $(3^2 + 3^{-2})(3^2 - 3^{-2})$　　　　　**8.** $\dfrac{3^2 + 3^{-2}}{3^2 - 3^{-2}}$

9. $10^0 + 6 \cdot 10^{-1} + 4 \cdot 10^{-2}$　　　**10.** $\left(\dfrac{1}{3}\right)^{-2}$

11. $\left(\dfrac{2}{5}\right)^{-4}$　　　　　　**12.** $(.1)^{-3}$　　　　　　**13.** $(.2)^{-2}$

In Problems 14–20 write the expression in a form that does not involve negative exponents and simplify.

14. $(8a^4 + 16a^3)(2a)^{-3}$ $(a \neq 0)$

15. $a^2(1 + a^{-1})(1 - a^{-1})$ $(a \neq 0)$

16. $(5ab)^3(a^2b^3 + a^3b^2)^{-1}$ $(a \neq 0, b \neq 0, a \neq -b)$

17. $(a^2b)^3[(a + b)^2 - (a - b)^2]^{-3}$ $(a \neq 0, b \neq 0)$.

18. $\dfrac{(6a)^{-3}}{(15a)^{-2}}$ $(a \neq 0)$

19. $[(a + b)^{-2} + (a - b)^{-2}](a^2 - b^2)^2$ $(a \neq b, a \neq -b)$

20. $(2a^{-3}b^2)^2\left(\dfrac{1}{2}a^2b^{-3}\right)^3$ $(a \neq 0, b \neq 0)$

21. Find the interest that you owe at the end of one year if you borrow \$3250 at 12%.

22. Mr. Evans borrowed \$850. A year later he repaid the bank \$956.25. What rate of interest was he charged?

†23. If you invest your money in a long-term savings account at 8.45% (see Example 3), approximately how many years does it take to double your money? to triple it? Does the amount of your original investment matter?

†24. An investor deposits \$12,500 in a savings account at 8% per year, leaving principal and interest on deposit.
   (a) What is his account worth after 3 years?
   (b) After 6 years?
   (c) After 9 years?

†25. An investor deposits \$7500 in a savings account, leaving principal and interest on deposit for 2 years. At the end of the 2-year period his account was worth \$8427. Find the annual interest rate.

†26. If a savings account is worth \$855.64 after 2 years at 8% (no withdrawals), what was the original amount invested?

27. Bacteria in a culture are reproducing in such a way that the number triples every hour. Thus if there are 1000 bacteria at the beginning of the experiment, there will be 3000 at the end of one hour. How many bacteria will there be in the culture at the end of 4 hours?

†28. According to a 1980 census, the population of a certain country was 25 million. It has been observed that the population in recent years is increasing by about 2% per year so that an estimate of the 1981 population would be 25 + (.02) 25 million = 25.5 million. Assuming that the population continues to grow at 2% per year,

(a) estimate the population of the country in 1990 and in 2000.

(b) approximately when will the population reach 50 million?

29. An automobile depreciates at a rate of 30% per year so that its value at the end of the year is 70% of its value at the beginning. Determine the value of a 1980 automobile initially costing $6500 after

(a) 1 year.

(b) 2 years.

(c) 3 years.

(d) Did the automobile lose 90% of its value after 3 years?

†30. If the cost of buying a small house is now $25,000 and the inflation rate remains at 12%, find the cost of buying such a house

(a) 1 year from now.

(b) 3 years from now.

(c) 6 years from now.

31. A "chain letter" is sent to individuals in a certain city. First, it is sent to five persons. At the end of one week each person is to send a similar letter to five other persons. At the end of another week each of these new persons is to send the letter to five others, and so on.

(a) How many letters have been sent after 3 weeks? 4 weeks? 5 weeks?

(b) Assuming nobody receives the letter more than once, how many weeks will it take before everybody in the city (population 400,000) receives the letter?

†32. (a) The half-life of the isotope Ra^{226} is about 1600 years. Express this half-life in seconds, using scientific notation.

(b) The half-life of Ra^{215} is 1.6×10^{-3} second. Express this half-life in years, using scientific notation.

(c) How many times longer is the half-life of Ra^{226} than that of Ra^{215}?

†33. The distance traveled by a ray of light in one year is called a light year. Astronomers tell us this distance is about 9.5×10^{12} kilometers. If it takes 7 million years for light to travel from a certain star to the earth, how far away is that star? Express your answer in kilometers, using scientific notation.

†34. If a cubic inch of gas contains about 4.42×10^{20} molecules, how many molecules are in a cubic foot? Express your answer in scientific notation.

FOR THE ENTHUSIASTS

35. The *Handbook of Chemistry and Physics*[5] gives data concerning the mass and the radius of each of the nine planets as shown in the table on the next page. The term following the plus or minus sign in each case indicates maximum error; thus the table tells us that the mass of Mercury is between 3.071×10^{23} and 3.291×10^{23} kilograms.

(a) Arrange the planets in order of increasing radius as best you can from the given information, taking into account the possible errors. (Is the radius of Uranus larger than that of Neptune?)

[5]*CRC Handbook of Chemistry and Physics*, CRC Press, Inc., 61st ed., 1980, Boca Raton, FL 33431.

Planet	Mass (kilograms)	Average Radius (kilometers)
Mercury	$(3.181 \pm 0.110) \times 10^{23}$	$2{,}433.0 \pm 35.0$
Venus	$(4.883 \pm 0.034) \times 10^{24}$	$6{,}053.0 \pm 7.8$
Earth	$(5.979 \pm 0.004) \times 10^{24}$	$6{,}371.315 \pm 0.437$
Mars	$(6.418 \pm 0.024) \times 10^{23}$	$3{,}380.0 \pm 20.0$
Jupiter	$(1.901 \pm 0.003) \times 10^{27}$	$69{,}758.0 \pm 139.0$
Saturn	$(5.684 \pm 0.022) \times 10^{26}$	$58{,}219.0 \pm 262.0$
Uranus	$(8.682 \pm 0.031) \times 10^{25}$	$23{,}470.0 \pm 732.0$
Neptune	$(1.027 \pm 0.014) \times 10^{26}$	$22{,}716.0 \pm 869.0$
Pluto	$(1.08 \pm 1.00) \times 10^{24}$	$5{,}700.0 \pm 413.0$

(b) Arrange the planets in order of increasing mass as best you can taking into account the possible error. (Where does Pluto fit in?)

1-3 POWERS OF A BINOMIAL

An expression consisting of a sum of two terms is called a **binomial**. Often we must deal with the positive integral powers of a binomial. In Eq. (2) of Section 1-1 we presented the formula for the second power (square) of a binomial. Using the basic properties of real numbers, we can develop the following list.

$$(u + v)^1 = u + v$$
$$(u + v)^2 = (u + v)(u + v) = (u + v)u + (u + v)v$$
$$= u^2 + 2uv + v^2$$
$$(u + v)^3 = (u + v)(u + v)(u + v) = (u + v)(u + v)^2$$
$$= u(u + v)^2 + v(u + v)^2$$
$$= u(u^2 + 2uv + v^2) + v(u^2 + 2uv + v^2)$$
$$= u^3 + 2u^2v + uv^2 + u^2v + 2uv^2 + v^3$$
$$= u^3 + 3u^2v + 3uv^2 + u^3$$
$$(u + v)^4 = (u + v)(u + v)(u + v)(u + v)$$
$$= (u + v)(u + v)^3$$
$$= u(u + v)^3 + v(u + v)^3$$
$$= u(u^3 + 3u^2v + 3uv^2 + v^3) + v(u^3 + 3u^2v + 3uv^2 + v^3)$$
$$= u^4 + 3u^3v + 3u^2v^2 + uv^3 + u^3v + 3u^2v^2 + 3uv^3 + v^4$$
$$= u^4 + 4u^3v + 6u^2v^2 + 4uv^3 + v^4$$

To obtain $(u + v)^5$, we can multiply both sides of the last equation by $u + v$; after simplifying, we get

$$(u + v)^5 = u^5 + 5u^4v + 10u^3v^2 + 10u^2v^3 + 5uv^4 + v^5.$$

Continuing in this way, we can obtain the "expansion" of $(u + v)^n$ for any positive integer n.

This pattern suggests the following properties of the expansion of $(u + v)^n$.

---

1. There are $n + 1$ terms after simplification.
2. The first and last terms are, respectively, u^n and v^n.
3. Each term consists of a coefficient times $u^{n-j}v^j$, where j is an integer from 0 to n.
4. The coefficient of $u^j v^{n-j}$ is the same as the coefficient of $u^{n-j}v^j$.
5. The coefficient of $u^{n-1}v^1$ is n.

For $n = 1, 2, 3, 4$, and 5 the coefficients in $(u + v)^n$ can be arranged in the following array, known as **Pascal's triangle**.

	$j = 0$	$j = 1$	$j = 2$	$j = 3$	$j = 4$	$j = 5$
$n = 1$	1	1				
$n = 2$	1	2	1			
$n = 3$	1	3	3	1		
$n = 4$	1	4	6	4	1	
$n = 5$	1	5	10	10	5	1

Notice that each internal entry in this array (i.e., those other than the 1's at the beginning or the end of a row) is the sum of two adjacent entries above it. It can be proved that if this pattern is continued, then the nth row will give the correct coefficients for the binomial expansion of $(u + v)^n$ for any positive integer n. The row of Pascal's triangle corresponding to $n = 6$, for example, is obtained by adding the adjacent entries in the fifth row and placing a 1 at the beginning and end.

$$1 \quad 6 \quad 15 \quad 20 \quad 15 \quad 6 \quad 1$$

Thus for $n = 6$ we have the expansion

$$(u + v)^6 = u^6 + 6u^5v + 15u^4v^2 + 20u^3v^3 + 15u^2v^4 + 6uv^5 + v^6.$$

Example 1

Expand $(x + 2y)^3$.

Solution. Let $u = x$, $v = 2y$ in the expansion of $(u + v)^3$ above. We get

$$(x + 2y)^3 = x^3 + 3x^2(2y) + 3x(2y)^2 + (2y)^3$$
$$= x^3 + 6x^2y + 12xy^2 + 8y^3.$$

The Algebra and Geometry of Real Numbers   Chap. 1

Example 2

Expand $\left(t - \dfrac{1}{t}\right)^3$.

Solution. Let $u = t$ and $v = -1/t$ in the expansion of $(u + v)^3$. The result is

$$\left(t - \frac{1}{t}\right)^3 = t^3 + 3t^2\left(-\frac{1}{t}\right) + 3t\left(-\frac{1}{t}\right)^2 + \left(-\frac{1}{t}\right)^3$$

$$= t^3 - \frac{3t^2}{t} + \frac{3t}{t^2} - \frac{1}{t^3}$$

$$= t^3 - 3t + \frac{3}{t} - \frac{1}{t^3}.$$

Using negative exponents this can be written

$$(t - t^{-1})^3 = t^3 - 3t + 3t^{-1} - t^{-3}.$$

It is valid for $t \neq 0$.

Example 3

Expand $(2t + 3)^4$.

Solution. Let $u = 2t$ and $v = 3$ in the expansion of $(u + v)^4$. We have

$$(2t + 3)^4 = (2t)^4 + 4(2t)^3 \cdot 3 + 6(2t)^2 \cdot 3^2 + 4(2t)3^3 + 3^4.$$

But $(2t)^4 = 2^4 t^4 = 16t^4$, $(2t)^3 = 8t^3$, and $(2t)^2 = 4t^2$; so

$$(2t + 3)^4 = 16t^4 + 96t^3 + 216t^2 + 216t + 81.$$

Example 4

Expand $(\sqrt{x} + \sqrt{y})^4$.

Solution. Let $u = \sqrt{x}$ and $v = \sqrt{y}$ in the expansion of $(u + v)^4$. We get

$$(\sqrt{x} + \sqrt{y})^4 = (\sqrt{x})^4 + 4(\sqrt{x})^3\sqrt{y} + 6(\sqrt{x})^2(\sqrt{y})^2$$

$$+ 4\sqrt{x}(\sqrt{y})^3 + (\sqrt{y})^4.$$

Because $(\sqrt{x})^2 = x$ and $(\sqrt{y})^2 = y$, we have $(\sqrt{x})^4 = x^2$ and $(\sqrt{y})^4 = y^2$ whereas $(\sqrt{x})^3 = x\sqrt{x}$ and $(\sqrt{y})^3 = y\sqrt{y}$. Thus

$$(\sqrt{x} + \sqrt{y})^4 = x^2 + 4x\sqrt{x}\sqrt{y} + 6xy + 4\sqrt{x}\,y\sqrt{y} + y^2.$$

This result is valid when x and y are nonnegative.

Example 5

Expand $\left(x + \dfrac{1}{x} - 1\right)^2$.

Solution. Let $u = x + \dfrac{1}{x}$ and $v = -1$ in the expansion of $(u + v)^2$. Then

$$\left(x + \frac{1}{x} - 1\right)^2 = \left(x + \frac{1}{x}\right)^2 + 2\left(x + \frac{1}{x}\right)(-1) + (-1)^2$$

$$= x^2 + 2 + \frac{1}{x^2} - 2x - \frac{2}{x} + 1$$

$$= x^2 - 2x + 3 - \frac{2}{x} + \frac{1}{x^2}.$$

Using negative exponents, we have
$$(x + x^{-1} - 1)^2 = x^2 - 2x + 3 - 2x^{-1} + x^{-2}.$$
This equation is valid for $x \neq 0$.

EXERCISES 1-3

In Problems 1–15 apply the binomial expansion.

1. $(a + 2b)^2$ **2.** $(2s - 3t)^2$ **3.** $\left(x + \dfrac{1}{x}\right)^3$

4. $\left(t - \dfrac{1}{2t}\right)^3$ **5.** $\left(3t + \dfrac{2}{t}\right)^4$

6. $(1 + \sqrt{x})^4$ where $x \geq 0$ **7.** $(\sqrt{t} - 2)^3$ where $t \geq 0$

8. $\left(\dfrac{5}{3} + x\right)^4$ **9.** $\left(\dfrac{x - 2y}{3}\right)^3$ **10.** $\left(\dfrac{x + y}{3}\right)^5$

11. $(x - y + 2)^2$ **12.** $(x^2 + y^2 + 1)^2$ **13.** $(3t + 2s - 1)^2$

14. $\left(\dfrac{x + 2}{x}\right)^3$ $(x \neq 0)$ **15.** $\left(\dfrac{1 - x}{x}\right)^4$ $(x \neq 0)$

In Problems 16–24 simplify, leaving your answer in terms of $\sqrt{2}$, $\sqrt{3}$, or $\sqrt{5}$.

16. $(3 + 2\sqrt{3})^2$ **17.** $\left(1 - \dfrac{\sqrt{3}}{2}\right)^2$ **18.** $(1 + \sqrt{3})^3$

19. $(1 - 2\sqrt{3})^3$ **20.** $(2 - \sqrt{3})^4$ **21.** $(1 + \sqrt{2})^4$

22. $(1 - \sqrt{2})^5$ **23.** $(3 + 2\sqrt{3})^5$ **24.** $(4 - 3\sqrt{2})^4$

25. Use Pascal's triangle to find the coefficients in the expansions of $(u + v)^7$ and $(u + v)^8$ and write out these expansions.

FOR THE ENTHUSIASTS

26. Verify that for any integer n greater than 1 and for any real numbers a and b,
$$a^n - b^n = (a - b)(a^{n-1} + a^{n-2}b + a^{n-3}b^2 + \cdots + ab^{n-2} + b^{n-1}).$$
(*Hint:* Multiply out the product on the right and simplify.)

27. Verify that for any *odd* integer n greater than 1 and for any real numbers a and b,
$$a^n + b^n = (a + b)(a^{n-1} - a^{n-2}b + a^{n-3}b^2 - \cdots + b^{n-1}).$$
(*Hint:* Replace b by $-b$ in Problem 26. Why does this work only if n is odd?)

In Problems 28–32 use the results of Problems 26 and 27 to factor the given expression.

28. $a^4 - b^4$ **29.** $x^4 - 16y^4$ **30.** $8t^3 + 27s^3$

31. $9 - x^4$ **32.** $u^5 + 1$

33. Verify the formula
$$1 + r + r^2 + \cdots + r^k = \frac{1 - r^{k+1}}{1 - r}$$
for $r \neq 1$ and k any positive integer.
(*Hint:* Use Problem 26.)

In Problems 34–38 evaluate the given sum, using the formula in Problem 33.

34. $1 + 2 + 4 + 8 + \cdots + 2^9$

35. $1 + \dfrac{1}{2} + \dfrac{1}{4} + \dfrac{1}{8} + \cdots + \dfrac{1}{2^9}$

36. $1 - \dfrac{1}{3} + \dfrac{1}{9} - \dfrac{1}{27} + \dfrac{1}{81} - \dfrac{1}{243}$

37. $1 + \dfrac{2}{3} + \dfrac{4}{9} + \dfrac{8}{27} + \cdots + \dfrac{64}{729}$

38. $1 - \dfrac{3}{4} + \dfrac{9}{16} - \dfrac{27}{64} + \dfrac{81}{256} - \dfrac{243}{1024}$

1-4 THE GEOMETRY OF REAL NUMBERS

A geometric way of thinking about the real numbers is made possible by assigning coordinates to the points on a line. Given a line, we first decide on a **positive direction**, then select a point O as **origin**, and choose a **unit of length**, or **scale**, for measuring distances. The point O is assigned the coordinate zero. The point one unit on the positive side of O is assigned the coordinate 1. The point two units on the positive side of O is assigned the coordinate 2 and so on. Negative numbers correspond to points on the negative side of O.

If P is any point on the line, the coordinate a of P is the directed distance from O to P. This directed distance is positive when P is on the positive side of O, negative when P is on the negative side of O. The horizontal line in Fig. 1-1 is directed positively to the right, as indicated by the arrow. The coordinate of the point P shown there is a positive number a; we estimate that a is between 3 and 4.

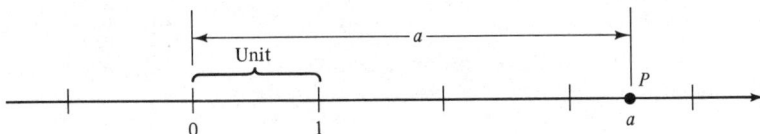

Figure 1-1

Each point on the line determines exactly one real number (its coordinate) and each real number is the coordinate of exactly one point. Because of this correspondence, we are free to think of numbers as points or of points as numbers—whichever is convenient. Thus we can refer to the point whose coordinate is 5 simply as "the point 5."

A line, together with its system of coordinates, is called a **number line**. It serves as a geometric picture of the system of all real numbers. *Ordinarily we draw a number line horizontally, choosing the positive direction to the right.* The choice of origin and scale depends on the numbers to be represented.

The points with coordinates $-3/2$, $-1/2$, $1/4$, $\sqrt{2}$, $5/3$, and π are plotted on the number line in Fig. 1-2(a). With the scale given in Fig. 1-2(a), it would not be convenient to plot larger numbers, such as 1375 or 5280; but by making the scale smaller, as in Fig. 1-2(b), we can plot these numbers conveniently.

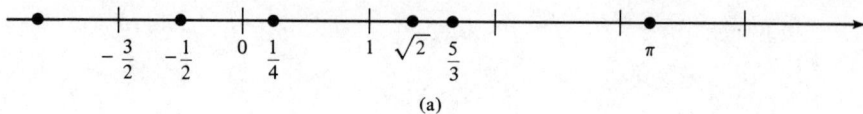

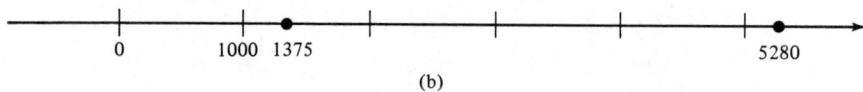

(a)

(b)

Figure 1-2

The **directed distance** from a point a to a point b is obtained by subtracting a from b: $b - a$.

Example 1

The directed distance from $2\frac{3}{4}$ to $3\frac{1}{2}$ is $3\frac{1}{2} - 2\frac{3}{4} = \frac{3}{4}$. The directed distance from $3\frac{1}{2}$ to $2\frac{3}{4}$ is $2\frac{3}{4} - 3\frac{1}{2} = -\frac{3}{4}$.

The directed distance from a to b is positive if b is on the positive side (to the right) of a; it is negative if b is on the negative side (to the left) of a.

Inequalities

We will say that a is **less than** b, written $a < b$, if b is to the right of a on the number line—that is, if the directed distance $b - a$ is a positive number.

Definition.   If a and b are real numbers, then $a < b$ means $b - a$ is positive.

Figure 1-3(a) shows points a and b such that $a < b$ whereas Fig. 1-3(b) shows points with $b < a$.

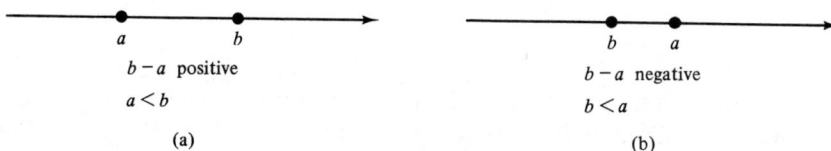

(a)

(b)

Figure 1-3

The statement $a < b$ is called an **inequality** and the symbol $<$ an **inequality sign**.

Example 2

$5 < 6$ because $6 - 5 = 1$, which is positive.

Example 3

$-4 < -2$ because $-2 - (-4) = 2$, which is positive.

Figure 1-4 shows the ordering of the numbers in Examples 2 and 3.

The inequality statement $a < b$ can also be written $b > a$, which is read "b is

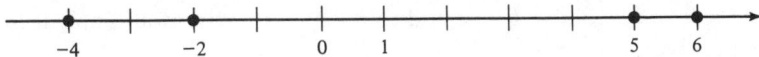

Figure 1-4

greater than a." So $b > a$ is just another way of saying that $b - a$ is positive or that b is to the right of a on the number line. The inequalities in Examples 2 and 3 could be expressed as $6 > 5$ and $-2 > -4$, respectively. The inequality signs $<$ and $>$ are said to have the "opposite sense."

The positive numbers are just those numbers that are greater than zero whereas the negative numbers are those that are less than zero. The number 0 itself is neither positive nor negative.

---

x is positive if and only if $x > 0$

x is negative if and only if $x < 0$.

Properties of Inequalities

If we compare inequalities with equations we will find some interesting similarities and differences. The "transitive" property of equality, for example, states that if $a = b$ and $b = c$, then $a = c$. There is a similar property for inequalities.

Transitive Property.   If $a < b$ and $b < c$, then $a < c$.

Proof. It is easy to understand this property geometrically on the number line. If a is to the left of b and b is to the left of c, then certainly a is to the left of c.

Here is an algebraic proof. Because $a < b$ and $b < c$, we know that $b - a$ and $c - b$ are positive. Therefore $(c - b) + (b - a)$ is also positive. But $(c - b) + (b - a) = c - a$. So $c - a$ is positive and $a < c$.

Because an inequality can be written in either of two ways (e.g., $2 < 3$ or $3 > 2$), the transitive property could just as well have been stated: "If $a > b$ and $b > c$, then $a > c$."

If the same number is added to both sides of an equation, the equality sign still holds. In fact, the new equation is equivalent to the original. The situation with inequalities in similar. The same number can be added to both sides of an inequality and the new inequality is equivalent to the original.

Addition Property.   The inequality $a < b$ is equivalent to $a + c < b + c$.

Proof. This property is easily understood geometrically if we just observe that the points $a + c$ and $b + c$ are obtained from a and b by shifting through the same directed distance c. The algebraic argument is just as easy: $a < b$ means $b - a$ is positive whereas $a + c < b + c$ means $(b + c) - (a + c)$ is positive. However,

$(b + c) - (a + c) = b - a$ and so $b - a$ is positive if and only if $(b + c) - (a + c)$ is positive.

Example 4

The inequality $1 < x - 4$ is equivalent to $5 < x$. We have added 4 to both sides.

We can subtract any number from both sides of an inequality, because subtracting c is the same as adding $-c$.

Example 5

The inequality $x + 5 < 2$ is equivalent to $x < -3$. We have subtracted 5 from both sides.

If the same number is used as a multiplier on both sides of an equation, then the equality sign still holds. With inequalities the situation is not quite the same. If the multiplier is positive, then the resulting inequality holds. If the multiplier is negative, however, then the sense of the inequality is reversed.

Multiplication Property.
1. If c is positive, then $a < b$ is equivalent to $ca < cb$.
2. If c is negative, then $a < b$ is equivalent to $ca > cb$.

The proof of the multiplication property is left as an exercise (Problem 28).

Example 6

Starting with $-4 < 6$, we can multiply both sides by 3, obtaining $-12 < 18$. This is an application of part (1) of the multiplication property because 3 is positive.

Example 7

Starting with $-4 < 6$ and multiplying both sides by the negative number -1, we can use part (2) of the multiplication property to obtain $(-1)(-4) > (-1)6$ or $4 > -6$. (Notice that failure to reverse the sense of the inequality in this case leads to the absurd result $4 < -6$.)

If d is a number different from 0, division by d is the same as multiplication by $1/d$. Hence the multiplication property enables us to divide an inequality on both sides by d. If d is positive, then its reciprocal $1/d$ is also positive and the sense of the inequality is not changed. If d is negative, however, its reciprocal is also negative and so the sense of the inequality is reversed.

Example 8

The inequality $2x > 4$ is equivalent to $x > 2$. We have divided both sides by 2.

Example 9

The inequality $-2x < 5$ is equivalent to $x > -5/2$. We have divided both sides by -2 and reversed the inequality sign.

EXERCISES 1-4

1. On a number line plot the points corresponding to the following real numbers: 0, 1, 2, 5, -4, $3\frac{1}{3}$, 11/4, 7/2, $-16/3$, $\sqrt{3}$, $(1 - \sqrt{2})/2$.

2. On a number line plot the points corresponding to the following real numbers: 200, 500, -400, 330, 270, -365, $60\sqrt{10}$, 490, -350.

3. On a number line plot the points corresponding to the following real numbers: $-.03$, $.5$, $2/3$, -1.7, $.345$, $\sqrt{2} - 1$, $\sqrt{3}$, $2 - (\sqrt{2}/2)$, $3 - \pi$.

In Problems 4–9 find the directed distance from P_1 to P_2.

4. $P_1 = 2, P_2 = 3\frac{1}{3}$
5. $P_1 = 2, P_2 = -4$
6. $P_1 = 3\frac{1}{3}, P_2 = \dfrac{11}{4}$

7. $P_1 = -\dfrac{16}{3}, P_2 = 3\frac{1}{3}$
8. $P_1 = -4, P_2 = \sqrt{3}$
9. $P_1 = \dfrac{1 - \sqrt{2}}{2}, P_2 = 0$

In Problems 10–15 insert an inequality sign, $<$ or $>$, between the given numbers in such a way as to make a true statement.

10. 3   5
11. 5   1
12. -3 4

13. -3 -4
14. $2 + \sqrt{2}$ π
15. $\sqrt{2} - 2$ $-\dfrac{3}{5}$

In Problems 16–22 insert (if possible) an inequality sign, $<$ or $>$, between the given expressions in such a way as to make the inequality true for all real values of x. If not possible, explain why.

16. $x + 4$ $x + 2$
17. $4x - 1$ $4x + 5$
18. $3 - x$ $4 - x$

19. 0 $x - 1$
20. 0 $x + 1$
21. $x + 4$ $2x + 2$

22. $4x - 1$ $3x + 5$

23. The **average** of a and b is defined to be the number $m = (a + b)/2$. Show that if $a < b$, then $a < m$ and $m < b$. Find the average of each of the following pairs of numbers. Plot a, m, and b on a number line.
   (a) $a = 2, b = 4$
   (b) $a = -2, b = 10$
   (c) $a = 0, b = 3$
   (d) $a = -6, b = -3$
   (e) $a = \frac{2}{3}, b = \frac{4}{5}$
   (f) $a = -\frac{1}{3}, b = \frac{5}{8}$

FOR THE ENTHUSIASTS

24. Show that if a and b are positive and if $a < b$, then $a^2 < b^2$.

25. Show that if a and b are positive and if $a < b$, then $1/a > 1/b$.

26. Show that $a^2 + b^2 \geq 2ab$ for all real numbers a and b.

27. Show that $a^2 + b^2 + c^2 \geq ab + ac + bc$ for all real numbers a, b, c.

28. Verify the multiplication property for inequalities.

29. (a) Describe a geometrical method of finding the exact location of the point on the number line corresponding to $\sqrt{2}$.
   (b) Describe how you could then find $\sqrt{3}$, $\sqrt{5}$, and so on.

30. Show that if p is a positive real number and if n is a positive integer greater than 1, then $(1 + p)^n > 1 + np$.

To solve an inequality like

$$2x + 3 < 9$$

means to find all values of the unknown x that satisfy the inequality—that is, that make the inequality a true statement. Using the addition and multiplication properties, we can work with an inequality much as if it were an equation. Adding -3 to both sides of the preceeding inequality, we get

$$2x < 6;$$

and multiplying by $\frac{1}{2}$, we obtain

$$x < 3.$$

This inequality is equivalent to the original. The set of all solutions, or **solution set**, consists of the shaded portion of the number line in Fig. 1-5; the number 3 itself is not included in this set, but all numbers to the left of 3 are included. We have labeled the number line with an "x"—the name of the variable being considered.

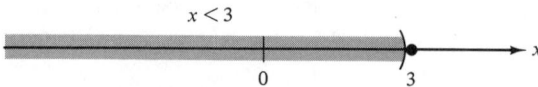

Figure 1-5

The following rules for solving inequalities are based on the properties of inequalities established in Section 1-4.

Rules for Solving Inequalities.
1. On both sides of an inequality we can add or subtract the same real number without changing the direction of the inequality. The resulting inequality is equivalent to the original.
2. On both sides of an inequality we can multiply or divide by the same *positive* real number without changing the direction of the inequality. The resulting inequality is equivalent to the original.
3. On both sides of an inequality we can multiply or divide by the same *negative* number, provided that we reverse the sense of the inequality. The resulting inequality is equivalent to the original.

Example 1

Solve $x - 3 < 5$ and sketch the solution set on a number line.

Solution. Add 3 to both sides, obtaining

$$x < 5 + 3$$

or

$$x < 8.$$

This inequality is equivalent to the original. The solution set forms the shaded portion of the number line in Fig. 1-6.

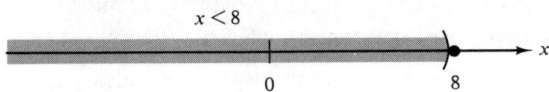

x < 8

0   8   x   **Figure 1-6**

Example 2

Solve $2 + 3x > 11$.

Solution. Subtract 2 from both sides and then divide by 3.

$$3x > 9$$
$$x > 3$$

The solution set appears in Fig. 1-7.

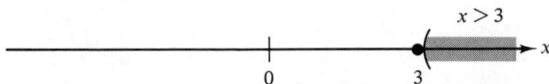

x > 3

0   3   x   **Figure 1-7**

Example 3

Solve $2 - 3t < 4 + 5t$.

Solution. Add $3t$ to both sides, subtract 4, and then divide by 8.

$$2 < 4 + 8t$$
$$-2 < 8t$$
$$-\frac{1}{4} < t$$

The solution set appears in Fig. 1-8, where the number line is labeled t.

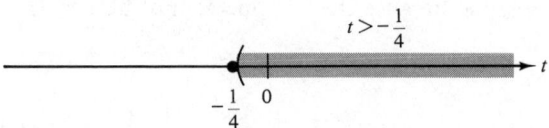

$t > -\frac{1}{4}$

$-\frac{1}{4}$ 0   t   **Figure 1-8**

Example 4

Solve $(3x - 2)/4 < (5x + 1)/3$.

Solution. To clear fractions, we first multiply both sides by 12 and then proceed as in earlier examples.

$$3(3x - 2) < 4(5x + 1)$$
$$9x - 6 < 20x + 4$$
$$-10 < 11x$$
$$-\frac{10}{11} < x$$

The solution set consists of all numbers greater than $-10/11$.

Sec. 1-5   Solving Inequalities   **35**

Example 5

A book salesman must travel from Hamilton to Albany (100 miles) in less than 3 hours. Allowing 45 minutes for lunch on the way, how fast must he travel?

Solution. Let x denote the salesman's average speed in miles per hour. His traveling time is then $100/x$ hours and so the total time for the trip is $(100/x) + \frac{3}{4}$ hours. We see that x must satisfy the inequality

$$\frac{100}{x} + \frac{3}{4} < 3.$$

This is equivalent to

$$\frac{100}{x} < \frac{9}{4}.$$

Multiplying by x and dividing by $\frac{9}{4}$ we see that

$$\frac{400}{9} < x.$$

Thus the salesman must average more than 44.4 miles per hour, approximately. (It is reasonable to assume here that x is positive. How is this assumption used in the solution?)

Some inequalities do not have any solutions.

Example 6

The square of any real number x is positive if x is positive or negative, zero if x is zero. In any event, x^2 is never negative; so the inequality $x^2 < 0$ is not true for any number x.

Example 7

The inequality $3 + x^2 + x^4 < 2$ has no solutions. It is equivalent to $x^2 + x^4 < -1$, which is clearly impossible because the left side cannot be negative.

Simultaneous Inequalities

We saw that the book salesman in Example 5 must average more than 44.4 miles per hour in order to arrive on time. Suppose, however, that the old company automobile that he is driving is safe only at speeds less than 50 miles per hour. Then his average speed x must be between 44.4 and 50 miles per hour. We express this situation with inequalities as follows:

$$44.4 < x < 50.$$

Definition. The inequality statement

$$a < x < b$$

means that $a < x$ and $x < b$. If $a < x < b$ (or if $b < x < a$), we say that x is **between** a and b.

The statement $a < x < b$ means that x satisfies *two* conditions simultaneously: $a < x$ *and* $x < b$. Accordingly, the set of all solutions of $a < x < b$ is the *intersection* of the sets shown in Fig. 1-9. The statement $a < x < b$ can also be written $b > x > a$.

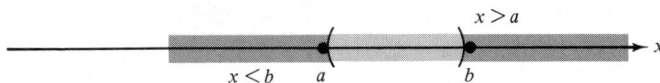

Figure 1-9

Example 8

Suppose that x must satisfy both inequalities $3 < x$ and $x < 4$ simultaneously. Then $3 < x < 4$ and x is between 3 and 4 (see Fig. 1-10).

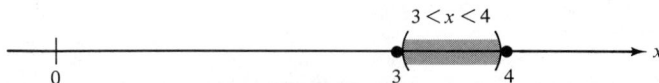

Figure 1-10

Example 9

Solve the simultaneous inequalities $2x + 3 > 0$ and $3 + 4x < 5$.

Solution. These inequalities are equivalent, respectively, to $-\frac{3}{2} < x$ and $x < \frac{1}{2}$. The solution set of the simultaneous inequalities consists of all x such that $-\frac{3}{2} < x < \frac{1}{2}$ (see Fig. 1-11).

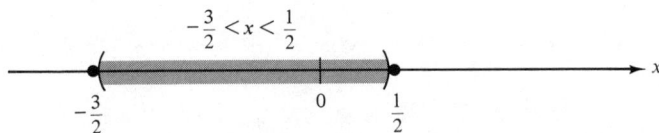

Figure 1-11

Example 10

Find all solutions of the simultaneous inequalities $5x - 3 < 0$ and $1 - 6x > 7$.

Solution. These inequalities are equivalent, respectively, to $x < \frac{3}{5}$ and $x < -1$. Because $-1 < \frac{3}{5}$, any x satisfying $x < -1$ automatically also satisfies $x < \frac{3}{5}$, as a result of the transitive property. Therefore the solution set consists of all x such that $x < -1$.

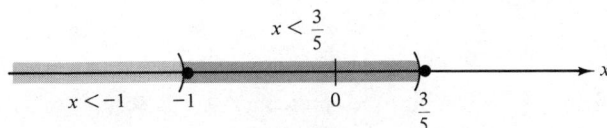

Figure 1-12

Example 11

Find all solutions of the simultaneous inequalities $3x + 4 > 7$ and $1 - 2x > 4$.

Solution. These inequalities are equivalent to $x > 1$ and $x < -\frac{3}{2}$, respectively. There are no values of x satisfying these two inequalities simultaneously.

Unions

Sometimes we will want to find those values of x that satisfy *at least one* of two inequalities.

Example 12

The set of all real numbers that satisfy at least one of the inequalities $x > 3$, $x < 2$ is the *union* of the sets shown in Fig. 1-13. It consists of all points on shaded parts of the line.

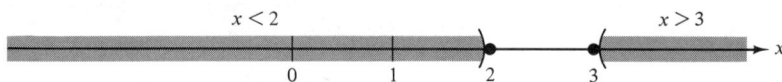

Figure 1-13

Example 13

The numbers that satisfy at least one of the inequalities $x > 2$, $x > 3$ are just those that satisfy $x > 2$. Any number that is greater than 3 is automatically also greater than 2 and so the second inequality does not give any additional solutions. The solution set appears in Fig. 1-14.

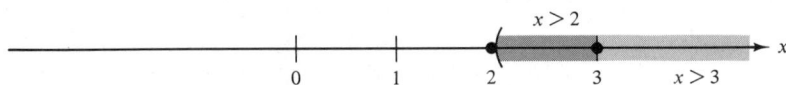

Figure 1-14

EXERCISES 1-5

In Problems 1–10 solve the given equations.

1. $x + 3 = 5$
2. $3 - 2y = 5$
3. $-1 + 2x = 5 - 2x$
4. $2 + \sqrt{2}\,x = 3 - \sqrt{2}\,x$
5. $-\sqrt{3}\,x + 4 = 1 - 2x$
6. $\frac{1}{2} - y = -\frac{2}{3}$
7. $\frac{1}{2} - \frac{x}{3} = \frac{3}{8} + \frac{2x}{3}$
8. $\frac{t+1}{2} = \frac{4-t}{3}$
9. $\frac{2}{3}y + \frac{1}{3} = 2y - 3$
10. $\frac{1}{5}x - \frac{1}{4} = \frac{13}{4} - \frac{6}{5}x$

In Problems 11–20 use the rules presented in this section to solve the given inequalities. Sketch the solution set on a number line.

11. $x + 3 < 5$
12. $3 - 2y < 5$
13. $-1 + 2x > 5 - 2x$
14. $2 + \sqrt{2}\,x > 3 - \sqrt{2}\,x$

15. $-x + 4 > 1 + 2x$

16. $\frac{1}{2} - y < -\frac{2}{3}$

17. $\frac{1}{2} - \frac{x}{3} > \frac{3}{8} + \frac{2x}{3}$

18. $\frac{t+1}{2} < \frac{4-t}{3}$

19. $\frac{2}{3}y + \frac{1}{3} < 2y - 3$

20. $\frac{1}{5}x - \frac{1}{4} > \frac{13}{4} - \frac{6}{5}x$

In Problems 21–24 find all solutions of the given inequalities if solutions exist.

21. $x^2 + 1 < 0$

22. $x(x + 1) < x(1 - x) - 4$

23. $x(x + 1) < x(x - 1)$

24. $u(u + 2) > u^2 + 3u + 4$

25. A train leaves one city at 8:00 A.M. destined for another city 365 miles away. It is due to arrive at 4:30 P.M. and must stop for 10 minutes at each of four intermediate cities. If the train averages less than 45 miles per hour while moving, can it arrive on time? Explain your conclusion.

In Problems 26–31 solve the simultaneous inequalities, and sketch the solution set on a number line.

26. $3x - 4 < 5$ and $2x > 1$

27. $3 + x < 4 - x$ and $1 + x < 3 - 2x$

28. $1 - 2y < 4$ and $3y < 0$

29. $4 + 2x > 12$ and $3x - 9 < 27 - 2x$

30. $t^2 + t + 1 < 2t + t^2$ and $3t < 2t + 5$

31. $\frac{1}{2}x + \frac{3}{5} < \frac{2}{3}x - \frac{3}{5}$ and $\frac{1}{5} - \frac{2}{3}x < \frac{2}{3}x - \frac{7}{5}$

In Problems 32–36 find all real numbers that satisfy at least one of the two inequalities and sketch on a number line.

32. $3x - 4 > 5, 2x > 1$

33. $2x < -10, 3x > 15$

34. $x < 2, x > 1$

35. $3x - 2 < x + 4, x < 0$

36. $\frac{x}{4} - 1 > \frac{3x}{8} + 2, \frac{2}{3}x + 1 > \frac{8}{3}$

37. Show that if $2 < x < 3$, then $5 < 2x + 1 < 7$.

38. Show that if $2 < x < 3$, then $-3 > 1 - 2x > -5$.

39. A temperature is recorded to be 20°C. The error in the thermometer reading is less than 0.1°C.
   (a) What inequalities does the Celsius temperature satisfy?
   (b) What inequalities does the Fahrenheit temeprature satisfy?
      (*Note:* $F = \frac{9}{5}C + 32$ is the formula that converts Celsius to Fahrenheit.)

40. The edge x of a square is recorded as 6 centimeters with an error less than .01 centimeter.
   (a) What inequalities does x satisfy?
   (b) What inequalities does the area satisfy?

1-6 INTERVALS AND ABSOLUTE VALUES

To express the fact that a is either less than or equal to b, we write $a \leq b$.

Definition. $a \leq b$ means $a < b$ or $a = b$.

The inequality statement $a \leq b$ is also written $b \geq a$, "b is greater than or equal to a."

It is easy to show that the transitive property, the addition property, and the multiplication property (Section 1-4) all hold if the symbol $<$ is replaced by $\leq$. So the rules for solving inequalities (Section 1-5) can readily be applied to situations involving $\leq$ or $\geq$.

Example 1

Solve $3x - 2 \leq 7$.

Solution. Adding 2 to both sides and then multiplying by $\frac{1}{3}$, we obtain the equivalent inequality

$$x \leq 3.$$

The solution set consists of the shaded portion of the line in Fig. 1-15.

Figure 1-15

To emphasize that the point 3 is included in the set shown in Fig. 1-15 we have used a bracket rather than a parenthesis in the diagram. A parenthesis at a point indicates the point is not included in the set, whereas a bracket indicates that it is included.

Open Intervals

An interval is a set of real numbers that can be described as a "connected" piece of the number line. There are several kinds of intervals, as we shall see.

Notation. Let a and b be real numbers. We will use the following notation.
1. $(a, +\infty)$ denotes the set of all real numbers x such that $x > a$.
2. $(-\infty, b)$ denotes the set of all real numbers x such that $x < b$.
3. If $a < b$, then (a, b) denotes the set of all real numbers x such that $a < x < b$.

Definition. A set of real numbers of one of types (1), (2), or (3) is called an **open interval**. The number a is the **left endpoint** of $(a, +\infty)$ while b is the **right endpoint** of $(-\infty, b)$. The numbers a and b are **left** and **right endpoints** of (a, b), respectively.

Example 2

The open intervals $(3, +\infty)$, $(-\infty, 4)$, and $(3, 4)$ are shown in Fig. 1-16. The interval $(3, +\infty)$ has a left endpoint 3 but no right endpoint. The interval $(-\infty, 4)$ has no left endpoint.

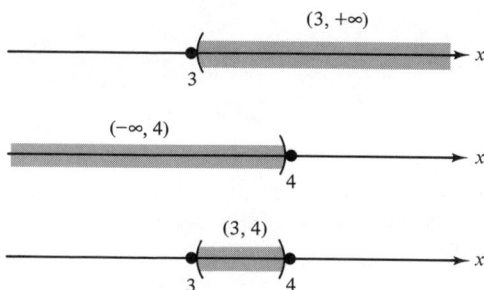

(3, +∞)

x

3

(−∞, 4)

x

4

(3, 4)

x

3   4

Figure 1-16

Example 3

The union of the open intervals (2, 4) and (3, 5) is the open interval (2, 5). The intersection of (2, 4) and (3, 5) is the open interval (3, 4). You should make a sketch verifying these statements.

Example 4

Solve the inequality $2x - 1 > 8 - x$ and express the solution set as an interval.

Solution. The given inequality is equivalent to $x > 3$ and so the solution set is the open interval $(3, +\infty)$.

Note that the endpoints of an open interval do not belong to the interval. The symbol ∞ is read "infinity"; $+\infty$ and $-\infty$ do not represent real numbers but simply tell us that the intervals extend infinitely far to the right or to the left.

Closed Intervals

An interval that does include its endpoints is called **closed**.

Notation. Let a and b be real numbers. We will use the following notation.

4. $[a, +\infty)$ denotes the set of all real numbers x such that $x \geq a$.
5. $(-\infty, b]$ denotes the set of all real numbers x such that $x \leq b$.
6. If $a \leq b$, then $[a, b]$ denotes the set of all real numbers x such that $a \leq x \leq b$.

Definition. A set of real numbers of one of the types (4), (5), or (6) is called a **closed interval**. The number a is the **left endpoint** of $[a, +\infty)$ while b is the **right endpoint** of $(-\infty, b]$. The numbers a and b are **left** and **right endpoints** of $[a, b]$, respectively.

The endpoints of a closed interval do belong to the interval. However, $+\infty$ and $-\infty$ do not stand for real numbers and so the notation $[a, +\infty]$ would not make sense.

Example 5

The closed intervals

$$[-1, +\infty)$$
$$(-\infty, 2]$$
$$[-1, 2]$$

are shown in Fig. 1-17. The use of brackets rather than parentheses indicates that the endpoints of each interval do belong to it.

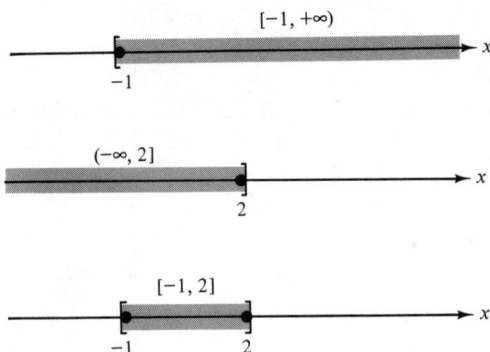

Figure 1-17

Example 6

The union of the two closed intervals $[-2, 4]$ and $[1, 5]$ is the closed interval $[-2, 5]$. The intersection of the two intervals is $[1, 4]$. Verify with a sketch.

Using the symbol $\cup$ for union and $\cap$ for intersection, we can express the results of Examples 3 and 6 as follows.

$$(2, 4) \cup (3, 5) = (2, 5)$$
$$(2, 4) \cap (3, 5) = (3, 4)$$
$$[-2, 4] \cup [1, 5] = [-2, 5]$$
$$[-2, 4] \cap [1, 5] = [1, 4]$$

Example 7

Solve the inequality $2x + 3 \leq 9$ and express the solution set as an interval.

Solution. The inequality is equivalent to

$$x \leq 3$$

and so the solution set is the closed interval $(-\infty, 3]$.

Example 8

Solve the simultaneous inequalities $2x - 3 \leq 5$ and $3 - 2x \leq 4 + x$ and express the solution set as an interval.

Solution. The first inequality is equivalent to $x \leq 4$ and the second is equivalent to $x \geq -\frac{1}{3}$. The solution set is the closed interval $[-\frac{1}{3}, 4]$.

The Algebra and Geometry of Real Numbers   Chap. 1

Other Intervals

> **Notation.** Let a and b be real numbers with $a < b$. We will use the following notation.
>
> 7. $[a, b)$ denotes the set of all real numbers x such that $a \leq x < b$.
> 8. $(a, b]$ denotes the set of all real numbers x such that $a < x \leq b$.
>
> The sets $[a, b)$ and $(a, b]$ are intervals but they are considered neither open nor closed.
>
> 9. $(-\infty, +\infty)$ denotes the set of all real numbers. The set $(-\infty, +\infty)$ is also considered an interval.

Example 9

The intervals $(-2, 3]$ and $[4, 5)$ are shown in Fig. 1-18.

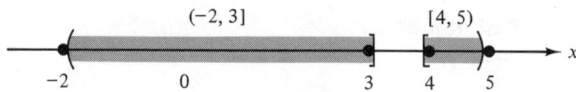

Figure 1-18

Example 10

Solve the simultaneous inequalities $2 - 3x < 8$ and $3 + x \leq 4$ and express the solution set as an interval.

Solution. Simplifying the given inequalities, we get $-2 < x$ and $x \leq 1$; so the solution set is the interval $(-2, 1]$.

Absolute Values

> **Definition.** The **absolute value** of a real number a is denoted by $|a|$ and is defined as follows.
>
> $$|a| = \begin{cases} a & \text{if } a \geq 0 \\ -a & \text{if } a < 0 \end{cases}$$

The absolute value of a nonnegative number is simply the number itself and hence is nonengative. The absolute value of a negative number has opposite sign and is therefore positive. Consequently, the absolute value of any real number is non-negative.

Example 11

$$|-3| = -(-3) = 3 \quad \text{because} \quad -3 < 0.$$

$$|3| = 3 \quad \text{because} \quad 3 > 0.$$

Example 12

Express $|2x - 3|$ without the absolute value notation.

Solution. It is easy to verify that $2x - 3 \geq 0$ for $x \geq \frac{3}{2}$, and that $2x - 3 < 0$ for $x < \frac{3}{2}$. So we have

$$|2x - 3| = \begin{cases} 2x - 3 & \text{for} \quad x \geq \frac{3}{2}, \\ -(2x - 3) = 3 - 2x & \text{for} \quad x < \frac{3}{2}. \end{cases}$$

Properties of Absolute Value.

$|a| \geq 0$ for every real number a.	(1-1)
$|ab| = |a||b|$ for all real numbers a, b.	(1-2)
$\left|\dfrac{a}{b}\right| = \dfrac{|a|}{|b|}$ for all real numbers a, b if $(b \neq 0)$.	(1-3)
$|a + b| \leq |a| + |b|$ for all real numbers a, b.	(1-4)

Although the absolute value of a product is always equal to the product of the absolute values, you should carefully note that the absolute value of a sum may be either less than or equal to the sum of the absolute values. It can be shown that the $<$ sign holds in (1-4) if and only if a and b have opposite signs.

Example 13

Some of the properties of absolute value listed are illustrated in these equations.

$$|3 \cdot (-2)| = |-6| = 6;$$
$$|3| \cdot |-2| = 3 \cdot 2 = 6; \quad \text{so} \quad |3 \cdot (-2)| = |3||-2|$$
$$|3 + 2| = |5| = 5; \quad |3| + |2| = 3 + 2 = 5; \quad \text{so} \quad |3 + 2| = |3| + |2|$$
$$|2 + (-3)| = |-1| = 1; \quad |2| + |-3| = 2 + 3 = 5; \quad \text{so} \quad |2 - 3| < |2| + |-3|$$

Example 14

It should be mentioned that $|-x| = |x|$. This result follows from (1-2) because

$$|-x| = |(-1)x| = |-1||x| = 1|x| = |x|.$$

Solution of Inequalities Involving Absolute Values

Geometrically $|b - a|$ gives the ordinary distance between a and b on the line. This observation can be helpful in solving certain inequalities.

Let a and d be fixed real numbers, $d > 0$, and consider the inequality $|x - a| < d$, where x is unknown. Now x satisfies this inequality if and only if its distance from a is less than d—that is, if and only if x is between $a - d$ and $a + d$ [Fig. 1-19(a)]. We see then that the solution set of $|x - a| < d$ is the open interval $(a - d, a + d)$.

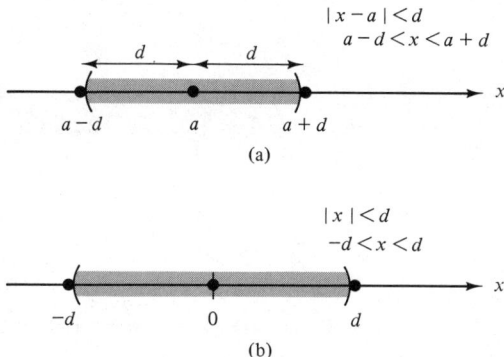

$$|x-a|<d$$
$$a-d<x<a+d$$

(a)

$$|x|<d$$
$$-d<x<d$$

(b)

Figure 1-19

The inequality ean be written without the absolute value symbol as

$$a - d < x < a + d.$$

An important special case occurs when $a = 0$. The inequality $|x| < d$ is satisfied if and only if the distance of x from the origin is less than d, that is, if and only if x is between $-d$ and d. The solution set of $|x| < d$ is the open interval $(-d, d)$ [Fig. 1-19(b)].

Example 15

(a) Solve $|x - 1| < 3$.

(b) Solve $|x| < 2$.

Solution. (a) We are looking for those points x whose distance from 1 is less than 3; these points are between $1 - 3$ and $1 + 3$. The solution set is the interval $(-2, 4)$ [Fig. 1-20(a)].

(b) We are looking for those points x whose distance from the origin is less than 2. The solution set is the interval $(-2, 2)$ [Fig. 1-20(b)].

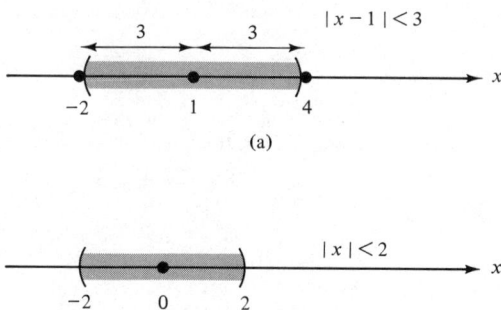

$$|x-1|<3$$

(a)

$$|x|<2$$

Figure 1-20

Example 16

Solve $|3 - x| < 1$.

Solution. This is equivalent to $|x - 3| < 1$. (Why?) The solution set is the open interval $(2, 4)$. You should make a sketch.

Example 17

Solve $|x + 4| < 2$.

Solution. Because $|x + 4| = |x - (-4)|$, the solutions include just those real numbers whose distance from -4 is less than 2. The solution set is the interval $(-6, -2)$.

Example 18

Solve $|x - 3| \leq 2$.

Solution. We want all numbers x whose distance from 3 is less than *or equal to* 2 (see Fig. 1-21). This includes all numbers from 1 to 5 inclusive; so the solution set is the closed interval $[1, 5]$.

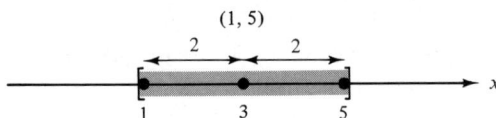

Figure 1-21

Example 19

Solve $|x - 3| > 2$.

Solution. We want all numbers x at a distance from 3 that is *greater* than 2. This is the union of the intervals $(-\infty, 1)$, $(5, +\infty)$ (see Fig. 1-22).

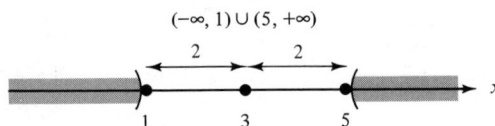

Figure 1-22

EXERCISES 1-6

In Problems 1–10 solve the given inequality and sketch the solution set on a number line.

1. $2x - 3 \leq 7$
2. $x + 1 \geq 2x - 5$
3. $\sqrt{2}\, x + 1 \leq x + \sqrt{2}$
4. $\frac{7}{3}y + \frac{2}{3} \geq -\frac{2}{3}y + 5$
5. $u + 1 \geq 2u + 3$
6. $1 + \sqrt{3}\, x > \sqrt{3} + 2x$
7. $\frac{1}{3} + \frac{3}{5}x \leq \frac{3}{5} - \frac{2}{3}x$
8. $3x + \frac{9}{4} \leq 0$
9. $9 - 6x \leq 4$
10. $\sqrt{5}\, x - 1 \geq 4 - \sqrt{5}\, x$

In Problems 11–25 sketch the interval on a number line marking any endpoint with the proper choice of bracket or parenthesis.

11. $(2, 5)$

12. $\left(-\frac{4}{3}, \frac{5}{6}\right)$

13. $(-\infty, -2)$

14. $(\sqrt{2}, +\infty)$

15. $\left[\frac{5}{8}, \frac{3}{4}\right]$

16. $[-1, \pi]$

17. $[1 - \sqrt{2}, 1 + \sqrt{2}]$

18. $[3, 3]$

19. $[3, +\infty)$

The Algebra and Geometry of Real Numbers   Chap. 1

20. $\left(-\infty, \frac{1}{2}\right]$ **21.** $[3, 7)$ **22.** $[-1, 4)$

23. $\left(0, \frac{1}{2\pi}\right]$ **24.** $(-\sqrt{2}, \sqrt{2}]$ **25.** $\left(-\frac{2}{7}, \frac{3}{10}\right]$

In Problems 26–34 find the intersection of the indicated intervals and express it in interval notation.

26. $(3, +\infty), (2, 5)$ **27.** $(-1, 0), (-2, \sqrt{2})$ **28.** $[0, \pi - 3], \left[\frac{1}{7}, 2\right]$

29. $[-\pi, \pi], [0, 2\pi]$ **30.** $(3, 5], [2, 4)$

31. $[1 - \sqrt{2}, +\infty), (-\infty, 1 + \sqrt{2})$ **32.** $\left(-\infty, \frac{1}{2}\right), \left(-\frac{1}{2}, +\infty\right)$

33. $\left(-\infty, -\frac{1}{2}\right], \left[\frac{1}{2}, +\infty\right)$ **34.** $(-\infty, +\infty), (-3, 4)$

In Problems 35–44 express the solution set in interval notation.

35. Problem 1   **36.** Problem 3   **37.** Problem 4

38. Problem 8   **39.** Problem 10   **40.** $3 - 2x < 7$

41. $\frac{1}{2}x + 5 < 3$ **42.** $1 - x < 1 + x$ **43.** $\sqrt{3} - x > 1 + 2x$

44. $5x + 3 > 12 - 3x$

In Problems 45–50 solve the simultaneous inequalities and express the solution set in interval notation.

45. $2x - 1 < x + 3$ and $3x > 0$ **46.** $2x < 1 + \sqrt{2}$ and $4 - x < 5$

47. $\frac{1}{3}y + 2 < \frac{4}{3}y - 1$ and $\frac{2}{3}y + 1 < 7$ **48.** $3u - \frac{1}{4} \geq u + \frac{7}{4}$ and $3u - 1 > 5$

49. $x + 1 \geq 2x - 5$ and $3x - 4 \leq 20$ **50.** $2x - 1 \leq 3x + 3$ and $x + 3 < \pi$

In Problems 51–54 find all real numbers that satisfy at least one of the two inequalities. Express this set as an interval or union of intervals.

51. $2x - 1 < x + 3, 3x > 0$ **52.** $2x - 1 \geq x + 3, 3x \leq 0$

53. $3x - 10 > 2 - x, 2x + 2 \leq 0$ **54.** $x + 1 > 2x - 5, 3x - 4 \geq 20$

In Problems 55–59 you are given values of a and b. Find $|ab|, |a + b|, |a| + |b|$, and $|a^2 - b^2|$.

55. $a = 3, b = 4$ **56.** $a = 3, b = -1$ **57.** $a = 2, b = -2$

58. $a = \sqrt{7}, b = \frac{3}{2}$ **59.** $a = 1 - \sqrt{2}, b = \frac{2}{3}$

In Problems 60–72 solve the given inequality or simultaneous inequalities.

60. $|x - 1| < 1$ **61.** $|x| < 2$ **62.** $|x + 4| < 5$

63. $|2x - 3| < 2$ **64.** $|1 - 3y| \leq 6$ **65.** $|\frac{1}{3} - \frac{2}{3}x| \leq \frac{5}{6}$

66. $|1 - \sqrt{2}x| < \sqrt{2}$ **67.** $|y - 1| \geq 1$ **68.** $|4x| > 5$

69. $|4x - 3| > \sqrt{32}$ **70.** $|x - 1| < 1$ and $|x - 2| < 1$

71. $|x - 1| < .5$ and $|x - 2| < .5$ **72.** $|x - 1| \leq .5$ and $|x - 2| \leq .5$

FOR THE ENTHUSIASTS

73. (a) Show that if x belongs to the open interval $(0, 1)$, then there is a number y in that interval such that $y > x$. In other words, there is no largest number in $(0, 1)$.
(b) Show that there is no smallest number in the open interval $(0, 1)$.

74. Generalize the result of Problem 73 by showing that an open interval (a, b) has no largest number and no smallest number.

1-7 INEQUALITIES OF SECOND DEGREE

Most inequalities solved so far have involved only the first power of the unknown x. In this section we consider second-degree inequalities, such as

$$x^2 - 5x + 6 > 0.$$

Consider instead the *equation*

$$x^2 - 5x + 6 = 0,$$

which we can solve by factoring the left side:

$$x^2 - 5x + 6 = (x - 3)(x - 2).$$

We know that the product $(x - 3)(x - 2)$ is zero if and only if one of the factors is zero. Hence either $x - 3 = 0$ or $x - 2 = 0$ and so the solutions of the equation are $x = 2$ and $x = 3$.

The method of factorization can also be applied to solving the inequality, which we write in the form

$$(x - 3)(x - 2) > 0.$$

We know that a product of two numbers is positive if and only if either

1. both factors are positive, or
2. both factors are negative.

The first case gives us

1. $x - 3 > 0$ and $x - 2 > 0$,

which is equivalent to $x > 3$. The second case gives us

2. $x - 3 < 0$ and $x - 2 < 0$,

which is equivalent to $x < 2$. The solution set therefore consists of all real numbers that are either less than 2 or greater than 3. This is the union of the open intervals $(-\infty, 2)$ and $(3, +\infty)$, shown in Fig. 1-13.

Example 1

Solve $x^2 + 3x + 2 < 0$.

Solution. Because $x^2 + 3x + 2 = (x + 1)(x + 2)$, the given inequality holds if and only if $x + 1$ and $x + 2$ are
1. both positive or
2. both negative.
Setting $x + 1 > 0$ and $x + 2 > 0$ gives $x > -1$. Setting $x + 1 < 0$ and $x + 2 < 0$ gives $x < -2$. The solution set is the union of $(-\infty, -2)$ and $(-1, +\infty)$ (see Fig. 1-23).

The Algebra and Geometry of Real Numbers   Chap. 1

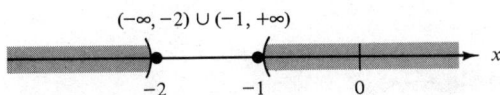

$(-\infty, -2) \cup (-1, +\infty)$

Figure 1-23

Example 2

Solve $4x^2 - 1 > 0$.

Solution. Factoring the left side, we have $4x^2 - 1 = (2x - 1)(2x + 1)$. Now this product is to be negative. A product is negative if and only if the two factors are of opposite sign. So two cases arise.
 1. $2x - 1 < 0$ and $2x + 1 > 0$
 2. $2x - 1 > 0$ and $2x + 1 < 0$
The first case gives us the intersection of $(-\infty, \frac{1}{2})$ with $(-\frac{1}{2}, +\infty)$, which is $(-\frac{1}{2}, \frac{1}{2})$. The second case gives us the intersection of $(\frac{1}{2}, +\infty)$ and $(-\infty, -\frac{1}{2})$, which is empty. The solution set is therefore just the open interval $(-\frac{1}{2}, \frac{1}{2})$ (see Fig. 1-24).

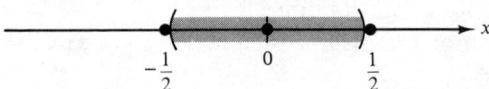

Figure 1-24

Example 3

Solve $2t^2 \leq t + 3$.

Solution. The given inequality is equivalent to $2t^2 - t - 3 \leq 0$. The left member factors: $2t^2 - t - 3 = (2t - 3)(t + 1)$. Because the product is to be less than or equal zero, we have two cases.
 1. $2t - 3 \leq 0$ and $t + 1 \geq 0$
 2. $2t - 3 \geq 0$ and $t + 1 \leq 0$
The first case yields $-1 \leq t \leq 3/2$ whereas the second case is impossible. The solution set is the closed interval $[-1, 3/2]$.

Example 4

Solve $u^2 > a^2$, where a is a positive real number.

Solution. The given inequality is equivalent to

$$u^2 - a^2 > 0$$

or
$$(u + a)(u - a) > 0.$$

It holds when
 1. $u + a > 0$ and $u - a > 0$, or
 2. $u + a < 0$ and $u - a < 0$.
The first case yields $u > a$ and the second case yields $u < -a$. The solution set of $u^2 > a^2$ is therefore $(-\infty, -a) \cup (a, +\infty)$, shown in Fig. 1-25. The two points $u = a$ and $u = -a$ are, of course, solutions of $u^2 = a^2$; and the remaining points, which satisfy $-a < u < a$, are solutions of inequality $u^2 < a^2$ (Problem 31 below).

Sec. 1-7   Inequalities of Second Degree   49

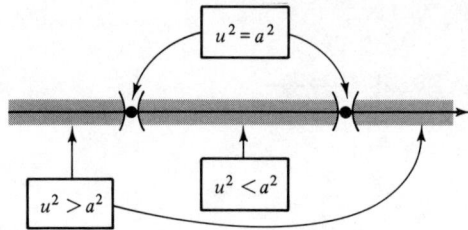

$$u^2 = a^2$$

$$u^2 > a^2 \qquad u^2 < a^2$$

Figure 1-25

You have probably already learned the usefulness of substitutions in solving equations. Consider, for example, the equation

$$x^4 - 3x^2 + 2 = 0.$$

We can substitute u for x^2. Because $u^2 = x^4$, this substitution enables us to rewrite the given equation as

$$u^2 - 3u + 2 = 0.$$

The new equation is easy to solve by factoring and we get $u = 2$ or $u = 1$. But $u = x^2$; so $x^2 = 2$ or $x^2 = 1$ and we have four solutions for x: $x = \pm\sqrt{2}$, $x = \pm 1$.

The method of substitution can also be used in solving inequalities.

Example 5

Solve $(x - 1)^2 > 9$.

Solution. Let us substitute u for $x - 1$. The resulting inequality looks simpler:

$$u^2 > 9$$

Now we can apply Example 4:

$$u < -3 \quad \text{or} \quad u > 3.$$

In terms of x,

$$x - 1 < -3 \quad \text{or} \quad x - 1 > 3.$$

Equivalently,

$$x < -2 \quad \text{or} \quad x > 4.$$

Thus the solution set is the union of $(-\infty, -2)$ and $(4, +\infty)$, shown in Fig. 1-26.

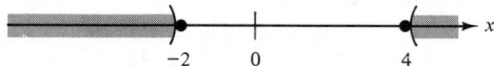

Figure 1-26

Example 6

Solve $x^4 + x^2 > 2$.

Solution. If we substitute u for x^2, the given inequality becomes

$$u^2 + u > 2$$

or

$$u^2 + u - 2 > 0$$

$$(u - 1)(u + 2) > 0.$$

The method of factoring yields the two cases:

1. $u > 1$ and $u > -2$; that is $u > 1$
2. $u < 1$ and $u < -2$; that is, $u < -2$.

In terms of x, these cases mean

1. $x^2 > 1$,
2. $x^2 < -2$.

The second case is clearly impossible. So the solution set is $(-\infty, -1) \cup (1, +\infty)$.

EXERCISES 1-7

In Problems 1–15 solve the given equations by factoring. In some cases, a substitution may help.

1. $x^2 + 2x - 15 = 0$ **2.** $x^2 = 4x + 12$ **3.** $-6y^2 + 13y = 5$

4. $12x^2 + x = 20$ **5.** $x(x + 2) = 24$ **6.** $(y + 1)(y - 1) = 3$

7. $(x^2 - 4)(x^2 - 9) = 0$ **8.** $2x^3 + x^2 = x$ **9.** $\frac{1}{2}x^2 - x - 4 = 0$

10. $4 - x = 5x^2$ **11.** $2x^2 + \frac{19}{15}x - \frac{4}{15} = 0$ **12.** $5y^2 - 9 = 0$

13. $x^4 - 5x^2 + 4 = 0$ **14.** $y^4 - 7y^2 + 12 = 0$ **15.** $2t^4 + 3t^3 - 5t^2 = 0$

In Problems 16–30, solve the given inequalities by factoring. Express the solution set as an interval or union of intervals and sketch.

16. $x^2 - x - 12 > 0$ **17.** $x^2 + x > 2$ **18.** $x^2 + 2x - 8 < 0$

19. $u^2 + 2u < 3$ **20.** $15x^2 + 14x - 8 > 0$ **21.** $12x^2 + 4x \leq 0$

22. $15y^2 - 8 \leq 14y$ **23.** $x^2 + 3x + \frac{9}{4} \geq 0$ **24.** $\frac{2}{9}x^2 - \frac{1}{3}x - 1 \geq 0$

25. $14t^2 - 17t - 6 \leq 0$ **26.** $3x^2 + x \geq 0$ **27.** $x^2 - 5x \geq -6$

28. $1 - x^2 \leq \frac{1}{4}$ **29.** $x(x + 2) \leq 24$ **30.** $4 - x \geq 5x^2$

31. Show that the solution set of $u^2 < a^2$ (where $a > 0$) is the interval $(-a, a)$.

In Problems 32–37 use the results of Example 4 and Problem 31 to solve the following inequalities.

32. $x^2 > 25$ **33.** $x^2 - 9 < 0$ **34.** $3x^2 + 1 > 13$

35. $(x + 2)^2 < 4$ **36.** $(3x - 2)^2 < 9$ **37.** $(x + 1)^2 > 5$

FOR THE ENTHUSIASTS

As you would expect, solving inequalities of degree higher than 2 gets more difficult. Here is a sample solution of a third-degree inequality. (We will return to solutions of higher-degree inequalities in Chapter 5.)

Example 7

Solve $x^3 - 16x < 0$.

Solution. The given inequality can be factored:

$$x(x^2 - 16) > 0.$$

We consider two cases.

    1. $x > 0$ and $x^2 - 16 > 0$
    2. $x < 0$ and $x^2 - 16 < 0$

1. Because the solutions of $x^2 - 16 > 0$ are $x < -4$, $x > 4$ (Example 4), we want the intersection of the interval $(0, +\infty)$ with the union of $(-\infty, -4)$ and $(4, +\infty)$. This intersection is just $(4, +\infty)$.

2. By Problem 31, the solution of $x^2 < 16$ is the interval $(-4, 4)$. So in this case we are looking for the intersection of $(-\infty, 0)$ with $(-4, 4)$. It is $(-4, 0)$.

    Therefore the solution set of the original inequality is the union of the intervals found in (1) and (2):

$$(-4, 0) \cup (4, +\infty).$$

In Problems 38–43 find all solutions of the given inequalities. Express in interval notation.

38. $x(x^2 - 4) > 0$ **39.** $x^3 - 9x \geq 0$ **40.** $16x^3 + x < 0$

41. $16x^4 - 16x^2 + 3 > 0$ **42.** $5x^2 - x^4 \leq 6$ **43.** $y^4 + y^2 \geq 20$

44. Show that the inequality $|a| < |b|$ is equivalent to $a^2 < b^2$ and that $|a| \leq |b|$ is equivalent to $a^2 \leq b^2$.

Use the properties derived in Problem 44 to solve the inequalities given in Problems 45–52 by squaring both sides.

45. $|x - 1| < |x|$ **46.** $|x + 1| < |x + 4|$ **47.** $|4x - 3| \leq |3x - 4|$

48. $|2x + 1| \leq |x - 4|$ **49.** $|\frac{4}{3}x + 1| > |\frac{2}{3}x + \frac{1}{3}|$ **50.** $|\frac{1}{5} + \frac{2}{3}x| < |\frac{1}{3} - \frac{2}{3}x|$

51. $|1 + \sqrt{2}x| \geq |1 - \sqrt{2}x|$ **52.** $|\sqrt{5} + x| \leq |\sqrt{5} - x|$

SUMMARY

In Chapter 1 you should have reviewed

1. the basic algebraic properties of real numbers and the meaning and properties of integral exponents.

2. the way we represent the real numbers as points on a number line.

You should have learned

3. applications of exponents to problems involving interest, population growth, decay, and scientific notation.

4. to expand a binomial to the nth power.

5. the meaning and properties of the inequality symbols $<, >, \leq, \geq$.

6. the meaning and properties of absolute value.

7. the notation for various kinds of intervals and their representation on a number line.

You should be able to solve an inequality of first or second degree, express the solution set as an interval or union of intervals, and sketch the solution set on a number line.

You should recognize $|x - a|$ as the distance between x and a and be able to use it to find geometric solutions of inequalities involving $|x - a|$.

REVIEW EXERCISES

In Problems 1–10 multiply the given expressions and simplify.

1. $(a + 2)(a - 3)$
2. $(t + 1)(t - 1)$
3. $(2a + 3b)(a - 4b)$
4. $(3x + 2)(4x - 5)$
5. $(\frac{2}{3}x - 5)(\frac{3}{4}x + 1)$
6. $(\sqrt{5} - 3x)(\sqrt{5} + 3x)$
7. $(t + 3)(t + 2)(t - 3)$
8. $(3t^2 - 2t + 1)(2t - 5)(3t + 4)$
9. $(a^2 - b^2)(a^4 + a^2b^2 + b^4)$
10. $(x^2 + 2y^2)(x^4 - 2x^2y^2 + 4y^4)$

In Problems 11–23 use the binomial expansion to evaluate.

11. $(3y + 2)^2$
12. $(1 - 2t)^2$
13. $(5x + 2y)^3$
14. $(x^2y^{-1} + xy)^2$
15. $(a - 2b^{-2})^3$
16. $(\sqrt{3} + 4t)^3$
17. $\left(1 + \frac{x}{2}\right)^4$
18. $(2x - y)^4$
19. $(2\sqrt{2}t + 2\sqrt{2}t^2)^2$
20. $(x + y + 1)^2$ (*Hint:* First let $u = x + y$.)
21. $(ax + by + c)^2$
22. $(ab^{-2} + a^{-2}b)^3(ab)^3$
23. $(t^2 + 3t^{-1})^2(t^2 - 3t^{-1})^2$

In Problems 24–36 factor the given expression.

24. $9a^2 - b^2$
25. $8p^2q - 16p^3$
26. $5abc + 10bcd$
27. $16x^{-2} - 4y^{-2}$
28. $x^2 + 4xy + 4y^2$
29. $7uv + v^2 + 10u^2$
30. $u^{-2} - 7u^{-1} + 12$
31. $10x^2 + 7x + 1$
32. $x^3 - 3x^2 + 2x$
33. $8a^3 - 27b^3$
34. $p^{-3} + 125q^{-3}$
35. $a^4 - 16b^4$
36. $\frac{1}{16}y^4 - \frac{1}{81}x^2$

In Problems 37 and 38 express the number as an ordinary decimal.

37. 4.68×10^9
38. 4.68×10^{-9}

In Problems 39–42 express the number in scientific notation.

†39. $(4.68 \times 10^{-9}) \times (2.34 \times 10^7)$
†40. $(4.68 \times 10^{-9}) \div (2.34 \times 10^7)$
†41. $(4.68 \times 10^{-9}) \div (2.34 \times 10^{-7})$
†42. $(4.68 \times 10^{-9})^3$

In Problems 43–54 solve the given inequality, express the solution in interval form, and sketch on a number line.

43. $2x - 3 > 5x + 6$
44. $x + 1 < 2x - 3$
45. $y^2 + 3y + 2 > 0$
46. $4x^2 + 15 \le 16x$
47. $|x - 2| < 5$
48. $|x + 1| < 3$
49. $\frac{1}{2}y + 3 \ge \frac{3}{2}y - 3$
50. $1 - x \le 1 + x$
51. $6y^2 - y \le 1$
52. $4y^2 \ge 9$
53. $|3 - x| \le 4$
54. $|2x + 3| > 6$

In Problems 55–58 solve the simultaneous inequalities, express the solutions in interval form, and sketch on a number line.

55. $3x > 7$ and $2x < 9$

56. $4y - 1 \le 5y - 6$ and $3y - 2 < 19$

57. $\frac{x}{3} - 1 \ge \frac{2x}{3} + 1$ and $1 - x < 11$

58. $\frac{x+1}{2} \le \frac{x+1}{3}$ and $\frac{2x+1}{2} \ge \frac{2x-1}{3}$

In Problems 59–64 sketch the interval on a number line. Which intervals are closed? Which are open?

59. $(-1, 4)$ **60.** $(-1, 3]$ **61.** $(-\infty, 3)$

62. $[1, 4]$ **63.** $(\pi - 3, \pi + 3]$ **64.** $[1 - \sqrt{2}, +\infty)$

In Problems 65–69 determine whether the statement is true or false.

65. $|ab| = |a| b$ for all real numbers a, b.

66. $|a + b| < a + |b|$ for all real numbers a, b.

67. $|a - b| < |a| - |b|$ for all real numbers a, b.

68. $|a^2| = a^2$ for all real numbers a.

69. $\sqrt{a^2} = |a|$ for all real numbers a.

70. Find the error in the following "solutions" of inequalities.

    (a) $x(x + 1) > x(2x - 4)$
$$x + 1 > 2x - 4$$
$$5 > x$$

    (b) $(x + 2)(2x + 3) \le (x + 2)(x - 2)$
$$2x + 3 \le x - 2$$
$$x \le -5$$

71. A truck driver is required to drive his rig a distance of 360 miles in less than 8 hours. If he obeys the speed limit of 55 mph and arrives on time, describe his average speed with inequalities.

†72. A spherical tank, initially full of water, is drained. The volume of the water removed is found to be 36π cubic feet. As much as 1% of the water, however, may have evaporated before the volume was determined. What inequalities does the radius of the tank satisfy? (Recall that the volume of a sphere is $\frac{4}{3}\pi$ times the radius cubed.)

73. Show that if a and b are positive real numbers, then their **geometric mean** $\sqrt{ab}$ is less than or equal to their average $\frac{1}{2}(a + b)$.

74. Show that the sum of any positive real number and its reciprocal is at least 2.

Basic coordinate geometry

2-1 RECTANGULAR COORDINATES

In a given plane we choose a fixed point O and construct two perpendicular number lines, one horizontal (usually called the x axis), the other vertical (usually called the y axis). Each line has its origin at O. The positive direction on the x axis is to the right. The positive direction on the y axis is upward.

We shall give each point in the plane a pair of coordinates as follows. (See Fig. 2-1.) If P is a point, let x be the directed distance from O to the foot P' of the perpendicular from P to the x axis. This directed distance is positive if P' is to the right of O, negative if P' is to the left of O. Then let y be the directed distance from O to the foot P'' of the perpendicular from P to the y axis. This directed distance is positive if P'' is above O, negative if P'' is below O. The numbers x and y are called, respectively, the **abscissa** (or x coordinate) and the **ordinate** (or y coordinate) of P. In this way, each point P determines an ordered pair (x, y) of real numbers. Conversely, each

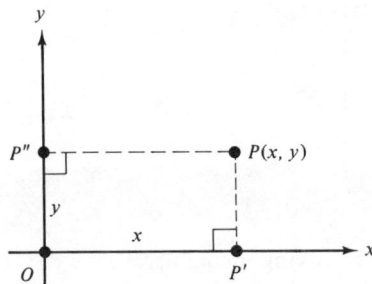

Figure 2-1

ordered pair (x, y) of real numbers determines a point $P(x, y)$ having x and y as its coordinates. We seldom bother to distinguish between a point and the corresponding pair of real numbers—for example, we say "the point $(1, 2)$" rather than "the point whose coordinates are $(1, 2)$."

The system of coordinates just described is known as a **rectangular coordinate system**.

Several points are plotted in Fig. 2-2. Compare the location of $(1, 2)$ with that of $(2, 1)$. Unless $a = b$, the ordered pairs (a, b) and (b, a) give different points; so the order in which the coordinates of a point are named is important.

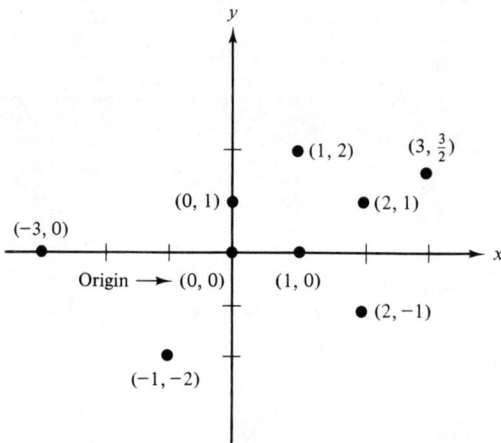

Figure 2-2

Example 1

The upper right quarter of the plane, Q_I (shaded in Fig. 2-3), is the set of points (x, y) for which x and y are both positive; Q_I is called the "first quadrant." There are three other quadrants.

Q_{II} is the set of points (x, y), where $x < 0$ and $y > 0$.
Q_{III} is the set of points (x, y), where $x < 0$ and $y < 0$.
Q_{IV} is the set of points (x, y), where $x > 0$ and $y < 0$.

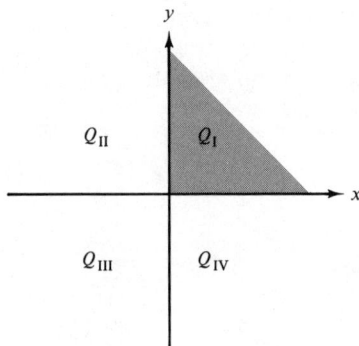

Figure 2-3

Example 2

The set of points (x, y) satisfying inequality $1 < x < 2$ is shaded in Fig. 2-4.

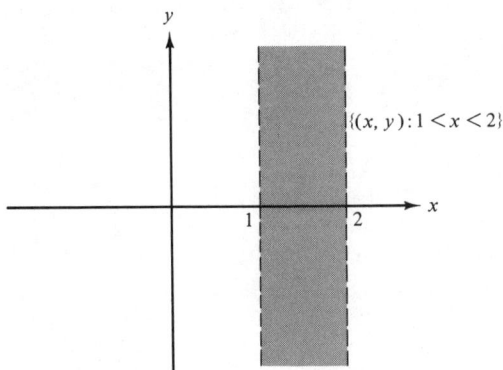

{(x, y): 1 < x < 2}

Figure 2-4

Example 3

The set of points (x, y) satisfying the simultaneous inequalities $1 < x < 2$ and $\frac{1}{2} \le y \le 1$ is shaded in Fig. 2-5. Points on the solid lines are included in the set whereas points on the dashed lines are not.

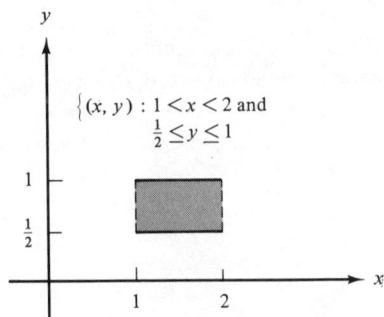

$\{(x, y) : 1 < x < 2 \text{ and } \frac{1}{2} \le y \le 1\}$

Figure 2-5

In this discussion we have chosen x and y to label our coordinate axes. Although we will usually continue to do so, it is sometimes natural or desirable to use other letters for this purpose.

Example 4

In a physics experiment you are asked to measure the length l of a spring when various weights w are attached. Suppose that your results are given in Table 2-1.

TABLE 2-1

w (grams)	l (centimeters)
1	7.2
2	11.5
3	15.7
4	20.0
5	24.2

These points (w, l) are plotted in a coordinate system (Fig. 2-6) in which the horizontal axis is labeled w and the vertical axis l.

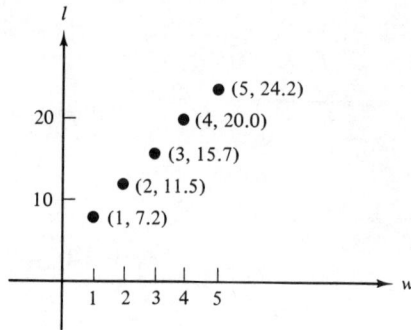

Figure 2-6

Slope of a Line Segment

Let $P_1(x_1, y_1)$ and $P_2(x_2, y_2)$ be two points in a coordinate plane. Consider the directed line segment *from P_1 to P_2*, which we denote by P_1P_2 (see Fig. 2-7). From P_1 to P_2 the x coordinate has changed from x_1 to x_2, a change of $x_2 - x_1$. The y coordinate has undergone a change of $y_2 - y_1$. We denote these changes by Δx and Δy, in which the Greek letter Δ (delta) stands for "difference" or "change."

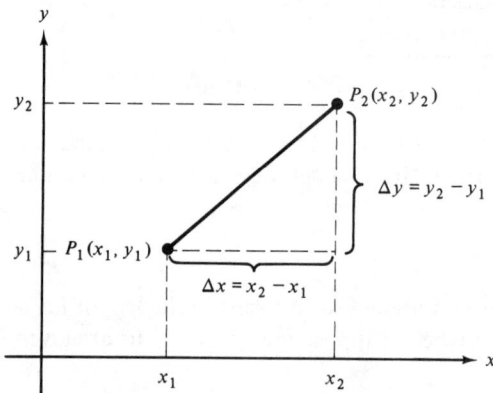

Figure 2-7

Notation.

$\Delta x = x_2 - x_1$ is the change in x from $P_1(x_1, y_1)$ to $P_2(x_2, y_2)$.

$\Delta y = y_2 - y_1$ is the change in y from $P_1(x_1, y_1)$ to $P_2(x_2, y_2)$.

If Δx is not zero, the ratio of Δy to Δx is called the **slope** of P_1P_2 and is often denoted by the letter m.

Basic Coordinate Geometry   Chap. 2

The diagrams in Fig. 2-8 show various arrangements of P_1 and P_2; in each case, Δy, Δx, and m are calculated. You should verify the following observations.

1. A positive slope occurs when Δx and Δy are both positive, as in Fig. 2-8(a) and (b). An increase in x produces an increase in y. The segment P_1P_2 rises to the right but more slowly in (a), where the slope is smaller.

2. In Fig. 2-8(c) Δx and Δy are both negative, but the slope is positive.

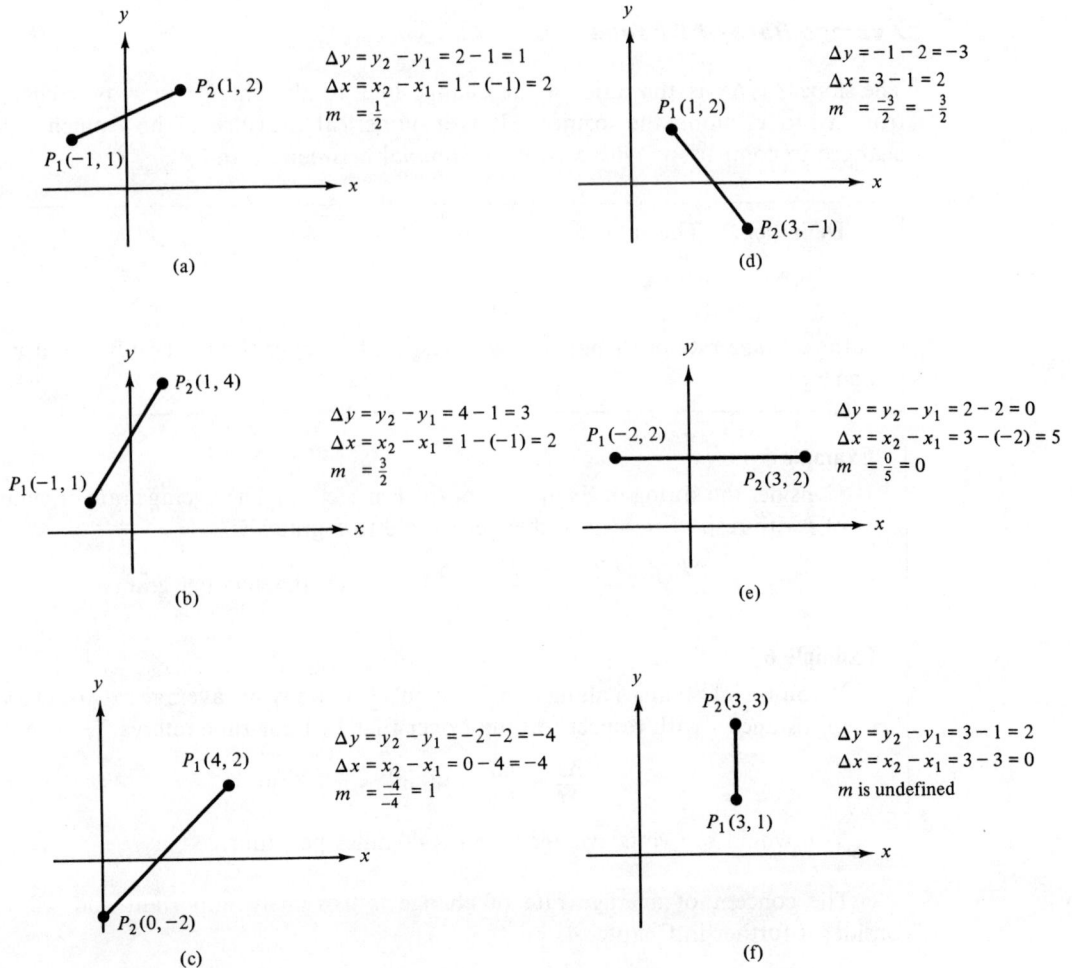

$\Delta y = y_2 - y_1 = 2 - 1 = 1$
$\Delta x = x_2 - x_1 = 1 - (-1) = 2$
$m = \frac{1}{2}$

(a)

$\Delta y = -1 - 2 = -3$
$\Delta x = 3 - 1 = 2$
$m = \frac{-3}{2} = -\frac{3}{2}$

(d)

$\Delta y = y_2 - y_1 = 4 - 1 = 3$
$\Delta x = x_2 - x_1 = 1 - (-1) = 2$
$m = \frac{3}{2}$

(b)

$\Delta y = y_2 - y_1 = 2 - 2 = 0$
$\Delta x = x_2 - x_1 = 3 - (-2) = 5$
$m = \frac{0}{5} = 0$

(e)

$\Delta y = y_2 - y_1 = -2 - 2 = -4$
$\Delta x = x_2 - x_1 = 0 - 4 = -4$
$m = \frac{-4}{-4} = 1$

(c)

$\Delta y = y_2 - y_1 = 3 - 1 = 2$
$\Delta x = x_2 - x_1 = 3 - 3 = 0$
m is undefined

(f)

Figure 2-8

From (1) and (2) we note that a positive slope occurs when Δx and Δy have the same sign.

3. A negative slope occurs when Δy and Δx have opposite signs [Fig. 2-8(d)]; an increase in x produces a decrease in y. The segment P_1P_2 falls to the right.
4. The slope is zero when $\Delta y = 0$—that is, when y does not change at all; in this case [Fig. 2-8(e)], the segment P_1P_2 is horizontal.
5. The slope of P_2P_1 is the same as the slope of P_1P_2. In calculating a slope, the order in which the points are taken makes no difference (although in calculating Δx and Δy, the order does make a difference).
6. Figure 2-8(f) shows a vertical line segment, for which the slope is undefined. Here $x_1 = x_2 = 3$ so that $\Delta x = 0$; the formula for m is meaningless because we cannot divide by zero.

Average Rate of Change

The slope $\Delta y/\Delta x$ is the ratio of the change in y to the change in x as x changes from x_1 to x_2 along the segment. It is a numerical measure of how much y has changed in comparison with x over the interval between x_1 and x_2.

Definition.   The ratio

$$\frac{\Delta y}{\Delta x} = \frac{y_2 - y_1}{x_2 - x_1}$$

is the **average rate of change** of y with respect to x over the interval between x_1 and x_2.

Example 5

Consider the spring in Example 4 of the last section. The average rate of change of l with respect to w as w changes from 3 to 5 grams is

$$\frac{\Delta l}{\Delta w} = \frac{24.2 - 15.7}{5 - 3} = \frac{8.5}{2} = 4.25 \text{ centimeters per gram.}$$

Example 6

If you travel 60 miles along a highway in $1\frac{1}{2}$ hours, your average rate of change of distance s with respect to time t over that $1\frac{1}{2}$-hour time interval is

$$\frac{\Delta s}{\Delta t} = \frac{60}{1.5} = 40 \text{ miles per hour.}$$

You would say your average speed is 40 miles per hour.

The concept of average rate of change is extremely important and we will consider it further in Chapter 4.

EXERCISES 2-1

1. In Fig. 2-9 points P, Q, R, and S are shown. Estimate their coordinates.

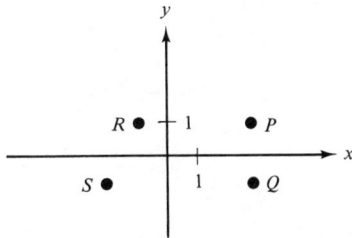

Figure 2-9

2. Draw coordinate axes in a plane and plot these points. $(1, 1)$, $(2, 2)$, $\left(\frac{5}{2}, \frac{5}{2}\right)$, $(1, -1)$, $(-2, 2)$, $(0, 3)$, $(3, 0)$, $(\sqrt{3}, 1 - \sqrt{3})$, $\left(\frac{1}{2}\sqrt{7}, 4\right)$, and $(1.3, 4)$

In Problems 3–13 shade the region in the plane consisting of all those points (x, y) that satisfy the given condition.

3. $x > 0$	4. $y \geq 0$	5. $x < 0$ and $y \geq 0$
6. $x \geq 0$ and $y < 0$	7. $\lvert x \rvert < 4$	8. $\lvert x \rvert > \sqrt{2}$
9. $-2 < x < 3$	10. $y \geq -1$	11. $\sqrt{2} < x < \sqrt{3}$
12. $x < y$	13. $0 < x < 1$ and $0 < y < 2$	

14. Plot a point (a, b) anywhere in the plane. On the same sketch plot the points

$$(a, -b), (-a, b), (-a, -b), (b, a), (a + 1, b - 2).$$

In Problems 15–26 plot P_1 and P_2 and find the slope of $P_1 P_2$.

15. $P_1(3, 2)$; $P_2(5, 7)$ 16. $P_1(-1, 4)$; $P_2(4, -7)$

17. $P_1\left(\frac{2}{3}, \frac{1}{6}\right)$; $P_2\left(\frac{1}{3}, \frac{5}{6}\right)$ 18. $P_1\left(\frac{2}{3}, \frac{4}{5}\right)$; $P_2\left(\frac{1}{5}, \frac{5}{6}\right)$

19. $P_1\left(-\frac{2}{3}, \frac{4}{9}\right)$; $P_2\left(\frac{1}{2}, \frac{3}{4}\right)$ 20. $P_1(\sqrt{2}, 0)$; $P_2(0, \sqrt{6})$

21. $P_1(\sqrt{2}, \sqrt{3})$; $P_2(1 + \sqrt{2}, -\sqrt{3})$

22. $P_1(\sqrt{5}, 1 + \sqrt{3})$; $P_2(1 + \sqrt{5}, 1 - \sqrt{5})$

23. $P_1(3 + \sqrt{2}, 1)$; $P_2(2, 2 - \sqrt{2})$ 24. $P_1\left(\frac{2}{\sqrt{5}}, \frac{1}{\sqrt{3}}\right)$; $P_2\left(\frac{\sqrt{5}}{2}, \frac{\sqrt{3}}{4}\right)$

25. $P_1\left(\frac{1}{\sqrt{3}}, \frac{2}{\sqrt{3}}\right)$; $P_2\left(\frac{1}{\sqrt{5}}, \frac{3}{\sqrt{5}}\right)$ 26. $P_1\left(\frac{1}{\sqrt{2}}, -\frac{1}{\sqrt{3}}\right)$ $P_2\left(\frac{1}{\sqrt{3}}, -\frac{1}{\sqrt{2}}\right)$

27. Find the slope of $P_1 P_2$ for P_1 and P_2 as given. Simplify your answer if possible.
    (a) $P_1(a, b)$; $P_2(b, a)(b \neq a)$
    (b) $P_1(a, a^2)$; $P_2(a + b, (a + b)^2)$ $(b \neq 0)$
    (c) $P_1\left(a, \frac{1}{a}\right)$; $P_2\left(a + b, \frac{1}{a + b}\right)$ (a and b are positive)

In Problems 28–33 determine a so that the slope of $P_1 P_2$ is the given number m.

28. $P_1(2, 3)$, $P_2(5, a)$; $m = 4$ 29. $P_1(a, -1)$, $P_2(3, 5)$; $m = -2$

30. $P_1(a, 1)$, $P_2(2, 2a)$; $m = 4$

31. $P_1(a, a + 6)$, $P_2(a - 6, 6)$; $m = 6$

32. $P_1\left(\frac{2a}{3}, 1\right)$, $P_2\left(-3, \frac{4a}{3}\right)$; $m = 4$

33. $P_1(\sqrt{7}\,a, 0)$, $P_2(1, a)$; $m = \sqrt{7}$

34. You take a trip in your car along the turnpike, starting at noon at milepost 0. At 1:00 you see that you are just passing milepost 40. At 1:45 you have traveled a total of 79 miles. At 2:30 you have traveled a total of 124 miles. At 3:00 you have traveled a total of 147 miles.

(a) Complete the table showing t, the number of hours after the start, and s, the total number of miles traveled then.

t	s
0	0
1	40
1.75	79
2.5	
3	

(b) Plot the five points (t, s) in a coordinate system.

(c) Enlarge the table to include a column for Δt, the time change in hours between measurements, and Δs, the distance traveled during each time interval. Also include a column for $\Delta s / \Delta t$, the average velocity over each time interval.

t	s	Δt	Δs	$\Delta s / \Delta t$
0	0			
		1	40	40
1	40			
		.75	39	52
1.75	79			
2.5				
3				

(d) Connect the five points successively with four line segments. Indicate the slope of each segment. Remember that the slope of each segment is equal to the average velocity over the corresponding time interval.

(e) Estimate your position at 1:30; at 2:15; at 2:45.

(f) At approximately what time were you a distance of 100 miles from the start?

35. An amateur meteorologist keeps a detailed record of the early morning temperature in Hamilton, New York. Suppose that on January 6 he records the temperatures every half hour from 5:30 A.M. through 8:00 A.M. The temperatures (°C) are (in order) -29, -32, -31, -31, -29, -26.

(a) Make a table like the one in Problem 34 showing time t (in hours after 5:30 A.M.) and temperature T (°C) at that time.

(b) Plot these six points (t, T) in a coordinate system.

(c) Enlarge the table to include a column for Δt, the time change in hours between measurements, and ΔT, the corresponding change in temperature.

(d) Connect the six points on your graph successively with five line segments and indicate the slope of each segment. What does this slope represent?

(e) Estimate the temperature at 5:45; at 6:45; at 7:50.

(f) Estimate the times at which the temperature was $-30°\text{C}$.

36. Every 10 years the U.S. Government conducts a census to determine the population of the country. Here are some recent results.

1930	122,775,046
1940	131,669,275
1950	151,325,798
1960	179,323,175
1970	203,211,926

(a) Make a table like the one in Problem 34, showing time t (in years since 1930) and population P (to the nearest million people) at that time.

(b) Plot these points (t, P) in a coordinate system.

(c) Enlarge the table to include a column for Δt, the time change in years between measurements, and ΔP, the corresponding change in population.

(d) Connect the six points on your graph successively with five line segments and indicate the slope of each segment. What does this slope represent?

(e) What was the average rate of change of population with respect to time from 1940 to 1950?

(f) Do you suppose the increase in population in the year from 1949 to 1950 was more or less than 1,900,000? Why?

37. Referring to the spring in Example 4, how many centimeters would you estimate the spring to measure when a 6-gram weight is attached? Estimate the length of the spring when no weight is attached.

2-2 DISTANCE FORMULA, CIRCLES

In setting up a coordinate system for a plane, we have not assumed that the units of length on the two coordinate axes are equal. If we wish to represent points (x, y), where x is time and y is temperature, there is no reason to make the scale on the x axis equal to the scale on the y axis.

But to deal with distances between points in the plane, we need a common unit of length for measuring *all* distances. In particular, the scales on the two axes must be equal. In the following discussion—and in any discussion in which the geometric concepts of distance, area, or angle are involved—we shall assume that the coordinate axes have equal scales.

Distance

We wish to find the distance d between two points $P_1(x_1, y_1)$ and $P_2(x_2, y_2)$. Consider the right triangle P_1QP_2 (Fig. 2-10), where Q is the point (x_2, y_1). The hypotenuse of this right triangle is P_1P_2, so its length d is the desired distance. If P_1 and P_2 are arranged as Fig. 2-10 suggests, with $x_2 > x_1$ and $y_2 > y_1$, then the lengths of P_1Q

and QP_2 are $\Delta x = x_2 - x_1$ and $\Delta y = y_2 - y_1$, respectively. By the Pythagorean theorem,

$$d^2 = (x_2 - x_1)^2 + (y_2 - y_1)^2$$

and because d is not negative,

$$d = \sqrt{(x_2 - x_1)^2 + (y_2 - y_1)^2}. \tag{2-1}$$

Although we derived this formula under the assumption that P_1 and P_2 are arranged as in Fig. 2-10, it can be shown that the formula holds regardless of the arrangement of P_1 and P_2.[1]

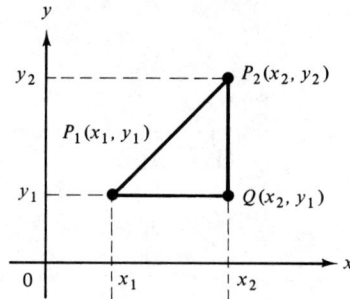

Figure 2-10

Example 1

Find the distance between $(3, 4)$ and $(7, 5)$.

Solution. $d = \sqrt{(7 - 3)^2 + (5 - 4)^2} = \sqrt{16 + 1} = \sqrt{17}$.

Example 2

Find the distance between $(3, 4)$ and the origin.

Solution. $d = \sqrt{(3 - 0)^2 + (4 - 0)^2} = \sqrt{25} = 5$.

Example 3

For what values of a is $(a, 4)$ at a distance 5 from $(3, 0)$?

Solution. We are given

$$\sqrt{(a - 3)^2 + (4 - 0)^2} = 5$$

so that

$$a^2 - 6a + 9 + 16 = 25$$
$$a^2 - 6a = 0$$
$$a(a - 6) = 0.$$

The solutions are $a = 0$ and $a = 6$.

---

[1] In all cases, the lengths of the legs of the right triangle $P_1 Q P_2$ are $|x_2 - x_1|$ and $|y_2 - y_1|$. Because $|x_2 - x_1|^2 + |y_2 - y_1|^2 = (x_2 - x_1)^2 + (y_2 - y_1)^2$, we arrive at the same formula for d.

Circles

Recall that a circle is defined as the set of points in a plane at a fixed distance (the **radius**) from a given fixed point (the **center**). Now consider the circle C with center at the origin O and radius 5 (see Fig. 2-11). If $P(x, y)$ is any point on that circle, then the distance between P and O must must be 5. Hence x and y satisfy the equation.

$$\sqrt{(x - 0)^2 + (y - 0)^2} = 5.$$

Squaring both sides, we obtain

$$x^2 + y^2 = 25. \tag{2-2}$$

We have seen that each point P on the circle has coordinates (x, y) that satisfy Eq. (2-2). Moreover, it is easy to see that if the coordinates (x, y) of a point satisfy Eq. (2-2), then that point is at a distance of 5 from the origin and must lie on C. We say that C is the **graph** of Eq. (2-2), and Eq. (2-2) is an equation of C.

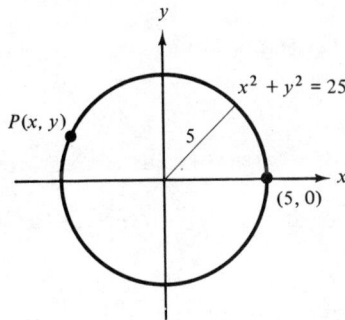

Figure 2-11

In the same way, we can derive an equation for any circle if we know its radius r and the coordinates of its center (h, k). A point $P(x, y)$ lies on this circle (Fig. 2-12)

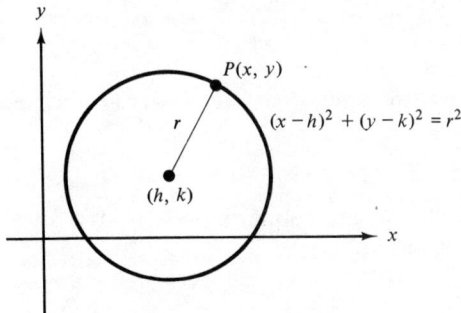

Figure 2-12

if and only if the distance between (x, y) and (h, k) is equal to r. This condition can be expressed by the equation

$$\sqrt{(x - h)^2 + (y - k)^2} = r,$$

which is equivalent to

$$(x - h)^2 + (y - k)^2 = r^2. \tag{2-3}$$

Equation (2-3) is called the "standard" equation of the circle; it displays the coordinates of the center as well as the radius.

Example 4

(a) The standard equation of the circle with center at $(1, -2)$ and radius 3 is

$$(x - 1)^2 + (y + 2)^2 = 9.$$

Squaring the terms on the left and simplifying, we get

$$x^2 + y^2 - 2x + 4y - 4 = 0.$$

This is also an equation of the given circle.

(b) The graph of the equation

$$(x - 2)^2 + (y - 3)^2 = 25$$

is a circle with center $(2, 3)$ and radius 5.

(c) The graph of the equation

$$x^2 + (y + 4)^2 = 12$$

is a circle with center $(0, -4)$ and radius $\sqrt{12} = 2\sqrt{3}$.

If we simplify the standard equation (2-3) as in Example 4(a), we obtain

$$x^2 + y^2 - 2hx - 2ky + h^2 + k^2 - r^2 = 0$$

or

$$x^2 + y^2 + Dx + Ey + F = 0 \tag{2-4}$$

where $D = -2h$, $E = -2k$, $F = h^2 + k^2 - r^2$. Equation (2-4) is called the "general" equation of a circle; unlike the standard equation (2-3), it does not display the center and radius. To find the center and radius from the general equation, we apply the process of completing the square.

Completing the Square

The technique of "completing the square" is based on the formula for squaring a binomial:

$$(x + a)^2 = x^2 + 2ax + a^2. \tag{2-5}$$

A familiar situation in which this technique is employed is in solving quadratic equations.

Example 5

Solve the quadratic equation $x^2 + 6x = 7$.

Solution.   Let us add 9 to both sides of the equation obtaining

$$x^2 + 6x + 9 = 16.$$

Notice that by Eq. (2-5) the left side is now a square:

$$x^2 + 6x + 9 = (x + 3)^2;$$

so the given equation is equivalent to

$$(x + 3)^2 = 16$$

and we get

$$x + 3 = \pm\sqrt{16} = \pm 4$$

$$x = -3 \pm 4$$

or

$$x = -7, 1.$$

How do we know what number to add to both sides in the first step of the solution of Example 5? The objective is to make the left side the square of a binomial. Starting with

$$x^2 + 6x + \ ?$$

and looking at the right side of Eq. (2-5), we equate the coefficients of x:

$$6 = 2a$$

$$a = 3.$$

So the number to be added is a^2 or 9. We have divided the coefficient of x by 2 and then squared.

Completing the Square. The number that must be added to $x^2 + cx$ to make a perfect square is $(c/2)^2$, or $c^2/4$. Then $x^2 + cx + (c^2/4)$, is the square of $x + (c/2)$:

$$x^2 + cx + \frac{c^2}{4} = \left(x + \frac{c}{2}\right)^2.$$

Example 6

Solve $x^2 - 10x - 2 = 0$.

Solution. First, add 2 to both sides, obtaining

$$x^2 - 10x = 2.$$

To complete the square on the left side, we add $(-10/2)^2 = 25$. This results in

$$x^2 - 10x + 25 = 2 + 25 = 27.$$

Hence

$$(x - 5)^2 = 27.$$

$$x - 5 = \pm\sqrt{27}.$$

Solutions are

$$x = 5 \pm \sqrt{27} = 5 \pm 3\sqrt{3}.$$

Example 7

Solve $3x^2 + 6x = 22$.

Solution. Because the coefficient of x^2 is not 1, we first factor out the coefficient 3 on the left: $3(x^2 + 2x)$. To complete the square for $x^2 + 2x$, we add 1. Replacing $x^2 + 2x$ by $x^2 + 2x + 1$, we see that we have actually added 3 to the left side of the equation:

$$3(x^2 + 2x + 1) = 3(x^2 + 2x) + 3.$$

Hence

$$3(x + 1)^2 = 22 + 3$$

$$(x + 1)^2 = \frac{25}{3}$$

$$x + 1 = \pm\sqrt{\frac{25}{3}}$$

$$x = -1 \pm \sqrt{\frac{25}{3}} = -1 \pm \frac{5\sqrt{3}}{3}.$$

Finding the Radius and Center of a Circle

We return to the problem of identifying a circle from its general equation.

Example 8

Find the center and radius of the circle whose equation is

$$x^2 + y^2 + 6x - 2y + 6 = 0.$$

Solution. Subtract 6 from both sides; then collect the terms involving x and those involving y:

$$(x^2 + 6x) + (y^2 - 2y) = -6.$$

Completing the squares, we obtain

$$(x^2 + 6x + 9) + (y^2 - 2y + 1) = -6 + 9 + 1 = 4$$

or

$$(x + 3)^2 + (y - 1)^2 = 4.$$

This is the standard equation of a circle with center $(-3, 1)$ and radius 2.

Example 9

Given that

$$9x^2 + 9y^2 - 6x + 12y + 2 = 0$$

represents a circle, find the center and radius.

Solution. We can make the coefficients of x^2 and y^2 equal to one by dividing both sides by 9:

$$x^2 + y^2 - \frac{2}{3}x + \frac{4}{3}y + \frac{2}{9} = 0.$$

Now we collect terms involving x and terms involving y and subtract 2/9 on both sides:

$$\left(x^2 - \frac{2}{3}x\right) + \left(y^2 + \frac{4}{3}y\right) = -\frac{2}{9}.$$

Completing the squares, we obtain

$$\left(x^2 - \frac{2}{3}x + \frac{1}{9}\right) + \left(y^2 + \frac{4}{3}y + \frac{4}{9}\right) = -\frac{2}{9} + \frac{1}{9} + \frac{4}{9}$$

$$\left(x - \frac{1}{3}\right)^2 + \left(y + \frac{2}{3}\right)^2 = \frac{1}{3}.$$

This is the standard equation of a circle with center $(1/3, -2/3)$ and radius $\sqrt{1/3} = \sqrt{3}/3$.

Not every equation of the form (2-4) gives us a circle.

Example 10

The equation

$$x^2 + y^2 - 2x + 2y + 4 = 0$$

is of the form (2-4). Completing the squares as in the preceding examples, we obtain

$$(x - 1)^2 + (y + 1)^2 = -2.$$

There are no values of x and y that satisfy this equation because the sum of squares on the left side cannot be negative. The graph is empty.

Inside and Outside of a Circle

A point is **inside** a circle if its distance from the center of the circle is less than the radius. It is **outside** the circle if its distance from the center is greater than the radius. In case of equality, of course, the point is on the circle.

Example 11

Is $(-2, 3)$ inside, outside, or on the circle whose equation is

$$x^2 + y^2 + 12x - 10y + 45 = 0?$$

Solution. You should verify that the center of this circle is $(-6, 5)$ and its radius is 4. The distance between $(-2, 3)$ and the center is $\sqrt{4^2 + 2^2} = \sqrt{20} = 2\sqrt{5}$. Because this number is greater than 4, the point $(-2, 3)$ is outside the given circle.

We can shorten our work by comparing the *square* of the distance between P and the center with the *square* of the radius. In Example 11 the square of the distance between $(-2, 3)$ and the center is 20, which is greater than 16, the square of the radius. So $(-2, 3)$ is outside the circle.

EXERCISES 2-2

In Problems 1–7 find the distance between the given points P_1 and P_2.

1. $P_1(2, 3)$; $P_2(3, 5)$

2. $P_1(-2, 3)$; $P_2(-3, -5)$

3. $P_1(0, 4)$; $P_2(3, 0)$

4. $P_1(\frac{1}{2}, 2)$; $P_2(\frac{1}{3}, \frac{1}{2})$

5. $P_1(\sqrt{2}, 3)$; $P_2(3, -\sqrt{2})$ **6.** $P_1(0, -1)$; $P_2(\sqrt{2}, \sqrt{3})$

7. $P_1(\sqrt{2}, \sqrt{3})$; $P_2(-\sqrt{2}, -\sqrt{3})$

In Problems 8–10 find the values of a for which the distance between P_1 and P_2 is equal to $\sqrt{10}$.

8. $P_1(a, 0)$, $P_2(1, 3)$ **9.** $P_1(-1, -1)$, $P_2(2, a)$

10. $P_1(a, 0)$, $P_2(0, a)$

In Problems 11–16 the center and radius of a circle are given. Find the standard equation (2–3) and then the general equation (2–4).

11. $(0, 0)$; 3   **12.** $(-1, 4)$; 1   **13.** $(2, -3)$; 4

14. $(-4, -5)$; $\dfrac{3}{\sqrt{10}}$ **15.** $\left(\dfrac{1}{2}, \dfrac{1}{3}\right)$; $\dfrac{5}{6}$ **16.** $\left(\dfrac{1}{\sqrt{2}}, -\dfrac{1}{\sqrt{2}}\right)$; 1

In Problems 17–26 solve the equations by completing the squares.

17. $x^2 + 20x = 44$ **18.** $x^2 - 12x = 13$ **19.** $x^2 + x - \dfrac{15}{4} = 0$

20. $3x + x^2 = \dfrac{27}{4}$ **21.** $5x - x^2 = 2$ **22.** $5 - x^2 + 3x = 0$

23. $4x^2 + 8x = 5$ **24.** $\dfrac{1}{3}x + \dfrac{1}{2}x^2 - \dfrac{2}{9} = 0$ **25.** $x^2 + 2\sqrt{2}x = 7$

26. $3x^2 + 5\sqrt{5}x + \dfrac{1}{3} = 0$

In Problems 27–35 each equation has a circle for its graph. By completing squares, find the center and radius. Draw the circle.

27. $x^2 + y^2 - 6x - 4y + 9 = 0$ **28.** $x^2 + y^2 - 4x + 4y - 1 = 0$

29. $x^2 + y^2 + 6x + 4y - 12 = 0$ **30.** $x^2 + y^2 - 10y = 0$

31. $x^2 + y^2 - 2x + 2y = 10$ **32.** $x^2 + y^2 - \dfrac{2}{3}x + \dfrac{4}{3}y + \dfrac{4}{9} = 0$

33. $x^2 + y^2 = x + y$ **34.** $8x^2 + 8y^2 + 12x - 4y + 3 = 0$

35. $x^2 + y^2 - 2\sqrt{2}x - 2\sqrt{2}y + 2 = 0$

In Problems 36–43 determine whether P is inside, outside, or on the circle whose equation is given.

36. $P(4, 1)$; $(x - 3)^2 + (y - 2)^2 = 4$ **37.** $P(1, 4)$; $(x - 3)^2 + (y - 2)^2 = 4$

38. $P(3, 4)$; $(x - 3)^2 + (y - 2)^2 = 4$ **39.** $P(2, 2)$; $x^2 + y^2 + 4x - 2y = 11$

40. $P\left(\dfrac{1}{2}, -\dfrac{5}{6}\right)$; $\left(x - \dfrac{1}{3}\right)^2 + \left(y + \dfrac{2}{3}\right)^2 = \dfrac{1}{17}$

41. $P(1, \sqrt{2})$; $x^2 + y^2 = 3$

42. $P(\sqrt{2}, \sqrt{3})$; $x^2 + y^2 - 2x - 2y + 1 = 0$

43. $P(1, 1 + 3\sqrt{3})$; $x^2 + y^2 + 4\sqrt{5}x - 6\sqrt{3}y + 16 = 0$

SUMMARY

In Chapter 2 you should have learned how to

1. plot points in a coordinate plane and represent data by such points.
2. find the slope of a nonvertical line segment and interpret this slope.
3. find the distance between two points.

4. find an equation for a circle with given center and radius.

5. find the center and radius of a circle whose equation is given. (This is likely to involve completing the square.)

6. determine whether a given point lies inside, outside, or on a circle whose equation is given.

REVIEW EXERCISES

1. (a) The following table shows the maximum temperature T (degrees Fahrenheit) each day for the first five days in August.

Day t	Maximum Temperature T (degrees Fahrenheit)
1	78
2	78
3	83
4	86
5	72

Plot the points (t, T) for $t = 1, 2, 3, 4, 5$ in a coordinate plane, using a horizontal t axis and a vertical T axis.

(b) Find the slopes of the four line segments connecting consecutive points on the graph in part (a). Interpret the slopes.

2. (a) For a certain chemical substance, the weight w (grams) that will dissolve in one liter of water at various temperatures T (degrees Celsius) is given in the table shown.

T	w
30	30
40	32
50	34
60	36

Plot the points (T, w) for $T = 30, 40, 50, 60$, using a horizontal T axis and a vertical w axis.

(b) Find the slopes of the three line segments connecting consecutive points on the graph in part (a). Interpret the slopes.

In Problems 3–6 find the distance between the points.

3. $(1, 1), (3, 2)$

4. $(1, 3), (-2, -1)$

5. $(1, 0), (0, -3)$

6. $(-2, 6), (3, 9)$

In Problems 7–11 find the general equation for the circle described.

7. center at $(0, 0)$, radius 3

8. center at $(0, 2)$, radius 2

9. center at $(4, -3)$, radius 5

10. center at $(-1, 1)$, passing through $(3, 0)$

11. center at $(-2, 1)$, passing through $(2, 4)$.

In Problems 12–17 find the center and radius of the circle whose general equation is given.

12. $x^2 + y^2 + 2x - 2y - 2 = 0$ **13.** $x^2 + y^2 - 6x - 4y + 4 = 0$

14. $x^2 + y^2 + 6x - 1 = 0$ **15.** $9x^2 + 9y^2 - 6x + 12y - 4 = 0$

16. $9x^2 + 9y^2 - 12x + 6y + 4 = 0$ **17.** $x^2 + y^2 + 2\sqrt{2}x - 2\sqrt{3}y = 0$

In Problems 18 and 19 determine whether the given point P is inside, outside, or on the circle whose equation is given.

18. $P(3, 1)$; $x^2 + y^2 - 8x + 12y + 3 = 0$

19. $P(0, -1)$; $x^2 + y^2 - 12x + 5y + 4 = 0$

Graphs

3-1 GRAPHS OF EQUATIONS

In Section 2-2 we found that the graph of the equation $x^2 + y^2 = 25$ is a circle with center at the origin and radius 5. We said that this meant two things: (a) every point whose coordinates (x, y) satisfy the given equation lies on that circle and (b) every point on the circle has coordinates (x, y) that satisfy the equation.

> **Definition.** A set of points in the coordinate plane is called the **graph** of an equation in x and y, provided that
> 1. every point whose coordinates (x, y) satisfy the equation belongs to the set.
> 2. every point belonging to the set has coordinates (x, y) that satisfy the equation.

One of our major objectives is to learn how to draw the graph of an equation. Sometimes a graph will have a familiar geometric shape, such as a circle or a line. Other times its shape can best be indicated by a sketch. Ordinarily the graph of an equation will be a line or curve of some sort.

Example 1

It is easy to see that the graph of the equation $x = 3$ is the line parallel to the y axis and three units to the right of it (see Fig. 3-1). Every point (x, y) whose coordinates satisfy $x = 3$ lies on this line and every point on the line has coordinates (x, y) such that $x = 3$. Similarly, the graph of $y = 2$ is a line parallel to the x axis, two units above it.

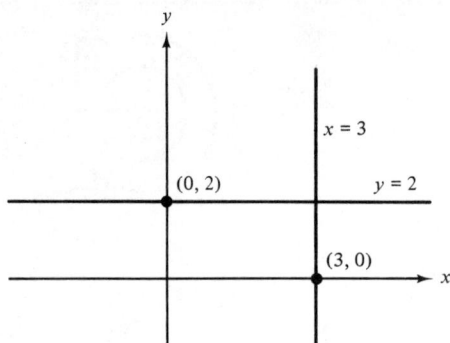

Figure 3-1

We can generalize Example 1 as follows. The graph of the equation $x = c$, where c is a constant, is a vertical line (parallel to the y axis). It passes through $(c, 0)$. The graph of the equation $y = c$ is a horizontal line (parallel to the x axis). It passes through $(0, c)$.

A simple-minded method of obtaining a graph is to plot a number of points whose coordinates satisfy its equation. To be sure, we cannot obtain all points on the graph in this way, but we may be able to get a fairly good picture if we select our points well.

Example 2

Sketch the graph of $y = 2x + 3$ by plotting points.

Solution. In Fig. 3-2 we have tabulated a number of pairs that satisfy the equation and plotted the corresponding points. Obviously the plotted points do not constitute the entire graph, but they appear to lie on a straight line. We shall show in the next section that the graph of this equation actually is a straight line.

$y = 2x + 3$

x	y
-3	-3
-2	-2
-1	1
0	3
1	5
2	7
3	9

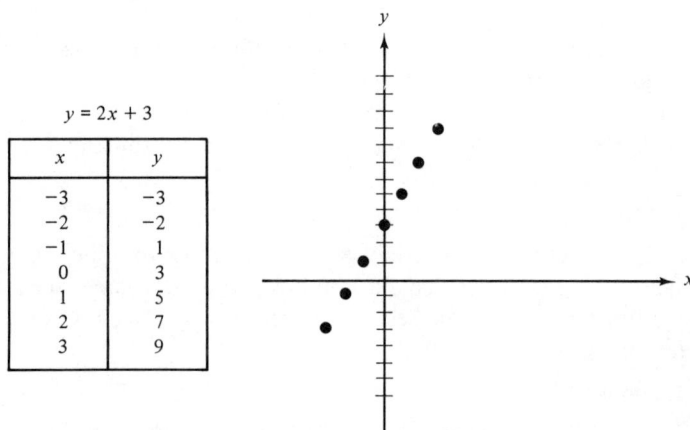

Figure 3-2

In addition to selecting a good variety of points to plot, we can improve our ability to draw a good graph by looking for algebraic clues in the equation.

Symmetry

Example 3

Sketch the graph of $y = x^2$.

Solution. First, we consider some points on the graph with nonnegative x coordinates: $(0, 0)$, $(1, 1)$, $(2, 4)$, $(3, 9)$ [see Fig. 3-3(a)]. Then we choose some

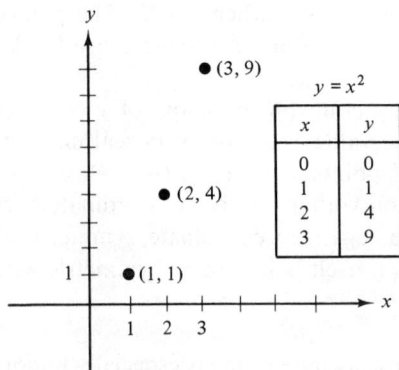

$y = x^2$

x	y
0	0
1	1
2	4
3	9

(a)

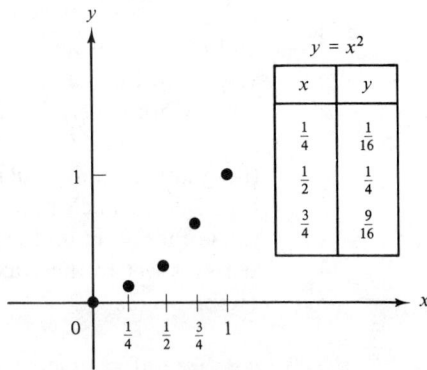

$y = x^2$

x	y
$\frac{1}{4}$	$\frac{1}{16}$
$\frac{1}{2}$	$\frac{1}{4}$
$\frac{3}{4}$	$\frac{9}{16}$

(b)

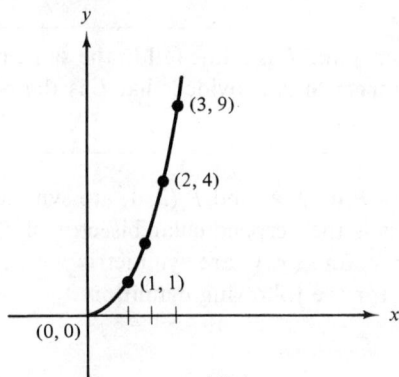

(c)

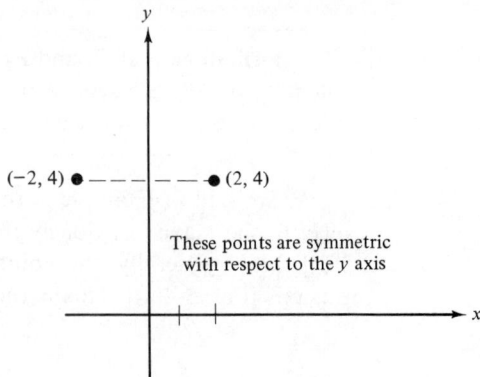

These points are symmetric with respect to the y axis

(d)

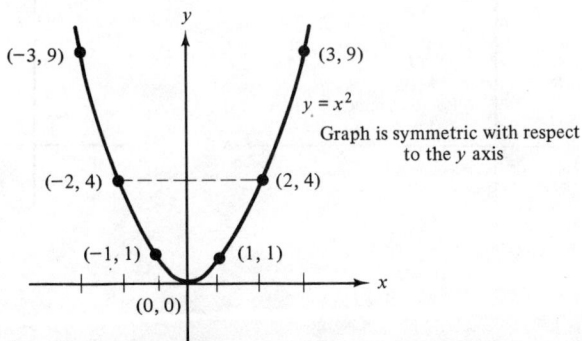

$y = x^2$

Graph is symmetric with respect to the y axis

(e)

Figure 3-3

additional points with x coordinates between 0 and 1:

$$\left(\frac{1}{4}, \frac{1}{16}\right), \quad \left(\frac{1}{2}, \frac{1}{4}\right), \quad \left(\frac{3}{4}, \frac{9}{16}\right).$$

These are shown in Fig. 3-3(b). Continuing in this fashion, we find that the graph takes the shape shown in Fig. 3-3(c).

Now let us consider negative values of x. For example, when $x = -2$, y has the value 4; this is the same value y has when $x = 2$. The points $(-2, 4)$ and $(2, 4)$ both lie on the graph of $y = x^2$ and are symmetrically placed with respect to the y axis [Fig. 3-3(d)].

More generally, if (x, y) is any point on the graph of $y = x^2$, then so is $(-x, y)$ because $(-x)^2$ has the same value as x^2 for every real number x. Thus the graph contains such additional points as $(-1, 1)$, $(-3, 9)$, and so on [see Fig. 3-3(e)]. Each point on the graph with a positive x coordinate corresponds to another point on the graph with a negative x coordinate, symmetrically placed with respect to the y axis. The graph itself is said to be symmetric with respect to the y axis.

The role of symmetry as a tool for drawing graphs is especially important. Let us consider symmetry more closely.

Definition.   If P_1 and P_2 are points and L is a line (all in the same plane), then P_1 and P_2 are **symmetric with respect to L** provided that L is the perpendicular bisector of the segment $P_1 P_2$.

As we said in Example 3, the points $P_1(-2, 4)$ and $P_2(2, 4)$ are symmetric with respect to the y axis; obviously the y axis is the perpendicular bisector of $P_1 P_2$ [Fig. 3-3(d)]. More generally, the points $(-x, y)$ and (x, y) are symmetric with respect to the y axis [Fig. 3-4(a)]. This is the basis for the following definition.

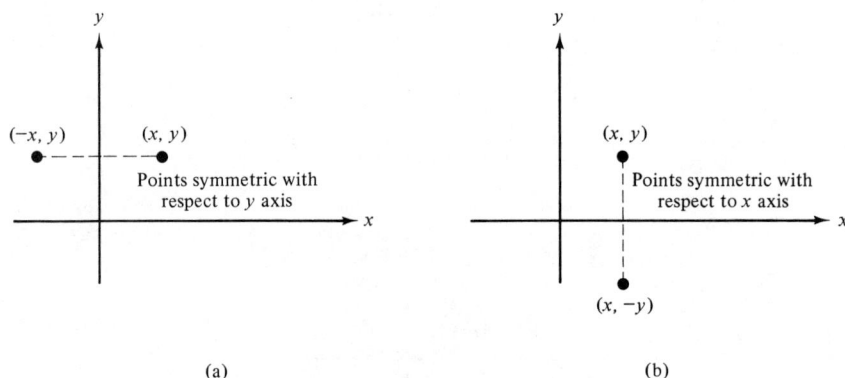

(a)   (b)

Figure 3-4

> **Definition.** The graph of an equation in x and y is said to be **symmetric with respect to the y axis** provided that the point $(-x, y)$ belongs to it whenever (x, y) does.

Algebraically, symmetry with respect to the y axis means that if x is replaced by $-x$ in the given equation, then the resulting equation is equivalent to the original. Consider the equation $y = x^2$ in Example 3. If x is replaced by $-x$, the resulting equation is $y = (-x)^2$, which is equivalent to $y = x^2$ because $(-x)^2 = x^2$. Some other equations whose graphs have symmetry with respect to the y axis are

$$y = x^4$$
$$y = |x| + 5$$
$$x^2 y = 1$$
$$x^2 + y^2 = 25.$$

The payoff for discovering symmetry is this: A graph that is known to be symmetric with respect to the y axis can be obtained by first drawing the part corresponding to nonnegative values of x and then reflecting that portion through the y axis [as in Fig. 3-3(c) and (e)].

The definition of symmetry with respect to the x axis is analogous.

> **Definition.** The graph of an equation in x and y is said to be **symmetric with respect to the x axis** provided that the point $(x, -y)$ belongs to it whenever (x, y) does.

The basis for this definition is the fact that $(x, -y)$ and (x, y) are symmetric with respect to the x axis [Fig. 3-4(b)]. Algebraically, symmetry with respect to the x axis means if y is replaced by $-y$ in the given equation, then the resulting equation is equivalent to the original. Some equations whose graphs have this symmetry are

$$x + y^2 = 1$$
$$x|y| = -1$$
$$3x^2 + 4y^2 = 12.$$

A graph that is known to be symmetric with respect to the x axis can be obtained by first drawing the part corresponding to nonnegative values of y and then reflecting that portion through the x axis.

Example 4

Draw the graph of $x + y^2 = 1$.

Solution. This graph is symmetric with respect to the x axis because the replacement of y by $-y$ results in the equivalent equation $x + (-y)^2 = 1$. Selecting some nonnegative values for y and calculating $x = 1 - y^2$, we obtain the following points on the graph:

$$(1, 0), \quad \left(\frac{3}{4}, \frac{1}{2}\right), \quad (0, 1), \quad (-3, 2), \quad \left(-\frac{21}{4}, \frac{5}{2}\right).$$

These points suggest the curve in Fig. 3-5(a); the remainder of the graph is obtained by reflection through the x axis as in Fig. 3-5(b).

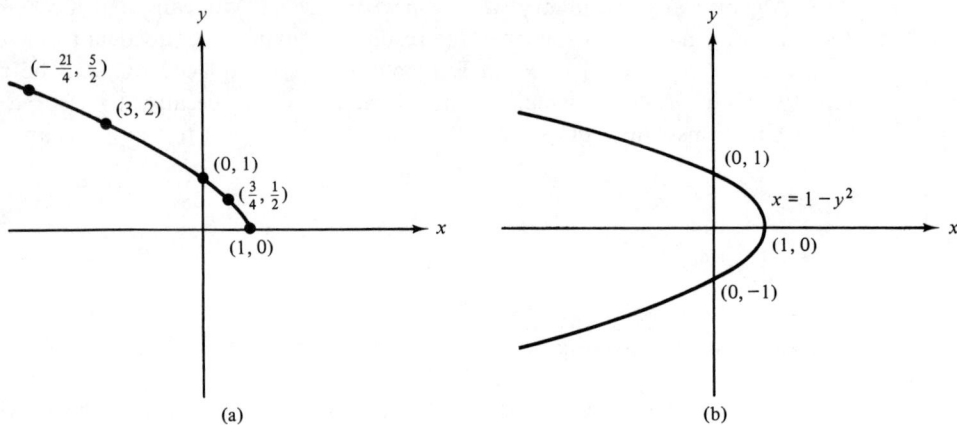

Figure 3-5

A third kind of symmetry is symmetry with respect to the origin.

Definition. Two points P_1 and P_2 are **symmetric with respect to a point O** provided that O is the midpoint of the segment P_1P_2.

If O is the origin of our coordinate system, then the points (x, y) and $(-x, -y)$ are symmetric with respect to O (Fig. 3-6).

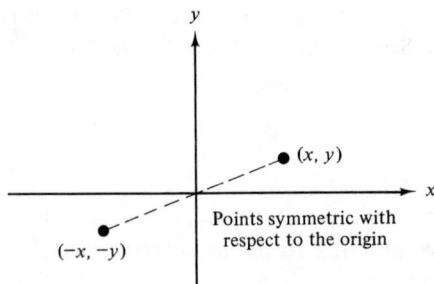

Points symmetric with respect to the origin

Figure 3-6

Definition. The graph of an equation in x and y is said to be **symmetric with respect to the origin**, provided that $(-x, -y)$ belongs to the graph whenever (x, y) does.

Algebraically, symmetry with respect to the origin means that if x and y are replaced by $-x$ and $-y$, respectively, *at the same time*, then the resulting equation is equivalent to the original.

Graphs Chap. 3

Example 5

Show that the equation $y = x^3$ does not have symmetry with respect to the x axis or y axis but that it does have symmetry with respect to the origin.

Solution. If we replace x by $-x$, we get $y = (-x)^3 = -x^3$, which is *not* equivalent to $y = x^3$. This shows that the graph is not symmetric with respect to the y axis. Replacing y by $-y$, we have $-y = x^3$ and conclude that the graph is not symmetric with respect to the x axis. But if we make the *simultaneous* replacement of x by $-x$ and y by $-y$, we obtain

$$-y = (-x)^3$$
$$-y = -x^3$$

and the latter is equivalent to $y = x^3$.

Some other equations whose graphs are symmetric with respect to the origin are

$$y = x$$
$$xy = 1$$
$$y^3 = 3x$$
$$x^2 + y^2 = 25.$$

A graph that is known to be symmetric with respect to the origin can be obtained by first drawing the part corresponding to nonnegative values of x and then reflecting that portion through the origin.

Example 6

Draw the graph of $y = x^3$.

Solution. The graph has symmetry with respect to the origin. We obtain the points $(0, 0)$, $(1, 1)$, $(2, 8)$, $(\frac{1}{2}, \frac{1}{8})$, suggesting a curve of the shape in Fig. 3-7(a).

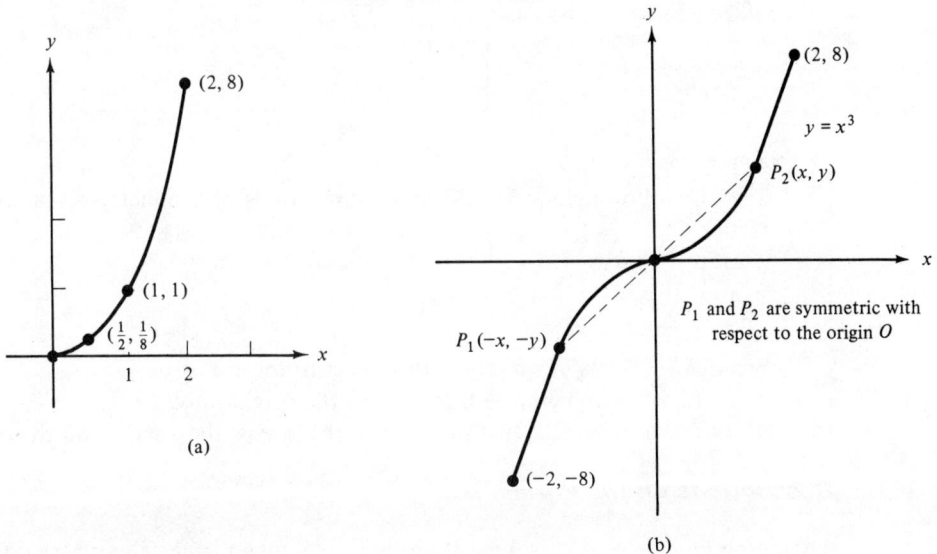

Figure 3-7

To obtain the reflection of this part, connect each point to the origin with a line segment and then extend this segment through the origin an equal distance on the other side [Fig. 3-7(b)].

The three kinds of symmetry are summarized in Table 3-1.

TABLE 3-1

Symmetry	Typical graph	Test
y axis	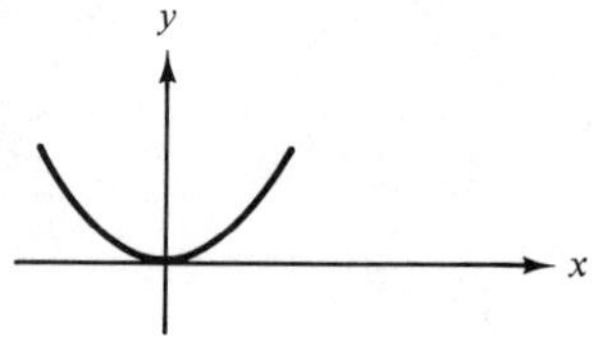	Replace x by $-x$.
x axis	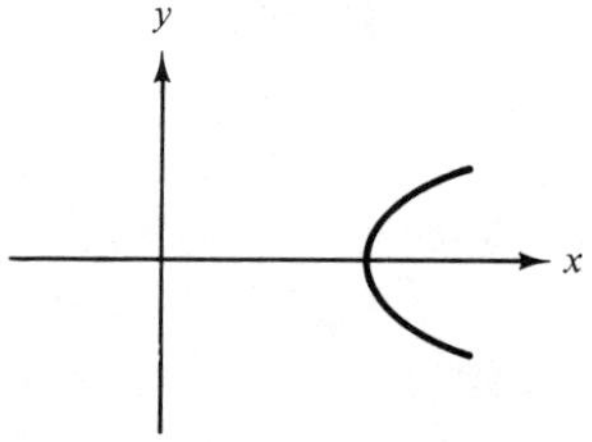	Replace y by $-y$.
Origin	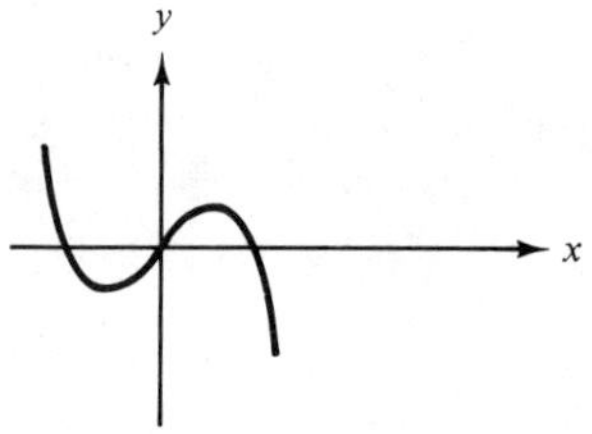	Replace x by $-x$ *and y by $-y$.*

Example 7

Test the following equations for the three kinds of symmetry discussed in this section.

(a) $y = 1 - x^2$
(b) $x + y = 0$
(c) $x^2 - y^2 = 4$

Solution. (a) has symmetry with respect to the y axis only.

(b) has symmetry with respect to the origin only.

(c) has symmetry with respect to the x axis, the y axis, and the origin.

Restrictions on the Variables

The graph of $x + y^2 = 1$ in Fig. 3-5(b) indicates that $x \leq 1$. This restriction on the values that x can take is due to the fact that $y^2 \geq 0$ and, consequently,

$$x = 1 - y^2 \leq 1 - 0 = 1.$$

Another way to see that $x \leq 1$ is to solve for y:

$$y = \pm\sqrt{1 - x}.$$

Because the square root exists only when $1 - x \geq 0$, we must have $x \leq 1$. There is no restriction on y, however, because for any real number y we can find an x—namely, $x = 1 - y^2$—such that (x, y) is on the graph.

Our discussion suggests a way of finding restrictions on the variables: solve for each one in terms of the other and then eliminate any values that would lead to an undefined expression.

Example 8

Determine any restrictions on x and y if $x^2 + 4y^2 = 4$ and sketch the graph of this equation.

Solution. Solving for x,

$$x = \pm\sqrt{4 - 4y^2},$$

we must have $4 - 4y^2 \geq 0$ or $-1 \leq y \leq 1$. Similarly,

$$y = \pm\frac{1}{2}\sqrt{4 - x^2};$$

so we must have $4 - x^2 \geq 0$ or $-2 \leq x \leq 2$. The graph is therefore contained within the rectangle in Fig. 3-8(a), whose sides lie on the lines $x = -2$, $x = 2$,

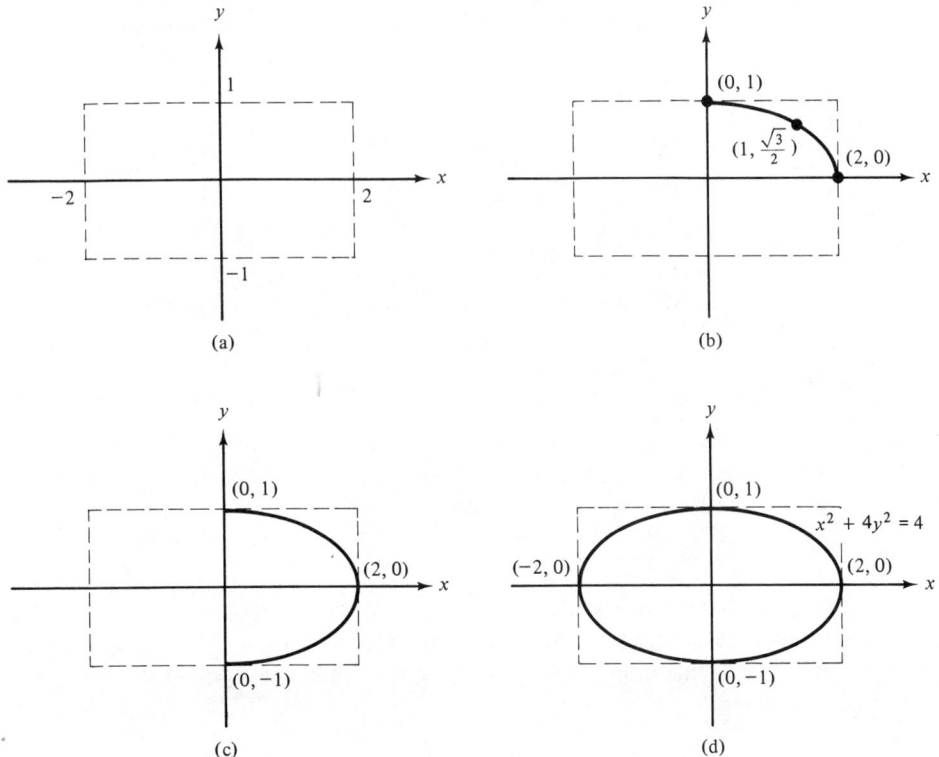

(a)

(b)

(c)

(d)

Figure 3-8

$y = -1$, $y = 1$. There is symmetry with respect to both axes as well as the origin. Plotting various points, such as $(0, 1)$, $(1, \sqrt{3}/2)$, $(2, 0)$, we obtain the curve in the first quadrant shown in Fig. 3-8(b). Using the x axis symmetry, we can reflect this piece through the x axis and obtain the curve in Fig. 3-8(c). Finally, we can use the y axis symmetry to obtain the graph in Fig. 3-8(d). (This curve is known as an ellipse; ellipses will be discussed in more detail in Chapter 11.)

EXERCISES 3-1

In Problems 1–7 you are given an equation in x and y and a particular value of x. Find any values of y such that (x, y) lies on the graph of the equation.

1. $y = 2x - 7 \ (x = 5)$ 2. $y = x^2 - 10 \ (x = 3)$

3. $y = \frac{2}{4}x - \frac{1}{2} \ (x = \frac{4}{3})$ 4. $x + y = 3 \ (x = \frac{1}{5})$

5. $x^2 + y^2 = 9 \ (x = -2)$ 6. $y^2 - 2x^2 = \frac{1}{9} \ (x = 1)$

7. $y^2 + 2x^2 = \frac{1}{9} \ (x = 1)$

In Problems 8–11 you may assume that the graph is a straight line. Draw the graph by finding two points on it and connecting them by a line.

8. $y = 3x - 1$ 9. $x - y = 0$

10. $x + y = 0$ 11. $2x + y = 4$

In Problems 12–23 determine which of the three symmetries the graph of the given equation has. Are x and y restricted to certain intervals? If so, specify. Sketch the graph.

12. $y = x^2 + 1$ 13. $y = 2x^2$ 14. $y = 2x^2 - 3$

15. $x = y^2$ 16. $x = y^2 - 1$ 17. $xy = 3$

18. $x = y^3$ 19. $9x^2 + y^2 = 36$ 20. $x^2 + 3y^2 = 6$

21. $x + 7y^2 = 8$ 22. $x^2 = y^2 + 1$ 23. $y(x^2 + 1) = 1$

24. In the same coordinate system sketch the graphs of the equations

$$y = x, \quad y = x^2, \quad y = x^3, \quad y = x^4,$$

letting x vary from -2 to 2. What points belong to all four graphs?

In Problems 25–29 determine the values of x such that $(x, 0)$ is a point on the graph. Draw the graph, labeling these points.

25. $y = x^2 - 1$ 26. $y = 6x + 16$ 27. $7x + 5y = 10$

28. $x^2 + y^2 = 10$ 29. $2x^2 + y^2 = 25$

In Problems 30–33 show that the given equation is not satisfied by any pair (x, y). The "graph" of such an equation is empty.

30. $x^2 + y^2 = -1$ 31. $x^2 + y^2 + 4 = 0$

32. $4y^2 = -1 - 3|x|$ 33. $4 + |x| = 2 - |y|$

34. Consider the graph of $y = x(x - 1)$. For which intervals on the x axis is y positive? negative? [Solve the inequalities $x(x - 1) > 0$, $x(x - 1) < 0$.]

35. Consider the graph of $y = x^2 + 3x$. For which intervals on the x axis is y positive? negative?

36. Consider the graph of $y = x^3 - x$. For which intervals on the x axis is y positive? negative?

37. Consider the graph of $y = x^3 + x$. For which intervals on the x axis is y positive? negative?

38. Show that if the graph of an equation has any two of the three symmetries discussed in this section, it must also have the third.

In Problems 39–46 find any symmetries and any restrictions on the variables for the graph of the given equation. Sketch the graph.

39. $|x| + |y| = 1$ 40. $|x| + |2y| = 2$ 41. $|x| - |y| = 1$

42. $|3x| - |4y| = 1$ 43. $|x + y| = 1$ 44. $|x - y| = 1$

45. $|x - 2y| = 1$ 46. $|3x + 2y| = 6$

47. Devise a test for symmetry with respect to the line $y = 2$.

48. Devise a test for symmetry with respect to the line $x = -3$.

3-2 EQUATION OF A LINE

In Section 2-2 we discussed the slope of a *line segment*. We now extend the slope idea to *lines*.

Slope of a Line

If L is a line in the x, y coordinate plane that is not parallel to the y axis and if we choose two points $P_1(x_1, y_1)$ and $P_2(x_2, y_2)$ on L such that $x_1 \neq x_2$, then the slope of the segment P_1P_2 is given by

$$m = \frac{y_2 - y_1}{x_2 - x_1}.$$

If we should choose two other points $P_3(x_3, y_3)$ and $P_4(x_4, y_4)$ on L such that $x_3 \neq x_4$, then the slope of the segment P_3P_4 is

$$m' = \frac{y_4 - y_3}{x_4 - x_3}.$$

Referring to Fig. 3-9, we see that the right triangles $P_1Q_1P_2$ and $P_3Q_2P_4$ are similar, for both are right triangles and the angles at P_1 and P_3 are equal. Because of this similarity,

$$\frac{y_2 - y_1}{x_2 - x_1} = \frac{y_4 - y_3}{x_4 - x_3}$$

and so $m = m'$.

The result of the preceding discussion is that the slope of *any* segment connecting a pair of distinct points on L is *always the same number*. This number is called the **slope** of the line L.

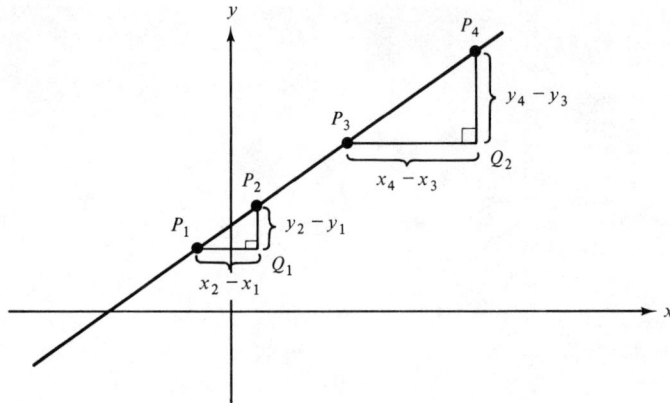

Figure 3-9

Example 1

It can be shown that the points $(0, 3)$, $(1, 5)$, $(2, 7)$, $(3, 9)$, and so on, tabulated in Fig. 3-2, all lie on a line. Observe that the slope of the segment connecting $(0, 3)$ and $(1, 5)$ is 2. You will get the same result if you calculate the slope of the segment connecting *any* pair of points in this list.

The slope m can be thought of as measuring the steepness of the line. The steepness does not change as we move along the line. If $m > 0$, the line rises to the right; if $m < 0$, it falls to the right. The larger the absolute value of m, the steeper the rise or fall of the line.

A line parallel to the y axis is vertical. Such a line has no slope, because the value of x is the same at all points and the denominator of the expression

$$m = \frac{y_2 - y_1}{x_2 - x_1}$$

is always zero. For a vertical line the slope is undefined.

A horizontal line has slope $m = 0$ because the numerator of the expression for m is always zero (y is constant on a horizontal line). You must take care to distinguish between having slope zero and having no slope at all.

Point-slope Equation of a Line

We can now show that a nonvertical line is the graph of a certain type of equation.

Theorem 3-1. The line passing through $P_0(x_0, y_0)$ with slope m is the graph of

$$y - y_0 = m(x - x_0). \tag{3-1}$$

Proof. Let L be the line described in the theorem; L is nonvertical because it has a slope m. To show that L is the graph of Eq. (3-1), we must verify that the coordinates of any point on L satisfy Eq. (3-1) and that any point whose coordinates satisfy Eq. (3-1) is on L.

The point $P_0(x_0, y_0)$ on L obviously satisfies Eq. (3-1). If $P_1(x_1, y_1)$ is any other point on L, then $x_1 \neq x_0$ (L is not vertical) and the slope of $P_0 P_1$ is equal to m:

$$\frac{y_1 - y_0}{x_1 - x_0} = m.$$

Consequently,

$$y_1 - y_0 = m(x_1 - x_0) \tag{3-2}$$

so that the coordinates of P_1 do satisfy Eq. (3-1).

Conversely, if the coordinates of $P_1(x_1, y_1)$ satisfy (3-1) then (3-2) holds. If $x_1 = x_0$, then $y_1 = y_0$ also, and the point P_1 coincides with P_0. If $x_1 \neq x_0$, let $P_1'(x_1, y_1')$ be that point on L whose x coordinate is x_1; such a point exists since L is nonvertical. Because P_1' is on L,

$$\frac{y_1' - y_0}{x_1 - x_0} = m.$$

Hence

$$y_1' - y_0 = m(x_1 - x_0).$$

But by Eq. (3-2), $m(x_1 - x_0) = y_1 - y_0$; so $y_1' - y_0 = y_1 - y_0$ and it follows that $y_1' = y_1$. The point $P_1(x_1, y_1)$ coincides with $P_1'(x_1, y_1')$ and hence is on L.

Equation (3-1) is called the **point-slope** equation of L because it displays the slope m and the coordinates of a point (x_0, y_0) on L.

Example 2

The point-slope equation of the line in Example 1 can be written in various equivalent ways by using, respectively, the points $(0, 3)$, $(1, 5)$, and $(3, 9)$:

$$y - 3 = 2(x - 0)$$
$$y - 5 = 2(x - 1)$$
$$y - 9 = 2(x - 3).$$

Each equation can be expressed simply as

$$y = 2x + 3.$$

Slope-intercept Equation of a Line

We can rewrite Eq. (3-1) as

$$y = m(x - x_0) + y_0$$
$$= mx - mx_0 + y_0.$$

Letting b denote $-mx_0 + y_0$, we see that an equation of L is

$$y = mx + b. \tag{3-3}$$

Clearly b is the value of y when $x = 0$; so $(0, b)$ is a point on L. It is called the **y intercept** of L (see Fig. 3-10). Equation (3-3) is called the **slope-intercept** equation of L.

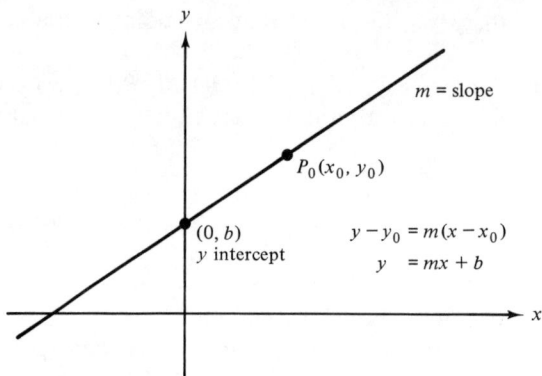

Figure 3-10

Example 3

Consider a line L having slope 4 and passing through $(-1, 2)$. Its point-slope equation is

$$y - 2 = 4(x + 1).$$

Simplifying this equation, we obtain the slope-intercept equation

$$y = 4x + 6.$$

The y intercept of L is $(0, 6)$.

Example 4

The two points $(5, -1)$ and $(4, 2)$ determine a line. Its slope is easily found to be -3. Using the first point, $(5, -1)$, the point-slope equation is $y + 1 = -3(x - 5)$. Using the second point, $(4, 2)$, the point-slope equation turns out to be $y - 2 = -3(x - 4)$. These equations are equivalent and both yield the slope-intercept equation $y = -3x + 14$. The y intercept of this line is $(0, 14)$.

Intercepts

The y intercept of a nonvertical line L is the point at which L crosses the y axis. It can be found by setting $x = 0$ and solving for y. Similarly, the x intercept of L is the point at which L crosses the x axis. It can be found by setting $y = 0$ and solving for x. In Example 3 the x intercept is obtained by setting

$$4x + 6 = 0$$

$$x = -\frac{3}{2}$$

Thus the x intercept is the point $(-3/2, 0)$. In Example 4 the x intercept is found to be $(14/3, 0)$.

A good way of graphing a line is to plot its two intercepts[1] and connect them by using a straightedge.

_____

[1] If the line passes through the origin or is parallel to one of the coordinate axes, it does not have two intercepts and this method does not work.

Example 5

Consider the line whose equation is $y = 3x - 5$. The y intercept $(0, -5)$ and x intercept $(5/3, 0)$ are connected by a line in Fig. 3-11.

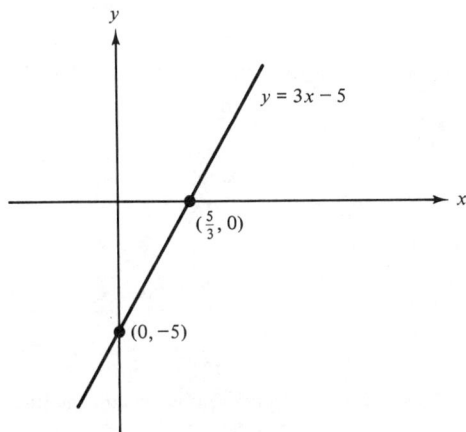

Figure 3-11

General Linear Equation

We have shown that any line in the plane is the graph of an equation of first degree in x and y. If the line is nonvertical, it has an equation $y = mx + b$, where m and b are constants; if the line is vertical, it has an equation $x = c$, where c is constant.

Conversely, the graph of any equation of first degree in x and y is a line. Consider the general first degree equation

$$Ax + By + C = 0 \tag{3-4}$$

in which A, B, and C are constants, and A and B are not both zero. Such an equation is expressible in one of the forms

$$y = -\frac{A}{B}x - \frac{C}{B} \qquad \text{(if } B \neq 0\text{)}$$

$$x = -\frac{C}{A} \qquad \text{(if } B = 0\text{)}.$$

In the first case, the graph is a line of slope $-A/B$ and y intercept $(0, -C/B)$. In the second case, the graph is a vertical line.

The general equation (3-4) is called **linear** because its graph is a line.

Example 6

The linear equation $4x + 3y - 2 = 0$ can be written $y = -\frac{4}{3}x + \frac{2}{3}$. Its graph is the line having slope $-\frac{4}{3}$ and y intercept $(0, \frac{2}{3})$.

EXERCISES 3-2

In Problems 1–10 find an equation of the line passing through the two points.

1. $(4, 1), (3, 2)$ **2.** $\left(\frac{1}{2}, 1\right), \left(0, \frac{1}{3}\right)$ **3.** $(-2, 1), (1, 2)$

4. $(3, 0)$, $(4, 0)$ **5.** $(\sqrt{2}, -\sqrt{2})$, $(1, 1)$ **6.** $\left(\frac{2}{3}, \frac{4}{5}\right)$, $\left(\frac{1}{5}, -\frac{5}{6}\right)$

7. $\left(\frac{4}{7}, \frac{11}{12}\right)$, $\left(\frac{4}{7}, \frac{5}{17}\right)$ **8.** $(0, 1)$, $(0, 4)$

9. (a, c), (b, c), where $a \neq b$ **10.** (a, b), (b, a), where $a \neq b$

In Problems 11–20 find the slopes and x and y intercepts of the given lines.

11. Problem 1   **12.** Problem 2   **13.** Problem 3   **14.** Problem 4

15. Problem 5   **16.** Problem 6   **17.** Problem 7   **18.** Problem 8

19. Problem 9   **20.** Problem 10

In Problems 21–32 draw the lines whose equations are given.

21. $y = 2x$ **22.** $3y - 2x + \sqrt{5} = 0$ **23.** $x = 3y$

24. $3x + 3y - 1 = 0$ **25.** $5x + 4y - 20 = 0$ **26.** $x = 2y + \sqrt{3}$

27. $y = -2$ **28.** $x = \sqrt{7}$ **29.** $\frac{x}{2} + \frac{y}{3} = 1$

30. $\frac{x}{2} - \frac{y}{3} = 1$ **31.** $\frac{3x}{4} - \frac{y}{4} = 1$ **32.** $\frac{3x}{4} - \frac{5y}{6} = 0$

In Problems 33–39 find a value of C such that the given point lies on the line whose equation is $3x - 5y + C = 0$.

33. $(2, 4)$ **34.** $(0, 0)$ **35.** $\left(\frac{1}{2}, \frac{1}{4}\right)$

36. $\left(-\frac{1}{7}, 0\right)$ **37.** $\left(3, \frac{1}{\sqrt{3}}\right)$ **38.** $(\sqrt{2}, \sqrt{2} - 1)$

39. $(0, 1)$

40. Fahrenheit (F) and Celsius (C) temperatures are related by a linear equation. Given that 0°C corresponds to 32°F and that 100°C corresponds to 212°F, determine a linear equation relating F and C. Graph this equation, using a vertical F axis and horizontal C axis.

FOR THE ENTHUSIASTS

41. (a) Show that if a line has two intercepts $(a, 0)$ and $(0, b)$, where a and b are nonzero, then its equation can be written in the form

$$\frac{x}{a} + \frac{y}{b} = 1.$$

This is called the **intercept** equation of the line.

(b) Find the intercept equation of the line whose intercepts are $(0, 4)$ and $(-3, 0)$.

(c) Find the intercept equation of the line having a point-slope equation

$$y - 3 = \frac{1}{3}(x + 4).$$

(d) Find the area of the triangle bounded by the coordinate axes and the line

$$\frac{x}{a} + \frac{y}{b} = 1.$$

42. Three (or more) points are said to be **collinear** if they all lie on one line. Show that the given points are not collinear.

(a) $\left(1, \frac{4}{3}\right)$ $(2, 2)$ $(4, 3)$

(b) $(1, 2)$ $(11, 3)$ $(34, 5)$

3-3 INTERSECTIONS OF GRAPHS

A point belongs to the intersection of two graphs if and only if its coordinates satisfy the equations of both graphs. Therefore to find the points of intersection, we solve the equations simultaneously. It is wise to check each point obtained to ensure that it satisfies both equations.

Example 1

Find the points of intersection of the graphs of $y = x^2$ and $x + 2y = 1$.

Solution. We can substitute x^2 for y in the second equation: $x + 2(x^2) = 1$. Doing so gives us the equation $2x^2 + x - 1 = 0$, which we can solve by factoring:

$$(2x - 1)(x + 1) = 0$$

$$x = \frac{1}{2} \qquad x = -1.$$

When $x = \frac{1}{2}$, $y = \frac{1}{4}$; when $x = -1$, $y = 1$. By substitution, it is easy to verify that points $(\frac{1}{2}, \frac{1}{4})$ and $(-1, 1)$ do have coordinates that satisfy both equations; therefore they are points of intersection (see Fig. 3-12) of the two graphs.

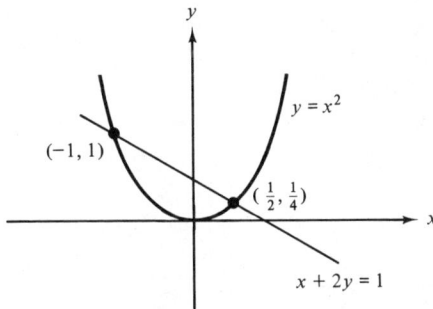

Figure 3-12

Intersection of Two Lines

You know from geometry that two lines in a plane either are parallel or intersect in just one point.

Example 2

Find the point of intersection of the lines $3x - y = 5$ and $5x + y = 3$.

Solution. (a) *Method of elimination.* It is easy to eliminate y by adding the given equations

$$3x - y = 5$$
$$5x + y = 3,$$

obtaining

$$8x = 8$$

so that

$$x = 1.$$

We see by substituting 1 for x in the second equation that $5 + y = 3$ and $y = -2$. So $x = 1$ and $y = -2$ and the point of intersection is $(1, -2)$.

(b) *Method of substitution.* First, use one of the equations to solve for one unknown in terms of the other. For example, the second equation gives $y = 3 - 5x$. Substituting this result in the first equation, we obtain

$$3x - (3 - 5x) = 5$$
$$8x = 8$$
$$x = 1.$$

Then $$y = 3 - 5 = -2.$$

The two lines and their point of intersection $(1, -2)$ are shown in Fig. 3-13.

Example 3

Consider the lines $3x + 2y = 0$ and $6x + 4y = 10$. If (x, y) satisfies the first equation, then $y = -\frac{3}{2}x$; substituting this expression in the second equation leads to the false statement $0 = 10$. These lines do not intersect and hence they are parallel (Fig. 3-14).

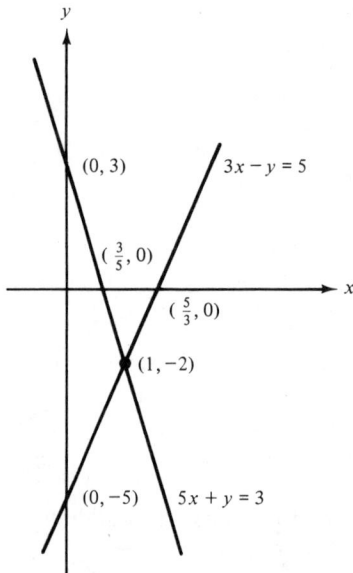

Figure 3-13

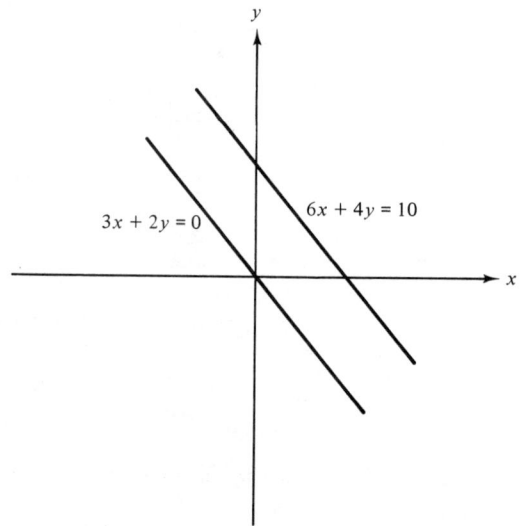

Figure 3-14

Example 4

The equations $3x + 2y = 0$ and $6x + 4y = 0$ are equivalent; so their graphs coincide.

Graphs That Do Not Intersect

Here is another example of graphs that do not intersect; this time the graphs are not lines.

Example 5

Find the points of intersection of $y = x^2 + 1$ and $y = \frac{1}{2}x^2$.

Solution. Substituting the value of y from the second equation into the first gives

$$\tfrac{1}{2}x^2 = x^2 + 1$$

or

$$0 = \tfrac{1}{2}x^2 + 1.$$

But there are no real numbers x satisfying this equation, for the right-hand side is at least one for any value of x and the left-hand side is zero. Thus there are no points of intersection (see Fig. 3-15).

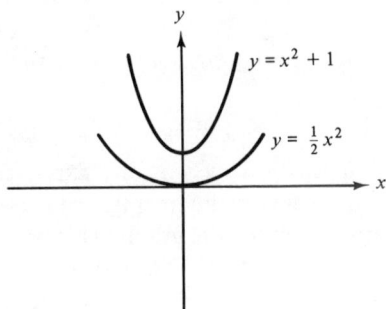

Figure 3-15

Extraneous Solutions

Sometimes the process of solving two equations simultaneously leads to points whose coordinates do not satisfy both equations. These so-called extraneous solutions do not represent points of intersection at all and must be rejected.

Example 6

Find the points of intersection of the graphs of $y = \sqrt{x}$ and $x + 2y = 3$.

Solution. Substituting $\sqrt{x}$ for y in the second equation, we obtain

$$x + 2\sqrt{x} = 3$$

or

$$2\sqrt{x} = 3 - x.$$

Squaring both sides,

$$4x = 9 - 6x + x^2$$

or

$$x^2 - 10x + 9 = 0.$$

The left side is easily factored and we find that the solutions of this equation are $x = 1$ and $x = 9$. Because $y = (3 - x)/2$, we obtain $(1, 1)$ and $(9, -3)$. But the second point *cannot* lie on the graph of $y = \sqrt{x}$, for its y coordinate is negative and $\sqrt{x}$ is nonnegative. The only point of intersection is $(1, 1)$ (see Fig. 3-16).

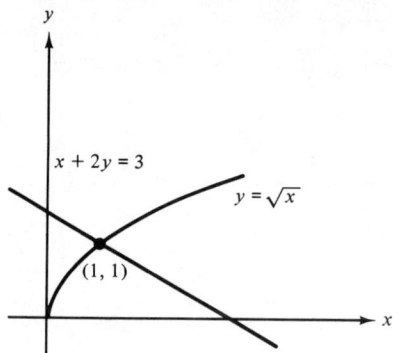

$x + 2y = 3$

$y = \sqrt{x}$

$(1, 1)$

Figure 3-16

Note. The procedure of squaring both sides of an equation is not a reversible one. The resulting equation may not be equivalent to the original even though it is implied by it. (Recall the discussion of implication and equivalence at the end of Section 1-1.) When solving equations, you should always be on guard when you use the squaring operation and should check your solutions in the original equations.

Intercepts of a Graph

A point of intersection of a graph with the x axis or y axis is called an **intercept**. The x axis has equation $y = 0$; so to find x intercepts, we solve the equation $y = 0$ simultaneously with the equation given for the graph. This process amounts to setting $y = 0$ in the given equation and solving for x. To find the y intercepts, we set $x = 0$ and solve for y.

Example 7

Find the x and y intercepts of $x^2 + 4y^2 = 16$.

Solution. Setting $y = 0$, we have $x^2 = 16$; so the x intercepts are $(4, 0)$, $(-4, 0)$. Setting $x = 0$, we obtain the y intercepts $(0, 2)$, $(0, -2)$.

EXERCISES 3-3

In Problems 1–15 find the points of intersection. Sketch the graphs and label the points of intersection.

1. $y = x^2$, $y = 3x - 2$
2. $x^2 + y^2 = 25$, $y = x + 1$
3. $y^2 = x + 4$, $x + y = 2$
4. $x^2 + y^2 = 10$, $y = x$
5. $x^2 + 2y^2 = 1$, $y = \frac{2}{3}x$
6. $y = 3x^2 + 4$, $y = x^2 - 1$
7. $x - 2y = 3$, $x + 2y = 7$
8. $2x + 3y = 1$, $x + 4y = 3$
9. $2x + 7y = 4$, $2x + 3y = 6$
10. $3x + 1 = 0$, $5y = 10x + 1$
11. $x = \sqrt{y}$, $y = 8 - 2x$
12. $y = \sqrt{3}(x + 1)$, $y = -\sqrt{3}(x - 1)$
13. $y = \sqrt{x - 1}$, $x + 2y = 1$
14. $x = 4y - 3$, $2x = 5 + 8y$
15. $x = 2y^2$, $y = 3x - 5$

In Problems 16–21 find the x and y intercepts. Sketch the graphs and label the intercepts.

16. $x^2 + y^2 = 25$
17. $x^2 - y = 25$
18. $x^2 + 4y^2 = 12$

19. $x^2 - y^2 = 0$ **20.** $y = x(x - 1)$ **21.** $y = x^3 - 15x$

22. The sides of a triangle lie on the lines

$$x - y - 1 = 0$$
$$x + y - 3 = 0$$
$$y = 2.$$

Find its vertices.

3-4 PARALLEL, PERPENDICULAR LINES

Parallel Lines

It is easy to tell whether two lines are parallel by comparing their slopes. Suppose that L_1 and L_2 are nonvertical parallel lines. As suggested by Fig. 3-17, triangles $P_1Q_1R_1$ and $P_2Q_2R_2$ are similar and so the ratios $Q_1R_1 : P_1Q_1$ and $Q_2R_2 : P_2Q_2$ are

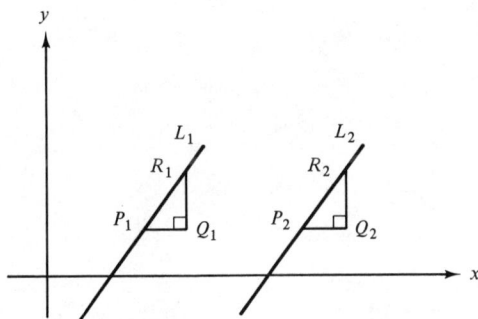

Figure 3-17

equal. But these ratios are the slopes of L_1 and L_2; thus we see that parallel lines have equal slopes. The converse is also true: if the slopes are equal, then the lines are parallel.

Theorem 3-2.   Two nonvertical lines are parallel if and only if they have the same slope.

Theorem 3-2 can also be proved algebraically. You are asked to do so later in Problem 30.

Example 1

The lines $3x + 2y = 0$ and $6x + 4y = 10$ (Example 3, Section 3-3) are parallel by Theorem 3-2, for each has slope $-3/2$.

Example 2

Find an equation for the line parallel to $2x + 3y = 6$ and passing through point $(2, 1)$.

Solution. Because the slope of the parallel line must be $-2/3$, its point-slope equation is $y - 1 = -(2/3)(x - 2)$ or $2x + 3y = 7$.

Perpendicular Lines

It is also possible to use slopes to determine whether two nonvertical lines are perpendicular. Because the geometric concept of angle is involved here, we must assume that we are working in a coordinate system in which the two axes have equal scales.

> **Theorem 3-3.**   Two nonvertical lines are perpendicular if and only if the product of their slopes is -1.

Proof. Let L_1 and L_2 be nonvertical lines with slopes m_1, m_2 respectively. Construct L_1' through the origin with slope m_1 and construct L_2' through the origin with slope m_2, as in Fig. 3-18. Clearly $L_1 \perp L_2$ if and only if $L_1' \perp L_2'$.

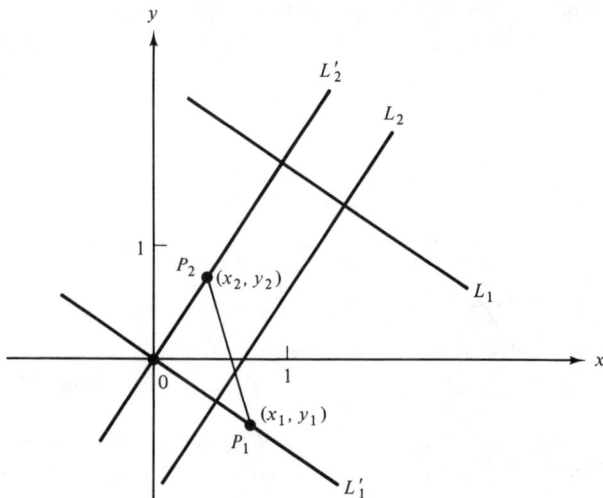

Figure 3-18

Choose points $P_1(x_1, y_1)$ on L_1' and $P_2(x_2, y_2)$ on L_2' such that neither point is on the y axis. Observe that $m_1 = y_1/x_1$ and $m_2 = y_2/x_2$.

Look at triangle OP_1P_2. Now $L_1 \perp L_2$ if and only if $L_1' \perp L_2'$ and $L_1' \perp L_2'$ if and only if OP_1P_2 has a right angle at O. But by the Pythagorean theorem OP_1P_2 has a right angle at O if and only if $\overline{OP_1}^2 + \overline{OP_2}^2 = \overline{P_1P_2}^2$. Applying the distance formula to each term of this equation, we find that $L_1 \perp L_2$ if and only if

$$(x_1^2 + y_1^2) + (x_2^2 + y_2^2) = (x_2 - x_1)^2 + (y_2 - y_1)^2.$$

This equation is equivalent to

$$0 = -2x_1x_2 - 2y_1y_2$$

or

$$y_1y_2 = -x_1x_2$$

or

$$\frac{y_1}{x_1} \cdot \frac{y_2}{x_2} = -1.$$

Because $m_1 = y_1/x_1$ and $m_2 = y_2/x_2$, we have proved that $L_1 \perp L_2$ if and only if $m_1 m_2 = -1$.

Two numbers whose product is -1 are called **negative reciprocals** of each other. Theorem 3-3 says that two nonvertical lines are perpendicular if and only if their slopes are negative reciprocals.

Example 3

The lines $2x + 3y = 6$ and $3x - 2y = 4$ are perpendicular because their slopes are negative reciprocals ($-\frac{2}{3}$ and $\frac{3}{2}$, respectively).

Example 4

Find an equation for the line perpendicular to $4x + 5y = 6$ and passing through (2, 1).

Solution. The slope of the perpendicular line must be the negative reciprocal of $-\frac{4}{5}$, which is $\frac{5}{4}$. So its point-slope equation is

$$y - 1 = \frac{5}{4}(x - 2)$$

or $$5x - 4y - 6 = 0.$$

Applications to Geometry

Theorems 3-2 and 3-3 enable us to give algebraic proofs of geometric theorems.

Example 5

Show that the diagonals of a square are perpendicular.

Solution. For any square we can set up a coordinate system so that the origin is at one vertex of the square and the coordinate axes contain two of its sides, as shown in Fig. 3-19.

If s is the length of a side of the square, then the coordinates of the other three vertices are $(s, 0)$, (s, s), $(0, s)$. The diagonal D_1 has slope $m_1 = (s - 0)/(s - 0) = 1$; the diagonal D_2 has slope $m_2 = (s - 0)/(0 - s) = -1$. Because the product of these slopes is -1, the diagonals are perpendicular.

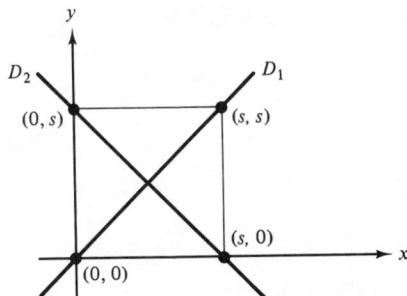

Figure 3-19

Example 6

Show that an angle inscribed in a semicircle is a right angle.

Solution. For any semicircle we can choose an origin at its center and let the x axis pass through its endpoints A and B (Fig. 3-20). These points then have coordinates $(-r, 0)$, $(r, 0)$, where r is the radius. Because the semicircle is the upper half of the circle $x^2 + y^2 = r^2$, its equation can be written $y = \sqrt{r^2 - x^2}$. For any point P on the semicircle consider the inscribed angle APB. If the x coordinate of P is a $(-r < a < r)$, then its y coordinate is $\sqrt{r^2 - a^2}$. The slope of AP is $\sqrt{r^2 - a^2}/(a + r)$ whereas the slope of PB is $-\sqrt{r^2 - a^2}/(r - a)$. The product of these slopes is easily seen to be -1, showing that $AP \perp PB$, and so the inscribed angle APB is a right angle.

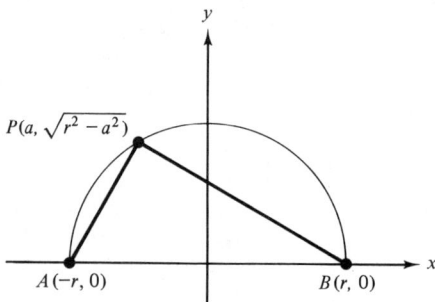

Figure 3-20

The Tangent to a Circle

Let P be a point on a circle whose center is C. The tangent to the circle at P is defined in geometry to be the line through P perpendicular to CP (see Fig. 3-21).

Example 7

Find an equation of the tangent to the circle $x^2 + y^2 = 25$ at $P(3, 4)$.

Solution. Observe that $P(3, 4)$ does lie on this circle. The center is $C(0, 0)$ and so the slope of CP is $\frac{4}{3}$. Hence the slope of the tangent is $-\frac{3}{4}$ and its equation is

$$y - 4 = -\frac{3}{4}(x - 3)$$

or

$$y = -\frac{3}{4}x + \frac{25}{4}.$$

(See Fig. 3-22.)

Example 8

Find an equation of the tangent to $x^2 + y^2 - 4x + 6y - 37 = 0$ at $P(-3, 2)$.

Solution. Observe that $(-3, 2)$ does lie on the circle. By completing the squares, we can express the given equation as

$$(x - 2)^2 + (y + 3)^2 = 50.$$

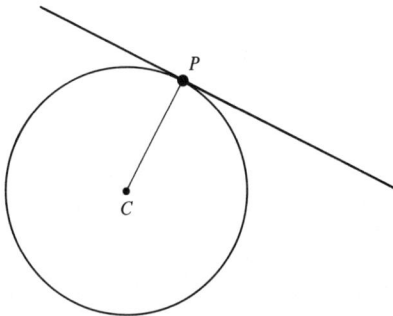

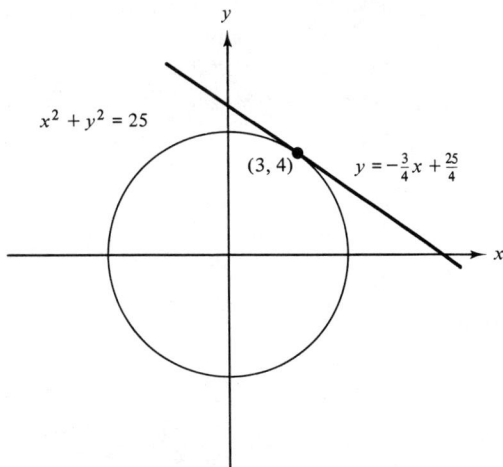

$x^2 + y^2 = 25$

$(3, 4)$ $y = -\frac{3}{4}x + \frac{25}{4}$

Figure 3-21   **Figure 3-22**

Hence the center is $(2, -3)$. The slope of CP is -1 and so the slope of the tangent is 1. The desired equation is easily found to be $y = x + 5$.

EXERCISES 3-4

In Problems 1–7 find an equation for the line parallel to the given line and passing through the given point.

1. $5x - 3y = 6$; $(-1, 2)$ **2.** $5x + 4y = 10$; $(1, 1)$ **3.** $y = 3$; $(1, 4)$

4. $\frac{x}{2} + \frac{y}{3} = 1$; $(0, 0)$ **5.** $x = 4$; $(2, -3)$ **6.** $6x - 2y = 3$; $(-1, 1)$

7. $2x - 5y = 10$; $(0, 0)$

In Problems 8–14 find an equation for the line perpendicular to the given line and passing through the given point.

8. $5x - 3y = 6$; $(-1, 2)$ **9.** $5x + 4y = 9$; $(1, 1)$ **10.** $y = 3$; $(1, 4)$

11. $\frac{x}{2} + \frac{y}{3} = 1$; $(0, 0)$ **12.** $x = 4$; $(2, -3)$ **13.** $6x - 2y = 3$; $(-1, 1)$

14. $2x - 5y = 0$; $(0, 0)$

15. Show that $(-4, -1)$, $(1, -3)$, $(5, 7)$ are the vertices of a right triangle.

16. Show that

$$\left(\frac{x_1 + x_2}{2}, \frac{y_1 + y_2}{2} \right)$$

is the midpoint of the segment connecting (x_1, y_1), (x_2, y_2).

In Problems 17–20, you may use the result of Problem 16.

17. Find the point of intersection of the diagonals of the square in Fig. 3-19. Show that these diagonals bisect one another.

18. Show that the line segment connecting the midpoints of two sides of a triangle is parallel to the third side and half as long.

Sec. 3-4   Parallel, Perpendicular Lines   **97**

19. Show that the line segments joining the midpoints of the sides of any quadrilateral form a parallelogram.

20. Find an equation of the circle having the ends of its diameter at (1, 3) and (3, 1).

In Problems 21–28 verify that the given point P lies on the circle and then find an equation of the tangent to the circle at P.

21. $x^2 + y^2 - 8x + 10y = 32; P(1, 3)$

22. $x^2 + y^2 + 6x + 6y = 11; P(2, -1)$

23. $x^2 + y^2 + 5x - 9y = 16; P(-2, -2)$

24. $x^2 + y^2 + x - 4y = -3; P\left(\dfrac{1}{2}, \dfrac{3}{2}\right)$

25. $x^2 + y^2 - 3x - 7y = \dfrac{7}{2}; P\left(-\dfrac{3}{2}, \dfrac{1}{2}\right)$

26. $x^2 + y^2 + 6y = 0; P(0, 0)$

27. $x^2 + y^2 + 4x = 0; P(-4, 0)$

28. $x^2 + y^2 = 13; P(2, -3)$

FOR THE ENTHUSIASTS

29. The perpendicular bisector of $P_1 P_2$ can be described as the set of all points that are equidistant from P_1 and P_2. Use this fact to find an equation for the perpendicular bisector of $P_1 P_2$.
(a) $P_1(2, 1), P_2(4, 3)$
(b) $P_1(3, -1), P_2(-3, 1)$
(c) $P_1(0, 0), P_2(4, 4)$
(d) $P_1(a, b), P_2(b, a)$, where $b \neq a$

30. In geometry two lines in a plane are defined to be parallel provided that they never intersect. With this definition you can prove Theorem 3–2 algebraically. Show that if L_1 and L_2 are two different lines having slopes m_1 and m_2, respectively, then their equations have no simultaneous solution if and only if $m_1 = m_2$.

31. The midpoint formula in Problem 16 can be proved in either of two ways: by using similar triangles or by using the distance formula. Is one of these methods to be preferred over the other and if so, why?

SUMMARY

In Chapter 3 you should have learned

1. the meaning of the graph–equation relationship. (Remember, this is a two-way relationship!)

2. the meaning of symmetry with respect to a line or with respect to a point and how to test an equation for symmetry with respect to the x axis, y axis, or origin.

3. how to determine any restrictions on the variables in an equation.

4. how to use symmetry and restrictions in drawing graphs.

5. how to recognize or find the equation of a line in various forms (point slope, slope-intercept, general linear).

6. how to determine the points of intersection of two graphs from their equations.

7. slope criteria for lines to be parallel or perpendicular.

8. how to apply slopes and distances to prove propositions in plane geometry.

9. methods for finding an equation of a line parallel (or perpendicular) to a given line and passing through a given point.

10. how to find an equation of the tangent to a circle at a point.

REVIEW EXERCISES

In Problems 1–12 test the equation to see whether its graph has symmetry with respect to the x axis, the y axis, or the origin.

1. $4x^2y = 1$ 2. $x = \sqrt{y^2 + 1}$ 3. $x^2 = y^2 + 1$ 4. $x^3y = 1$

5. $x = |y + 4|$ 6. $x = |y| + 4$ 7. $(x - y)^2 = 9$ 8. $x^4 + y^4 = 16$

9. $y = x + \dfrac{1}{x}$ 10. $y = x|x|$ 11. $y = \dfrac{|x|}{x}$ 12. $xy = 0$

In Problems 13–17 determine any restrictions on the variables.

13. $x = \sqrt{y^2 + 1}$ 14. $x^2 = y^2 + 1$ 15. $x = |y| + 4$

16. $x^4 + y^6 = 16$ 17. $9x^2 + 4y^2 = 36$

In Problems 18–23 draw the graph.

18. $x = \sqrt{y^2 + 1}$ 19. $x^2 = y^2 + 1$ 20. $y = \dfrac{x|x|}{x}$

21. $y = \dfrac{|x|}{x}$ 22. $9x^2 + 4y^2 = 36$ 23. $x = |y| + 4$

In Problems 24–28 find all points of intersection.

24. $x^2 + y^2 = 25, 3y - 4x = 0$

25. $x^2 + y^2 - 2x = 3, y = x + 1$

26. $x^2 + y^2 + x - 5y = 6, x^2 + y^2 + 2x - 6y = 7$

27. $x^2 + y^2 + 5y = 0, x^2 + y^2 + 4x + 7y = 6$

28. $x^2 + y^2 + 2x + 4y + 4 = 0, x^2 + y^2 + 3x + 3y + 1 = 0$

In Problems 29–32 determine whether the lines are perpendicular.

29. $3y + 4x = 5, y = \frac{3}{4}x + 2$

30. $y = 2x + 3, x = 2y - 3$

31. $y = 2, x = 3$

32. the line through $(3, 1)$ and $(2, 4)$; $x = 3y + 1$

In Problems 33–35 find an equation of the tangent to the circle at the given point.

33. $x^2 + y^2 = 25; (4, 3)$ 34. $x^2 + y^2 + 4x - 6y + 3 = 0; (1, 2)$

35. $x^2 + y^2 - 5x + 4y + 9 = 0; (2, -1)$

Elementary functions and coordinate geometry

TWO

Functions

4-1 RELATED VARIABLES

Relationships between variable quantities are at the heart of calculus. In this section we give a number of examples of related variables. We show how each relation can be expressed mathematically and then pictured in a graph.

Example 1

Suppose that your automobile has run out of gasoline just as you arrive at a service station. You begin filling the tank at 4:05, noticing that the pump supplies gasoline at a constant rate and that 12 gallons are pumped into the tank between 4:05 and 4:07.

Let y denote the amount of gasoline (in gallons) pumped into the tank at time x (in minutes after 4:05). Obviously the variables y and x are related. It takes exactly 2 minutes to pump 12 gallons, which means that y increases by 6 every minute. Therefore

$$y = 6x.$$

If your tank has a capacity of 20 gallons, then it will be full when

$$20 = 6x$$

or

$$x = \frac{10}{3}.$$

Thus the tank is full 10/3 minutes after 4:05—that is, 20 seconds after 4:08.

The graph of $y = 6x$ for $0 \le x \le 10/3$ shows the relationship between y and x (Fig. 4-1).

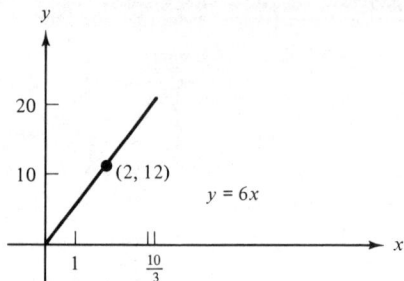

Figure 4-1

> **Proportion.** When two variables x and y are related in such a way that there is a certain constant k for which
>
> $$y = kx,$$
>
> then we say that y is **proportional** to x; k is called the **proportionality constant**.

In Example 1 y is proportional to x and the proportionality constant is $k = 6$.

Example 2 (Motion of a Falling Object)

Galileo observed that if an object is dropped, the distance s fallen by the object is proportional to the square of the time t of the fall. Therefore there is some constant k such that $s = kt^2$. Experimental evidence can be used to determine the proportionality constant k approximately. We would find, for example, that an object takes about 10 seconds to fall from a height of 1600 feet so that we can estimate the constant k from the equation

$$1600 = k \cdot 10^2,$$

obtaining $k = 16$. So we have

$$s = 16t^2,$$

where s is measured in feet and t in seconds. A graph of the relation between s and t for $0 \le t \le 10$ appears in Fig. 4-2.

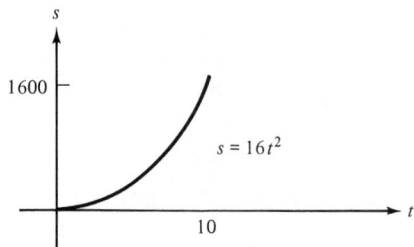

Figure 4-2

Example 3

Because of heavy traffic, an airplane is required to circle an airport for a long time. Suppose that the pilot is instructed to maintain a constant altitude of 4000 feet and a constant distance of 5000 feet from the control center. (See Fig. 4-3.) His path is a circle of radius 3000 feet. If we choose an x, y coordinate system at

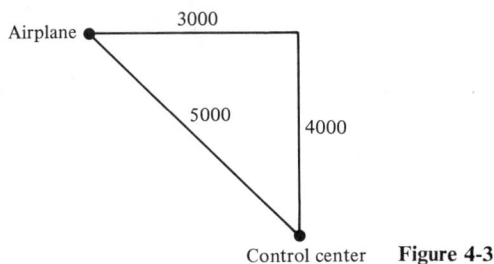

Airplane

3000

5000

4000

Control center   **Figure 4-3**

the center of this circle with the y axis directed north and the x axis directed east, then any point on the path of the plane has coordinates (x, y) (both measured in feet) that are related by the equation

$$x^2 + y^2 = (3000)^2.$$

(See Fig. 4-4.) Thus when the plane is 2000 feet east of the control center (i.e., when $x = 2000$), its y coordinate is $\pm\sqrt{(3000)^2 - (2000)^2} = \pm1000\sqrt{5}$ so that the plane is either $1000\sqrt{5}$ north of the control center or $1000\sqrt{5}$ south of it.

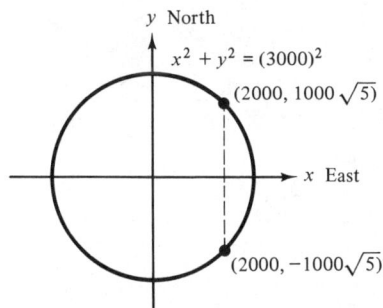

y North

$x^2 + y^2 = (3000)^2$

$(2000, 1000\sqrt{5})$

x East

$(2000, -1000\sqrt{5})$

Figure 4-4

Inverse Proportion.   When two variables x and y are related in such a way that there is a certain constant k for which

$$y = \frac{k}{x},$$

then we say that y is **inversely proportional** to x; k is called the **proportionality constant**.

Example 4 (Boyle's Law)

When a gas is held at a constant temperature in a flexible container, the pressure p is inversely proportional to the volume v. Therefore $p = k/v$ for some constant k. If by experiment we find that the pressure of a certain gas is 20 pounds per square inch when the volume of the container is 800 cubic inches, then $k = 16,000$ and we see that the pressure and volume satisfy the equation

$$p = \frac{16,000}{v}$$

in which p is measured in pounds per square inch and v in cubic inches.

Using this equation, it is easy to show that the pressure is 40 pounds per square inch when the volume is 400 cubic inches; if the volume is reduced to 100 cubic inches, the pressure is increased to 160 pounds per square inch. The relation between p and v is graphed in Fig. 4-5.

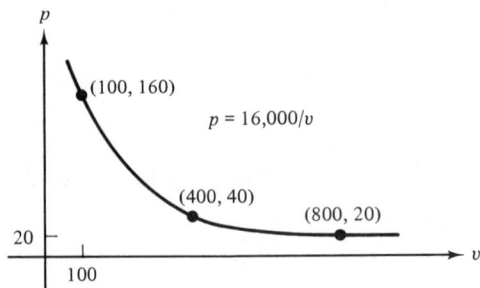

Figure 4-5

EXERCISES 4-1

1. Referring to Example 1 and assuming that gasoline costs $1.75 per gallon, determine the total cost C of the gasoline pumped into your tank x minutes after you start pumping. Graph the relation between x and C for $0 \leq x \leq 10/3$. What would the cost have been if you had shut off the pump at 4:08?

Problems involving falling objects
Use the result of Example 2 to work Problems 2–5.

2. A boy drops a stone from a bridge into a brook 18 feet below. After how many seconds does the stone hit the water?

3. An apple on a tree falls from a height of 25 feet. How long is it in the air?

4. A parachutist jumps from a plane at an altitude of 7500 feet. His parachute opens 5 seconds after he jumps. How far above the ground is he when the parachute opens?

5. If the parachutist in Problem 4 must open his parachute at an altitude of no less than 2600 feet, how long does he have after jumping until he must pull the cord?

6. Water is flowing out of a tank at a constant rate of 5 gallons per minute. If there were 300 gallons in the tank initially, determine the number g of gallons remaining after t minutes.
   (a) Graph the relation between t and g for appropriate values of t.
   (b) After how many minutes are there exactly 75 gallons left in the tank?
   (c) When will the tank be empty?

Fahrenheit and Celsius scales
7. The linear equation relating Fahrenheit temperature (F) to Celsius (C) is

$$F = \frac{9}{5}C + 32.$$

   (a) What value of F corresponds to a temperature of 30°C?
   (b) What value of C corresponds to a temperature of 75°F?
   (c) Find the temperature at which a Fahrenheit thermometer and a Celsius thermometer give exactly the same reading.

Fuel consumption

8. The consumption C of fuel (in gallons per hour) by a boat is proportional to the square of the speed v.

   (a) If 50 gallons per hour is required to maintain a speed of 20 knots, how many gallons per hour would be required to maintain a speed of 30 knots? v knots?

   (b) Graph the relation between C and v for $10 \leq v \leq 40$.

Intensity of illumination

9. The intensity of illumination from a source of light (measured in candlepower) is inversely proportional to the square of the distance from the source. At a distance of 10 meters the intensity of illumination of a certain street light is 20 candlepower. Find the intensity of illumination I at distance d meters. Graph the relation between d and I for $5 \leq d \leq 15$.

Period of a pendulum

10. The time required for one oscillation of a pendulum is called its **period**. It is known that the period of a pendulum is proportional to the square root of its length. A 16-inch pendulum is timed and it is found that an oscillation requires 1.3 seconds, approximately.

   (a) Use thus information to determine the period T of a pendulum in terms of its length L and graph this equation.

   †**(b)** How much time is required for one oscillation of a pendulum of length 50 inches?

4-2 FUNCTIONS

In the last section we discussed several relations between variables. A certain kind of relation has great importance and is given the special name "function." We shall now describe what we mean by a function, explain the notation and terminology used in connection with functions, and then look at a number of examples.

Definition. Suppose that two variables x and y are related in the following way:

1. The first variable x may assume any value in a specified set of numbers D.
2. There is a rule f that determines for each x in D exactly one value of the second variable y. The value of y that the rule associates with x will be denoted by the symbol $f(x)$.
3. R is the set of all values of y that are associated by the rule with some x in D.

Then:

4. The rule f is called a **function**.
5. The first variable is called the **independent** variable. The set D of all values of the independent variable is called the **domain** of f.
6. The second variable is called the **dependent** variable. The set R of all values of the dependent variable is called the **range** of f.

Briefly a function is a *rule* that associates with each number in its domain *exactly one* number in its range. We shall usually assume that the domain and range are sets of real numbers.

The notation $f(x)$ should be understood to be the result of applying the rule f to x; it does not mean multiplication of f by x. We call $f(x)$ the **value** of f at x.

The **graph** of a function f with domain D is the graph of the equation $y = f(x)$ for x belonging to D.

Other letters, such as $g, h, \ldots, F, G, H, \ldots$, are also commonly used to represent functions.

Let us return to the examples of Section 4-1 and determine which give rise to functions.

Example 1

Here the relation between x and y was found to be $y = 6x$ for $0 \leq x \leq 10/3$. The rule f: "multiply x by 6," associates with each x in the interval $[0, 10/3]$ exactly one value of y—namely, $y = 6x$. The set of all values of y consists of the interval $[0, 20]$. Thus f is a function with domain $D = [0, 10/3]$ and range $R = [0, 20]$. The independent variable is x, the dependent variable y. The graph of f is the graph of $y = 6x$ for $0 \leq x \leq 10/3$. (See Fig. 4-1.)

Example 2

The relation between t and s was given by $s = 16t^2$ for $0 \leq t \leq 10$. The rule g: "square t and then multiply by 16," associates with each t in the interval $[0, 10]$ exactly one value of s in the interval $[0, 1600]$. Thus g is a function with domain $D = [0, 10]$ and range $R = [0, 1600]$. The independent variable is t, the dependent variable s. The graph of g appears in Fig. 4-2.

Example 3

The relation between x and y was given by $x^2 + y^2 = (3000)^2$. This equation does not determine a unique y for each value of x. For example, if $x = 2000$, *two* values of y are determined—namely, $1000\sqrt{5}$ and $-1000\sqrt{5}$. Furthermore, this equation does not determine a unique x for each value of y. Therefore a function is not determined by this equation.

Example 4

The relation between v and p was given

$$p = \frac{16{,}000}{v} \quad \text{for } v > 0$$

and the rule h: "divide 16,000 by v," associates with each $v > 0$ exactly one value of p. Thus h is a function; its domain consists of all positive numbers and so does its range. The independent variable is v, the dependent variable p. The graph of h appears in Fig. 4-5.

If f is any function and x is an element of its domain, then the symbol $f(x)$ denotes the number associated with x by the function f. If f is the function of Example

1, we have

$$f(0) = 6 \cdot 0 = 0$$
$$f(1) = 6 \cdot 1 = 6$$
$$f(3) = 6 \cdot 3 = 18$$
$$f\left(\frac{10}{3}\right) = 6 \cdot \frac{10}{3} = 20$$

and, in general, $f(x) = 6x$ for $0 \le x \le 10/3$.

If g denotes the function of Example 2, we have

$$g(0) = 16 \cdot 0^2 = 0$$
$$g(1) = 16 \cdot 1^2 = 16$$
$$g(3) = 16 \cdot 3^2 = 144$$

and so on, and, in general, $g(t) = 16t^2$ for $0 \le t \le 10$.

Representing a Function and Finding Its Values. A function can be represented by
1. specifying its domain D and
2. giving a rule that indicates how to find the value of $f(x)$ for each x in D.
The value $f(x)$ at any particular x may be obtained by applying the rule to that x. If the rule is in the form of an equation for $f(x)$, we substitute for x in the given equation to obtain the desired function value.

Example 5

A function f is defined by $f(x) = 2x + 1$ for $0 \le x \le 1$. Its domain is the interval $[0, 1]$. The rule is given by the equation $f(x) = 2x + 1$. To find $f(\frac{1}{2})$, for example, we substitute $\frac{1}{2}$ for x, obtaining

$$f\left(\frac{1}{2}\right) = 2 \cdot \frac{1}{2} + 1 = 2.$$

Similarly,

$$f(1) = 2 \cdot 1 + 1 = 3.$$

The graph of this function appears in Fig. 4-6.

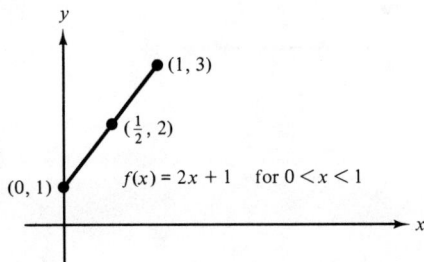

Figure 4-6

Example 6 (Square Function)

The **square function** is defined for all real numbers by $f(x) = x^2$. Then $f(0) = 0^2 = 0$, $f(3) = 3^2 = 9$, $f(1 + a) = (1 + a)^2 = 1 + 2a + a^2$, and so on. The graph of the square function is the graph of the equation $y = x^2$, shown in Fig. 3-3(e), and reproduced below.

Example 7 (Square Root Function)

The **square root function** is defined for $x \geq 0$ by $h(x) = \sqrt{x}$, where the symbol $\sqrt{}$ means, as usual, the nonnegative square root. Thus $h(1) = 1$, $h(\frac{1}{4}) = \frac{1}{2}$, and $h(-3)$ is undefined. The graph of the square root function appears in Fig. 4-7.

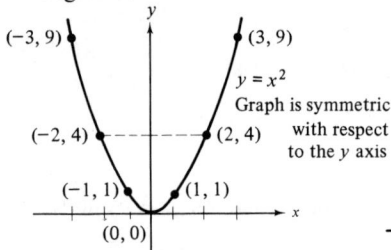

Figure 3-3(e)

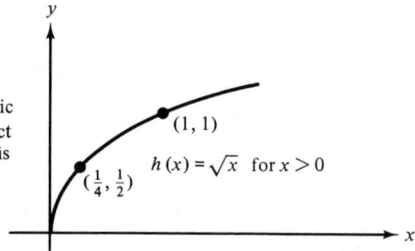

Figure 4-7

Example 8 (Reciprocal Function)

The **reciprocal function** is defined for $x \neq 0$ by $F(x) = 1/x$. Notice that if we let x increase through values $1, 2, 3, 4, \ldots$, then $F(x)$ decreases: $F(1) = 1$, $F(2) = \frac{1}{2}$, $F(3) = \frac{1}{3}$, $F(4) = \frac{1}{4}, \ldots$. If x decreases through values $\frac{1}{2}, \frac{1}{3}, \frac{1}{4}, \ldots$, then $F(x)$ increases: $F(\frac{1}{2}) = 2$, $F(\frac{1}{3}) = 3$, $F(\frac{1}{4}) = 4, \ldots$. $F(0)$ is undefined; we exclude zero from the domain, for $1/x$ is meaningless when $x = 0$. The graph of the reciprocal function appears in Fig. 4-8; it is called a **hyperbola**.

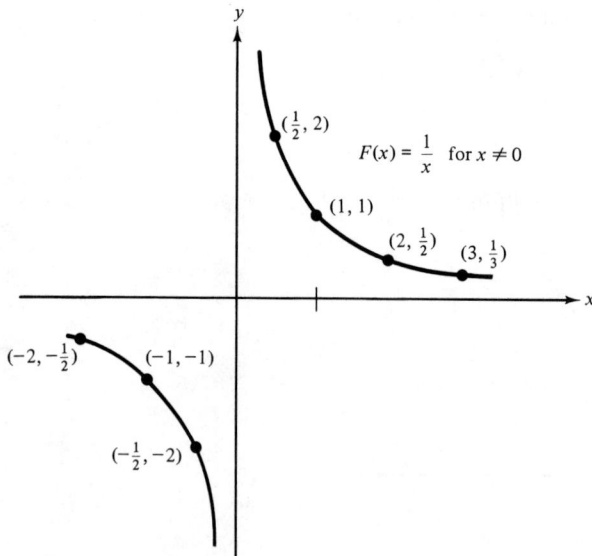

Figure 4-8

Functions   Chap. 4

It is common practice to represent a function f simply by stating the rule for finding $f(x)$ without specifying the domain of f. It is left for the reader to determine the domain D with the tacit understanding that D is the set of all real numbers for which the rule makes sense.

Example 9

If a function f is represented by the equation $f(x) = x^2 - 1$ and no domain is explicitly specified, we shall assume that D consists of all real numbers because the expression $x^2 - 1$ is meaningful for all real values of x.

Values of x that lead to divison by zero, to square roots of negative numbers, and so forth must be excluded when dealing with functions whose domains and ranges contain only real numbers.

Example 10

If we are given $F(x) = 1/x$ and no domain is specified, we shall assume that the domain of F is all real numbers except zero: $F(x) = 1/x$ for $x \neq 0$.

Example 11

If we are given $h(x) = \sqrt{5x - x^2}$ and no domain is specified, we shall assume that the domain of h is the set of all real numbers x for which $5x - x^2 \geq 0$. Solving this inequality, we obtain $0 \leq x \leq 5$ and so the domain of h is the interval $[0, 5]$.

EXERCISES 4-2

1. In each of the following statements a relation between variables x and y is described. Does this relation determine a function f with x as independent variable and y as dependent variable? If so, what is the rule and what are the sets D and R?
   (a) $y = 2x + 3$ for $-2 \leq x \leq 1$
   (b) $x + 2y = 5$ for $0 \leq x \leq 2$
   (c) $x + 2y = 5$ for real numbers x
   (d) $x^2 + y^2 = 16$ for $0 \leq x \leq 4$
   (e) $x^2 + y^2 = 16$ and $y \geq 0$ for $0 \leq x \leq 4$
   (f) $xy = 2$ for $x \neq 0$ and $y \neq 0$
   (g) $y = 16(x + 3)^2$ for $x \geq 0$

2. A function is defined for all real numbers $x \neq -1$ by

$$f(x) = \frac{1-x}{1+x}.$$

   Find each of the following values of this function.
   (a) $f(1)$ (b) $f(0)$

   (c) $f(3)$ (d) $f\left(\frac{1}{2}\right)$

   (e) $f\left(-\frac{2}{5}\right)$ (f) $f(1 + \sqrt{3})$

   (g) $f(1 - 2\sqrt{3})$ (h) $f\left(\frac{1}{\sqrt{3}}\right)$

(i) $f\left(\dfrac{1}{a}\right)$, where $a \neq -1$ (j) $f(f(a))$, where $a \neq -1$

(k) $[f(a)]^2$, where $a \neq -1$ (l) $f(a^2)$

3. Let f be the square function, defined in Example 6.
   (a) Evaluate $f(-2)$, $f(-3)$, $f(-5)$, $f(6)$, $f(2/3)$, $f(a^2 + 1)$, $f(\sqrt{a^2 + 1})$.
   (b) Is $f(a + b) = f(a) + f(b)$ for every pair of real numbers a, b?
   (c) Is $f(ab) = f(a)f(b)$ for every pair of real numbers a, b?
   (d) Is $f(a/b) = f(a)/f(b)$ for every pair of real numbers a, b such that $b \neq 0$?
   (e) Show that $f(a + b) + f(a - b) = 2f(a) + 2f(b)$ for every pair of real numbers a, b.

4. Let h be the square root function, defined in Example 7.
   (a) Evaluate $h(100)$, $h(9)$, $h(3)$, $h(1/2)$, $h(4/9)$, $h(0)$, $h(-4)$, $h(a^2)$.
   (b) Is $h(a + b) = h(a) + h(b)$ for every pair of positive real numbers a, b?
   (c) Is $h(ab) = h(a)h(b)$ for every pair of positive real numbers a, b?
   (d) Find a number x such that $h(x) = h(1 - x)$.

5. Let F be the reciprocal function, defined in Example 8.
   (a) Evaluate $F(2)$, $F(3)$, $F(5)$, $F(2/3)$, $F(a^2 + 1)$.
   (b) Is $F(a + b) = F(a) + F(b)$ for every pair of nonzero real numbers a, b?
   (c) Is $F(a/b) = F(a)/F(b)$ for every pair of nonzero real numbers a, b?
   (d) If $x \neq 0$, what is $F(F(x))$?
   (e) Show that $F(-x) = -F(x)$ for every $x \neq 0$.

6. For which of the following functions f is it true that

$$f(a + b) = f(a) + f(b)$$

   for every pair of real numbers a, b in the domain of f?
   (a) $f(x) = 2x$ for all real x
   (b) $f(x) = \frac{1}{2}x$ for all real x
   (c) $f(x) = x + 1$ for all real x
   (d) $f(x) = \sqrt{x}$ for $x \geq 0$
   (e) $f(x) = x^3$ for all real x

In Problems 7–25 a function is represented by stating a rule. The domain is the set of all real numbers for which the rule makes sense. Find the domain.

7. $f(x) = \dfrac{2}{3x - 6}$ 8. $g(x) = \dfrac{2x + 4}{3x + 8}$ 9. $h(x) = \dfrac{1}{x^2 - 1}$

10. $f(x) = \dfrac{2x + 1}{x^2 - 5x + 6}$ 11. $f(x) = \dfrac{x + 3}{x^2 - 4}$ 12. $g(x) = \dfrac{1}{x^2 + 4}$

13. $h(x) = \dfrac{x}{x^2 + 6x + 8}$ 14. $f(x) = \sqrt{x - 1}$ 15. $g(x) = \sqrt{4x - 3}$

16. $f(x) = \sqrt{1 - x^2}$ 17. $F(x) = \sqrt{x^2 - 4}$ 18. $g(x) = \sqrt{3x^2 - 6}$

19. $h(x) = \sqrt{x(x - 1)}$ 20. $G(x) = \sqrt{3x^2 + 6x}$ 21. $f(x) = \sqrt{x^2 + 3}$

22. $g(x) = \dfrac{1}{\sqrt{2x + 3}}$ 23. $h(x) = \dfrac{1}{\sqrt{x^2 + 5x}}$ 24. $F(x) = \dfrac{3x}{\sqrt{4x - x^2}}$

25. $f(x) = \dfrac{x - 1}{\sqrt{x^2 - 3x - 4}}$

26. A ball is thrown so that its height t seconds after it is thrown is given by $96t - 16t^2$ feet.
    (a) How high is the ball after 2 seconds?
    (b) When does the ball hit the ground?
    (c) State the rule and give the domain of a function h that describes this motion.

4-3 THE STATEMENT, "y IS A FUNCTION OF x"

When two variables x and y are related as in the definition in Section 4-2, it is customary to say that "y is a function of x." (A more precise statement would be, "There is a function f such that $y = f(x)$.") We are often required to express one of two variables as a function of the other—in which case, what we must do is determine the function that relates them. The resulting function shows explicitly how the dependent variable y depends on the independent variable x.

Example 1

Express the area y of a square as a function of its edge x.

Solution. From geometry we know that $y = x^2$. Because x can be any nonnegative real number, the function that we want is defined by $f(x) = x^2$ for $x \geq 0$.

Example 2

Express the area A of a circle (a) as a function of its radius r, (b) as a function of its diameter d, and (c) as a function of its circumference C.

Solution. (a) $A = \pi r^2$, where $r \geq 0$. Thus $A = g(r)$, where g is defined by $g(r) = \pi r^2$ for $r \geq 0$.

(b) Because $r = d/2$ we have $A = \pi \cdot (d/2)^2 = \pi d^2/4$ so that $A = h(d)$, where h is defined by $h(d) = \pi d^2/4$ for $d \geq 0$.

(c) The circumference C is $2\pi r$ and so $r = C/2\pi$. Hence $A = \pi r^2 = \pi(C/2\pi)^2 = C^2/4\pi$ so that $A = G(C)$, where $G(C) = C^2/4\pi$ for $C \geq 0$.

Notice in Example 2 that (a), (b), and (c) result in *three different* functions. These functions all represent the area A of a circle, but they represent A in terms of three different independent variables, r, d, and C.

Example 3

Express the radius r of a circle (a) as a function of its area A and (b) as a function of its circumference C.

Solution. (a) The equation $A = \pi r^2$ can be solved for r: $r = \sqrt{A/\pi}$. Because r and A are nonnegative, we have $r = F(A) = \sqrt{A/\pi}$ for $A \geq 0$.

(b) Because $C = 2\pi r$, we have $r = H(C)$, where $H(C) = C/2\pi$ for $C \geq 0$.

In Example 2(a) we expressed the area of a circle as a function of its radius, the rule being

$$A = g(r) = \pi r^2.$$

In Example 3(a) we expressed the radius as a function of the area, by means of the rule

$$r = F(A) = \sqrt{A/\pi}.$$

These rules were both derived from the same equation, $A = \pi r^2$. The first rule tells how to obtain A from r; A is dependent whereas r is independent. The second rule tells how to obtain r from A; here r is dependent, A is independent. Obviously the two rules are different.

When one variable y is a function of another variable x, we cannot say, in general, whether x is a function of y. It depends on the rule involved and the values the given variables are permitted to assume.

> **Example 4**
>
> Consider the equation $v = u^2 + 4$. This equation defines v as a function of u:
>
> $$v = f(u) = u^2 + 4 \qquad \text{for } u \text{ real.}$$
>
> It does not define u as a function of v, however. If we try to determine u from v, we are led to
>
> $$u^2 = v - 4$$
> $$u = \pm\sqrt{v - 4},$$
>
> which is ambiguous because of the plus or minus sign.

> **Example 5**
>
> Consider again the equation $v = u^2 + 4$. Suppose, however, that u has been restricted to nonnegative values. Then we can determine u from v by means of the equation
>
> $$u = \sqrt{v - 4}.$$
>
> Under these circumstances, u is a function of v, given by
>
> $$u = g(v) = \sqrt{v - 4} \qquad \text{for } v \geq 4.$$

EXERCISES 4-3†

1. Express the volume v of a cube as a function of its edge e. Graph the function.
2. Express the volume V of a sphere as a function
   (a) of its radius r.
   (b) of its diameter d.
   (c) of its surface area S.
3. (a) Express the surface area S of a sphere as a function of its radius r.
   (b) Is r a function of S? If so, find the function.
4. A rectangular flag is $1\frac{1}{2}$ times as long as it is wide. Express the area as a function of its width.
5. A rectangular vegetable garden is to be enclosed with 400 feet of fencing. Express the area of the garden as a function of its width. (Notice that neither the width nor the length can exceed 200 feet.) Graph the function.
6. A farmer has a field bordered along one side by a straight stone wall. He wants to plant a small rectangular vegetable garden within this field, using 400 feet of fencing for three sides of the garden and letting the stone wall serve as the fourth side. Express area A of the garden as a function of length x of the side perpendicular to the wall. (Notice that x cannot exceed 200 feet.)

†Some important geometric formulas are listed on the inside cover.

7. A topless rectangular box has a square base with edge x and height $\frac{2}{3}x$, where x is measured in meters and $2 \le x \le 4$. Express the volume and the surface area of the box as functions of x.

8. Suppose that the base of the box in Problem 7 must be specially reinforced and costs $1.50 per square meter whereas the sides cost $.75 per square meter. Express the cost of the material as a function of x.

9. A conical drinking cup full of water has a radius of 3 centimeters and a height of 8 centimeters. A hole is punched in the bottom so that the water leaks out. Find the volume of water in the cup when the depth of the water is h centimeters. Graph this function.

10. Sand being poured from a truck forms a conical pile in which the radius of the base is always twice the altitude. The pile is 4 feet high when the truck is emptied.
    (a) As the altitude increases from 0 to 4 feet, the volume of the growing pile of sand is a function of the altitude. Find this function.
    (b) Is the altitude a function of the volume? If so, find this function.

11. A cylindrical can is to hold 25 cubic units. The manufacturer wants to minimize the total surface area S. Express S as a function of the altitude h of the can, assuming that $1 \le h \le 5$.

12. A rectangle is inscribed in a circle of radius 6 units. Express the area of the rectangle as a function of its length.

13. A painter stands at the top of a 20-foot ladder that is leaning against a vertical wall. The base of the ladder slides away from the wall. Express the height of the painter above the ground as a function of the distance of the bottom of the ladder from the wall. Graph the function.

14. A spherical balloon is being inflated in such a way that its volume V is proportional to time t and increases by 64 cubic inches every minute. Assuming it is fully inflated when $t = 4$ minutes, express the radius r of the balloon as a function of time for $0 \le t \le 4$.

15. In the following problems you are given an equation relating the variables x and y, with some restrictions on their values. Determine whether y is a function of x and whether x is a function of y.
    (a) $x^2 + y^2 = 25$, $y \ge 0$
    (b) $x^2 + y^2 = 25$, $x < 0$
    (c) $x^2 + y^2 = 25$, $x \ge 0$, $y \le 0$
    (d) $x^2 - y^2 = 36$, $x > 0$, $y > 0$
    (e) $xy = 1$, $x \ne 0$, $y \ne 0$
    (f) $x^2y^2 = 1$, $x > 0$

4-4 PIECEWISE DEFINED FUNCTIONS

Sometimes a function is defined by a rule that is not simply an equation. The following examples illustrate this fact.

Example 1 (Absolute Value Function)

The **absolute value function** is defined for all real numbers by the rule

$$g(x) = \begin{cases} x & \text{for } x \ge 0 \\ -x & \text{for } x < 0 \end{cases}.$$

Applying this rule to $x = -3$, we get $g(-3) = -(-3) = 3$ because $-3 < 0$.

But if $x = 3$, we get $g(3) = 3$. Of course, we ordinarily write $g(x)$ as $|x|$, but you should recognize the vertical bar notation as a convenient way of abbreviating the foregoing rule.

The graph of the absolute value function (Fig. 4-9) consists of two half lines, $y = x$ for $x \geq 0$ and $y = -x$ for $x < 0$. They meet at the origin, making a right angle.

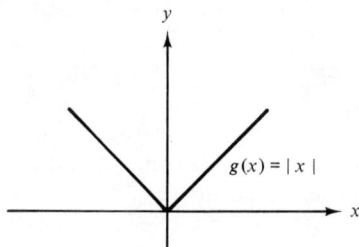

Figure 4-9

Properties of the Absolute Value Function. The absolute value function g has the properties

$$g(x) \geq 0 \quad \text{for all real numbers.}$$

$$g(xy) = g(x)g(y) \quad \text{for all real numbers } x \text{ and } y.$$

$$g(x + y) \leq g(x) + g(y) \quad \text{for all real numbers } x \text{ and } y.$$

These properties appeared in the vertical bar notation in Section 1-6.

Example 2

A function f is defined for real numbers by the rule

$$f(x) = \begin{cases} 1 + x & \text{for } x \geq 2 \\ 5 - x & \text{for } x < 2 \end{cases}.$$

For example, $f(9/2) = 1 + (9/2) = 11/2$; $f(1) = 5 - 1 = 4$. The graph of f appears in Fig. 4-10.

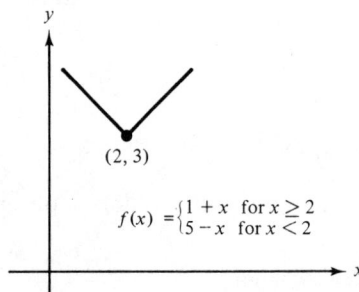

$$f(x) = \begin{cases} 1 + x & \text{for } x \geq 2 \\ 5 - x & \text{for } x < 2 \end{cases}$$

Figure 4-10

Examples 1 and 2 illustrate functions that are defined piecewise—that is, they are given by different equations on different intervals. In these two examples the "pieces" fit together to form a connected graph. In the next example we have a function defined piecewise whose graph consists of disconnected parts.

Example 3 (Postage Function)

The postage required for a first-class letter is a function of its weight w. According to 1982 regulations, a letter weighing an ounce or less requires 20¢ postage; for each additional ounce, or part of an ounce, the additional postage is 17¢. Thus for values of w up to 5 ounces, the required postage P is given by

$$P = f(w) = \begin{cases} 20 & 0 < w \le 1. \\ 37 & 1 < w \le 2 \\ 54 & 2 < w \le 3 \\ 71 & 3 < w \le 4 \\ 88 & 4 < w \le 5. \end{cases}$$

The graph of f appears in Fig. 4-11. The solid dot at point $(1, 20)$ means that $(1, 20)$ is on the graph; the hollow dot at $(1, 37)$ means that $(1, 37)$ is not on the graph.

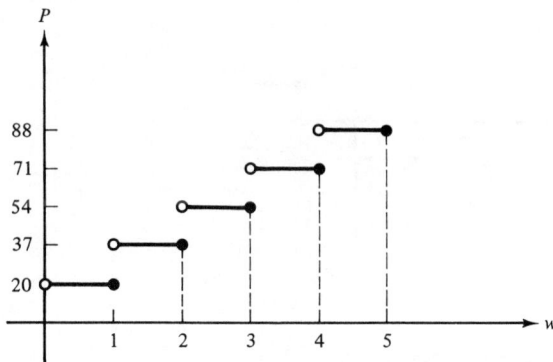

Figure 4-11

Example 4 (Income Tax Function)

State income tax for residents of a certain state is computed according to a table, part of which is given here. A function T is determined by this information.

TABLE 4-1

Taxable income		
Over	But not over	
$5,000	$ 7,000	$160 plus 5% of excess over $5,000
7,000	9,000	$260 plus 6% of excess over $7,000
9,000	11,000	$380 plus 7% of excess over $9,000

Algebraically, T may be given as follows, where x (representing taxable income) and $T(x)$ (representing tax) are given in dollars.

$$T(x) = \begin{cases} 160 + .05(x - 5,000) & \text{for } 5,000 < x \le 7,000 \\ 260 + .06(x - 7,000) & \text{for } 7,000 < x \le 9,000 \\ 380 + .07(x - 9,000) & \text{for } 9,000 < x \le 11,000 \end{cases}$$

The graph of T is interesting and you are asked to draw it in Problem 4.

Example 5 (Greatest Integer Function)

A function G is defined for all real numbers x by the rule

$G(x)$ is the greatest integer that is less than or equal to x.

The function G is called the **greatest integer function**. Its values are customarily denoted by using square brackets: $G(x) = [x]$. For instance, $G(2\frac{1}{2}) = [2\frac{1}{2}] = 2$, $G(.75) = [.75] = 0$, $G(\sqrt{125}) = [\sqrt{125}] = 11$, $G(-8/5) = [-8/5] = -2$, $G(4) = 4$. The graph of G for $-2 \le x \le 2$ appears in Fig. 4-12.

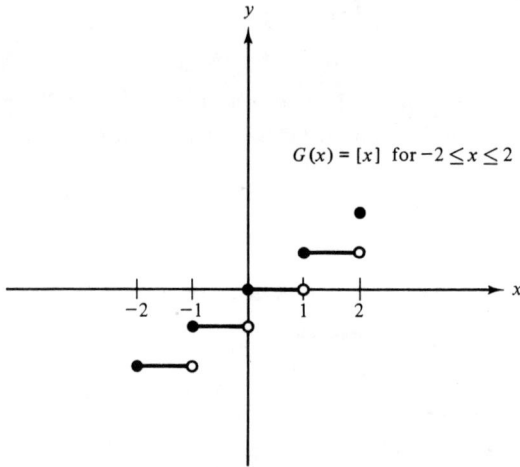

$G(x) = [x]$ for $-2 \le x \le 2$

Figure 4-12

EXERCISES 4-4

1. Draw the graph of each function.

   (a) $f(x) = \begin{cases} 2 + x & \text{for } x \le 1 \\ 4 - x & \text{for } x > 1 \end{cases}$

   (b) $f(x) = \begin{cases} x & \text{for } 0 \le x \le 1 \\ x^2 & \text{for } x > 1 \end{cases}$

   (c) $g(x) = \begin{cases} x^2 & \text{for } 0 \le x \le 1 \\ 1 & \text{for } x > 1 \end{cases}$

   (d) $h(x) = \begin{cases} -1 & \text{for } x < 0 \\ 0 & \text{for } x = 0 \\ 1 & \text{for } x > 0 \end{cases}$

2. (a) Let $f(x) = |x - 2| + 3$. Show that f is the function in Example 2. [*Hint:* What is $f(x)$ for $x \ge 2$? What is $f(x)$ for $x < 2$?]

   (b) Let $g(x) = |x + 2| - 1$. What is $g(x)$ for $x \ge -2$? What is $g(x)$ for $x < -2$? Draw the graph of g.

   (c) Let $h(x) = 4 - |2x - 6|$. Draw the graph of h.

3. The fare T charged by a taxi company is $1 for the first quarter mile and $.50 for each additional quarter mile or part thereof. Express T as a function of the number m of miles traveled, for $0 < m \le 1.5$. Graph the function.

4. Draw the graph of the income tax function defined in Example 4.

5. The weight W of an object depends on its distance d from the center of the earth. If d is less than or equal to the radius R of the earth, then W is proportional to d, but if d is greater than or equal to R, then W is inversely proportional to d^2. Assuming the weight of an object is w on the earth's surface, express W as a function of d for $d > 0$. Draw the graph.

6. Suppose that the gasoline tank of an automobile has a capacity of 20 gallons and that the automobile gets 15 miles per gallon on a trip. The owner takes a trip of 500 miles, starting with a full tank and stopping regularly every 150 miles to refill the tank. If g is the number of gallons in the tank after m miles, express g as a function of m for $0 \leq m \leq 500$. Draw the graph.

7. A company has a vacation policy for its employees. No vacation is earned until the employee has worked a full year. On completing a full year, a 2-week vacation (10 working days) is earned. Then for each additional year of work the employee's vacation increases by 2 days, except that no vacation longer than 20 working days is permitted.

    The number n of days of vacation is a function of t, the number of years of employment. Graph this function for $0 \leq t \leq 10$. How many days of vacation has an employee earned after 2 years and 9 months? After 10 years?

8. (a) The greatest integer function (Example 5) has the property that

$$[x + 1] = [x] + 1.$$

Verify this property for the following values of x.

$$1, \quad 2.5, \quad 0, \quad -.5, \quad -1, \quad -1.5$$

(b) Is the following equation true for all real numbers x and y?

$$[x + y] = [x] + [y]$$

4-5 AVERAGE RATE OF CHANGE

Example 1

A stone is dropped into a pond. The resulting ripples form circles with expanding radii. The function

$$y = f(x) = \pi x^2 \quad \text{for } x \geq 0$$

describes the relationship between radius x and area y of such a circle. Let us study the way in which y changes as x changes. Suppose that x increases from 2 to 5 so that $\Delta x = 3$ units. By how much has the area changed? When $x = 2$, $y = f(2) = 4\pi$; and when $x = 5$, $y = f(5) = 25\pi$, so that $\Delta y = 25\pi - 4\pi = 21\pi$. The ratio of these changes

$$\frac{\Delta y}{\Delta x} = \frac{f(5) - f(2)}{5 - 2} = \frac{21\pi}{3} = 7\pi$$

is called the average rate of change of y with respect to x between $x = 2$ and $x = 5$. Over that interval, *on the average*, y has changed by 7π square units for each unit change in x.

We emphasize the words "on the average" in Example 1. The change in y is not always 7π square units for a given unit change in x; sometimes it may be greater, sometimes less, as Table 4-2 shows.

TABLE 4-2

x changes from	Δy
2 to 3	$9\pi - 4\pi = 5\pi$
3 to 4	$16\pi - 9\pi = 7\pi$
4 to 5	$25\pi - 16\pi = 9\pi$
2.5 to 3.5	$12.25\pi - 6.25\pi = 6\pi$

Definition. Suppose that $y = f(x)$ is a function and that x_1 and x_2 ($x_1 \neq x_2$) are numbers such that the closed interval connecting x_1 and x_2 is contained in the domain of f. Then the **average rate of change** of $f(x)$ (or y) with respect to x between $x = x_1$ and $x = x_2$ is the ratio

$$\frac{\Delta y}{\Delta x} = \frac{f(x_2) - f(x_1)}{x_2 - x_1}.$$

Geometrically, the average rate of change of $f(x)$ with respect to x between $x = x_1$ and $x = x_2$ is the slope of the segment connecting the two points $(x_1, f(x_1))$ and $(x_2, f(x_2))$ on the graph of f (see Fig. 4-13). This slope does not depend on the order in which the two points are taken.

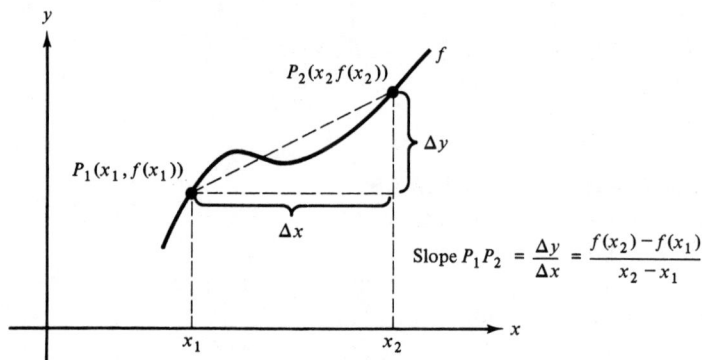

Figure 4-13

Example 2

Consider the function

$$p = h(v) = \frac{16{,}000}{v} \quad \text{for } v > 0,$$

which describes the pressure of a gas in a container having volume v (see Example 4, Section 4-1 and 4-2). The average rate of change of $h(v)$ with respect to v between $v = 400$ and $v = 500$ is given by

$$\frac{\Delta p}{\Delta v} = \frac{\frac{16{,}000}{500} - \frac{16{,}000}{400}}{500 - 400} = \frac{-8}{100}.$$

This means that over the interval from $v = 400$ to $v = 500$ an increase of 1 cubic inch in volume causes the pressure to decrease *on the average* by .08 pound per square inch.

An important special case of the average rate of change occurs when the position s of an object moving on a line is expressed as a function of time t, $s = f(t)$. Then the average rate of change of s with respect to t between $t = t_1$ and $t = t_2$

$$\frac{\Delta s}{\Delta t} = \frac{f(t_2) - f(t_1)}{t_2 - t_1}$$

is called the **average velocity** of the object over that time inverval.

Example 3

If you are at milepost 40 of a straight turnpike at noon and you arrive at milepost 79 at 12:45 P.M., then your average velocity over the time interval between noon and 12:45 is given by

$$\frac{\Delta s}{\Delta t} = \frac{79 - 40}{3/4} = 52 \text{ miles per hour}$$

Of course, your speedometer probably read less than 52 at some times during the trip and more than 52 at other times.

Example 4

An object falls from a height of 1600 feet. The distance s (feet) fallen after time t (seconds) is given by

$$s = f(t) = 16t^2 \quad \text{for } 0 \le t \le 10.$$

(See Example 2, Sections 4-1 and 4-2.)

The average velocity (feet per second) of the object over various time intervals is shown in Table 4-3.

TABLE 4-3

t_1	t_2	Δt	Δs	$\dfrac{\Delta s}{\Delta t}$
0	5	5	400	80
5	10	5	1200	240
0	10	10	1600	160
4	8	4	768	192

EXERCISES 4-5

1. Let $f(x) = \pi x^2$ for $x \ge 0$. Find the average rate of change of f with respect to x between $x = x_1$ and $x = x_2$ if
   (a) $x_1 = 2$, $x_2 = 4$.
   (b) $x_1 = 1$, $x_2 = 3$.
   (c) $x_1 = 0$, $x_2 = 5$.

2. Let $h(v) = 16,000/v$ for $v > 0$. Find the average rate of change of $h(v)$ with respect to v between $v = 500$ and $v = 200$.

3. An object falls from a height of 6400 feet. After how many seconds will it hit the ground? What was its average velocity during the first 5 seconds of fall? during the last 5 seconds of fall?

4. Let V be the volume of a sphere of radius r units. Find the average rate of change of V with respect to r

(a) between $r = 3$ and $r = 5$.

(b) between $r = 5$ and $r = 7$.

(c) between $r = 4$ and $r = 2$.

5. Find the average rate of change of the surface area of a sphere with respect to the radius x between $x = x_1$ and $x = x_2$.

6. The altitude h of a conical pile of sand is increasing. The shape of the cone is such that the radius of the base equals twice the altitude. Find the average rate of change of the volume with respect to the altitude between $h = 2$ meters and $h = 3$ meters.

7. Find the average rate of change of the circumference of a circle with respect to its diameter d between

(a) $d = 1$ and $d = 4$.

(b) $d = 3$ and $d = 9$.

(c) $d = d_1$ and $d = d_2$.

8. A company estimates that the profit P resulting from the manufacture of n units of a certain product can be expressed by the function

$$P = f(n) = 150n - n^2 \quad \text{for } 0 \le n \le 100.$$

Find the average rate of change of P with respect to n over the interval

(a) between $n = 0$ and $n = 50$.

(b) between $n = 50$ and $n = 75$.

(c) between $n = 50$ and $n = 100$.

9. Let T be the income tax function defined in Example 4, Section 4-4. Find the average rate of change of $T(x)$ with respect to x between $x = x_1$ and $x = x_2$ if

(a) $x_1 = 6000$ and $x_2 = 7000$.

(b) $x_1 = 7000$ and $x_2 = 9000$.

(c) $x_1 = 9000$ and $x_2 = 11,000$.

(d) $x_1 = 6000$ and $x_2 = 11,000$.

4-6 LINEAR FUNCTIONS

We showed in Section 3-2 that if m and b are constants, the graph of equation $y = mx + b$ is a nonvertical line having slope m and y intercept $(0, b)$. Accordingly, a function defined by such an equation is called linear.

> **Definition.** A **linear function** is a function f defined for all real numbers x by an equation of the form
> $$f(x) = mx + b. \tag{4-1}$$

Recall that slope m is the ratio of the change in y to the change in x, measured between any two points $P_1(x_1, y_1)$, $P_2(x_2, y_2)$ on the line:

$$m = \frac{\Delta y}{\Delta x} = \frac{y_2 - y_1}{x_2 - x_1}.$$

If m is positive, the line rises to the right. If m is negative, the line falls to the right. If $m = 0$, the line is horizontal. (See Sections 2-1 and 3-2, especially Fig. 2-8.)

Example 1

The equation

$$f(x) = 3x - 4$$

defines a linear function whose graph is a line with slope 3 and y intercept $(0, -4)$. It is easy to verify that the x intercept is $(\frac{4}{3}, 0)$. [See Fig. 4-14(a).]

Example 2

The equation

$$x + 3y = 9.$$

determines y as a linear function of x, for we can solve for y, obtaining

$$y = -\tfrac{1}{3}x + 3.$$

The slope of this line is $-\frac{1}{3}$, the intercepts $(0, 3)$ and $(9, 0)$. [See Fig. 4-14(b).]

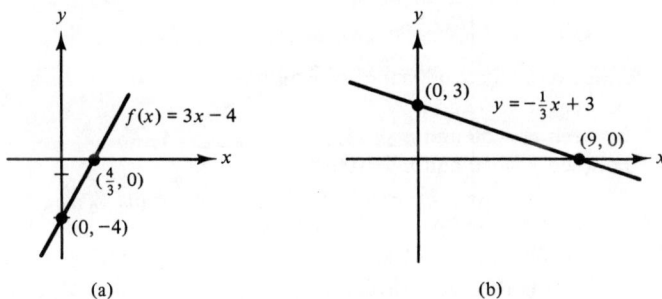

(a)

(b)

Figure 4-14

Rate of Change of a Linear Function

Let us calculate the average rate of change of a linear function $f(x) = mx + b$ between any two numbers x_1 and x_2.

$$\frac{\Delta y}{\Delta x} = \frac{f(x_2) - f(x_1)}{x_2 - x_1} = \frac{(mx_2 + b) - (mx_1 + b)}{x_2 - x_1} = \frac{m(x_2 - x_1)}{x_2 - x_1} = m.$$

The result is just the slope of the graph of f—the constant m. Because the average rate of change is constant, we call it simply the **rate of change** of f.

It follows that if y is a linear function of x, then y changes at a constant rate with respect to x. Conversely, if y changes at a constant rate with respect to x, then y can be expressed in terms of x by means of a linear function. We will show how to do so in some of the examples that follow. First, let us summarize the preceding discussion.

---

1. A function is linear if and only if it has a constant rate of change.
2. The rate of change of a linear function is the slope of its graph.

Example 3

Water is leaking out of a bucket at a constant rate of $\frac{1}{2}$ liter per minute. If the bucket contains 5 liters initially, we can express the number w of liters remaining after t minutes have elapsed by means of equation

$$w = 5 - \tfrac{1}{2}t.$$

Because the bucket is empty after 10 minutes, this formula applies when $0 \leq t \leq 10$. The function f defined by

$$f(t) = 5 - \tfrac{1}{2}t \qquad (0 \leq t \leq 10)$$

is part of the linear function graphed in Fig. 4-15. The rate of change of f is $-\frac{1}{2}$. The negative rate of change means that w is decreasing.

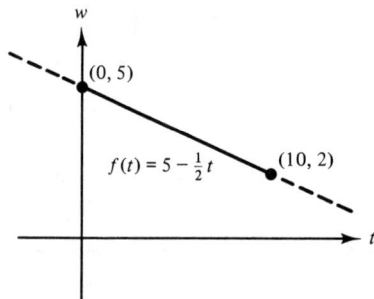

Figure 4-15

Example 4

An object is thrown straight up. The rate of change of its velocity v with respect to time t is known to be constant. Suppose that the velocity at time $t = 0$ is 50 kilometers per hour and that 4 seconds later it is 40 kilometers per hour. Express v as a function of t, find the velocity when $t = 10$ seconds, and find the time at which the velocity is zero.

Solution. Because the rate of change of v is constant, v is a linear function of t:

$$v = f(t) = mt + b$$

for certain constants m and b. We can determine m and b from the given information. When $t = 0$, $v = 50$; hence

$$50 = f(0) = b.$$

When $t = 4$, $v = 40$; so

$$40 = f(4) = 4m + 50.$$

Solving for m

$$m = -\frac{10}{4} = -\frac{5}{2}$$

and so, finally

$$v = f(t) = -\frac{5}{2}t + 50.$$

To find the velocity when $t = 10$, we evaluate $f(10)$:

$$f(10) = -\frac{5}{2} \cdot 10 + 50 = 25.$$

So at 10 seconds the velocity is 25 kilometers per hour. To determine when the velocity is zero, we set $f(t) = 0$, obtaining

$$0 = -\frac{5}{2}t + 50$$

$$t = 20.$$

The velocity is zero when $t = 20$ seconds.

Identity and Constant Functions

Here are two simple but important linear functions.

Example 5 (Identity Function)

The linear function $I(x) = x$ is called the **identity function**. It associates each real number with itself. Its graph appears in Fig. 4-16(a). The rate of change of I is one.

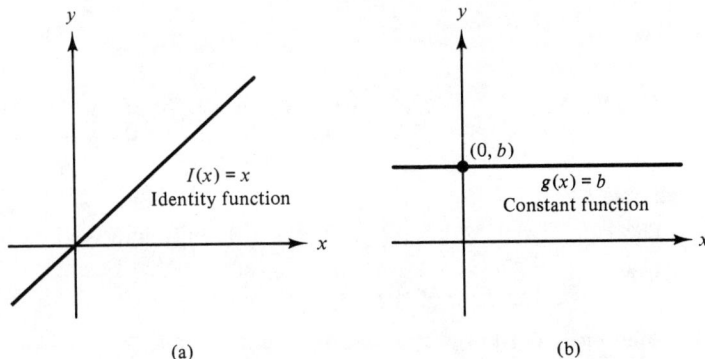

(a)

(b)

Figure 4-16

Example 6 (Constant Function)

If b is a fixed real number, the linear function $g(x) = b$ is called a **constant function**. It associates with each real number the constant b. The graph of such a function g appears in Fig. 4-16(b). The rate of change of a constant function is zero.

Hooke's Law for Springs

Hooke's law states that a spring will stretch from its natural length by an amount proportional to the weight hung on it. If w is the weight, then the amount of stretch is kw, where the proportionality constant k depends on the nature of the particular spring. Thus the extended length L of the spring is a linear function of w,

$$L = \text{natural length} + \text{stretch}$$

$$= L_0 + kw,$$

where L_0 is the natural length (see Fig. 4-17). Of course, Hooke's law holds only for limited values of w because a sufficiently heavy weight will break the spring.

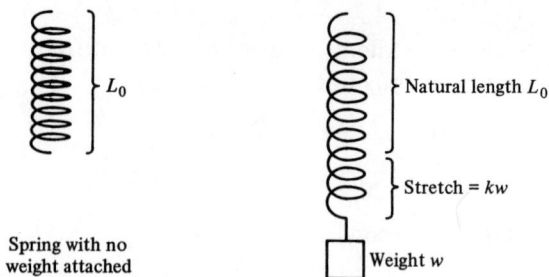

Spring with no weight attached

Natural length L_0

Stretch = kw

Weight w

Figure 4-17

Example 7

A spring is 10 centimeters long when no weight is attached, 13 centimeters long when a weight of 2 grams is attached. Express its extended length L as a function of the attached weight w grams and find the extended length when $w = 5$.

Solution. Because 2 grams produces a stretch of 3 centimeters, we can obtain the constant for this spring:

$$k \cdot 2 = 3$$
$$k = \tfrac{3}{2}.$$

Therefore $L = L_0 + kw = 10 + \tfrac{3}{2}w$. A weight of 5 grams produces a stretch of 7.5 centimeters, making the extended length 17.5 centimeters.

Example 8

The extended length of a spring is 10 centimeters when a 2-gram weight is attached, 15 centimeters when a 5-gram weight is attached. Express L for this spring as a function of the attached weight.

Solution. We know that Hooke's equation $L = L_0 + kw$ holds and that the pairs $w = 2$, $L = 10$; $w = 5$, $L = 15$ each satisfy this equation. Substituting, we obtain the simultaneous equations

$$10 = L_0 + 2k$$
$$15 = L_0 + 5k$$

whose solution is $k = 5/3$, $L_0 = 20/3$. Thus $L = 20/3 + (5/3)w$. This spring has a natural length 20/3 and its constant is 5/3.

Linear Function Through Two Points

Because two points are sufficient to determine a line, a linear function is completely determined by two points on its graph. This means that if we know the values of a linear function at two particular numbers, we can then find its value at any other number.

Example 9

Let f be a linear function such that $f(1) = 4$ and $f(2) = 7$. Find $f(\frac{3}{2})$.

Solution. Because the graph of f contains the two points $(1, 4)$ and $(2, 7)$, its slope is 3. Let $(\frac{3}{2}, y)$ be the point on the graph whose x coordinate is $\frac{3}{2}$. The slope of the segment connecting $(1, 4)$ and $(\frac{3}{2}, y)$ is also 3.

$$\frac{y - 4}{\frac{3}{2} - 1} = 3$$

$$y - 4 = \frac{1}{2} \cdot 3$$

$$y = 4 + \frac{3}{2} = \frac{11}{2}$$

Therefore

$$f\left(\frac{3}{2}\right) = \frac{11}{2}.$$

EXERCISES 4-6

1. Graph the following linear functions, labeling any intercepts.
   (a) $f(x) = 2x$
   (b) $f(x) = \sqrt{10}x + 4$
   (c) $h(x) = 4 + \frac{2}{3}x$
   (d) $f(x) = 10x - \sqrt{5}$
   (e) $f(x) = 1 - 3x$
   (f) $g(t) = 3 - \frac{1}{2}t$
   (g) $h(u) = \sqrt{5} - 2\sqrt{5}\,u$
   (h) $f(x) = \frac{2}{3} + \frac{1}{3}x$

2. Show that each equation determines y as a linear function of x. Find the rate of change of y with respect to x in each case.
   (a) $3y - 4x = 12$
   (b) $x + 3y = 3$
   (c) $\sqrt{2}x + \sqrt{3}\,y = 1$
   (d) $5y - 3 = 0$

3. Show that each of the equations in Problem 2(a), (b), (c) determines x as a linear function of y and find the rate of the change of x with respect to y. Show that the equation in Problem 2(d) does not determine x as a function of y.

4. The temperature in a mine shaft increases at a constant rate as the depth increases. For every 5 feet of depth the temperature increases by 1°C.
   (a) Given that the temperature at the surface is 5°C, express temperature T as a linear function of the depth d feet, for $d \geq 0$. Graph this function.
   (b) Find the rate of change of the function in part (a).
   (c) Find the temperature in the mine shaft at a depth of 220 feet.

5. The driver of an automobile applies her brakes at time $t = 0$ seconds. The speed of her automobile then decreases at a constant rate. If the speed at $t = 0$ was 40 miles per hour and if it takes 5 seconds to stop, express the speed v as a function of the time t for $0 \leq t \leq 5$.

6. The velocity v of a falling object increases at a constant rate in such a way that during each second v increases by 32 feet per second. Thus $v = f(t) = v_0 + 32t$, where v_0 is the initial velocity and t is time in seconds.

(a) What is the rate of change of v with respect to t?

(b) Suppose that an object is thrown downward with an initial velocity of 10 feet per second, reaching the ground at $t = 8$ seconds. Find the velocity at 4 seconds and find the impact velocity (i.e., the velocity when it hits the ground).

7. The extended length of a spring is 5 centimeters when a 2-gram weight is attached and is 7 centimeters when a 5-gram weight is attached. Find its natural length.

8. A spring has natural length 9 centimeters and measures 12 centimeters when a weight of 5 grams is attached. What weight would stretch this spring to 11 centimeters?

9. Express the circumference of a circle as a linear function of its diameter. What is the rate of change of this function?

10. The boiling point of water decreases with altitude. Suppose that at sea level the boiling point is 100°C while at an altitude of 5000 feet the boiling point is 94°C. Assuming that the relationship is linear, express the boiling point of water as a function of altitude h (in feet). What is the rate of change of this function?

11. Oil is being pumped at a constant rate into a vertical cylindrical tank 10 meters high. At 3:08 the depth of the oil in the tank is 4.5 meters. Five minutes later the depth is 4.7 meters.

(a) Show that the depth d of oil in the tank is a linear function of time t in minutes after 3:08 and find its rate of change.

(b) At what time will the tank be full?

(c) How long did it take to fill the tank if it was originally empty?

In Problems 12–16 assume that f is linear. Find the rate of change of f. Using the two given values of f, find the value of f at the indicated number. Sketch the graph, showing all three points.

12. $f(2) = 7, f(3) = 9; f(2.5) = ?$

13. $f(-1) = 3, f\left(\frac{1}{2}\right) = \frac{1}{3}; f\left(\frac{1}{3}\right) = ?$

14. $f(4) = 4, f(5) = 4; f(2) = ?$

15. $f(-2) = 0, f(6) = -3; f(1) = ?$

16. $f(-1) = \sqrt{3}, f(1) = -\sqrt{3}; f(\sqrt{12}) = ?$

4-7 APPROXIMATIONS BY LINEAR FUNCTIONS

Sometimes we are required to deal with a function whose values are not given by a convenient algebraic expression.

Example 1

A weather observer has recorded the temperature every hour from 12 noon to 5 P.M. on a certain afternoon. The result appears in Table 4-4 and the corresponding points in Fig. 4-18(a). The table does not show the temperature at 3:30, but we would guess that the temperature then was about 30°C. In any event, the relation

$$y = T(x) = \text{temperature at time } x \qquad \text{for } 0 \le x \le 5$$

defines a function even though no algebraic formula can be written for $T(x)$. A possible graph of T is represented by the curve in Fig. 4-18(b).

TABLE 4-4

x = time in hours after noon	y = temperature in degrees Celsius
0	26
1	28
2	28
3	29
4	31
5	27

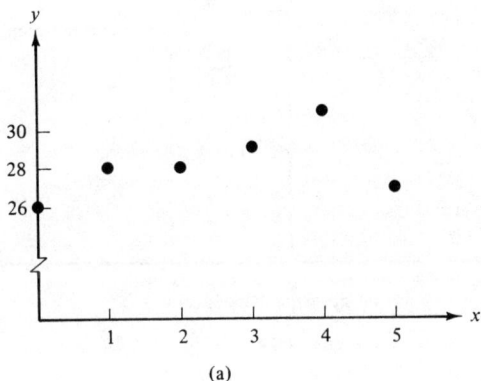

(a)

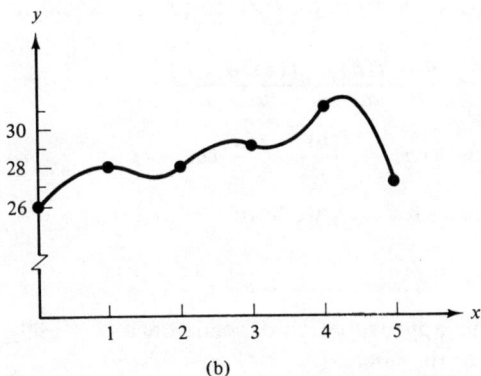

(b)

Figure 4-18

In Example 1, why did we guess that the temperature at 3:30 was approximately 30°? Since the temperature was 29° at 3:00 and increased by 2° over the next hour, its average rate of change over that interval was 2° per hour. *If the temperature were changing at a constant rate*, then the change in temperature over the half-hour from 3:00 to 3:30 would have been 1°, and the temperature at 3:30 would have been 30°. Our estimate was based on the assumption that the function T is approximately linear over a short interval.

Linear Interpolation

Suppose that we know the values of a function f at a and b and that we want to approximate the value of f at c, where c is between a and b. As in Example 1, we imagine that f changes at a constant rate between a and b—equivalently, that the graph over that interval is linear. In Fig. 4-19 we show the two points $(a, f(a))$ and $(b, f(b))$,

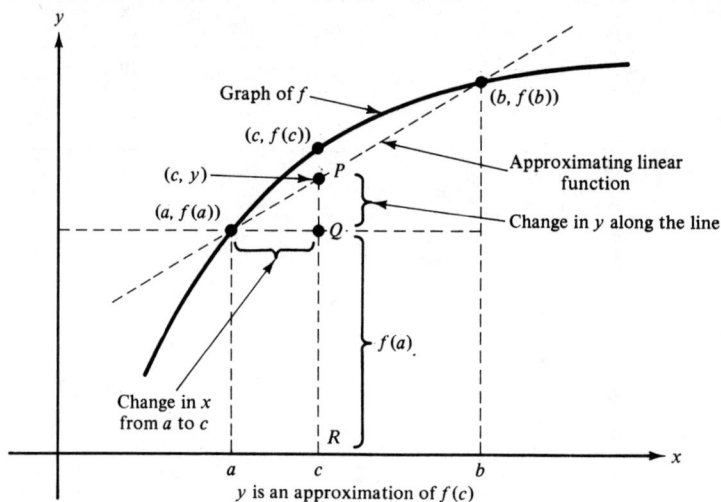

Figure 4-19

a possible graph of f, and the approximating linear function. If (c, y) is on the approximating line, then the slope from $(a, f(a))$ to (c, y) is equal to the slope from $(a, f(a))$ to $(b, f(b))$ and so

$$\frac{y - f(a)}{c - a} = \frac{f(b) - f(a)}{b - a}$$

$$y = f(a) + \frac{f(b) - f(a)}{b - a}(c - a).$$

Taking y as our approximation for $f(c)$, we have

$$f(c) \approx f(a) + \frac{f(b) - f(a)}{b - a}(c - a). \qquad (4\text{-}2)$$

Obviously the quality of this approximation depends on how closely the graph of f resembles a straight line over the interval.

This approximation technique is called **linear interpolation** because it makes the assumption that the function f is approximately linear.

Equation (4-2) can be interpreted as follows.

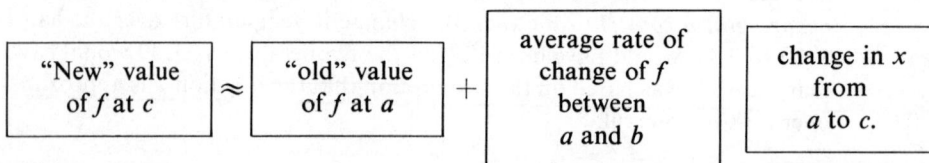

$$\boxed{\begin{array}{c}\text{``New'' value}\\ \text{of } f \text{ at } c\end{array}} \approx \boxed{\begin{array}{c}\text{``old'' value}\\ \text{of } f \text{ at } a\end{array}} + \boxed{\begin{array}{c}\text{average rate of}\\ \text{change of } f\\ \text{between}\\ a \text{ and } b\end{array}} \boxed{\begin{array}{c}\text{change in } x\\ \text{from}\\ a \text{ to } c.\end{array}}$$

The product on the right represents the change that the function f would undergo from $x = a$ to $x = c$ if f were changing at a constant rate. Adding this approximate change (QP in Fig. 4-19) to the "old" value $f(a)$ (RQ in Fig. 4-19) gives an approximation to the "new" value $f(c)$.

Example 2

Consider a function f, some of whose values are given in Table 4-5. The value

TABLE 4-5

x	$f(x)$
2	-2
4	6
6	15
8	25

of f at 3.75 is not listed, but we can approximate it by linear interpolation. We know that $f(2) = -2, f(4) = 6$. So using $c = 3.75$ in Eq. (4-2), we have

$$f(3.75) \approx f(2) + \frac{f(4) - f(2)}{4 - 2}(3.75 - 2)$$

$$= -2 + \frac{6 + 2}{2}(1.75)$$

$$= -2 + 7$$

$$= 5.$$

You can do this geometrically without using formula (4-2). Connect the points $(2, -2)$ and $(4, 6)$ with a line segment and determine the y coordinate of the point on that segment whose x coordinate is 3.75.

Square Roots by Interpolation

Square roots are often approximated by linear interpolation in conjunction with a table. A glance at the graph of $f(x) = \sqrt{x}$ (Fig. 4-7) suggests that, for large x, the graph can be approximated well by line segments; hence linear interpolation should work well. The first step is to locate the number whose square root we want between two consecutive entries in the table.

Example 3

A table of square roots shows that

$$\sqrt{23} \approx 4.796$$

$$\sqrt{24} \approx 4.899.$$

Use linear interpolation to approximate $\sqrt{23.4}$.

Solution. We let $f(x) = \sqrt{x}$. We know the values of f at 23 and 24 and we choose $c = 23.4$. Then (4-2) gives

$$\sqrt{23.4} = f(c) \approx f(23) + \frac{f(24) - f(23)}{24 - 23}(23.4 - 23)$$

$$\approx 4.796 + (.103)(.4)$$

$$\approx 4.796 + .041$$

$$\approx 4.837.$$

Figure 4-20 indicates that the interpolated value is a little less than the true value of $\sqrt{23.4}$.

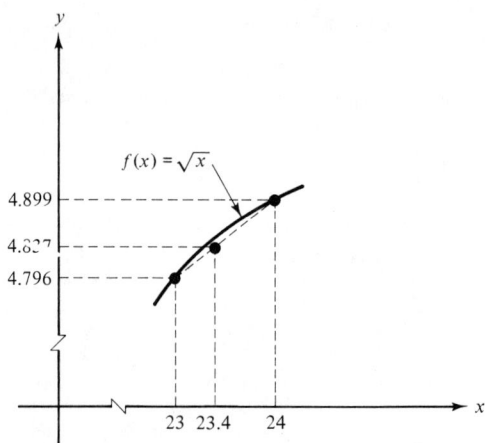

Figure 4-20

Interpolating for x, Given y

We have dealt with the problem of approximating the value of the dependent variable y for a certain value of the independent variable x. Occasionally it is necessary to do the opposite.

Example 4

Referring to Table 4-4, approximate the time between 4:00 and 5:00 when the temperature was 28°.

Solution. Let $(x, 28)$ be a point on the line segment connecting points $(4, 31)$ and $(5, 27)$ in Fig. 4-18(b). We get

$$\frac{28 - 31}{x - 4} = \frac{27 - 31}{5 - 4} = -4,$$

which yields

$$-3 = -4(x - 4)$$

$$x = 4\frac{3}{4}.$$

Thus the temperature was 28° at approximately 4:45.

EXERCISES 4-7

1. Using the method of linear interpolation, approximate the following values of f for the function in Table 4-5.
   (a) $f(2.5)$
   (b) $f(4.6)$
   (c) $f\left(\frac{15}{7}\right)$

2. *Linear extrapolation*
   Continuing with the function in Problem 1, devise a method for approximating the following values.
   (a) $f(8.6)$ [*Hint:* Extend the line segment connecting $(6,15)$ and $(8, 25)$.]
   (b) $f(1.75)$

3. (a) Let f be a function such that $f(-1) = 0$ and $f(3) = 8$. By linear interpolation, approximate $f(1)$.
   (b) Show that if $f(x) = x^2 - 1$, then f satisfies the conditions of part (a). What is the true value of $f(1)$? Why is the approximation poor?

4. Water in a container is evaporating during a chemistry experiment. A student observes the weight of the water at several times during the afternoon and tabulates her observations.

Time	Weight of water in grams
1:00	525
1:30	490
2:00	450
3:00	355

   (a) Approximate the weight of the water at 2:30 by linear interpolation.
   (b) Is the weight of the water a linear function of time?
   (c) Do you think your approximation in part (a) is greater or less than the true weight at 2:30?
   (d) Approximate the time at which the weight of water was 400 grams.

5. A ball is rolling along an inclined plane. The distance rolled is observed at several times.

Time (seconds)	Distance (meters)
0	0
1	1.4
2	2.8
3	4.2

   (a) Approximate the distance that the ball has rolled at 2.5 seconds.
   (b) Approximate the time at which the ball passes the 2-meter mark.

(c) If distance is a linear function of time, find the function.

†6. Use linear interpolation and the information in Example 3 to approximate $\sqrt{23.7}$. Check your answer with a calculator.

A table of square roots and cube roots lists the following entries.

x	$\sqrt{x}$	$\sqrt[3]{x}$
70	8.367	4.121
71	8.426	4.141

In Problems 7–12 use linear interpolation and the information above to approximate the given number. Check your answer with a calculator.

†7. $\sqrt{70.5}$ †8. $\sqrt{70.3}$ †9. $\sqrt[3]{70.7}$

†10. $\sqrt[3]{70.2}$ †11. $\sqrt{282.4}$ †12. $\sqrt[3]{560.8}$

4-8 TRANSLATION OF AXES

Our purpose here is to consider the effects of changing the coordinate system in a plane by moving the axes in a certain way.

Suppose that an x, y coordinate system is given and let C be the point with coordinates (h, k) in that system. We construct new coordinate axes, a u axis and a v axis, as follows.

1. The origin of the u, v coordinate system is at $C(h, k)$.
2. The u axis is parallel to the x axis, with the same direction and scale.
3. The v axis is parallel to the y axis, with the same direction and scale.

The "old" x, y axes and the "new" u, v axes are shown in Fig. 4-21. It is clear geometrically that if a point P has coordinates (u, v) in the new system, then its coordinates in the old x, y system satisfy

$$x = u + h$$
$$y = v + k$$
(4-3)

or equivalently

$$u = x - h,$$
$$v = y - k.$$
(4-4)

[These equations hold regardless of the position of P relative to the point (h, k) and regardless of the quadrant in which (h, k) lies.] Equations (4-3) and (4-4) are said to define a **translation of axes** to the point $C(h, k)$.

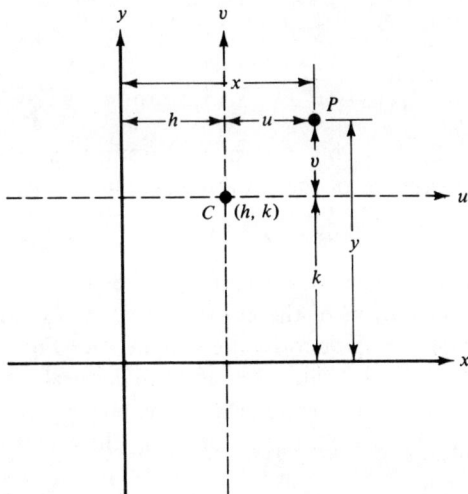

Figure 4-21

Example 1

Give the equations for translation to the point $(1, -2)$ and find the new coordinates of the point P whose old coordinates are $(5,1)$.

Solution. Because $h = 1$ and $k = -2$, formulas (4-3) and (4-4) become

$$x = u + 1, \qquad u = x - 1,$$
$$y = v - 2; \qquad v = y + 2.$$

With $x = 5$ and $y = 1$, we get $u = 4$, $v = 3$; hence P has new coordinates $(4, 3)$.

Example 2

Express the linear equation $y = 2x - 4$ in terms of u and v after translating the axes to the point $(1, -2)$.

Solution. The equations for this translation are given in Example 1. Substituting $x = u + 1$ and $y = v - 2$ in $y = 2x - 4$, we get

$$v - 2 = 2(u + 1) - 4$$

or
$$v = 2u.$$

Notice that this line passes through the origin of the new coordinate system. (See Fig. 4-22.)

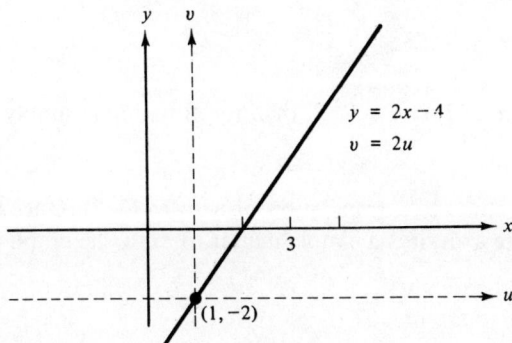

Figure 4-22

Example 3

Consider the circle

$$(x - 1)^2 + (y + 2)^2 = 9 \qquad (4\text{-}5)$$

(Example 4, Section 2-2). It seems natural to substitute $u = x - 1$ and $v = y + 2$ in this equation because of the resulting simplification:

$$u^2 + v^2 = 9. \qquad (4\text{-}6)$$

This substitution is actually just a translation of axes to $(1, -2)$. We can see quite clearly from the new form of the equation that it is a circle of radius 3 centered at the origin of the u, v coordinate system. (See Fig. 4-23.) Equation (4-6) belongs to the circle just as legitimately as (4-5) does except that it refers to a different coordinate system. The simplicity of the new equation is due to the fact that the new coordinate system has its origin at the center of symmetry of the circle.

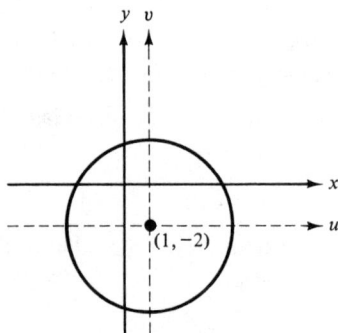

Figure 4-23

The last example suggests that translation of axes to a suitable point (h, k) may help us to draw a graph. If a substitution $u = x - h$, $v = y - k$ simplifies the equation, then a translation to (h, k) should be investigated.

Example 4

Graph the function $f(x) = |x - 3| + 2$.

Solution. Let us set $y = f(x)$ so that

$$y = |x - 3| + 2$$

or

$$y - 2 = |x - 3|.$$

If we substitute $u = x - 3$, $v = y - 2$, then the equation is simply

$$v = |u|.$$

This substitution amounts to a translation of axes to $(3, 2)$. Once we have set up the new coordinate axes, it is a simple matter to draw the graph (Fig. 4-24).

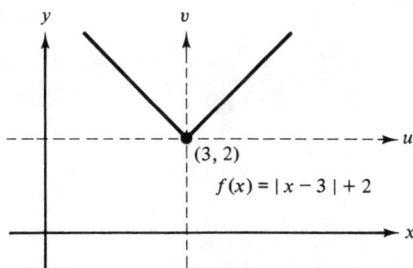

Figure 4-24

When an equation contains terms of second degree, it may be helpful to complete the square before deciding on which translation to make.

Example 5

Graph the function f defined for real x by $f(x) = x^2 - 4x + 2$.

Solution. Setting $y = f(x)$, we have

$$y = x^2 - 4x + 2;$$

and completing the square,

$$y = (x^2 - 4x + 4) - 2$$
$$y = (x - 2)^2 - 2.$$

or

$$y + 2 = (x - 2)^2.$$

We substitute $u = x - 2$, $v = y + 2$. This translation of axes to $(2, -2)$ gives us a new equation $v = u^2$, whose graph we can easily draw (Fig. 4-25).

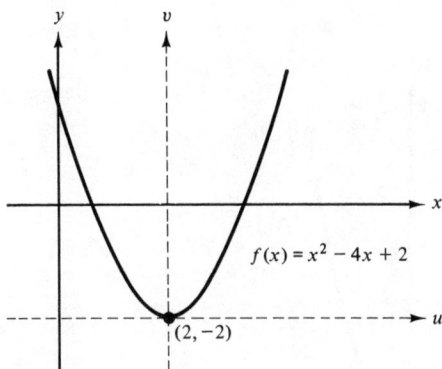

$f(x) = x^2 - 4x + 2$

$(2, -2)$

Figure 4-25

EXERCISES 4-8

In Problems 1–8 express the equation in terms of u and v after translating the axes to the given point.

1. $x + y = 3$; $(1, 2)$

2. $y = 4x + 5$; $(-1, 1)$

3. $x - y = 2$; $(3, 0)$

4. $x^2 - 2x - y = 4$; $(1, -5)$

5. $x^2 + y^2 - 4x + 6y + 9 = 0$; $(2, -3)$

6. $x^2 - y^2 + 2x + 6y = 9$; $(-1, 3)$

7. $xy = 1$; $(2, 3)$

8. $x^2 + y^2 - 2\sqrt{3}x + \sqrt{5}y + 2 = 0$; $\left(\sqrt{3}, -\dfrac{\sqrt{5}}{2}\right)$

In Problems 9–20 use a suitable translation of axes to graph the given function.

9. $f(x) = |x - 3| + 1$ **10.** $g(x) = |x + 2| - 5$ **11.** $f(x) = |2x + 4| + 1$

12. $f(x) = x^2 + 6x + 10$ **13.** $g(x) = \sqrt{x + 3} - 2$ **14.** $h(x) = x^2 - 4x$

15. $g(x) = x^2 - 5x + 2$ **16.** $f(x) = 3 + \sqrt{x - 5}$ **17.** $f(x) = 6x - x^2$

18. $f(x) = \dfrac{1}{x - 3} + 1$ **19.** $f(x) = \dfrac{2}{x + 1} - 3$ **20.** $g(x) = 2|x - 3| - 4$

4-9 QUADRATIC FUNCTIONS

> **Definition.** A **quadratic function** is a function f defined for all real numbers x by an equation of the form
>
> $$f(x) = Ax^2 + Bx + C \tag{4-7}$$
>
> in which A, B, C are constants and $A \neq 0$.

The simplest quadratic function is defined by $f(x) = x^2$. Here $A = 1$ whereas B and C are both zero. The graph of $y = f(x) = x^2$ was discussed in Section 3-1 and is reproduced in Fig. 4-26.

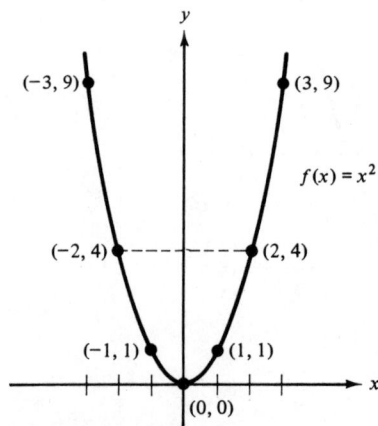

Figure 4-26

Next, consider the case in which $f(x) = Ax^2$. The graph of $y = Ax^2$ is similar to that in Fig. 4-26 except that the y coordinate of every point is multiplied by the constant A. Several examples are shown in Fig. 4-27. Of course, the graph of $f(x) = Ax^2$ is symmetric with respect to the y axis. Multiplication by a negative value of A causes the graph to be reflected through the x axis [Fig. 4-27(c)].

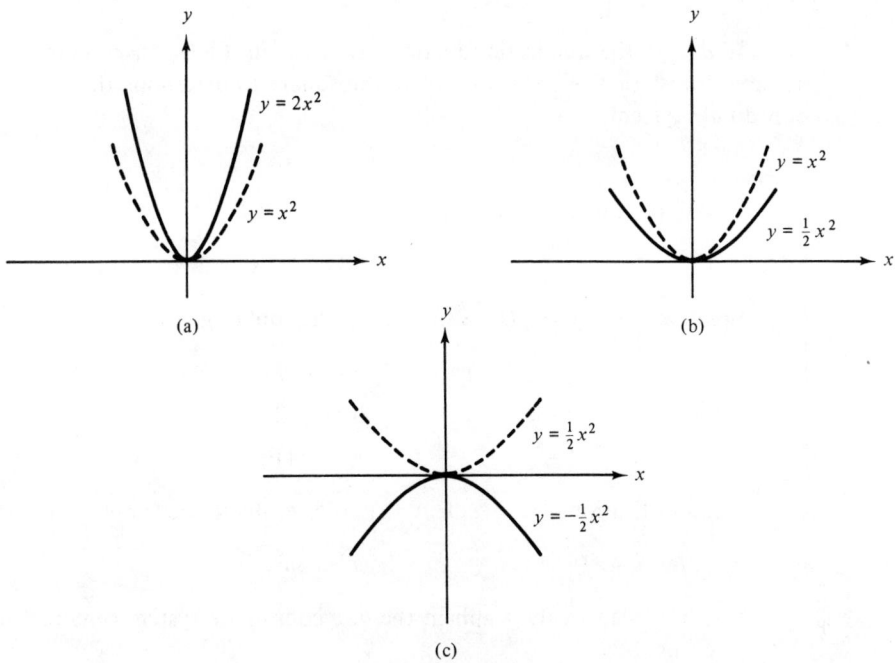

(a)

(b)

(c)

Figure 4-27

Because addition of a constant C shifts the graph vertically, we can obtain the graph of a quadratic function of the form $f(x) = Ax^2 + C$, as indicated in Fig. 4-28. This is just a vertical translation.

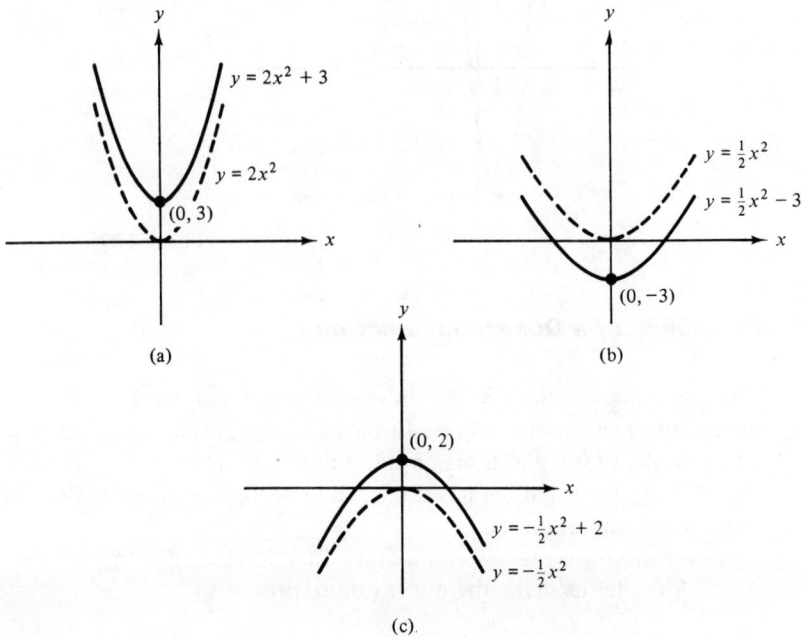

(a)

(b)

(c)

Figure 4-28

If $B \neq 0$, the quadratic function given by Eq. (4-7) can be graphed by using a suitable translation of axes. Complete the square to determine the center of the new coordinate system.

Example 1

Graph the quadratic function

$$f(x) = 3x^2 + 6x - 2.$$

Solution. Let $y = f(x)$. Completing the square gives

$$y \quad\;\; = 3(x^2 + 2x + \;\;) - 2$$
$$y + 3 = 3(x^2 + 2x + 1) - 2$$
$$y + 5 = 3(x + 1)^2.$$

The substitution $u = x + 1$, $v = y + 5$ results in the simpler equation

$$v = 3u^2,$$

which we can easily graph in the u, v coordinate system centered at $(-1, -5)$ (see Fig. 4-29).

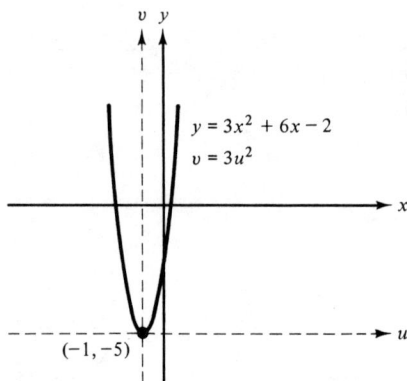

$$y = 3x^2 + 6x - 2$$
$$v = 3u^2$$

$(-1, -5)$

Figure 4-29

Extremum of a Quadratic Function

The quadratic functions in the previous examples all have "extreme" points—that is, minimum or maximum points. In terms of the graph, a minimum point is a lowest point on the graph and a maximum point is a highest point. The function shown in Fig. 4-28(b), for example, has a minimum at $(0, -3)$ whereas the one in Fig. 4-28(c) has a maximum at $(0, 2)$.

We will show that every quadratic function has either a minimum or a maximum point. First, let us make these ideas more precise.

Definition.

1. A function f assumes a **minimum** at $x = c$, provided that

$$f(x) \geq f(c)$$

for every x in the domain of f. The number $f(c)$ is the **minimum value** of f. The point $(c, f(c))$ is a **minimum point** on the graph of f.

2. A function f assumes a **maximum** at $x = d$, provided that

$$f(x) \leq f(d)$$

for every x in the domain of f. The number $f(d)$ is the **maximum value** of f. The point $(d, f(d))$ is a **maximum point** on the graph of f.

Example 2

The function $f(x) = \frac{1}{2}x^2 - 3$ assumes a minimum at $x = 0$ because $f(0) = -3$ and, for every real x,

$$f(x) = \tfrac{1}{2}x^2 - 3 \geq 0 - 3 = f(0).$$

The point $(0, -3)$ is a minimum point on the graph of f [Fig. 4-28(b)]. The minimum value of f is -3.

Example 3

The function $f(x) = -\frac{1}{2}x^2 + 2$ assumes a maximum at $x = 0$ because $f(0) = 2$ and, for every real x,

$$f(x) = -\tfrac{1}{2}x^2 + 2 \leq 0 + 2 = f(0).$$

The point $(0, 2)$ is a maximum point on the graph of f [Fig. 4-28(c)]. The maximum value of f is 2.

The term **extremum** is used to indicate either a maximum or a minimum without specifying which.

Example 4

Find the extremum of the function of Example 1.

Solution.   In graphing $y = f(x) = 3x^2 + 6x - 2$, we first obtained

$$y + 5 = 3(x + 1)^2.$$

Because $3(x + 1)^2 \geq 0$, we know that

$$f(x) + 5 \geq 0$$

or

$$f(x) \geq -5$$

for every real number x. But letting $x = -1$, we get $f(-1) = -5$. Hence

$$f(x) \geq f(-1)$$

for every real number x. This shows that $(-1, -5)$ is a minimum point on the graph of f (see Fig. 4-29).

We will now show that every quadratic function

$$f(x) = Ax^2 + Bx + C \qquad (A \neq 0) \tag{4-7}$$

has an extremum of one type or the other. The argument is a generalization of the one in Example 3. We start by rewriting (4-7) as

$$f(x) = A\left(x^2 + \frac{B}{A}x + \right) + C.$$

Next, add $A \cdot (B^2/4A^2)$ to both sides so as to complete the square on the right:

$$f(x) + \frac{B^2}{4A} = A\left(x^2 + \frac{B}{A}x + \frac{B^2}{4A^2}\right) + C.$$

Subtract $B^2/4A$ from both sides, obtaining

$$f(x) = A\left(x + \frac{B}{2A}\right)^2 + \left(C - \frac{B^2}{4A}\right).$$

The term $D = C - B^2/4A$ is the value of f at $x = -B/2A$. Hence

$$f(x) = A\left(x + \frac{B}{2A}\right)^2 + f\left(-\frac{B}{2A}\right). \tag{4-8}$$

Now if $A > 0$, the term $A(x + B/2A)^2$ is greater than or equal to zero for all real x. Therefore $f(x) \geq f(-B/2A)$ for all real x, showing that $(-B/2A, D)$ is a minimum point. If $A < 0$, then $A(x + B/2A)^2$ is less than or equal to zero and so $f(x) \leq f(-B/2A)$ for all real x, showing that $(-B/2A, D)$ is a maximum point. The following theorem summarizes these results.

Theorem 4-1. The quadratic function

$$f(x) = Ax^2 + Bx + C \qquad (A \neq 0)$$

assumes an extremum at $x = -B/2A$. If $A > 0$, this extremum is a minimum; whereas if $A < 0$, it is a maximum. In either case, the extreme value is given by

$$D = f\left(-\frac{B}{2A}\right) = C - \frac{B^2}{4A}$$

and the extreme point is $(-B/2A, D)$.

Example 5

Use Theorem 4-1 to determine the extremum of $f(x) = 6x - x^2 + 10$.

Solution. Because $A = -1$, we have a maximum at

$$x = -\frac{B}{2A} = -\frac{6}{2 \cdot (-1)} = 3.$$

The maximum value of f is $D = f(3) = 18 - 9 + 10 = 19$. (Alternatively, we could use the formula $D = C - B^2/4A = 10 - 36/(-4) = 19$.) The maximum point is $(3, 19)$.

Theorem 4-1 can best be understood geometrically by thinking in terms of a translation of axes to point $(-B/2A, D)$. Equation (4-8) can be written as follows, with $f(x)$ replaced by y and $f(-B/2A)$ replaced by D:

$$y - D = A\left(x + \frac{B}{2A}\right)^2.$$

Using the translation equations $u = x + (B/2A)$ and $v = y - D$, we get a new equation

$$v = Au^2,$$

which obviously has its extremum at the new origin. It is not necessary to memorize the formulas for the coordinates of the new origin, for they can be found directly by completing the square.

Example 6

Determine the extremum of $f(x) = 6x - x^2 + 10$ by completing the square.

Solution. We rewrite the equation in the form

$$y = -x^2 + 6x + 10$$
$$y - 9 = -(x^2 - 6x + 9) + 10$$
$$y - 19 = -(x - 3)^2.$$

The extremum occurs at the origin of the u, v coordinate system, where $u = x - 3$ and $v = y - 19$. Because the second-degree term has a negative coefficient, we see that $(3, 19)$ is a maximum point.

EXERCISES 4-9

In Problems 1–12 draw the graph of the given quadratic function. If necessary, use a translation of axes.

1. $f(x) = 3x^2$

2. $g(x) = \frac{1}{3}x^2$

3. $f(x) = 1 - x^2$

4. $g(u) = 2u^2 - 5$

5. $h(x) = \frac{1}{5}(x^2 + 1)$

6. $f(x) = 3x^2 + 6x + 5$

7. $f(x) = 2x^2 - 4x + 5$

8. $g(t) = 5t - t^2$

9. $h(x) = 1 - 3x + 2x^2$

10. $g(z) = 3 + 5z + 2z^2$

11. $h(x) = 3x^2 + 2\sqrt{3}x + 4$

12. $f(t) = \frac{1}{3}t^2 - \frac{1}{3}t + \frac{5}{12}$

In Problems 13–21 use Theorem 4-1 to find the extreme point. Say whether this extremum is a maximum or a minimum.

13. $f(x) = x^2 + 6x + 13$

14. $g(x) = 90 - 18x + x^2$

15. $f(t) = 4t^2 - 4t + 3$

16. $f(x) = 4x - x^2$

17. $g(x) = 3 - x^2$

18. $f(t) = 1 - 12t - 3t^2$

19. $g(u) = 3u^2 + 10u + 3$

20. $F(x) = 3x^2 + 2\sqrt{3}x + 1$

21. $g(t) = 5t^2 - 4t + 8$

In Problems 22–25 find the extremum by completing the square. Check your answer by using Theorem 4-1.

22. $f(x) = 2x^2 + 4x + 5$

23. $g(x) = 6x - x^2$

24. $f(t) = \sqrt{5}\,t^2 + \sqrt{10}\,t$

25. $f(x) = \frac{2}{3}x^2 + \frac{8}{9}x + \frac{17}{27}$

26. The distance s (feet) fallen by an object thrown downward with an initial velocity of v_0 feet per second is given by

$$s = 16t^2 + v_0 t,$$

where t is the number of seconds of fall.

(a) If $v_0 = 10$ feet per second, how far will the object fall after 3 seconds?

(b) If $v_0 = 10$ feet per second and the object strikes the ground at 5 seconds, from what height was the object thrown?

(c) If $v_0 = 12$ feet per second, how many seconds will it take the object to fall from a height of 1404 feet?

(d) If the object falls 1072 feet after 8 seconds, what was its initial velocity?

27. If a bullet is fired into the air from the ground, its altitude h (feet) at time t (seconds) satisfies an equation of the form

$$h = kt - 16t^2,$$

where k is a constant that depends on the angle of elevation of the gun and the speed with which the bullet leaves the muzzle of the gun.

(a) If the altitude is 144 feet after 1 second, find k.

(b) When does the bullet hit the ground?

(c) Express h as a function of t over the interval when the bullet is in the air.

(d) When does the bullet reach its greatest altitude and what is that altitude?

FOR THE ENTHUSIASTS

28. Show that if $f(x) = mx + b$ for x real, and if $m \neq 0$, then f does not assume an extreme value at any point.

29. Show that the function $f(x) = (x^2 - 1)^2$ assumes a minimum at two different values of x.

4-10 ZEROS OF A FUNCTION

The solutions of the equation $f(x) = 0$ play an important role in drawing the graph of f. They determine the x intercepts and, as we shall see in Chapter 5, they are helpful in determining the intervals on which $f(x)$ is positive or negative.

> **Definition.** A **zero** of a function f is a number c in its domain such that $f(c) = 0$.

Example 1

Find the zeros of the function $f(x) = x^2 - 5x + 6$ and sketch the graph.

Solution. The zeros of f are the solutions of equation $x^2 - 5x + 6 = 0$. By factoring, we obtain solutions $x = 2$ and $x = 3$. Using the two intercepts $(2, 0)$ and $(3, 0)$, together with the minimum point $(\frac{5}{2}, -\frac{1}{4})$, we obtain the graph in Fig. 4-30.

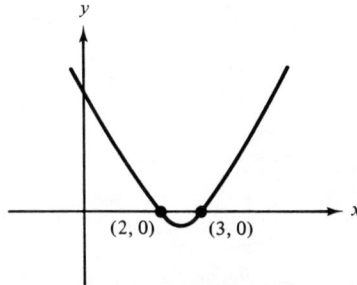

Figure 4-30

Example 2

Find the zeros of the function $f(x) = |4 - x| - 3$.

Solution. Because an absolute value occurs, we consider two cases:
   (a) $4 - x \geq 0$ (i.e., $x \leq 4$).
   (b) $4 - x < 0$ (i.e., $x > 4$).
By the definition of absolute value,

$$f(x) = \begin{cases} (4 - x) - 3 = 1 - x & \text{for } x \leq 4 \\ -(4 - x) - 3 = x - 7 & \text{for } x > 4 \end{cases}.$$

Can $f(x)$ take the value 0 for some $x \leq 4$? Yes, because $1 - x = 0$ when $x = 1$. Thus 1 is a zero of f. Considering values of $x > 4$, we see that $f(x)$ takes the value 0 when $x = 7$. Thus 7 is also a zero of f. A graph of f appears in Fig. 4-31.

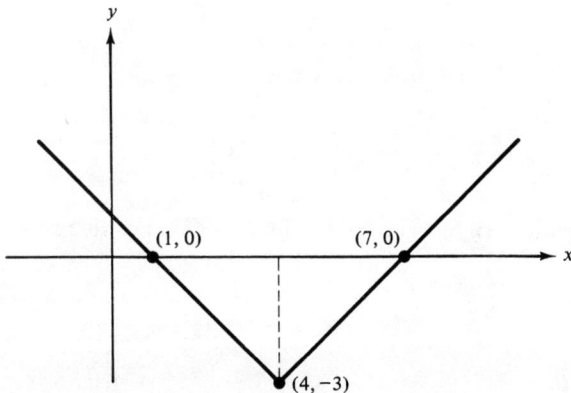

Figure 4-31

Example 3

Find the zeros of the function $g(x) = |4 - x| + 3$.

Solution. There are no zeros of g. One way to see this quickly is to observe that $|4 - x| \geq 0$; hence $g(x) \geq 3$ for all x.

If we follow the pattern of the previous example, we have

$$g(x) = \begin{cases} 7 - x & \text{for } x \leq 4 \\ x - 1 & \text{for } x > 4 \end{cases}.$$

There are no values of $x \leq 4$ for which $7 - x = 0$ and there are no values of $x > 4$ for which $x - 1 = 0$. So $g(x)$ cannot take the value 0 for any real number x.

Zeros of Linear and Quadratic Functions

By solving the equation $mx + b = 0$, it is easy to verify the following fact.

If $m \neq 0$, the linear function $f(x) = mx + b$ has exactly one zero, $-b/m$.

Example 4

The zero of $f(x) = 4x + 6$ is $-3/2$.

A quadratic function

$$f(x) = Ax^2 + Bx + C \qquad (A \neq 0).$$

may have two real zeros, one real zero, or no real zeros at all, We will show that the number of zeros of f and their values depend on the **discriminant** $B^2 - 4AC$.

Our method is to set $f(x) = 0$ and complete the square.

$$Ax^2 + Bx + C = 0$$

$$A\left(x^2 + \frac{B}{A}x + \phantom{\frac{B^2}{4A^2}}\right) = -C$$

$$A\left(x^2 + \frac{B}{A}x + \frac{B^2}{4A^2}\right) = \frac{B^2}{4A} - C$$

Now dividing both sides by A,

$$\left(x + \frac{B}{2A}\right)^2 = \frac{B^2}{4A^2} - \frac{C}{A} = \frac{B^2 - 4AC}{4A^2}. \qquad (4\text{-}9)$$

If $B^2 - 4AC$ is positive, then the right-hand side of (4-9) is also positive and has two square roots. Thus

$$x + \frac{B}{2A} = \pm\sqrt{\frac{B^2 - 4AC}{4A^2}} = \pm\frac{1}{2A}\sqrt{B^2 - 4AC}$$

$$x = \frac{-B}{2A} \pm \frac{1}{2A}\sqrt{B^2 - 4AC} = \frac{-B \pm \sqrt{B^2 - 4AC}}{2A}.$$

This means that there are *two* zeros of f:

$$x = \frac{-B + \sqrt{B^2 - 4AC}}{2A} \qquad x = \frac{-B - \sqrt{B^2 - 4AC}}{2A}.$$

If $B^2 - 4AC$ is equal to zero, Eq. (4-9) is equivalent to

$$\left(x + \frac{B}{2A}\right)^2 = 0$$

and, in this case, f has just one zero,

$$x = -\frac{B}{2A}.$$

If $B^2 - 4AC$ is negative, it is not possible to find a real number x satisfying Eq. (4-9), for the left-hand side is a square and the right-hand side is negative. In this case, there are no real zeros. These facts are summarized in the following theorem.

Theorem 4-2 (Quadratic Formula). Let f be the quadratic function

$$f(x) = Ax^2 + Bx + C \qquad (A \neq 0).$$

1. If $B^2 - 4AC > 0$, then f has two real zeros, given by

$$x = \frac{-B + \sqrt{B^2 - 4AC}}{2A} \quad \text{and} \quad x = \frac{-B - \sqrt{B^2 - 4AC}}{2A}.$$

2. If $B^2 - 4AC = 0$, then f has just one real zero, given by

$$x = -\frac{B}{2A}.$$

3. If $B^2 - 4AC < 0$, then f has no real zeros.

Example 5

Find the zeros of $f(x) = 3x^2 - 6x + 2$.

Solution. The discriminant is

$$B^2 - 4AC = (-6)^2 - 4 \cdot 3 \cdot 2$$
$$= 36 - 24 = 12 > 0.$$

So there are two zeros:

$$x = \frac{-(-6) \pm \sqrt{12}}{2 \cdot 3} = \frac{6 \pm 2\sqrt{3}}{6} = 1 \pm \frac{\sqrt{3}}{3}.$$

Example 6

Find the zeros of $f(x) = 3x^2 - 6x + 3$.

Solution. Here the discriminant is

$$B^2 - 4AC = (-6)^2 - 4 \cdot 3 \cdot 3 = 0$$

and so there is just one zero:

$$x = -\frac{B}{2A} = \frac{6}{6} = 1.$$

Example 7

Find the zeros of $f(x) = 5 + x + 2x^2$.

Solution. For this function, $A = 2$, $B = 1$, $C = 5$ and the discriminant is

$$B^2 - 4AC = 1 - 4 \cdot 2 \cdot 5 < 0.$$

Therefore f has no real zeros.

Example 8

Find the zeros of the function $f(x) = 3x^3 + 6x^2 - 8x$.

Solution. This function is not quadratic, but the equation $f(x) = 0$ is equivalent to

$$x(3x^2 + 6x - 8) = 0.$$

This equation is satisfied if and only if $x = 0$ or $3x^2 + 6x - 8 = 0$; so the zeros of $f(x)$ consist of $x = 0$, together with the two solutions of

$$3x^2 + 6x - 8 = 0.$$

The latter are $x = -1 \pm \sqrt{33}/3$.

We will consider "imaginary" zeros of quadratics and other functions in Chapter 6.

EXERCISES 4-10

In Problems 1–6 find the zeros of the given function and make a sketch of the graph, showing the intercepts on the horizontal axis.

1. $f(x) = 3x - 4$ **2.** $f(x) = 2 - x$ **3.** $f(t) = (2t + 5)^2$

4. $g(x) = x$ **5.** $h(u) = u^2 - 4$ **6.** $f(x) = 1 - x^2$

In Problems 7–22 find the zeros of the given function.

7. $F(x) = x^2 - 3x + 2$ **8.** $g(x) = x^2 + 3x^2 + 2x$ **9.** $f(t) = t^4 - 6t^2 + 5$

10. $f(y) = 4y^2 + 2y - 12$ **11.** $g(z) = 6z^2 + z - 1$ **12.** $f(x) = \dfrac{1}{x}$

13. $f(x) = \dfrac{4}{x - 1}$ **14.** $f(x) = x^2 - \dfrac{16}{x^2}$ **15.** $F(x) = \dfrac{x^2 - 9}{x^2 - 16}$

16. $g(t) = |t - 3| - 4$ **17.** $h(t) = |t - 3| + 4$ **18.** $f(x) = |3x - 5| - 7$

19. $f(x) = |4 + 3x| + 2$ **20.** $g(x) = \sqrt{6x - x^2}$ **21.** $f(x) = 1 - \sqrt{1 + x^2}$

22. $h(u) = 1 + \sqrt{1 - u^2}$

23. Graph the function g in Example 3.

24. Graph the function f in Example 7.

In Problems 25–37 use the quadratic formula to find the zeros of the given function.

25. $f(x) = 2x^2 + x - 1$ **26.** $f(x) = 5x^2 + 2x - 3$

27. $g(x) = 10x + 5x^2 + 6$ **28.** $f(x) = x^2 + 2\sqrt{5}\,x + 1$

29. $h(x) = 10x + 5x^2 - 6$

30. $f(t) = 2t^2 + 3\sqrt{3}\,t + 3$

31. $f(x) = 1 + 5x + 8x^2$

32. $g(x) = \frac{1}{2}x^2 + x + \frac{1}{4}$

33. $g(w) = 3w^2 + 2w + 4$

34. $F(t) = 4t^2 + 12t + 9$

35. $f(x) = \frac{2}{3}x^2 + \frac{1}{3}x + 1$

36. $f(w) = w^3 + 2w^2 + w$

37. $g(t) = \frac{1}{2}t^3 + t^2 + \frac{1}{4}t$

In Problems 38–41 sketch the graph of the given function, labeling the intercepts and extreme point.

38. $f(x) = 6 - 6x - x^2$

39. $g(x) = 1 + x - 2x^2$

40. $f(u) = 1 + 4u + 3u^2$

41. $F(x) = x^2 + 2\sqrt{2}\,x + 1$

42. What are the zeros of the linear function $f(x) = mx + b$ if $m = 0$?

43. A ball is $192 + 64t - 16t^2$ feet high t seconds after it is thrown. What is its time of impact?

FOR THE ENTHUSIASTS

44. Let $f(x) = 4x^2 + 5x + C$. Determine all values of C for which f has
  (a) two real zeros.
  (b) one real zero.
  (c) no real zeros.

45. Let $f(x) = Ax^2 + 5x + 3$. Determine all values of A for which f has
  (a) two real zeros.
  (b) one real zero.
  (c) no real zeros.

46. Show that if $f(x) = Ax^2 + Bx + C$ $(A \neq 0)$ has two real zeros, then
  (a) the sum of the zeros is $-B/A$.
  (b) the product of the zeros is C/A.

47. Show that if a quadratic function f has two intercepts at $(x_1, 0)$, $(x_2, 0)$, then the extremum of f occurs at $x = \frac{1}{2}(x_1 + x_2)$.

48. Sketch each graph by finding the intercepts and then locating the extremum by the method suggested in Problem 47.
  (a) $f(x) = x^2 - 6x + 8$
  (b) $f(x) = 2x^2 - 3x - 2$
  (c) $g(x) = 5x^2 - 7x + 2$
  (d) $h(x) = 3x^2 - 5x - 12$

4-11 APPLICATIONS

Some important applications of mathematics require that an extremum of a function be found. A manufacturer, for instance, wants to *maximize* his profits or a consumer wants to *minimize* her costs. In the following example a farmer wants to maximize the area of his vegetable garden.

Example 1

A farmer has a field bordered along one side by a straight stone wall. He wants to plant a small rectangular vegetable garden within this field. He has 400 feet of fencing to use for three sides of the garden; the wall is to serve as the fourth side. How can he use his fencing material so as to enclose the maximum area and what is the maximum area?

Solution. The situation is diagrammed in Fig. 4-32, where we denote by x and y the dimensions (in feet) of the field; note that y is the dimension of the side parallel to the wall. The area of the rectangular garden is given by

$$A = xy \tag{4-10}$$

and we want to choose x and y so as to make A as large as possible.

Now the variables x and y are not independent of each other. Because the farmer has just 400 feet of fencing, x and y are related by the equation

$$2x + y = 400. \tag{4-11}$$

Thus $y = 400 - 2x$; and returning to (4-10) with this value of y, we have

$$A = x(400 - 2x) = 400x - 2x^2.$$

Let us consider the values of x that are permissible in this discussion. Certainly $x \geq 0$ because x is a length. But also $y \geq 0$ and this requires

$$400 - 2x \geq 0$$

or
$$x \leq 200.$$

Hence

$$0 \leq x \leq 200.$$

At this point we have expressed the area A as a function of x:

$$A = f(x) = 400x - 2x^2 \quad \text{for} \quad 0 \leq x \leq 200 \tag{4-12}$$

The domain of f is the interval $0 \leq x \leq 200$ and the rule for determining $f(x)$ for each x in the domain is $f(x) = 400x - 2x^2$.

The graph of the function f appears in Fig. 4-33. It is part of the graph of a quadratic function, limited to the interval $0 \leq x \leq 200$. From Theorem 4-1 we know that the graph has a maximum point where

$$x = -\frac{B}{2A} = -\frac{400}{2 \cdot (-2)} = 100.$$

When $x = 100$, we see that $y = 200$. So to get the maximum area from his 400 feet of fencing, the farmer should construct his rectangular garden with dimensions 100 feet by 200 feet, using 200 feet parallel to the stone wall. The maximum area is therefore 20,000 square feet.

There are two important steps in the solution of Example 1: (a) setting up the function that we want to maximize and (b) actually finding the maximum value. In the first step we expressed the area A as a function of x, eliminating y by means of Eq. (4-11). We determined the domain of this function by considering the possible

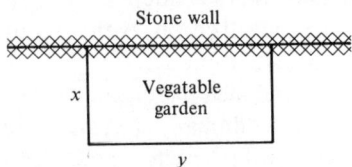

Figure 4-32

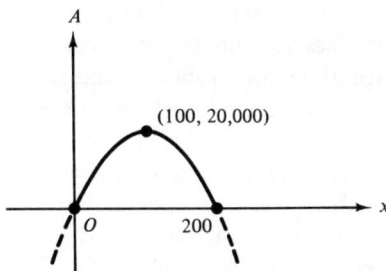

Figure 4-33

values of the variables in the problem. In step (b) we used Theorem 4-1 to find where the maximum occurs.

Example 2

The Rotary Club is planning a tour of the Pennsylvania Dutch country for some of its senior citizens. The bus company will provide a bus with a capacity of 80 seats and stipulates that at least 50 people must go. If only 50 people go, the charge will be $100 per person; for each person in excess of 50, the fare for everybody will be reduced by $1. (So, for example, if 52 people go, the fare will be $98 per person.) Find the number of people that will maximize the bus company's revenue.

Solution. Let us denote by n the number of persons in excess of 50 who go on the tour, so that the total number going is $50 + n$. Then the fare paid per person is $100 - n$ dollars and the bus company's revenue is given by

$$P(n) = (50 + n)(100 - n)$$
$$= 5000 + 50n - n^2.$$

This function is actually defined only for $n = 0, 1, 2, \ldots, 30$ and we could answer the question by calculating $P(n)$ for those values of n. Doing so would be tedious.

Let us consider instead the graph of the corresponding quadratic function, defined for real x by

$$f(x) = 500 + 50x - x^2.$$

This quadratic is shown in Fig. 4-34; it assumes a maximum when $x = 25$. Because 25 is a number in the domain of P, we have found the point where P assumes its maximum.

The number of tourists that maximizes the bus company's proceeds is $50 + 25 = 75$.

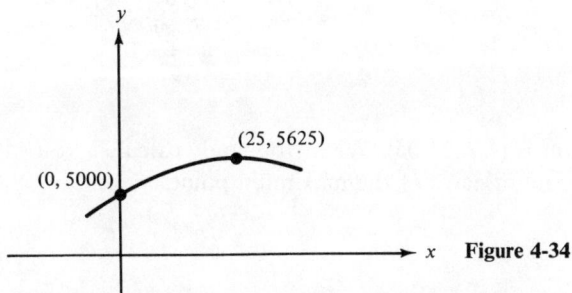

Figure 4-34

Note. We were lucky to have quadratic functions in Examples 1 and 2. Calculus provides techniques for maximizing or minimizing other types of functions; in the meantime, for a function that is not quadratic, the best we can do is to draw a careful graph and estimate where an extremum occurs (if there is one).

In Example 2 we made use of a convenient idealization. Although the values of x should be nonnegative integers to apply to the situation described, we were able to get an answer more quickly by ignoring this fact and maximizing a quadratic function. Had the value of x that maximizes f not been in the domain of P, we might still have used f. If the point where f assumed its maximum had been $x = 25.7$, for instance, we would have checked the values at $x = 25$ and $x = 26$.

Example 3

Suppose that you are given a square piece of cardboard 10 inches by 10 inches. By cutting equal squares out of the corners and folding up the remaining cardboard as shown in Fig. 4-35 you obtain a rectangular box without a top. If you want to obtain the maximum volume, how long should you make the edges of the corner squares?

Solution. Let us consider the variables that arise in this problem. The volume V of the box depends on the edge x of the corner squares. The box has a height of x inches and its base is a square $10 - 2x$ inches by $10 - 2x$ inches. Hence its volume V is $x(10 - 2x)^2$. Because $10 - 2x \geq 0$ and $x \geq 0$, we can express V as a function of x:

$$V = g(x) = x(10 - 2x)^2 \quad \text{for} \quad 0 \leq x \leq 5.$$

A table of values and a graph of g are shown in Fig. 4-36. It appears that a maximum point occurs somewhat to the left of point $(2, 72)$. Let us tabulate some additional values of g between 1.5 and 2. An approximation of the maxi-

x	$g(x)$
1.5	73.50
1.6	73.98
1.7	74.05 ←
1.8	73.73
1.9	73.04
2.0	72.00

mum point is $(1.7, 74.05)$. When you study calculus, you will learn how to find the exact coordinates of the maximum point.

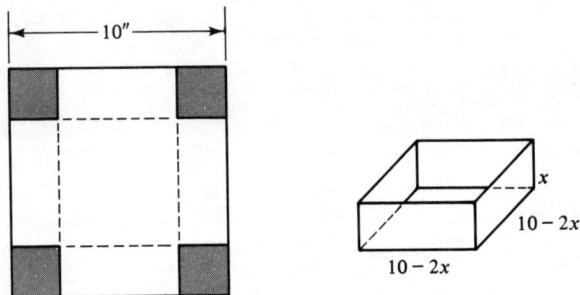

Figure 4-35

x	$g(x)$
0	0
1	64
2	72
3	48
4	16

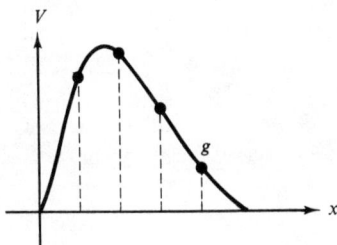

Figure 4-36

EXERCISES 4-11

1. Suppose that the farmer in Example 1 has a 3-foot gate that he wants to insert into his fence along the side parallel to the stone wall. Find the dimensions of the garden of maximum area that he can obtain.

2. If the farmer in Example 1 has 400 feet of fencing material with which to enclose a rectangular plot on all four sides (no stone wall, no gate), find the dimensions that he should use to get a plot of maximum area.

3. The sum of two positive numbers is 25. What are the two numbers if
   (a) their product is maximum?
   (b) the sum of their squares is minimum?
   (c) the sum of the first number and the square of the second number is minimum?

4. A rectangle is to have its sides along the coordinate axes, one vertex at the origin, and the diagonally opposite vertex in the first quadrant, on the line $3x + 4y = 12$. Find the coordinates of the vertex in the first quadrant that maximizes the area.

5. A newspaper publisher can sell 6000 copies of her weekly if the price is 12¢ per copy, but she can sell 1500 additional copies for each 1¢ reduction in the price per copy. Suppose that she reduces the price by n¢, where $0 < n < 12$. Express the total weekly revenue as a function of n. Graph this function. What value of n gives the greatest revenue?

6. The manufacturers of Sweetie Gum Drops now sell their candy at 75¢ per pound. At the current price they can sell 12,750 pounds of gum drops each day. Marketing analysts predict that, for every penny reduction in price, sales will increase by 250 pounds per day. What should the selling price be if the Sweetie Gum Drop Company is to maximize its total daily income?

†7. Suppose that a box without a top is to be made from a rectangular piece of cardboard 8 inches by 12 inches by cutting out corner squares and folding up the remainder (as in Example 3). Express the volume V of the box as a function of the length x of the edge of the corner squares. Tabulate some values of this function, graph it, and estimate the maximum volume such a box can have.

8. A farmer now has a crop of 200 bushels of potatoes. The current price is $5 per bushel. Each week for the next 8 weeks the crop will increase by 20 bushels, but the weekly market price will drop 30¢ per bushel. Express the value V of the farmer's crop of potatoes as a function of the number n of weeks he waits before harvesting them. Determine the value of n that gives the farmer maximum income.

†9. A truck carrying frozen food is destined for a city 500 miles away. Suppose that the truck consumes fuel at a rate proportional to the square of its speed and that at 50 miles per hour its rate of fuel consumption is $30 per hour. Other costs of operating the truck amount to $24 per hour. Express the cost of trucking the frozen food to its destination as a function of the speed r of the truck, assuming that r is constant and that $40 \leq r \leq 55$. Estimate the speed that makes the cost a minimum.

SUMMARY

The basic theme of Chapter 4 is the function concept.

1. You should be able to express verbally stated relationships between variables by means of equations. In particular, you should understand the mathematical meaning of "proportion" and "inverse proportion."

2. You should be able to say whether a given mathematical relation determines a function and, if it does, to give the rule and the domain.

3. You should understand what is meant by the "graph" of a function and be familiar with the graphs and properties of such basic functions as the square function, the square root function, the reciprocal function, the absolute value function, the identity function, and a constant function.

4. You should be able to express geometric, physical, and other relationships by functions; this includes stating the rule and giving the domain.

5. You should know the geometric formulas inside the front cover of the text, because they are the source of many functional relationships.

6. You should understand that not every function can be defined with a single formula; many common relationships involve "piecewise" defined functions.

7. You should understand the concept of average rate of change of a function.

8. You should recognize the linear function by its various equations and should know that a linear function is one for which the rate of change is constant. This rate of change is the slope of the graph of the function.

9. You should understand the role of linear functions in approximating unknown values of other functions by means of linear interpolation.

10. You should be able to use translation of axes to draw graphs.

11. You should know the definition of quadratic function and be able to sketch the graph of a quadratic function, by means of a translation of axes if necessary. You should be able to locate the maximum or minimum point on the graph of such a function.

12. You should be able to find the zeros of functions—in particular, to apply the quadratic formula to find the zeros of a quadratic function. You should be able to tell from the discriminant whether a quadratic function does not have any real zeros.

13. You should be able to set up the functions required to solve maximum/minimum problems and find, either exactly or approximately, the coordinates of any maximum and/or minimum points.

REVIEW EXERCISES

1. The velocity v of an object falling from rest is proportional to the square root of the distance s fallen. An object is dropped from a height of 900 feet and is falling at 96 feet per second when it has fallen just 144 feet. Express v as a function of s, stating the domain and range of this function.

2. The "strength" of a wooden beam is the maximum load that can be safely placed at the midpoint of the beam when the ends of the beam are supported at equal heights. For a beam of fixed width and depth, this strength is inversely proportional to the length. If the strength of a 6-foot beam cut from a long piece of lumber 4 inches deep and 10 inches wide is 1000 pounds, what is the strength of a 10-foot beam cut from the same piece? Express the strength S (in pounds) as a function of the length x (in feet).

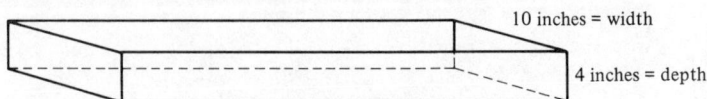

10 inches = width

4 inches = depth

3. Draw the graph of the functions defined for real x by
   (a) $f(x) = x + |x|$.   (b) $g(x) = x|x|$.

4. Let $]x[$ denote the smallest integer greater than or equal to x. Draw the graph of the function $h(x) =]x[$ for $-2 \le x \le 2$.

5. Let $f(x) = \sqrt{x^2 + 4}$ for real x. Evaluate:
   (a) $f(0)$ (b) $f(2)$

   (c) $f(-2)$ (d) $f\left(\dfrac{1}{a}\right)$ for $a \ne 0$

   (e) $f\left(a - \dfrac{1}{a}\right)$ for $a > 0$ (f) $f\left(a - \dfrac{1}{a}\right)$ for $a < 0$

   (g) $f\left(a - \dfrac{1}{a}\right)$ for $a \ne 0$

6. Let $f(x) = x + (1/x)$ for $x \ne 0$. Show that, for $x \ne 0$, $[f(x)]^2 = f(x^2) + 2$.

7. Find the average rate of change of volume v of a cube with respect to its edge x over the interval between $x = 1$ and $x = 6$.

8. Give a formula for $f(x)$, where f is a linear function, if
   (a) the slope of the graph of f is 2/3 and $f(1) = 1$.
   (b) the intercepts of the graph of f are $(2, 0)$ and $(0, -3)$.
   (c) the rate of change of f is 3/4 and f has a zero at 1/2.
   (d) $f(2) = 1$ and $f(1) = 2$.

9. (a) The speed V of sound in air varies with the temperature T of the air. At $T = 0°C$, this speed is about 332 meters per second whereas in air at $100°C$ it is about 392 meters per second. Assuming that the relationship is linear, express V as a function of T.
   (b) Watching a fireworks display from a distance one July 4 evening when the temperature was $25°C$, I counted 7 seconds between the flash of light and the sound of an exploding rocket. How far was I from the point where the rocket exploded?

†10. Use linear interpolation to estimate $f(3.142)$, given that $f(3) = 8.167$ and $f(3.5) = -1.324$.

11. Graph the following functions, using a translation of axes.
   (a) $f(x) = 5 + |x - 1|$ (b) $g(x) = 2 - \sqrt{x + 4}$
   (c) $h(x) = 4x^2 - 8x + 3$

12. Find and identify the extremum points of the following quadratics.
   (a) $h(x) = 4x^2 - 8x + 3$ (b) $f(x) = 9x - x^2$
   (c) $g(x) = 1 + x + 2x^2$

13. Find the zeros of the quadratic functions in Problem 12.

14. The sum of two nonnegative numbers is 8. What are the numbers if the sum of the first number and the square of the second is to be a minimum?

†15. A ball is thrown upward from a height of 5 feet in such a way that its height in feet at time t seconds is given by

$$s = -16t^2 + 128t + 5.$$

(a) What is the maximum height reached by the ball?
(b) When does the ball hit the ground?

Polynomial and rational functions

5-1 THE ALGEBRA OF POLYNOMIAL FUNCTIONS

The linear and quadratic functions we studied in Chapter 4 belong to a special class known as polynomial functions. The expressions

$$2x + 3$$

$$8 - 5x$$

$$5x^2 - 9$$

$$x^2$$

$$\frac{5}{3} + \frac{2}{3}x + \frac{1}{3}x^2$$

have the following property in common: Each is a term or a sum of terms consisting of a constant multiplied by a nonnegative integral power of x.

Definition. A **polynomial function** is a function P defined for real numbers x by an equation of the form

$$P(x) = A_n x^n + A_{n-1} x^{n-1} + \cdots + A_1 x + A_0, \tag{5-1}$$

where n is a nonnegative integer and $A_0, A_1, \ldots, A_n$ are real constants, $A_n \neq 0$. The nonnegative integer n is called the **degree** of P. The coefficient A_n is the **leading coefficient** of P. The coefficient A_0 is the **constant term** of P.

For convenience, the terms are arranged in Eq. (5-1) in order of descending powers of x; they need not always be ordered this way.

Example 1

The following functions, defined for all real numbers, are examples of polynomial functions.

$P(x) = 5x^3 - 2x^2 + 7x - 3$; degree 3, leading coefficient 5, constant term -3

$Q(t) = 3t - \dfrac{1}{2}t^5$; degree 5, leading coefficient $-\dfrac{1}{2}$, constant term 0

$f(x) = \dfrac{1}{4}x^7$; degree 7, leading coefficient $\dfrac{1}{4}$, constant term 0

$D(x) = 1 + \sqrt{2}\,x$; degree 1, leading coefficient $\sqrt{2}$, constant term 1

$R(u) = \dfrac{u^2 - 1}{4} + \dfrac{u^2 + 1}{3}$; degree 2, leading coefficient $\dfrac{7}{12}$, constant term $\dfrac{1}{12}$

A polynomial function of degree zero is a function of the form $P(x) = C$, where C is a nonzero constant. The equation $P(x) = 2$, for instance, defines a polynomial function of degree zero. The constant function $P(x) = 0$ is also considered a polynomial function, but its degree is undefined.

Example 2

The following are not polynomials functions.

$$f(x) = x^2 + 2x + x^{-1} \qquad \text{(the term } x^{-1} \text{ is not of the right form)}$$
$$g(t) = 1 + \sqrt{t} + 3t^2 \qquad \text{(the term } \sqrt{t} \text{ is not of the right form)}$$

The following definition indicates a natural extension of the algebraic operations—addition, subtraction, multiplication, and division—to functions.

Definition. Let f and g be functions and D the intersection of the domains of f and g. Then

1. the **sum** $f + g$ is the function s defined for x in D by $s(x) = f(x) + g(x)$.
2. the **difference** $f - g$ is the function d defined for x in D by $d(x) = f(x) - g(x)$.
3. the **product** fg is the function p defined for x in D by $p(x) = f(x)g(x)$.
4. the **quotient** f/g is the function q defined by $q(x) = f(x)/g(x)$ for all those x in D for which $g(x) \neq 0$.

These operations can be applied to various kinds of functions, but for the present we are concerned with polynomial functions.

Example 3

Consider the polynomial functions $P(x) = 2x - 3$ and $Q(x) = x^2 + x - 6$. Then the sum s, difference d, and product p are defined for all real x by the rules

$$s(x) = P(x) + Q(x) = (2x - 3) + (x^2 + x - 6)$$
$$= x^2 + 3x - 9$$
$$d(x) = P(x) - Q(x) = (2x - 3) - (x^2 + x - 6)$$
$$= -x^2 + x + 3$$
$$p(x) = P(x)Q(x) \quad = (2x - 3)(x^2 + x - 6)$$
$$= 2x(x^2 + x - 6) - 3(x^2 + x - 6)$$
$$= (2x^3 + 2x^2 - 12x) - (3x^2 + 3x - 18)$$
$$= 2x^3 - x^2 - 15x + 18.$$

The functions s, d, and p are polynomial functions. The quotient is defined when $Q(x)$ is not zero:

$$q(x) = \frac{2x - 3}{x^2 + x - 6} \quad \text{for } x \neq -3, 2.$$

The function q is *not* a polynomial function.

Example 3 shows that, although the quotient of two polynomial functions is not necessarily a polynomial function, the sum, the difference, and the product are.

We shall use the term **polynomial** to refer to an expression having the form given in Eq. (5-1)—that is, an expression that defines a polynomial function. As illustrated in Example 3, the algebraic operations on polynomial functions are carried out by working with these expressions.

To add two polynomials, we proceed as in Example 3 by collecting terms with equal powers and adding their coefficients.

$$(x^5 - 3x^4 + x^2 - 7) + \left(\frac{1}{2}x^5 + x^3 - 2x^2 + 4x + 1\right)$$

$$= \left(1 + \frac{1}{2}\right)x^5 + (-3 + 0)x^4 + (0 + 1)x^3 + (1 - 2)x^2 + (0 + 4)x + (-7 + 1)$$

$$= \frac{3}{2}x^5 - 3x^4 + x^3 - x^2 + 4x - 6$$

Subtraction is performed similarly.

$$\left(x^4 + \frac{2}{3}x^3 - x^2 + 7\right) - \left(\frac{3}{2}x^5 + x^3 + \frac{1}{2}\right)$$

$$= \left(0 - \frac{3}{2}\right)x^5 + (1 - 0)x^4 + \left(\frac{2}{3} - 1\right)x^3 + (-1 - 0)x^2 + (0 - 0)x + \left(7 - \frac{1}{2}\right)$$

$$= -\frac{3}{2}x^5 + x^4 - \frac{1}{3}x^3 - x^2 + \frac{13}{2}$$

To find the product of two polynomials, such as

$$(x^2 - x + 4)(x^3 + 3x^2 - 2),$$

we first multiply the right factor by each term in the left factor, using the distributive laws of algebra:

$$x^2(x^3 + 3x^2 - 2) + (-x)(x^3 + 3x^2 - 2) + 4(x^3 + 3x^2 - 2);$$

then multiply out each product, remembering that $x^m x^n = x^{m+n}$:

$$(x^5 + 3x^4 - 2x^2) + (-x^4 - 3x^3 + 2x) + (4x^3 + 12x^2 - 8).$$

Next, we collect equal powers and add, obtaining

$$x^5 + 2x^4 + x^3 + 10x^2 + 2x - 8.$$

The work we have just done can be arranged in the following way.

$$
\begin{array}{r}
x^3 + 3x^2 + 0 - 2 \\
x^2 - x + 4 \\
\hline
x^5 + 3x^4 + 0 - 2x^2 \\
- x^4 - 3x^3 + 0 + 2x \\
4x^3 + 12x^2 + 0 - 8 \\
\hline
x^5 + 2x^4 + x^3 + 10x^2 + 2x - 8
\end{array}
$$

For convenience, terms with similar powers of x are kept in a vertical column.

Division of polynomials is illustrated in the following example, in which we divide $P(x) = 5x^3 - 2x^2 + 7x - 3$ (the **dividend**) by $D(x) = x^2 + x + 2$ (the **divisor**).

$$
\begin{array}{r}
5x - 7 \\
x^2 + x + 2 \overline{)\; 5x^3 - 2x^2 + 7x - 3} \\
5x^3 + 5x^2 + 10x \\
\hline
- 7x^2 - 3x - 3 \\
- 7x^2 - 7x - 14 \\
\hline
4x + 11
\end{array}
$$

Notice that both polynomials are arranged in order of descending powers. The first term is found by dividing x^2 into $5x^3$, getting $5x$. This is then multiplied by the divisor and the result is subtracted from the dividend. We now divide x^2 into the first term of the difference. We continue the process until we reach a **remainder**, $4x + 11$, which has degree less than the degree of the divisor. The result is summarized in the equation

$$\frac{P(x)}{D(x)} = \frac{5x^3 - 2x^2 + 7x - 3}{x^2 + x + 2} = (5x - 7) + \frac{4x + 11}{x^2 + x + 2}. \qquad (5\text{-}2)$$

In general, a quotient of two polynomials is either a polynomial itself or it can be expressed as the sum of a polynomial and a fraction in which the degree of the numerator is less than the degree of the denominator.

Multiplying both sides of Eq. (5-2) by the divisor gives

$$5x^3 - 2x^2 + 7x - 3 = (5x - 7)(x^2 + x + 2) + (4x + 11);$$

that is, $P(x) = (5x - 7)D(x) + (4x + 11)$.

> **Division Process for Polynomials.** Let $P(x)$ and $D(x)$ be polynomials such that $D(x)$ is not the constant 0. Then there exist polynomials $Q(x)$ and $R(x)$ such that, for all real numbers x,
>
> $$P(x) = Q(x)D(x) + R(x) \qquad (5\text{-}3)$$
>
> and $R(x)$ either is the constant 0 or has degree less than the degree of $D(x)$.

In the preceding example we found $Q(x) = 5x - 7$ and $R(x) = 4x + 11$. Note that the degree of $R(x)$ is less than the degree of $D(x)$.

Example 4

Apply the division process to find $Q(x)$ and $R(x)$ if $P(x) = x^4 - 4x^2 + x + 2$ and $D(x) = x^2 - x + 1$.

Solution.

$$
\require{enclose}
\begin{array}{r}
x^2 + x - 4 \\
x^2 - x + 1 \enclose{longdiv}{x^4 + 0 - 4x^2 + x + 2} \\
\underline{x^4 - x^3 + x^2} \\
x^3 - 5x^2 + x \\
\underline{x^3 - x^2 + x} \\
-4x^2 + 2 \\
\underline{-4x^2 + 4x - 4} \\
-4x + 6
\end{array}
$$

Thus $Q(x) = x^2 + x - 4$ and $R(x) = -4x + 6$. This result may be expressed in the form (5-3) as follows:

$$x^4 - 4x^2 + x + 2 = (x^2 + x - 4)(x^2 - x + 1) + (-4x + 6).$$

EXERCISES 5-1

In Problems 1–7 determine whether the given expression defines a polynomial function. If it does, determine the degree, the leading coefficient, and the constant term.

1. $f(x) = 6x - x^2 + 1$ **2.** $g(u) = (1 + u)(1 - u)$

3. $h(x) = x(2x + 1)(3x + 2)$ **4.** $f(t) = 2t + \sqrt{t^2 - 1}$

5. $f(x) = x - \sqrt{1 - x^2}$ **6.** $g(t) = t + \dfrac{1}{t}$

7. $h(x) = \dfrac{x^2 - 1}{x^2 + 1}$

In Problems 8–11 find the sum and the product of the polynomials indicated.

8. $x^2 - 2x + 6,\ 5x^2 + 3x - 4$

9. $3x + 2\sqrt{2}x^2 + 4x^3,\ 1 - 3x + \sqrt{2}x^2 + x^3$

10. $x - 3x^2 - \dfrac{4}{3},\ \dfrac{3}{2}x^2 + \dfrac{1}{2}x + \dfrac{7}{6}$

11. $-u^2 + u + 1, u^2 + u + 1$

12. Suppose that a polynomial f has degree m, another polynomial g has degree n. What, if anything, can you conclude about the degree of the product? the sum? the difference?

In Problems 13–18 determine those values of x for which $f(x)/g(x)$ is *not* defined.

13. $f(x) = 2x, g(x) = x^2 - 1$

14. $f(x) = x^2 - 1, g(x) = 2x$

15. $f(x) = x^2 - 5x + 9, g(x) = x^2 - 9$

16. $f(x) = 3x^2 - 6x - 24, g(x) = x^2 - x - 12$

17. $f(x) = x^2 + 16, g(x) = 2x^2 + 5x - 4$

18. $f(x) = 2x^2 + 5x - 4, g(x) = x^2 + x + 3$

In Problems 19–27 use the division process to find polynomials $Q(x)$ and $R(x)$ satisfying Eq. (5-3).

19. $P(x) = 3x^4 - 4x^3 + x + 1; D(x) = x^2 - 1$

20. $P(x) = x^4 - 2x^2 + 1; D(x) = x^3 - 5x^2 + 6x + 4$

21. $P(x) = 4x^4 + 3x^3; D(x) = 2x^2 + 4x + 2$

22. $P(x) = \frac{3}{2}x^3 - \frac{2}{3}x + \frac{1}{3}; D(x) = \frac{1}{3}x + \frac{2}{3}$

23. $P(x) = 1 + 3\sqrt{2}x + x^2; D(x) = \sqrt{2}x + 3$

24. $P(x) = x^2 - 3; D(x) = x + \sqrt{3}$

25. $P(x) = x^2 - 16; D(x) = \frac{1}{2}x^2 + 4$

26. $P(x) = x^2 + 4; D(x) = x^2 - 2$

27. $P(x) = x + 2; D(x) = -x^2 + x + 3$

5-2 FACTORS AND ZEROS OF POLYNOMIALS

Factors

If $D(x)$ divides "evenly" into $P(x)$ (i.e., with zero remainder), then $D(x)$ is called a factor of $P(x)$.

Definition. A polynomial $D(x)$ is a **factor** of the polynomial $P(x)$ if there exists a polynomial $Q(x)$ such that, for all real numbers x,

$$P(x) = Q(x)D(x).$$

Example 1

(a) $x - 1$ is a factor of $x^2 - 1$ because

$$x^2 - 1 = (x + 1)(x - 1).$$

(b) x^2 is a factor of $x^4 - x^3 + 2x^2$ because

$$x^4 - x^3 + 2x^2 = (x^2 - x + 2)x^2.$$

Zeros

By the definition in Section 4-10, a zero of a polynomial function P is a real number r such that $P(r) = 0$.[1] We shall also refer to such an r as a zero of the polynomial $P(x)$. To find these zeros, we set the polynomial $P(x)$ equal to zero and solve.

Example 2

The polynomial function P defined for real x by $P(x) = x^2 - 9$ has two zeros, -3 and 3. They are obtained by solving the equation $x^2 - 9 = 0$.

Example 3

The function $Q(x) = 3x^2 + x + 4$ has no real zeros, as we can see by applying Theorem 4-2: $B^2 - 4AC = 1 - 48 < 0$.

The Connection Between Zeros and Factors

If the case of polynomial functions, the process of finding zeros is closely related to factorization. The first step in establishing this connection is a theorem conerning the division process.

Theorem 5-1 (Remainder Theorem). If the polynomial $P(x)$ is divided by the polynomial $D(x) = x - r$, where r is a constant, then the remainder is the constant $P(r)$.

Proof. We know that by applying the division process, we can find polynomials $Q(x)$ and $R(x)$ such that

$$P(x) = Q(x)(x - r) + R(x),$$

where $R(x)$ either is zero or has degree less than the degree of $x - r$. Because the degree of $x - r$ is one, we see in either case that $R(x)$ is a constant. Letting $R(x) = C$, we have

$$P(x) = Q(x)(x - r) + C.$$

But this equation holds for all real numbers x and so we can substitute $x = r$, obtaining

$$P(r) = Q(r) \cdot 0 + C = C,$$

showing that the constant remainder C is equal to $P(r)$.

Example 4

The remainder theorem tells us that the remainder obtained when $P(x) = x^2 + 2x + 5$ is divided by $x - 3$ is $P(3) = 9 + 6 + 5 = 20$. This result is confirmed by the division process, which gives us

$$x^2 + 2x + 5 = (x + 5)(x - 3) + 20.$$

[1] In Chapter 6 we will consider complex zeros of polynomial functions.

Example 5

Find the remainder when $P(x) = x^3 - 8x^2 - 15x + 2$ is divided by $x + 1$.

Solution. Because $x + 1 = x - (-1)$, we choose $r = -1$. According to the remainder theorem, the remainder is $P(-1) = 8$.

The remainder theorem can be expressed concisely by the equation

$$P(x) = Q(x)(x - r) + P(r),$$

which holds for all real numbers x. If r is a zero of P then $P(r) = 0$ and we have the following.

Corollary. If r is a zero of the polynomial $P(x)$, then there is a polynomial $Q(x)$ such that, for all real x,

$$P(x) = Q(x)(x - r).$$

The corollary states that if r is a zero of $P(x)$, then $x - r$ is a factor of $P(x)$. Since the converse also holds (Problem 18), we obtain a theorem that shows the relationship between zeros and factors.

Theorem 5-2 (Factor Theorem). The real number r is a zero of a polynomial $P(x)$ if and only if $x - r$ is a factor of $P(x)$.

Example 6

Let $P(x) = x^2 - 3x + 2$. Because $x^2 - 3x + 2 = (x - 1)(x - 2)$, we see that $x - 1$ and $x - 2$ are factors. Therefore 1 and 2 are zeros.

Example 7

Let $P(x) = x^3 - 8$. It is easy to verify that 2 is a zero. By Theorem 5-2, $x - 2$ must be a factor. Dividing $x - 2$ into $P(x)$, we obtain the factorization

$$P(x) = (x - 2)(x^2 + 2x + 4).$$

Suppose that we have found two different zeros, r_1 and r_2, of a polynomial $P(x)$. First, because r_1 is a zero, we have

$$P(x) = (x - r_1)Q(x),$$

where $Q(x)$ is a polynomial. Next, because r_2 is a zero of $P(x)$, and $r_2 \neq r_1$, we have

$$0 = P(r_2) = (r_2 - r_1)Q(r_2).$$

Dividing both sides of this equation by $r_2 - r_1$, we see that $Q(r_2) = 0$ so that r_2 is a zero of $Q(x)$. But this implies that $x - r_2$ is a factor of $Q(x)$. Therefore

$$Q(x) = (x - r_2)S(x),$$

where $S(x)$ is a polynomial, and so

$$P(x) = (x - r_1)Q(x) = (x - r_1)(x - r_2)S(x).$$

If we have additional zeros, r_3, r_4, and so on, all different, then we can extend this argument, obtaining

$$P(x) = (x - r_1)(x - r_2)(x - r_3) \cdots (x - r_k)T(x), \qquad (5\text{-}4)$$

where $T(x)$ is a polynomial. Notice that the degree of $T(x)$ is lowered by one for each factor of the form $x - r$. The reduction in degree often makes it possible to find new factors of $T(x)$ and, with them, additional zeros of $P(x)$.

Example 8

Let $P(x) = x^4 - 5x^2 + 4$. It is easy to see that -1 and 1 are zeros, indicating a factorization of the form

$$P(x) = (x - 1)(x + 1)T(x).$$

The division process shows that $T(x) = x^2 - 4$. But $T(x) = (x - 2)(x + 2)$; so

$$P(x) = (x - 1)(x + 1)(x - 2)(x + 2)$$

and P has the four zeros, $-2, -1, 1, 2$.

Equation (5-4) shows that the number of different zeros of a polynomial cannot exceed its degree; consequenty in Example 8 we obtained all possible zeros. But the number of zeros of a polynomial can be less than its degree.

Example 9

Let $Q(x) = x^4 - 1$. Then $Q(x) = (x - 1)(x + 1)(x^2 + 1)$ and we have two zeros, 1 and -1. Any other real zeros of $Q(x)$ would also be zeros of $x^2 + 1$. But $x^2 + 1$ has no real zeros; so -1 and 1 are the only real zeros of $Q(x)$.

Example 10

Suppose that, using trial and error, we find that -4 is a zero of the polynomial $P(x) = x^3 + 5x^2 + 7x + 12$. Then, by division, we obtain

$$P(x) = (x + 4)(x^2 + x + 3).$$

By Theorem 4-2, $x^2 + x + 3$ has no real zeros, so -4 is the only real zero of $P(x)$.

Example 11

Find the zeros of the function defined for real $x \neq 3/2$ by

$$f(x) = \frac{x^3 - 2x^2 + x}{3 - 2x}$$

Solution. Although f is not a polynomial, its numerator is, and we know a quotient can equal zero only if its numerator is zero. So we set $x^3 - 2x^2 + x = 0$, obtaining

$$x(x^2 - 2x + 1) = 0$$

$$x(x - 1)(x - 1) = 0.$$

There are just two real zeros, 0 and 1.

EXERCISES 5-2

In Problems 1–6 find the zeros of the given function.

1. $f(x) = x^5 - x$

2. $g(x) = x^3 - 4x^2 - 5x$

3. $f(x) = \dfrac{x^2 - 1}{x^2 + 1}$

4. $f(x) = \dfrac{2x^2 - 6}{x^3 + 8}$

5. $g(x) = \dfrac{x^3 - 2x^2 - x + 2}{x^2 + 9}$

6. $h(x) = \dfrac{x^3 - x^2 - 20x}{1 - x^2}$

In Problems 7–17 find the factors and the zeros of the given polynomial.

7. $x^3 - 3x^2 - 4x$

8. $x^3 - 3x^2 - 13x + 15$

9. $u^3 - 12u^2 + 24u - 8$

10. $x^3 + 4x^2 + 4x + 3$

11. $x^4 - 2x^3 - x^2 + 2x$

12. $t^4 - 2t^2 - 3$

13. $x^4 + 2x^3 + 6x^2$

14. $2z^3 - 2z^2 - z + 1$

15. $4x^3 + 8x^2 - 9x - 18$

16. $3x^3 - 2x^2 - 33x - 28$

17. $t^6 - 1$

FOR THE ENTHUSIASTS

18. Show that if $x - r$ is a factor of a polynomial $P(x)$, then r is a zero of $P(x)$.

19. If $(x - r)^2$ is a factor of polynomial $P(x)$, then r is said to be a **repeated** zero of $P(x)$. Show that if $B^2 - 4AC = 0$, then the quadratic function

$$P(x) = Ax^2 + Bx + C \qquad (A \neq 0)$$

has a repeated zero.

5-3 GRAPHING POLYNOMIAL FUNCTIONS

Intercepts

A first step in drawing the graph of a polynomial $y = P(x)$ is to locate its intercepts. Setting $x = 0$, we easily get the y intercept. The x intercepts are obtained by finding the zeros of P.

Example 1

Find all intercepts for these polynomials.
   (a) $P(x) = 5x^2 + 3x - 2$
   (b) $f(x) = x^4 - x^2$
   (c) $g(x) = x^3 - 3x$

Solution. (a) Because $P(0) = -2$, the y intercept is $(0, -2)$. Solving $5x^2 + 3x - 2 = 0$, we obtain $x = -1$ and $x = 2/5$ and so the x intercepts are $(-1, 0)$ and $(2/5, 0)$.

   (b) Because $f(0) = 0$, the y intercept is $(0, 0)$. The x intercepts are $(0, 0)$, $(-1, 0)$, and $(1, 0)$.

   (c) Because $g(0) = 0$, the y intercept is $(0, 0)$. The x intercepts are $(0, 0)$, $(-\sqrt{3}, 0)$, and $(\sqrt{3}, 0)$.

Even and Odd Functions

In Chapter 3 we saw how useful it could be to look for symmetry when drawing the graph of an equation. We now apply some of the methods developed there to the graphing of polynomial functions.

> **Definition.** A function f is said to be **even** provided that, for every x in its domain, $-x$ is also in its domain and
> $$f(-x) = f(x).$$

The graph of an even function is symmetric with respect to the y axis because the equation $y = f(-x)$ is equivalent to $y = f(x)$.

Example 2

The following functions, defined for all real numbers, are even.

(a) $f(x) = x^4 - x^2$, because $f(-x) = (-x)^4 - (-x)^2 = x^4 - x^2 = f(x)$.

(b) $g(x) = |x|$, because $g(-x) = |-x| = |x| = g(x)$.

(c) $h(x) = \dfrac{1}{1 + x^2}$, because $h(-x) = \dfrac{1}{1 + (-x)^2} = \dfrac{1}{1 + x^2} = h(x)$.

Any polynomial consisting only of terms with even powers of x defines an even function. An even function is not necessarily a polynomial function, as (b) and (c) show.

> **Definition.** A function f is said to be **odd** provided that, for every x in its domain, $-x$ is also in its domain and
> $$f(-x) = -f(x).$$

The graph of an odd function is symmetric with respect to the origin, because the equation $-y = f(-x)$ is equivalent to $y = f(x)$.

Example 3

Any polynomial consisting only of terms with odd powers of x defines an odd function. The function $h(x) = \sqrt[3]{x}$ is an example of an odd function that is not a polynomial.

Note. Not all functions are even or odd; for example, the function defined by $P(x) = 5x^2 + 3x - 2$ is neither even nor odd.

Example 4

Graph the polynomial $f(x) = x^4 - x^2$.

Solution. The intercepts were found in Example 1. With f even, we can draw the graph for positive x and reflect that through the y axis to get the remainder of the graph. The factorization

$$x^4 - x^2 = x^2(x^2 - 1)$$

shows that $f(x)$ is positive for $x > 1$, negative for $0 < x < 1$. The graph of $y = f(x)$ for $x \geq 0$ is shown in Fig. 5-1(a) and the reflection through the y axis in Fig. 5-1(b).

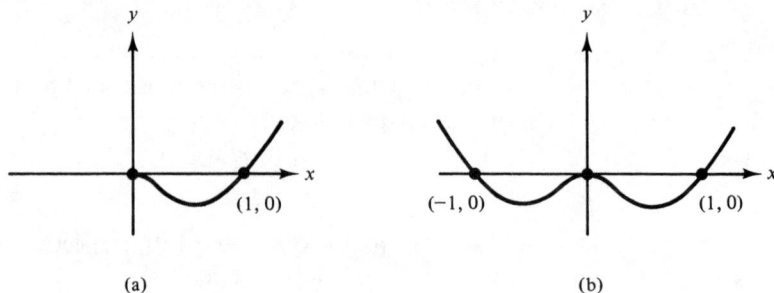

(a)   (b)

Figure 5-1

Example 5

Graph the polynomial $g(x) = x^3 - 3x$.

Solution. The intercepts were found in Example 1. Because g is odd, we begin by drawing its graph for positive x. The factorization $g(x) = x(x^2 - 3)$ shows that $g(x)$ is negative for $0 < x < \sqrt{3}$, positive for $x > \sqrt{3}$. The graph of g for positive values of x is shown in Fig. 5-2(a). Reflecting this through the origin, we obtain the graph shown in Fig. 5-2(b).

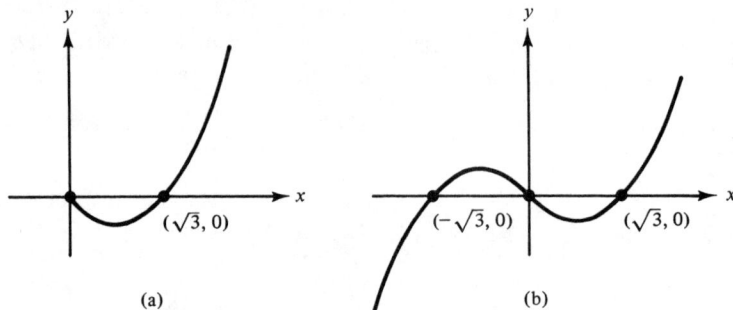

(a)   (b)

Figure 5-2

Behavior for Large x

No analysis of a function f is complete until we have determined how $f(x)$ behaves as x becomes very large. The following terminology will help express our ideas.

1. If a variable v becomes indefinitely large, without any upper bound on its values, then we say that v **approaches plus infinity**, written $v \to +\infty$.
2. If a variable v becomes indefinitely large in a negative sense, without any lower bound on its values, then we say that v **approaches minus infinity**, written $v \to -\infty$.

We want to know how $f(x)$ behaves as $x \longrightarrow +\infty$, and as $x \longrightarrow -\infty$. This is not hard to determine when f is a polynomial function.

Example 6

Consider $f(x) = x^2$. As x increases without bound, it is clear that the product $x^2 = x \cdot x$ also increases without bound. In fact, for $x > 1$, the value of x^2 is even greater than that of x. Therefore $f(x) \longrightarrow +\infty$ as $x \longrightarrow +\infty$. The graph of f rises indefinitely as x increases. (See Fig. 5-3.)

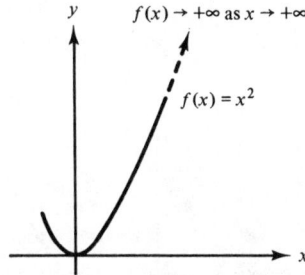

Figure 5-3

Example 7

Consider $g(x) = x - x^3$. This third-degree polynomial can be factored:

$$g(x) = x(1 - x^2).$$

As x increases without bound, the factor $1 - x^2$ approaches minus infinity; the product consists of x (large, positive) times $1 - x^2$ (large in absolute value but negative). The product is large in absolute value but negative; so $g(x)$ approaches minus infinity. In symbols, $g(x) \longrightarrow -\infty$ as $x \longrightarrow +\infty$.

Example 8

Determine how the polynomial $f(x) = x^3 + 3x^2$ behaves as $x \longrightarrow -\infty$.

Solution. We factor $f(x)$:

$$f(x) = x^2(x + 3).$$

As $x \longrightarrow -\infty$, the factor x^2 is positive (product of two negative factors) and becomes large. The factor $x + 3$ is negative but becomes large in absolute value. The product therefore approaches minus infinity: $f(x) \longrightarrow -\infty$ as $x \longrightarrow -\infty$.

Example 9

Let us return to the function $f(x) = x^4 - x^2$ of Example 4 and complete the graph by determining how $f(x)$ behaves as x becomes large. Because $f(x) = x^2(x^2 - 1)$, we see that $f(x) \longrightarrow +\infty$ as $x \longrightarrow +\infty$ and so the graph rises indefinitely as x increases. (See Fig. 5-4.) Because f is even, it follows that $f(x) \longrightarrow +\infty$ as $x \longrightarrow -\infty$.

Example 10

Consider the function $g(x) = x^3 - 3x$ of Example 5. By factoring, we see that $g(x) \longrightarrow +\infty$ as $x \longrightarrow +\infty$. Because g is odd, $g(x) \longrightarrow -\infty$ as $x \longrightarrow -\infty$. This behavior is shown in Fig. 5-5.

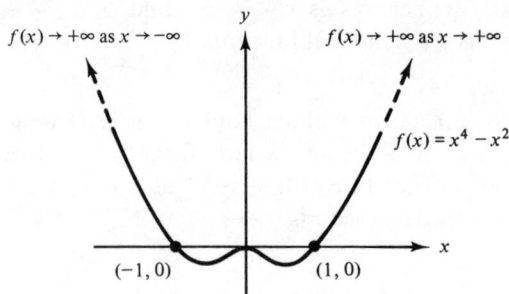

$f(x) \to +\infty$ as $x \to -\infty$ $f(x) \to +\infty$ as $x \to +\infty$

$f(x) = x^4 - x^2$

$(-1, 0)$ $(1, 0)$

Figure 5-4

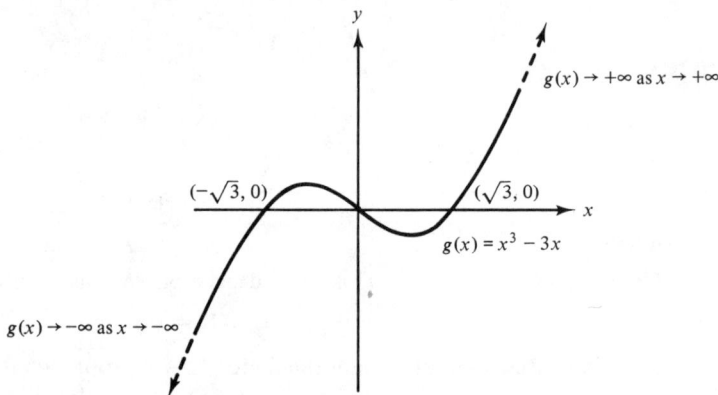

$g(x) \to +\infty$ as $x \to +\infty$

$(-\sqrt{3}, 0)$ $(\sqrt{3}, 0)$

$g(x) = x^3 - 3x$

$g(x) \to -\infty$ as $x \to -\infty$

Figure 5-5

It is not hard to see from our examples that a polynomial approaches plus or minus infinity, depending on whether its leading term approaches plus or minus infinity.

EXERCISES 5-3

In Problems 1–8 determine whether the function is even or odd (or neither).

1. $f(x) = 3x^2 + 4$

2. $g(x) = 3x^2 + 4x$

3. $f(x) = x + \dfrac{1}{x}(x \neq 0)$

4. $g(u) = \sqrt{25 - u^2}(-5 \leq u \leq 5)$

5. $f(x) = x|x|$

6. $F(t) = \left(t + \dfrac{1}{t}\right)^2$

7. $f(x) = |x - 2|$

8. $g(x) = |x^2 - 2|$

In Problems 9–14 sketch the graph of the given polynomial, specifying any intercepts and any symmetry. Determine the intervals on which the polynomial is positive.

9. $x^4 - 6x^2 + 8$

10. $x^4 - 16$

11. $x^5 - 3x^3$

12. $(x^2 - 1)^2$

13. $5x - x^3$

14. $12x - x^5$

In Problems 15–24 determine how the function behaves as $x \to +\infty$ and as $x \to -\infty$.

15. $f(x) = x(x - 1)$

16. $f(x) = x^2 - 3x + 2$

17. $g(x) = x^3 - 3x^2 + 2x$ 18. $f(x) = x(x + 1) + 4$
19. $g(x) = x(x + 1)(x + 2)$ 20. $f(x) = x^2 + x + 10$
21. $g(x) = x(1 - x^2)$
22. $f(x) = x^n$, where n is an even positive integer
23. $g(x) = x^m$, where m is an odd positive integer
24. $F(x) = x^n + x^m$, where m and n are positive integers and $n > m$.

In Problems 25–28 graph the function, showing its behavior as $x \longrightarrow +\infty$ and as $x \longrightarrow -\infty$.

25. The function in Problem 16.   26. The function in Problem 18.
27. The function in Problem 20.   28. The function in Problem 21.

FOR THE ENTHUSIASTS

29. Show that if f and g are even (odd), then $f + g$ is even (odd).
30. Fill in a "multiplication table" for even and odd functions, showing whether the product fg is even or odd. Verify your conclusions.

f \ g	even	odd
even		
odd		

31. Show that if f is odd, then the function $g(x) = f(x^3)$ is odd.

5-4 CONTINUITY AND THE SIGN OF A POLYNOMIAL

Continuous Functions

Polynomial functions, as well as certain other functions, possess an important property called **continuity**. To say that a function is **continuous** over an interval means that the function is defined over that interval and that its graph there is an unbroken curve. The mathematician and philosopher René Descartes is said to have described a continuous function as one whose graph can be drawn without lifting the pencil from the paper.

  Example 1

  The absolute value function (Example 1, Section 4-4) is continuous over every interval. Its graph can be drawn without lifting the pencil from the paper. If the interval contains zero, there is an abrupt change in the direction of the graph at $(0, 0)$, but the graph is not broken (Fig. 5-6).

A precise definition of continuity can be made only after the limit concept has

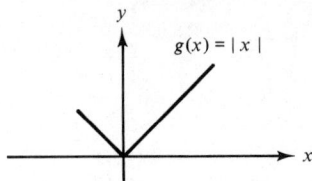

Figure 5-6

been studied in calculus. Because our description is informal, we must accept many facts about continuity without proof. The following theorem is a case in point.

Theorem 5-3. A polynomial function is continuous over every interval.

From Theorem 5-3 it follows that the linear and quadratic functions are all continuous over every interval. So also are $f(x) = x^3$, $g(x) = 5x^4 - 7x^3 + 8x^2 - 17x + 2$, and so on.

Points of Discontinuity

It is helpful to examine the idea of continuity by looking at its opposite: discontinuity. We have already encountered several examples of functions that are discontinuous.

Example 2

Consider the postage function described in Example 3, Section 4-4. This function takes a sudden jump from 20 to 37 as the variable w (weight) increases through the value 1, it then jumps again as w increases through the value 2 (see Fig. 5-7). The points $w = 1, w = 2, \ldots$, are called **points of discontinuity** of the function.

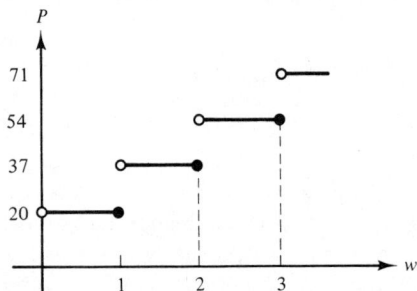

Figure 5-7

Example 3

The reciprocal function is defined by the rule $F(x) = 1/x$. This function has no value at zero, for division by zero is not possible. When we attempt to graph F for values of x near zero, we find that $F(x) \longrightarrow +\infty$ as x approaches zero through positive values, because the reciprocal of a small positive number becomes large. Moreover, $F(x) \longrightarrow -\infty$ as x approaches zero through negative values. There is a point of discontinuity at $x = 0$ where the graph is torn apart severely (see Fig. 5-8).

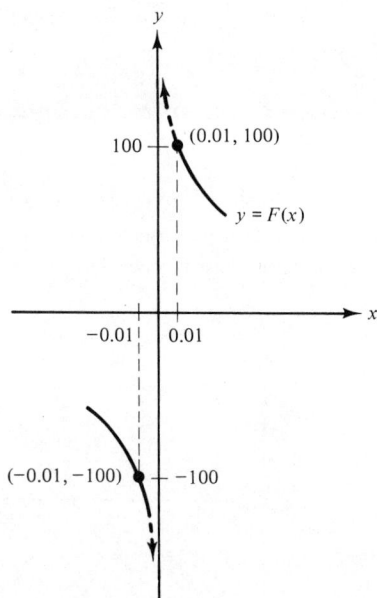

x	$F(x)$
0.1	10
0.01	100
0.001	1000
0.001	10,000
−0.01	−100
−0.001	−1000
−0.001	−10,000

Figure 5-8

Example 4

Let us define $f(x) = (x^2 - 4)/(x - 2)$ for $x \neq 2$. Because f is undefined at $x = 2$, the graph has a gap at that point. The point $x = 2$ is therefore a point of discontinuity of f. Because $f(x) = x + 2$ for $x \neq 2$, the graph consists of the line $y = x + 2$ with the point at $x = 2$ removed (see Fig. 5-9).

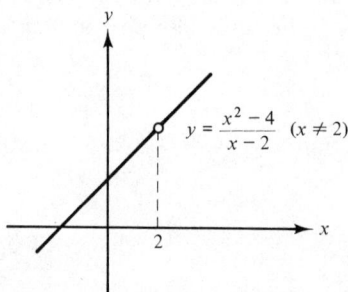

Figure 5-9

If we wish to find points of discontinuity of a function $y = f(x)$, the preceding discussion suggests we look for the following:

1. values of x at which the graph of f has a "jump."
2. values of x at which $f(x)$ approaches plus or minus infinity.
3. values of x at which $f(x)$ is undefined, such as where a denominator takes the value 0.

If f is a polynomial, of course, Theorem 5-3 assures us that there are no points of discontinuity.

Change of Sign of a Continuous Function

It is intuitively clear that the graph of a continuous function cannot go from a negative value at a to a positive value at b (or vice versa) without crossing the x axis at least once between a and b. A point where such a crossing takes place corresponds to a zero of the function (see Fig. 5-10).

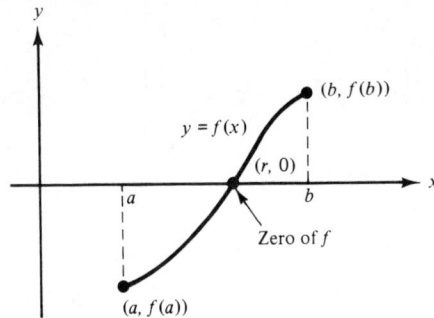

Figure 5-10

Theorem 5-4. If the function f is continuous over an interval containing a and b and if $f(a)$ and $f(b)$ have opposite signs, then there is at least one zero of f between a and b.

Despite its obviousness on an intuitive level, Theorem 5-4 is not an easy theorem to prove and we will not attempt to prove it here. As we shall see, however, this theorem can be very helpful in finding the zeros of a polynomial function and in analyzing its sign.

Example 5

The polynomial function $f(x) = x^3 + 5x - 7$ has the following values:

$$f(0) = -7$$

$$\left. \begin{array}{l} f(1) = -1 \\ f(2) = 11 \end{array} \right\} \text{ change of sign}$$

Because $f(x)$ changes sign between $x = 1$ and $x = 2$, we conclude that there is a zero of f between 1 and 2. If we wish to obtain a better approximation of this zero, we can evaluate $f(1.5)$, which turns out to be 3.875. Thus we have

$$\left. \begin{array}{l} f(1) \ \ = -1 \\ f(1.5) = 3.875 \end{array} \right\} \text{ change of sign}$$
$$f(2) \ \ = 11,$$

showing that a zero lies between 1 and 1.5. To improve our approximation to this zero, we can evaluate f at 1.25 obtaining 1.203. Because $f(1.25) > 0$, the zero is between 1 and 1.25. Evaluating f at 1.12, we get .0049, showing that the zero is between 1 and 1.12. Further calculation shows the zero is between 1.11 and 1.12. It is clear that we can obtain any desired degree of accuracy by continuing this process.

We will return to the approximation of zeros of polynomials in Section 5-6.

Sign of a Polynomial

Let us see how we can apply Theorems 5-3 and 5-4 to sketch a graph of a polynomial function that shows where the function is positive and where it is negative.

If f is a polynomial function, then f is continuous everywhere by Theorem 5-3. So, by Theorem 5-4, a change in the sign of $f(x)$ as x changes from a to b indicates a zero of f between a and b. Looking at it another way, on any interval in which there is no zero of f there can be no change of sign: we must have either $f(x) > 0$ for all x in the interval or $f(x) < 0$ for all x in the interval.

> On an interval containing no zeros of the polynomial function f the values of $f(x)$ all have the same sign.

Example 6

Let $f(x) = x^3 - 5x$. Determine where f is positive and where it is negative.

Solution. From the factorization

$$f(x) = x(x^2 - 5) = x(x + \sqrt{5})(x - \sqrt{5}) \qquad (5\text{-}5)$$

we obtain zeros $-\sqrt{5}, 0, \sqrt{5}$ and then the intercepts on the x axis. [See Fig. 5-11(a).] Now consider the sign of $f(x)$: for which x is $f(x)$ positive and for which x is it negative? The open intervals $(-\infty, -\sqrt{5}), (-\sqrt{5}, 0), (0, \sqrt{5})$, $(\sqrt{5}, +\infty)$ contain no zeros of f and so f is of constant sign on each interval. We can choose a representative point in each interval to determine the sign. Suppose, for instance, that we choose 10 to represent the interval $(\sqrt{5}, +\infty)$. Because $f(10)$ is positive, we know that $f(x)$ must be positive for all x in that interval. Proceeding similarly with the other intervals, we obtain the following data.

Interval	Representative point	Sign of $f(x)$
$(-\infty, -\sqrt{5})$	$x = -10$	$-$
$(-\sqrt{5}, 0)$	$x = -1$	$+$
$(0, \sqrt{5})$	$x = 1$	$-$
$(\sqrt{5}, +\infty)$	$x = 10$	$+$

Thus $f(x)$ is positive on the intervals $(-\sqrt{5}, 0), (\sqrt{5}, +\infty)$ and negative on $(-\infty, -\sqrt{5}), (0, \sqrt{5})$, as shown on the "sign chart."

Sign of $f(x)$ $+$ $-$ $+$

Values of x $-\sqrt{5}$ 0 $\sqrt{5}$

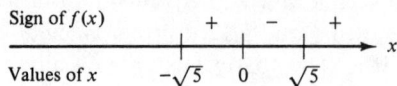

The graph of f appears in Fig. 5-11(b); it is easy to verify that $f(x) \rightarrow +\infty$ as $x \rightarrow +\infty$ and that $f(x) \rightarrow -\infty$ as $x \rightarrow -\infty$.

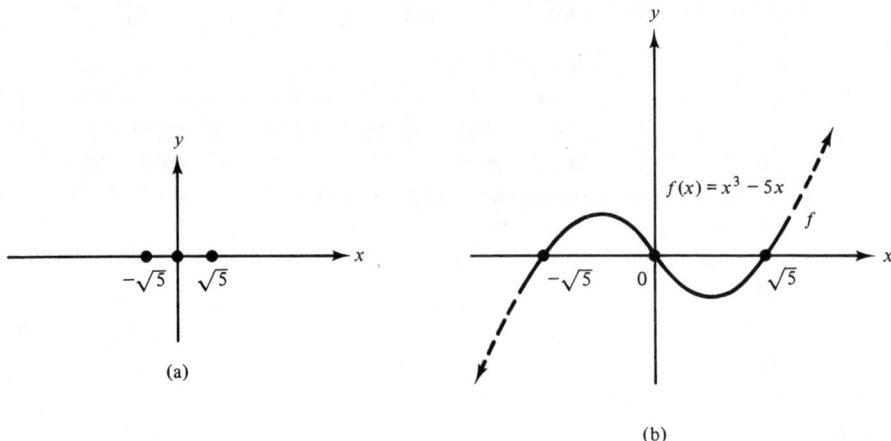

(a)

$f(x) = x^3 - 5x$

(b)

Figure 5-11

Another way to obtain the information in the sign chart of Example 6 is as follows. Choose a point (say 10) in the rightmost of the four intervals and determine the sign of f on that interval ($+$). Now as x decreases through $\sqrt{5}$, the factor $x - \sqrt{5}$ changes sign ($+$ to $-$) whereas none of the other factors in (5-5) changes sign. Thus $f(x)$ changes sign (from $+$ to $-$) as x decreases through $\sqrt{5}$. Notice that this information is obtained without a second evaluation of f. Next, as x decreases through 0, the factor x in (5-5) changes sign, but the other factors do not. Hence $f(x)$ changes sign ($-$ to $+$) as x decreases through 0. Finally, as x decreases through $-\sqrt{5}$, the factor $x + \sqrt{5}$ changes sign whereas no other factors of $f(x)$ change sign; so $f(x)$ changes sign ($+$ to $-$).

Example 7

Determine the intervals where the polynomial function defined by $f(x) = x(x - 1)(x - 2)(x^2 + 4)$ is positive and those where it is negative.

Solution. The zeros of f are $0, 1, 2$, which correspond to factors $x, x - 1$, $x - 2$, respectively. As x passes through each zero, the corresponding factor (and no others) changes sign. Hence $f(x)$ changes sign at each zero. Because $f(x)$ is clearly positive for large x, we obtain the following sign chart for $f(x)$, without any actual evaluation of f.

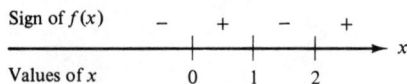

Sign of $f(x)$ $-$ $+$ $-$ $+$

Values of x 0 1 2

Polynomial and Rational Functions   Chap. 5

Care must be taken when applying the procedure of Example 7 to polynomials having repeated factors, as the following example shows.

Example 8

If $f(x) = x(x - 1)^2$, then as x passes through the value 1, the factor $x - 1$ changes sign, but $(x - 1)^2$ does *not*. So $f(x)$ does not change sign as x passes through 1. Thus we obtain the following sign chart for $f(x)$.

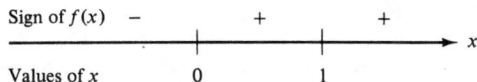

Whether $f(x)$ changes sign at a given zero depends on the "multiplicity" of the zero. A zero r of a polynomial $f(x)$ is said to be a **zero of multiplicity k**, provided that k is the exponent of the largest power of $x - r$ that is a factor of $f(x)$. Then

1. $f(x)$ does change sign as x passes through a zero of multiplicity k if k is *odd* $(k = 1, 3, \ldots)$.
2. $f(x)$ does *not* change sign as x passes through a zero of multiplicity k if k is *even* $(k = 2, 4, 6, \ldots)$.

Intercepts corresponding to zeros of odd multiplicity therefore represent points where the graph of f crosses the x axis and henceforth we will mark them with a cross. Intercepts corresponding to zeros of even multiplicity represent points where the graph touches but does not cross the x axis; we will mark them with a dot. After we have marked the intercepts properly, we choose any convenient value of x (not a zero) at which to calculate the sign of f. The sign of f on the remaining intervals can then be determined easily and we can construct a graph that shows correctly where f is positive and where it is negative.

Example 9

Determine the intervals where $f(x) = x(x - 1)^2(x - 2)^3(x - 5)^4$ is positive and those where it is negative.

Solution. The zeros are 0, 2 (odd multiplicity, crossings); 1, 5 (even multiplicity, no crossings). The intercepts are marked in Fig. 5-12(a). Now obviously $f(x)$ is positive for $x > 5$. Consequently, we have a graph like that in Fig. 5-12(b).

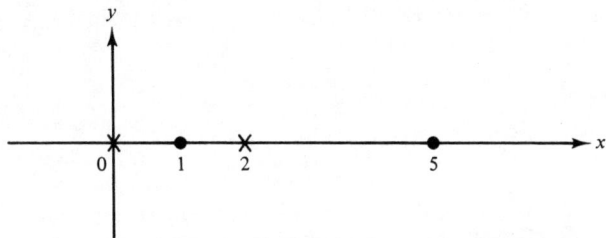

(a)

Figure 5-12 (a)

Notice that $f(x) \to +\infty$ as $x \to +\infty$ and $f(x) \to +\infty$ as $x \to -\infty$. The graph may not be accurate in some details, but it does show correctly where $f(x)$ is positive and where it is negative.

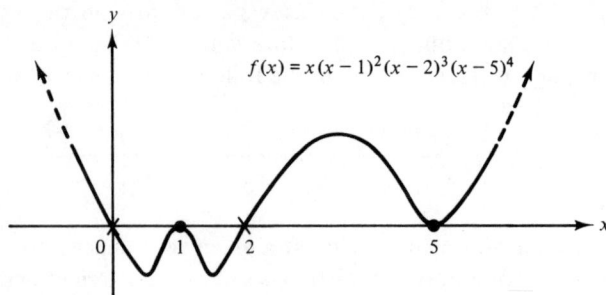

$$f(x) = x(x-1)^2(x-2)^3(x-5)^4$$

(b)

Figure 5-12 (b)

SUMMARY

To obtain a graph of f showing where $f(x) > 0$ and where $f(x) < 0$,

1. find the zeros of f and list the x intercepts of the graph.
2. identify each zero as having odd or even multiplicity and mark the corresponding intercept with a cross or dot accordingly.
3. find the sign of f at some convenient point and construct a sign chart.
4. draw a graph based on the sign chart.

EXERCISES 5-4

In Problems 1–10 graph the function and check for points of discontinuity.

1. $f(x) = \begin{cases} 1 & \text{for } x > 0 \\ 0 & \text{for } x = 0 \\ -1 & \text{for } x < 0 \end{cases}$

2. $F(t) = \begin{cases} 3t + 4 & \text{for } t > 1 \\ -2t + 8 & \text{for } t \le 1 \end{cases}$

3. $f(x) = \begin{cases} x + 1 & \text{for } x > 0 \\ x - 1 & \text{for } x \le 0 \end{cases}$

4. $g(u) = 3 + \dfrac{1}{u}$

5. $f(x) = \dfrac{1}{x - 2}$

6. $f(x) = 3x^2 + 4x^3$

7. $g(x) = |1 - x|$

8. $h(x) = \dfrac{1 + x}{1 - x^2}$

9. $f(x) = \dfrac{x}{x^2 - x - 12}$

10. $g(x) = \dfrac{x - 1}{x^3 - x^2}$

Polynomial and Rational Functions   Chap. 5

11. Show that each function has a zero between $x = 3$ and $x = 4$.
    (a) $g(x) = x^3 - 4x^2 + 7$
    (b) $h(x) = 7x - x^3 + 8$

†12. Use the method of Example 5 to find an approximation to a zero of the polynomial $P(x) = 2x^3 + 4x^2 - 7$.

In Problems 13–24 determine the intervals on which the given polynomial is positive.

13. $x(x^2 + 4)$ 14. $x(4 - x^2)$ 15. $x(x - 1)(x + 3)$

16. $(u^2 + u)(2 - u)$ 17. $x^2(x + 1)^3(x - 2)$ 18. $3t(t - 1)^4(t + 2)$

19. $3t^2 - 4t^3 - t^4$ 20. $x^4 - 6x^2 + 9$ 21. $6 + 5x - 2x^2 - x^3$

22. $x^4 + x^3 + x^2 + x$ 23. $12 + 4x - 3x^2 - x^3$ 24. $x^3 + 4x^2 - 3x - 18$

FOR THE ENTHUSIASTS

25. Show that if f is a quadratic function

$$f(x) = Ax^2 + Bx + C \quad (A \neq 0)$$

and if $B^2 - 4AC < 0$, then $f(x)$ is of constant sign, either positive for all x or negative for all x.

26. Show that if f is a polynomial function, if a and b are numbers such that $a < b$, and if, for some number m, $f(a) < m$ and $f(b) > m$, then there is a number c such that $a < c < b$ and $f(c) = m$.

5-5 INTEGRAL AND RATIONAL ZEROS

Guessing a Zero

If we can guess a zero r of a polynomial $P(x)$, we can then factor out the corresponding factor $x - r$, obtaining

$$P(x) = (x - r)Q(x).$$

Our search for zeros of $P(x)$ may then be limited to looking for zeros of $Q(x)$. Because $Q(x)$ is of lower degree, the chances are good that its zeros may be easier to find. The purpose of this section is to determine some reasonable clues for making the initial guess.

Often common sense rules out numbers of a certain size, as in the following example.

Example 1

Find the zeros of $P(x) = x^3 + 2x^2 + 4x + 8$.

Solution. Clearly no nonnegative number can be a zero of P and so we consider only negative numbers. Trying -1 gives $P(-1) = 5$. However, $P(-2) = 0$; so -2 is a zero and $x + 2$ a factor. By the division process, $P(x) = (x + 2)(x^2 + 4)$; and because $x^2 + 4$ has no real zeros, we see that $P(x)$ has only one real zero, -2.

Integral Zeros

On the optimistic assumption that $P(x)$ has an integral zero, we can try the numbers $0, \pm 1, \pm 2, \dots$. Fortunately, there are considerations that will help us reduce the number of possibilities.

Notice that in Example 1 the zero obtained (-2) is a factor of the constant term (8). This is not a coincidence at all but rather a consequence of the following theorem.

> **Theorem 5-5 (Integral Zeros Theorem).** Any integral zero of a polynomial of degree ≥ 1 with integral coefficients is a factor of the constant term of that polynomial.

Proof. Let r be an integral zero of the polynomial

$$P(x) = A_n x^n + A_{n-1} x^{n-1} + \cdots + A_1 x + A_0$$

whose degree is greater than or equal to one and whose coefficients $A_n, A_{n-1}, \dots, A_1, A_0$ are integers. Because r is a zero of P, we have

$$P(r) = A_n r^n + A_{n-1} r^{n-1} + \cdots + A_1 r + A_0 = 0$$

and it follows that

$$A_0 = -A_n r^n - A_{n-1} r^{n-1} - \cdots - A_1 r$$
$$A_0 = r(-A_n r^{n-1} - A_{n-1} r^{n-2} - \cdots - A_1)$$

The last equation shows that r is a factor of A_0, the constant term of $P(x)$.

If we are trying to find integral zeros of a polynomial with integral coefficients, Theorem 5-5 permits us to limit the search to the factors of the constant term.

Example 2

Find any integral zeros of the polynomial

$$P(x) = x^3 - 2x^2 - 2x - 3$$

and factor $P(x)$ accordingly.

Solution. The only possible integral zeros are the factors of -3—namely, $\pm 1, \pm 3$. Trying out these four numbers, we find that only 3 is a zero. We can then proceed to factor $P(x)$ by dividing out $x - 3$. The result is $P(x) = (x - 3)(x^2 + x + 1)$, in which the quadratic factor has no real zeros.

Example 3

Find any integral zeros of $P(x) = 12 - 4x - x^3$.

Solution. The constant term is 12, which has factors $\pm 1, \pm 2, \pm 3, \pm 4, \pm 6, \pm 12$. Obviously no negative number is a zero. Trying the positive factors of 12, we get $P(1) = 7$, $P(2) = -4$ and at this point it should be clear that no larger number is a zero. We conclude that P has no integral zeros.

Rational Zeros

Let us consider next the possibility that the given polynomial has rational[2] zeros. We begin with an example.

Example 4

Find the zeros of

$$P(x) = 6x^3 - x^2 - 23x + 14.$$

Solution. An attempt to find integral zeros fails; indeed, this polynomial has no integral zeros. We notice, however, that $P(x)$ changes sign between $x = 0$ and $x = 1$ and so, by Theorem 5-4, $P(x)$ has a zero between 0 and 1. Trying various rational numbers between 0 and 1—for instance, $\frac{1}{2}, \frac{3}{4}, \frac{3}{2}, \ldots$—we finally discover that $P(\frac{2}{3}) = 0$; hence $\frac{2}{3}$ is a zero. Dividing out the factor $x - \frac{2}{3}$, we get

$$P(x) = \left(x - \frac{2}{3}\right)(6x^2 + 3x - 21)$$

$$= 3\left(x - \frac{2}{3}\right)(2x^2 + x - 7).$$

The zeros of the quadratic factor are $(-1 \pm \sqrt{1 + 56})/4$. So P has the three zeros

$$\frac{2}{3}, \quad \frac{-1 + \sqrt{57}}{4}, \quad \frac{-1 - \sqrt{57}}{4}.$$

Notice that the rational zero $\frac{2}{3}$ obtained in Example 4 has the following property: the numerator 2 is a factor of the constant term (14) of $P(x)$ while the denominator 3 is a factor of the leading coefficient (6). Again, this is not a coincidence.

Theorem 5-6 (Rational Zeros Theorem). Let $P(x)$ be a polynomial of the form

$$P(x) = A_n x^n + A_{n-1} x^{n-1} + \cdots + A_1 x + A_0$$

with degree greater than or equal to 1 whose coefficients $A_n, A_{n-1}, \ldots, A_0$ are integers. If $r = p/q$ is a rational zero of P (in reduced form), then p is a factor of A_0 and q is a factor of A_n.

Proof. Because r is a zero of P, we know that $P(r) = 0$ and so

$$A_n r^n + A_{n-1} r^{n-1} + \cdots + A_1 r + A_0 = 0.$$

Substituting p/q for r gives

$$A_n \frac{p^n}{q^n} + A_{n-1} \frac{p^{n-1}}{q^{n-1}} + \cdots + A_1 \frac{p}{q} + A_0 = 0.$$

---

[2] A rational number is a real number that can be expressed as a quotient of two integers. Thus $\frac{2}{3}, \frac{5}{2}, -\frac{1}{5}$ are rational. A rational number p/q is "reduced" if p and q have no common factors greater than one.

Multiply by q^n on both sides and simplify.

$$A_n p^n + A_{n-1} p^{n-1} q + \cdots + A_1 p q^{n-1} + A_0 q^n = 0 \qquad (5\text{-}6)$$

Thus

$$A_n p^n = -q(A_{n-1} p^{n-1} + \cdots + A_1 p q^{n-2} + A_0 q^{n-1}).$$

The right-hand side clearly has q as a factor; therefore q is a factor of $A_n p^n$. But because q has no factors in common with p, the only possibility is that q is a factor of A_n.[3]

Returning to Eq. (5-6) and rewriting it now in a different way, we have

$$p(A_n p^{n-1} + A_{n-1} p^{n-2} q + \cdots + A_1 q^{n-1}) = -A_0 q^n.$$

The left-hand side has p as a factor; so p is a factor of $A_0 q^n$. But p has no factors in common with q, and so the only possibility is that p is a factor of A_0.

If follows from Theorem 5-6 that, for a given polynomial with *integer* coefficients, only certain rational numbers need be considered as possible zeros. They can be obtained as follows.

1. List for p all factors of the constant term.
2. List for q all factors of the leading coefficient.
3. Form all possible numbers p/q by dividing each number in the first list by those in the second list.

Any rational zero of the given polynomial must appear in the resulting list of numbers (3).

Example 5

Applying this procedure to the polynomial

$$P(x) = 6x^3 - x^2 - 23x + 14$$

in Example 4, we have
  (a) the factors of 14: $\pm 1, \pm 2, \pm 7, \pm 14$.
  (b) the factors of 6: $\pm 1, \pm 2, \pm 3, \pm 6$.
  (c) the numbers p/q: $\pm 1, \pm 2, \pm 7, \pm 14; \ \pm 1/2, \pm 7/2; \ \pm 1/3, \pm 2/3, \pm 7/3, \pm 14/3; \pm 1/6, \pm 7/6$.

Obviously not all numbers in our list (c) are zeros, but any rational zeros of $P(x)$ *must* appear in that list. In Example 4 we observed that P has a zero between 0 and 1. Looking at list (c), we see that the only possible *rational* zeros between 0 and 1 are $\frac{1}{2}, \frac{1}{3}, \frac{1}{6}, \frac{2}{3}$. And as we found out, $\frac{2}{3}$ is actually a zero.

Integral zeros are also rational, appearing as those fractions $r = p/q$ having $q = 1$. Ordinarily we would check for integral zeros as a first step in applying the

---

[3]The argument here depends on the fundamental theorem of arithmetic. See, for example, David M. Burton, *Elementary Number Theory* (Boston: Allyn and Bacon, Inc., 1976).

rational zeros theorem and it is easy to see that this step is tantamount to using the integral zeros theorem.

Example 6

Find any zeros of $P(x) = 6x^3 - 15x^2 - 2x + 5$.

Solution. Let us check for integral zeros first. By Theorem 5-5, the candidates are $\pm 1, \pm 5$ of which none are zeros; but evaluating P at these four numbers yields some valuable information: $P(1) < 0$ and $P(5) > 0$. So there is a zero between 1 and 5.

Candidates for rational zeros (other than the integers already mentioned) are $\pm 1/2, \pm 1/3, \pm 1/6, \pm 5/2, \pm 5/3, \pm 5/6$. If there is a rational zero between 1 and 5, it must be either $5/2$ or $5/3$. We find that $5/2$ is a zero and obtain the factorization $(x - 5/2)(6x^2 - 2)$. The other two zeros are $\pm \sqrt{3}/3$.

Example 7

Show that $P(x) = 3x^3 + 7x + 5$ has no rational zeros.

Solution. The candidates for p are $\pm 5, \pm 1$; those for q are $\pm 3, \pm 1$. And so the candidates for p/q are $\pm 5, \pm 1, \pm 5/3, \pm 1/3$. Clearly P has no positive zeros. Checking the values of P at $-5, -1, -5/3, -1/3$, we find that none of them is a zero, and conclude that P has no rational zeros.

EXERCISES 5-5

In Problems 1–7 use Theorem 5-5 to find all integral zeros of the given polynomial.

1. $P(x) = 2x^3 + 17x^2 + 41x + 30$ **2.** $P(x) = 5x^3 + 11x^2 - 2x - 8$

3. $Q(x) = x^3 + 3x^2 - 3x - 14$ **4.** $P(x) = 4x^3 + 12x^2 - x - 3$

5. $Q(y) = y^3 - 7y + 6$ **6.** $f(t) = t^3 + t^2 - 6t - 2$

7. $f(x) = x^4 + 15x^3 - 2x^2 - 34x - 60$

In Problems 8–14 find all real zeros of the given polynomials.

8. $P(x) = 8x^3 + 4x^2 - 18x - 9$ **9.** $P(x) = x^3 - x^2 - 4x - 6$

10. $Q(x) = 3x^3 - 8x^2 + x + 2$ **11.** $f(t) = 4t^3 - t^2 - 7t - 3$

12. $P(x) = 2x^3 + 5x^2 - 8x - 6$ **13.** $f(x) = 6x^4 - 8x^3 + 3x^2 - 4x$

14. $P(x) = x^4 + 6x^2 + 9$

In Problems 15–20 find a rational zero of the given polynomial or show that there is no rational zero.

15. $P(x) = 2x^3 - 3x^2 + 4x - 12$ **16.** $Q(x) = 3x^3 + 2x^2 + 15x + 10$

17. $P(x) = 5x^3 - x^2 + x + 4$ **18.** $f(x) = x^4 - 5x^2 + 6$

19. $f(x) = 2x^4 - 9x^2 - 5$ **20.** $P(t) = 2t^5 + t^4 - 3t^3 - t + 1$

21. Let $Q(y) = -2y^4 + 7y^3 + 2y^2 - 3y - 1$. Show that $-1/2$ is the only rational zero of Q and that there is an irrational zero of Q between $1/2$ and 1.

22. Show that if a polynomial with integral coefficients and leading coefficient 1 has a rational zero, then that zero must be an integer.

23. Show that $\sqrt[3]{2}$ is irrational by looking for rational solution of $x^3 = 2$.

24. Show that there is no rational number x such that the sum of the square of x and the reciprocal of x is 5.

5-6 APPROXIMATION OF IRRATIONAL ZEROS

Using the methods of the previous section, we can find all rational zeros of a polynomial with integral coefficients. In this section we develop a method for approximating an irrational root to any desired degree of accuracy. This method is based on Theorems 5-3 and 5-4, which tell us that if a polynomial changes sign from $x = a$ to $x = b$, then it has a zero between a and b.

Example 1

Locate a zero of the polynomial $P(x) = x^3 + x - 13$ between consecutive hundredths.

Solution. First, we observe that $P(x)$ has no rational zeros.

Checking $P(x)$ for some integral values of x, we find

x	$P(x)$	
0	-13	
1	-11	
2	-3	change of sign
3	$+17$	

We have located a zero between consecutive integers—namely, between 2 and 3.

Next, let us locate this zero between consecutive tenths. The procedure is simply to start calculating P at 2.1, 2.2, . . . , 2.9 and watch for a change of sign.

x	$P(x)$	
2.0	-3.000	
2.1	-1.639	
2.2	-0.152	change of sign
2.3	$+1.467$	
2.4		
2.5		
2.6		
2.7		
2.8		
2.9		
3.0	$+17$	

We need not complete this table, for we are now sure that there is a zero between 2.2 and 2.3. We have located a zero between consecutive tenths.

Next, let us locate this zero between consecutive hundredths.

x	$P(x)$	
2.20	-0.152	change of sign
2.21	$+0.004$	

We did not have to go far: The zero is now located between consecutive hundredths, 2.20 and 2.21. Clearly this process could be continued as long as we wish.

It can be shown by methods of calculus that the polynomial in Example 1 has only one zero. If there were others, they could be approximated in the same way.

Example 2

Locate a zero of the polynomial $P(x) = 5 + 10x - x^3$ between consecutive hundredths.

Solution. The table shows that there are three zeros.

x	$P(x)$	
-3	$+ 2$	change of sign
-2	$- 7$	
-1	$- 4$	change of sign
0	$+ 5$	
1	$+14$	
2	$+17$	
3	$+ 8$	change of sign
4	-19	

There are two negative zeros: one between -3 and -2, another between -1 and 0, and a positive zero between 3 and 4. We will locate the positive zero between consecutive hundredths.

Rather than check the value of P at each of the numbers 3.1, 3.2, . . . , 3.9, let us use linear interpolation to estimate where the graph of P crosses the x axis. Assuming the graph of P is approximately linear, we consider the line segment between points $(3, 8)$ and $(4, -19)$. (See Fig. 5-13.) Where does this segment cross the x axis? If r is the x coordinate of the intercept, we have

$$\frac{8 - 0}{3 - r} = \frac{8 - (-19)}{3 - 4} = \frac{27}{-1}.$$

Solving for r yields $r \approx 3.296$, indicating that the graph of P crosses the x axis near $x = 3.3$. Let us then begin by calculating $P(3.3)$. Because $P(3.3) \approx 2.063$ and $P(3)$ is also positive, we next try $P(3.4)$, which turns out to be $-.304$. Thus the zero is between 3.3 and 3.4.

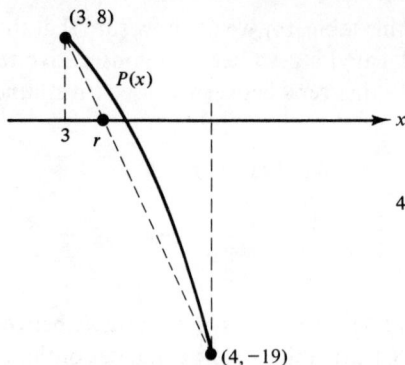

(3, 8)

P(x)

3 r

x

4

(4, −19)   Figure 5-13

To obtain the next approximation, we connect the points (3.3, 2.063) and (3.4, −.304) by a line segment and determine a new r in the same way as before (the reader should supply a sketch).

$$\frac{2.063 - 0}{3.3 - r} = \frac{2.063 - (-.304)}{3.3 - 3.4}$$

From this equation we obtain $r \approx 3.387$, which suggests that the zero is between 3.38 and 3.39. Indeed, we see that such is the case from the following values.

x	$P(x)$	
3.38	0.186	change of sign
3.39	−0.058	

EXERCISES 5-6

†1. Locate a zero of the given polynomial between consecutive hundredths.
   (a) $P(x) = x^3 + x - 28$ (b) $P(x) = 2x^3 + x - 7$
   (c) $P(x) = 8 - x - x^3$ (d) $Q(x) = x^3 + x^2 + x + 2$

†2. Locate $\sqrt[3]{25}$ between consecutive thousandths, using the method of this section.

†3. A rectangular box without a top is to be made from a square piece of cardboard 10 centimeters by 10 centimeters by cutting equal squares of edge x centimeters out of the corners and folding along the dotted lines as in Fig. 5-14. The resulting box must have volume 20 square centimeters. Show that there are two values of x between 0 and 5 for which the volume is 20 and locate each of them between consecutive hundredths.

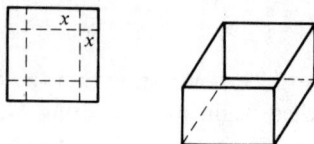

Figure 5-14

5-7 ALGEBRA OF RATIONAL FUNCTIONS

A function formed by dividing two polynomials is called a rational function.

> **Definition.** A **rational function** is one that can be expressed as the quotient of two polynomial functions.

If f is a rational function, then there exist polynomials

$$P(x) = A_0x^n + A_1x^{n-1} + \cdots + A_{n-1}x + A_0$$

and
$$Q(x) = B_0x^m + B_1x^{m-1} + \cdots + B_{m-1}x + B_0$$

such that

$$f(x) = \frac{P(x)}{Q(x)} = \frac{A_nx^n + A_{n-1}x^{n-1} + \cdots + A_1x + A_0}{B_mx^m + B_{m-1}x^{m-1} + \cdots + B_1x + B_0}. \tag{5-7}$$

An expression $P(x)/Q(x)$ that defines a rational function is called a **rational expression**; $P(x)$ is its **numerator** and $Q(x)$ is its **denominator**. The domain of the function f in (5-7) consists of all those real numbers x such that $Q(x) \neq 0$—that is, all real numbers except the zeros of Q.

Example 1

A rational function f is defined by

$$f(x) = \frac{x^2 - 2}{x^3 - x}.$$

Its domain consists of all real numbers x except $x = -1, 0, 1$, for these are the zeros of $x^3 - x$.

Any polynomial function is a rational function because we can take the denominator to be the constant 1. Also, the reciprocal of a polynomial function is a rational function because we can take the numerator to be 1.

You will recall from algebra that, in order to add two fractions, you first find a common denominator and express each fraction in a form having that denominator. Then the numerators can be added. This process is illustrated with rational expressions in the next three examples. These examples also show that a rational function may be given originally in a form other than a quotient.

Example 2

The expression

$$\frac{1}{x - 1} + \frac{3}{x + 1}$$

is algebraically the same as

$$\frac{(x + 1) \cdot 1}{(x + 1)(x - 1)} + \frac{3(x - 1)}{(x + 1)(x - 1)} = \frac{(x + 1) + 3(x - 1)}{(x + 1)(x - 1)} = \frac{4x - 2}{x^2 - 1}.$$

Thus the given expression defines a rational function. Its domain consists of all real numbers $x \neq -1, 1$.

Example 3

The expression

$$\frac{2x-3}{x^2+1} - \frac{3x-2}{x^3}$$

is equal to

$$\frac{(2x-3)x^3}{(x^2+1)x^3} - \frac{(x^2+1)(3x-2)}{(x^2+1)x^3}$$

$$= \frac{(2x-3)x^3 - (x^2+1)(3x-2)}{(x^2+1)x^3} = \frac{2x^4 - 6x^3 + 2x^2 - 3x + 2}{x^5 + x^3}.$$

This expression defines a rational function whose domain consists of all real numbers except 0. (Why?)

Example 4

The expression

$$\frac{2}{x(x+1)} + \frac{3}{x^2(x-1)}$$

is equal to

$$\frac{2x(x-1)}{x^2(x+1)(x-1)} + \frac{3(x+1)}{x^2(x-1)(x+1)} = \frac{2x^2 + x + 3}{x^2(x+1)(x-1)}.$$

This expression defines a rational function whose domain consists of all real numbers except $-1, 0, 1$.

As we saw in preceding examples, the sum and the difference of two rational functions are also rational functions. The same applies to the product and the quotient.

Example 5

Let

$$f(x) = \frac{3x+1}{x^2-4} \quad \text{and} \quad g(x) = \frac{3x}{x^2+4}.$$

Then the product and quotient are defined by

$$f(x)g(x) = \frac{(3x+1)(3x)}{(x^2-4)(x^2+4)} = \frac{9x^2 + 3x}{x^4 - 16}$$

and

$$\frac{f(x)}{g(x)} = \frac{(3x+1)(x^2+4)}{(x^2-4)(3x)} = \frac{3x^3 + x^2 + 12x + 4}{3x^3 - 12x}.$$

What are the domains of these functions?

Cancellation

When a polynomial occurs as a factor both in the numerator and in the denominator of a rational expression, it may be "canceled," as illustrated in the following example.

Example 6

Let

$$f(x) = \frac{x^2 + x - 6}{x^2 - 4}.$$

By factoring, we see that the polynomial $x - 2$ is a factor of the numerator and denominator.

$$\frac{x^2 + x - 6}{x^2 - 4} = \frac{(x + 3)(x - 2)}{(x + 2)(x - 2)}$$

For $x \neq 2$, we can cancel $x - 2$ because

$$f(x) = \frac{x + 3}{x + 2} \cdot \frac{x - 2}{x - 2} = \frac{x + 3}{x + 2} \cdot 1 = \frac{x + 3}{x + 2}.$$

If $x = 2$, however, the expression $(x - 2)/(x - 2)$ is meaningless and so the cancellation is not valid for $x = 2$. The equation

$$f(x) = \frac{x + 3}{x + 2}$$

does not hold when $x = 2$; the function f is undefined at 2 as well as -2.

As illustrated in Example 6, cancellation is invalid at any value of x for which the canceled factor is zero.

Example 7

Simplify

$$f(x) = \frac{x^4 + 2x^2 - 3}{x^4 - 1}.$$

Solution. Factoring the numerator and denominator gives

$$f(x) = \frac{(x^2 + 3)(x^2 - 1)}{(x^2 + 1)(x^2 - 1)},$$

showing that $x^2 - 1$ is a factor of both. So

$$f(x) = \frac{x^2 + 3}{x^2 + 1} \quad \text{for} \quad x \neq -1, 1.$$

Example 8

Simplify

$$f(x) = \frac{x^3 - 1}{x^2 + x + 1}.$$

Solution. The numerator factors:

$$x^3 - 1 = (x - 1)(x^2 + x + 1);$$

and we have

$$f(x) = \frac{(x - 1)(x^2 + x + 1)}{x^2 + x + 1} = x - 1.$$

Because $x^2 + x + 1$ has no real zeros, $f(x) = x - 1$ for all real x.

Zeros of a Rational Function

Because a quotient is zero only if its numerator is zero, the zeros of the rational function (5-7) can be found by solving $P(x) = 0$. Any solution of $P(x) = 0$ that belongs to the domain of f is a zero of f.

Example 9

Find the zeros of

$$f(x) = \frac{4x - 2}{x^2 - 1}.$$

Solution. The domain of f includes all real numbers except -1 and 1. Setting the numerator equal to zero, we get $4x - 2 = 0$, $x = \frac{1}{2}$. This number is in the domain and so it is a zero.

Example 10

Find the zeros of

$$f(x) = \frac{x^2 + x - 6}{x^2 - 4}.$$

Solution. The domain of f includes all real numbers except -2 and 2. The zeros of the numerator are -3 and 2. We reject 2, for it is not in the domain of f. The only zero of f is -3.

Continuity of a Rational Function

In subsequent sections we will study graphs of rational functions. The following theorem will be helpful in determining the intervals on which a rational function is continuous. On any interval of continuity the graph is an unbroken curve.

Theorem 5-5. A rational function f defined by $f(x) = P(x)/Q(x)$, where P and Q are polynomials, is continuous over every interval that does not contain a zero of $Q(x)$. The points of discontinuity of a rational function are the zeros of its denominator.

Example 11

Determine the intervals on which the function of Example 10 is continuous.

Solution. The points of discontinuity of

$$f(x) = \frac{x^2 + x - 6}{x^2 - 4}$$

are the zeros of $x^2 - 4$—namely, -2 and 2. Thus f is continuous on the intervals $(-\infty, -2)$, $(-2, 2)$, and $(2, +\infty)$.

Sign of a Rational Function

The methods of Section 5-4 enable us to determine the sign of a rational function.

Theorem 5-6. Let $P(x)$ and $Q(x)$ be polynomials. Then the rational function defined by

$$f(x) = \frac{P(x)}{Q(x)}$$

has the same sign as the polynomial

$$P(x)Q(x)$$

for every x in the domain of f.

Proof. Let x be a number in the domain of f so that $Q(x) \neq 0$. If $P(x)Q(x)$ is positive, we can divide both sides of the inequality

$$P(x)Q(x) > 0$$

by $[Q(x)]^2$ (which is positive), obtaining

$$\frac{P(x)}{Q(x)} > 0.$$

Thus $P(x)/Q(x)$ is positive whenever $P(x)Q(x)$ is positive. In the same way, if $P(x)Q(x)$ is negative, we can divide both sides of the inequality

$$P(x)Q(x) < 0$$

by $[Q(x)]^2$, obtaining

$$\frac{P(x)}{Q(x)} < 0.$$

Thus $P(x)/Q(x)$ is negative whenever $P(x)Q(x)$ is negative.

Example 12

Determine the intervals on which the rational function

$$f(x) = \frac{x^2 - 4}{x^3 + x}$$

is positive.

Solution. By Theorem 5-6, the sign of $f(x)$ is the same as that of the polynomial

$$g(x) = (x^2 - 4)(x^3 + x) = (x^2 - 4)x(x^2 + 1).$$

The zeros of g are $-2, 0, 2$ and $g(x)$ changes sign as x passes through each zero. Starting with the fact that $g(x) > 0$ for very large x, we construct a sign chart for g.

Sign of $g(x)$ $-$ $+$ $-$ $+$

Values of x -2 0 2

Since f has the same sign chart, we conclude that $f(x)$ is positive on the intervals $(-2, 0)$, and $(2, +\infty)$.

Example 13

Determine the intervals on which the function

$$f(x) = \frac{x^2 + x}{x^2 - 1}$$

is positive.

Solution. By Theorem 5-6, $f(x)$ is positive wherever $g(x) = (x^2 + x)(x^2 - 1)$ is positive. But $g(x) = x(x + 1)^2(x - 1)$; so the zeros of g are $-1, 0, 1$. Because -1 has multiplicity 2, $g(x)$ does not change sign as x passes through -1. Here is the sign chart of g.

Sign of $g(x)$	$+$	$+$	$-$	$+$

Values of x -1 0 1

It follows that $f(x)$ is positive on the intervals $(-\infty, -1), (-1, 0)$, and $(1, +\infty)$.

EXERCISES 5-7

In Problems 1–13 write the expression as a quotient of two polynomials.

1. $\dfrac{2}{3x + 1} + \dfrac{5}{x - 2}$

2. $\dfrac{3y}{y^2 - 4} - \dfrac{5}{y + 2}$

3. $(2x - 1)(x^2 - x)^{-1} + 3x(x^2 + 1)^{-1}$

4. $\left(\dfrac{1}{3}x - \dfrac{1}{2}\right)(6x + 12)^{-1} - \left(\dfrac{1}{2}x + \dfrac{2}{3}\right)(4x - 6)^{-1}$

5. $\left(5x + \dfrac{1}{2}\right)(2 - x)^{-1} + \left(\dfrac{3}{2} - \dfrac{3}{4}x\right)(10x + 4)^{-1}$

6. $\dfrac{t - 1 + \sqrt{5}}{t - 1 - \sqrt{5}} - \dfrac{t + 1 - \sqrt{5}}{t + 1 + \sqrt{5}}$

7. $\dfrac{3u}{u^2 - 4} \cdot \dfrac{2}{u + 5}$

8. $\dfrac{5x - 2}{x^2 + 3x + 2} \cdot \dfrac{4x + 3}{x^2 + 2}$

9. $\dfrac{3w}{w^2 - 4} \div \dfrac{2}{w + 5}$

10. $\dfrac{5x - 2}{x^2 + 3x + 2} \div \dfrac{x^2 + 2}{4x + 3}$

11. $\dfrac{\frac{1}{3}x - \frac{1}{2}}{6x + 12} \div \dfrac{\frac{1}{2}x + \frac{2}{3}}{4x - 6}$

12. $(x^{-1} - 3x^{-2}) \div (4x^{-1} - 5x^{-2})$

13. $\left(\dfrac{1}{x + 1} - \dfrac{1}{x - 1}\right) \div \left(\dfrac{1}{x + 1} + \dfrac{1}{x - 1}\right)$

In Problems 14–21 simplify the rational expression by canceling factors that occur in both numerator and denominator. Indicate any values of the variable for which the simplified expression is defined but where the cancellation is not valid (see Example 6).

14. $\dfrac{x(x - 3)}{x^2 - 4x + 3}$

15. $\dfrac{x^2 + x - 2}{x^2 + 5x + 6}$

16. $\dfrac{t^3 + t}{t^4 - 1}$

17. $\dfrac{4u^3 + 2u}{4u^4 - 1}$

18. $\dfrac{x^3 - 2x^2 + x}{x^3 - x^2 - x + 1}$

19. $\dfrac{x^3 + 4x^2 + 4x}{x^3 + x^2 - 8x - 12}$

20. $\dfrac{u^3 - 8}{u^2 + 2u + 4}$

21. $\dfrac{w^3 + 27}{w^2 - 3w + 9}$

In Problems 22–27 find the zeros of the expressions.

22. $\dfrac{x^2 + x - 6}{2x^2 - 5x}$

23. $\dfrac{x^3 - 3x^2 + 2x}{x^2 - 4x}$

24. $\dfrac{x^3 - 2x^2 + x}{x^3 - x^2 - x + 1}$

25. $\dfrac{u^3 - 8}{u^2 + 2u + 4}$

26. $\dfrac{2x(x - 2)}{x^2 - 1} - \dfrac{x}{x + 1}$

27. $\dfrac{2x - 1}{x^2 - x} - \dfrac{2x}{x^2 - 1}$

In Problems 28–32 find the points of discontinuity of the functions.

28. $f(x) = \dfrac{x + 4}{x^3 - x}$

29. $g(x) = \dfrac{x}{x^3 - 3x}$

30. $f(x) = \dfrac{2x - 3}{x^2 - 4}$

31. $g(x) = \dfrac{x - 2}{x^4 - 16}$

32. $f(x) = \dfrac{x^3 - 2x^2 + x}{x^3 - x^2 - x + 1}$

In Problems 33–42 determine the intervals on which the rational expression is positive.

33. $\dfrac{x}{x^2 - 1}$

34. $\dfrac{3x - 2}{x^2 - 4}$

35. $\dfrac{t + 1}{t - 1}$

36. $\dfrac{x^2 - 1}{x^2 + 1}$

37. $\dfrac{u(u + 1)}{u + 2}$

38. $\dfrac{x^2 - 6x + 8}{x^2 - 6x + 9}$

39. $\dfrac{t^2 - 1}{t^3 - 1}$

40. $\dfrac{1}{x} + \dfrac{x}{x^2 - 1}$

41. $\dfrac{3}{x + 4} - \dfrac{1}{x - 4}$

42. $\dfrac{x}{2x + 3} \div \dfrac{x - 1}{2x + 5}$

43. (a) Does the following procedure give the solution of $x^2 > 16x$?

$$x^2 - 16x > 0 \qquad \text{(addition property)}$$
$$x(x - 16) > 0 \qquad \text{(factoring)}$$
$$x - 16 > 0 \qquad \text{(dividing by } x\text{)}$$
$$x > 16 \qquad \text{(addition property)}$$

(b) Is it correct to multiply or divide both sides of an inequality by a variable expression?

(c) What justifies dividing both sides of an inequality by $[Q(x)]^2$ as in the proof of Theorem 5-6?

5-8 VERTICAL ASYMPTOTES

Suppose that we wish to graph the rational function

$$f(x) = \frac{4x}{x^2 - 1}$$

over the interval $-2 \leq x \leq 2$. First, we notice that there are two points inside the interval that do not belong to the domain of the function: $f(x)$ is undefined and hence discontinuous at $x = -1$ and $x = 1$. Let us draw dashed vertical lines through $x = -1$ and $x = 1$ and keep in mind that the graph of f will not cross these lines [see Fig. 5-15(a)].

We continue by looking for intercepts on the coordinate axes and for symmetry. We find that the only intercept is the point $(0, 0)$; and because f is odd, the graph of f is symmetric with respect to the origin:

$$f(-x) = \frac{4(-x)}{(-x)^2 - 1} = \frac{-4x}{x^2 - 1} = -f(x).$$

(a)

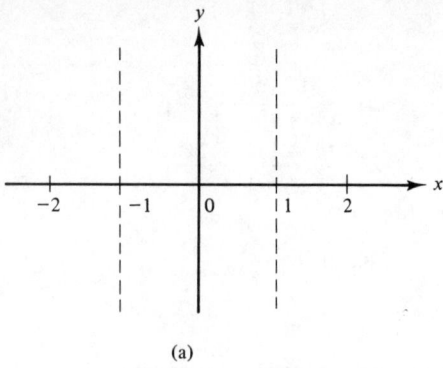

x	$f(x)$
0	0
$\frac{1}{2}$	$-\frac{8}{3}$
$\frac{3}{4}$	$-\frac{48}{7}$
$\frac{5}{4}$	$\frac{80}{9}$
$\frac{3}{2}$	$\frac{24}{5}$
2	$\frac{8}{3}$

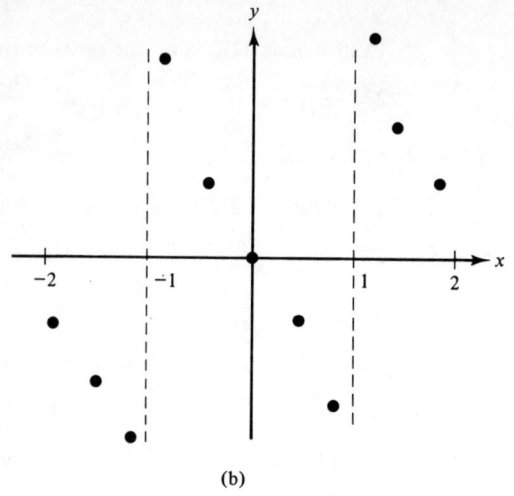

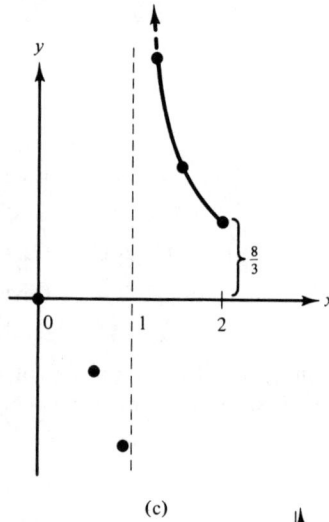

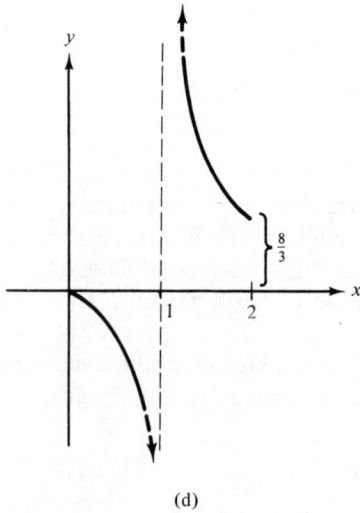

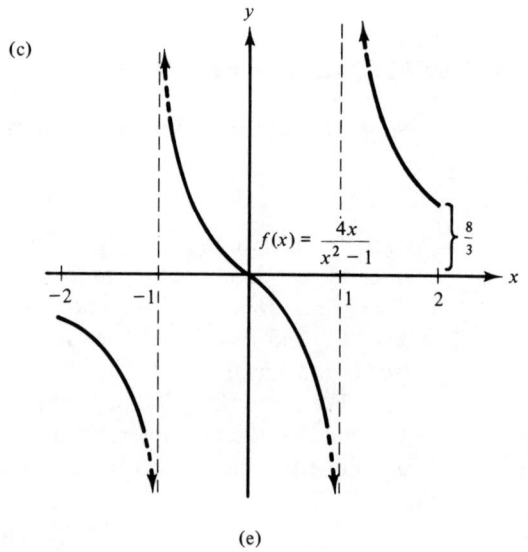

(b)

(c)

(d)

$$f(x) = \frac{4x}{x^2 - 1}$$

(e)

Figure 5-15

In Fig. 5-15(b) a short table of values for $x \geq 0$ is given and the corresponding points, together with the symmetric points, are plotted. How do we use these points to draw the graph? Let us allow x to approach 1 through values greater than 1. We will write "$x \rightarrow 1^+$" to mean that we let x become as close as desired to 1 but always greater than 1. Looking at the algebraic expression for $f(x)$, we can see that the denominator $x^2 - 1$ is a small positive number and becomes smaller as x gets closer to 1. Because the numerator $4x$ is near 4, the ratio $f(x) = 4x/(x^2 - 1)$ becomes larger and larger, approaching plus infinity as $x \rightarrow 1^+$. We will write

$$f(x) \longrightarrow +\infty \quad \text{as} \quad x \longrightarrow 1^+$$

to symbolize the situation. This behavior accounts for the tendency of the graph in Fig. 5-15(c) to turn upward sharply, just to the right of the line $x = 1$.

If we allow x to approach 1 through values *less* than 1 ($x \rightarrow 1^-$), the only difference is that $x^2 - 1$ becomes small and *negative*. This means that $f(x)$ is negative but becomes large in absolute value. In other words, $f(x)$ approaches minus infinity as x approaches 1 from the left:

$$f(x) \longrightarrow -\infty \quad \text{as} \quad x \longrightarrow 1^-.$$

The graph of f turns downward sharply just to the left of line $x = 1$ [see Fig. 5-15(d)].

Definition. The line $x = a$ is a **vertical asymptote** of the graph of a function f if any of the following conditions hold:
1. $f(x) \rightarrow +\infty$ as $x \rightarrow a^+$;
2. $f(x) \rightarrow -\infty$ as $x \rightarrow a^+$;
3. $f(x) \rightarrow +\infty$ as $x \rightarrow a^-$;
4. $f(x) \rightarrow -\infty$ as $x \rightarrow a^-$.

The line $x = 1$ is a vertical asymptote of the graph of $f(x) = 4x/(x^2 - 1)$ because, as we have seen, $f(x) \rightarrow +\infty$ as $x \rightarrow 1^+$ and $f(x) \rightarrow -\infty$ as $x \rightarrow 1^-$.

Because f is an odd function, we can now use symmetry to obtain the graph of $f(x)$ for $-2 \leq x \leq 2$ [see Fig. 5-15(e)]. Notice how the graph behaves near the line $x = -1$. Symmetry tells us that $x = -1$ is also a vertical asymptote. Indeed, it is easy to verify that $f(x) \rightarrow +\infty$ as $x \rightarrow -1^+$ and $f(x) \rightarrow -\infty$ as $x \rightarrow 1^-$.

Although not part of the graph itself, vertical asymptotes are important because they have a strong influence on the shape of the graph. So we want to find vertical asymptotes whenever they exist.

Definition. A **pole** of a rational function f is a point of discontinuity of f at which f becomes infinite—that is, a number a such that $x = a$ is a vertical asymptote of f.

The rational function $f(x) = 4x/(x^2 - 1)$ considered above has poles -1 and 1—the zeros of the denominator $x^2 - 1$. In general if P and Q are polynomial functions we look for the poles of $f = P/Q$ among the zeros of Q.

Example 1

The poles of the function

$$f(x) = \frac{x(x-1)(x+2)}{(x+4)(x-2)(x-1)}$$

are -4 and 2. The lines $x = -4$ and $x = 2$ are vertical asymptotes of the graph of f. Notice that we do not include 1 as a pole of $f(x)$ because the factor $x - 1$ can be eliminated by cancellation. Although 1 is a point of discontinuity, $f(x)$ does not approach $+\infty$ or $-\infty$ as $x \to 1^+$ or as $x \to 1^-$ and so $x = 1$ is not a vertical asymptote.

Example 2

The poles of the rational function

$$f(x) = \frac{x^3 + x^2 - 2x}{x^4 - x^2}$$

are 0 and -1 because the expression for $f(x)$ reduces to $(x + 2)/[x(x + 1)]$ after cancellation. The lines $x = 0$ and $x = -1$ are vertical asymptotes.

Example 3

There are no poles of the rational function $f(x) = x/(1 + x^2)$, for $1 + x^2$ has no zeros. Therefore there are no vertical asymptotes.

There are various ways in which the graph of a rational function can approach its vertical asymptotes, depending on the sign of the function.

Example 4

Let us see how the graph of the rational function in Example 2 approaches its vertical asymptotes. First, consider the asymptote $x = 0$. As $x \to 0^+$ the numerator and denominator of $(x + 2)/[x(x + 1)]$ are both positive; the numerator approaches 2 whereas the denominator is positive and becomes small. Hence $f(x) \to +\infty$ as $x \to 0^+$. As $x \to 0^-$ the denominator $x(x + 1)$ is negative (the factor x is negative while $x + 1$ is positive); the numerator approaches 2. So $f(x) \to -\infty$ as $x \to 0^-$ [see Fig. 5-16(a)].

Next, let us consider the asymptote $x = -1$. Letting $x \to -1^-$, we notice that the denominator consists of two factors, x and $x + 1$, both of which are negative because $x < -1$. Hence the denominator is positive and becomes small. Because the numerator approaches 1, $f(x) \to +\infty$. As $x \to -1^+$, the denominator is *negative* and becomes small so that $f(x) \to -\infty$.

Figure 5-16(b) shows both vertical asymptotes. It also shows a missing point in the graph at $x = 1$ because this number is not in the domain of the given function f.

In previous examples we saw rational functions that approach $+\infty$ on one side of an asymptote and $-\infty$ on the other. The next example shows that the graph of a rational function f can have a vertical asymptote such that $f(x)$ approaches $+\infty$ on both sides.

Polynomial and Rational Functions   Chap. 5

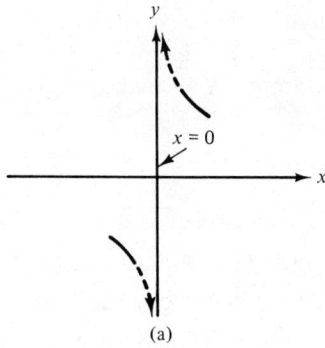

(a)

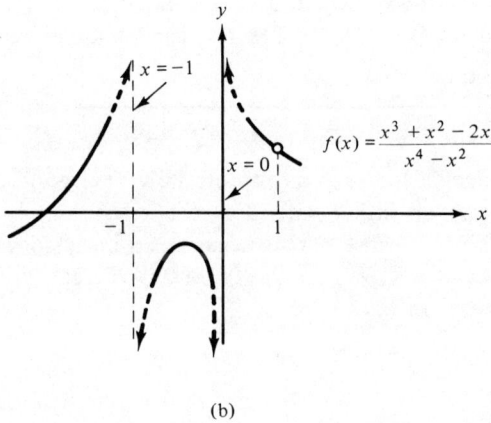

$$f(x) = \frac{x^3 + x^2 - 2x}{x^4 - x^2}$$

(b)

Figure 5-16

Example 5

The rational function $f(x) = x/(x-1)^2$ has $x = 1$ as a vertical asymptote. As x approaches 1 from either side, $(x-1)^2$ is positive and becomes small. Because the numerator x approaches 1, $f(x) \to +\infty$ as $x \to 1^+$ and as $x \to 1^-$ (see Fig. 5-17).

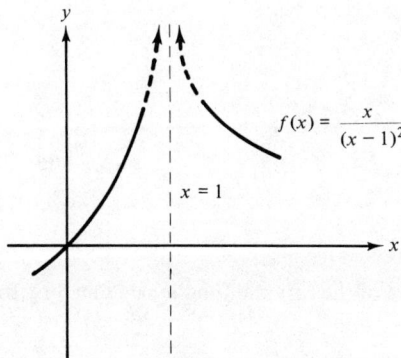

$$f(x) = \frac{x}{(x-1)^2}$$

Figure 5-17

The poles of a rational function are often easy to find, and when we know the poles we know the vertical asymptotes. To determine whether $f(x) \to +\infty$ or $f(x) \to -\infty$ on the two sides of an asymptote, we make a sign chart.

Example 6

Determine the poles of the function
$$f(x) = \frac{x^2 - 4x + 3}{x^3 - 4x^2 + 4x}$$

and show the corresponding vertical asymptotes in a graph.

Solution. The numerator and denominator of f can both be factored:
$$f(x) = \frac{(x-1)(x-3)}{x(x-2)^2}.$$

This rational function has two poles, 0 and 2. Remembering that the sign of $f(x)$ is the same as that of $(x-1)(x-3)x(x-2)^2$, we obtain the sign chart.

Sign of $f(x)$ $-$ $+$ $-$ $-$ $+$

Value of x 0   1   2   3 x

Now we can quickly decide for each asymptote how $f(x)$ approaches infinity (positively or negatively) on either side. For the vertical asymptote $x = 2$, $f(x) \to -\infty$ on both sides because $f(x)$ is negative on both sides of 2. For the vertical asymptote $x = 0$, $f(x) \to -\infty$ as $x \to 0^-$ whereas $f(x) \to +\infty$ as $x \to 0^+$. A sketch appears in Fig. 5-18.

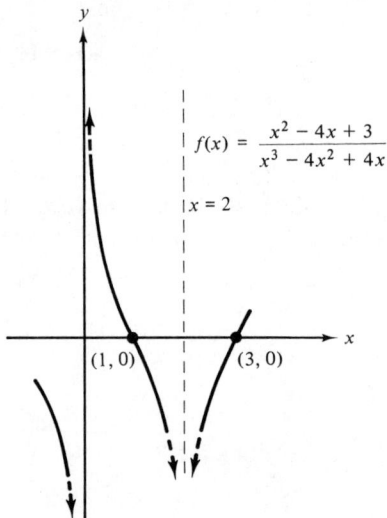

$$f(x) = \frac{x^2 - 4x + 3}{x^3 - 4x^2 + 4x}$$

$x = 2$

$(1, 0)$ $(3, 0)$

Figure 5-18

EXERCISES 5-8

In Problems 1–16 find the poles of the rational functions defined by the given expressions.

1. $\dfrac{2}{x-1}$

2. $\dfrac{4}{(x+2)^2}$

3. $\dfrac{1}{(3x+5)^2}$

4. $\dfrac{t}{4 - t^2}$

5. $\dfrac{2 - t^2}{2 + t^2}$

6. $\dfrac{3x - 1}{3x^2 + x + 4}$

7. $\dfrac{x^2 - 3x}{x^2 - 4x}$

8. $\dfrac{x^3 - 8}{x^2 - 2x}$

9. $\dfrac{x^2 - x + 1}{x^2(x - 1)^2}$

10. $\dfrac{x^2 + 2x + 4}{x(x + 1)^2(2x + 3)}$

11. $\dfrac{x^2 + 12}{x^2 - 12}$

12. $\dfrac{u^2 - 4}{u^2 + 2}$

13. $\dfrac{t^2 - t - 2}{t^2 + 3t + 2}$

14. $x - \dfrac{1}{x}$

15. $1 - \dfrac{1}{x^2}$

16. $\dfrac{1}{x + 2} - \dfrac{1}{x - 2}$

In Problems 17–21 find any vertical asymptotes and show in a sketch how the graph of the function approaches each of these asymptotes.

17. $g(x) = \dfrac{x}{4 - x^2}$

18. $f(x) = \dfrac{2}{(x - 1)^2}$

19. $h(t) = \dfrac{2 - t^2}{2 + t^2}$

20. $f(x) = x - \dfrac{1}{x}$

21. $f(x) = 1 - \dfrac{1}{x^2}$

FOR THE ENTHUSIASTS

In Problems 22–25 write an expression for a rational function whose graph has the given properties.

22. vertical asymptotes: $x = -1, x = 2$
x intercepts: $(0, 0), (1, 0)$

23. vertical asymptotes: $x = -1, x = 1$
x intercepts: $(-2, 0), (0, 0), (2, 0)$
symmetry with respect to the origin

24. vertical asymptotes: $x = -3, x = 3$
x intercepts: $(-2, 0), (0, 0), (2, 0)$
symmetry with respect to the y axis

25. vertical asymptotes: $x = -1, x = 0, x = 1$
x intercepts: $(-2, 0), (2, 0)$
graph does not cross x axis

5-9 HORIZONTAL ASYMPTOTES

In this section we consider the behavior of rational functions as $x \to +\infty$ and as $x \to -\infty$—briefly, as x becomes infinite.

Example 1

The rational function

$$f(x) = \dfrac{1}{1 + x^2}$$

is defined for all real numbers. As x increases without bound, so does $1 + x^2$. Because the reciprocal of a large number is small, it follows that $f(x)$ is near zero when x is large. Figure 5-19(a) shows the graph approaching the x axis as x becomes positively infinite:

$$f(x) \longrightarrow 0 \quad \text{as} \quad x \longrightarrow +\infty$$

It is also true that $f(x) \to 0$ as $x \to -\infty$. Noting the symmetry about the y axis we can complete the graph as in Fig. 5-19(b). There is a maximum at $(0, 1)$, as you may wish to show (Problem 30).

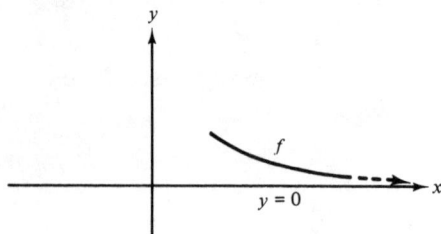

(a)

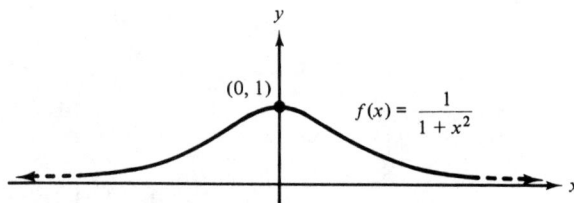

(b)

Figure 5-19

The line $y = 0$ (the x axis) plays a special role in determining the shape of the graph in Example 1; such a line is called a horizontal asymptote.

Definition. The line $y = c$ is a **horizontal asymptote** of the graph of f if $f(x) \to c$ as $x \to +\infty$ or $f(x) \to c$ as $x \to -\infty$.

Before we delve into methods for finding asymptotes of rational functions it is useful to examine the limiting behavior of a special type of function.

Example 2

Consider any function h of the form

$$h(x) = \frac{k}{x^p} \quad (x \neq 0)$$

where k is a constant and p is a positive integer. As x becomes infinite (positively or negatively) the denominator x^p becomes large in absolute value. Because the numerator is fixed, the quotient approaches zero. That is,

$$\frac{k}{x^p} \longrightarrow 0 \quad \text{as} \quad x \longrightarrow +\infty \quad \text{and} \quad \frac{k}{x^p} \longrightarrow 0 \quad \text{as} \quad x \longrightarrow -\infty.$$

Accordingly the x axis is a horizontal asymptote of the graph of h.

Example 3

Determine any horizontal asymptotes of the graph of

$$g(x) = \frac{x^2 - 1}{x^2 + 1}.$$

Solution. First consider positive x. What can we say about the behavior of $g(x)$ as $x \longrightarrow +\infty$? Dividing both the numerator and the denominator by x^2, we get, for $x \neq 0$,

$$g(x) = \frac{x^2\left(1 - \dfrac{1}{x^2}\right)}{x^2\left(1 + \dfrac{1}{x^2}\right)} = \frac{1 - \dfrac{1}{x^2}}{1 + \dfrac{1}{x^2}}.$$

As $x \longrightarrow +\infty$, the terms $1/x^2$ in numerator and denominator approach zero by Example 2, so that $g(x)$ approaches 1. Therefore the line $y = 1$ is a horizontal asymptote of the graph of g. Because $1 - 1/x^2$ is less than $1 + 1/x^2$, the value of $g(x)$ is less than 1 and the graph is below its asymptote. The fact that g is even means that $g(x)$ also approaches 1 as $x \longrightarrow -\infty$. There are intercepts at $(-1, 0)$, $(1, 0)$, $(0, -1)$. It can be shown that $(0, -1)$ is a minimum point of g (Problem 31). The graph appears in Fig. 5-20.

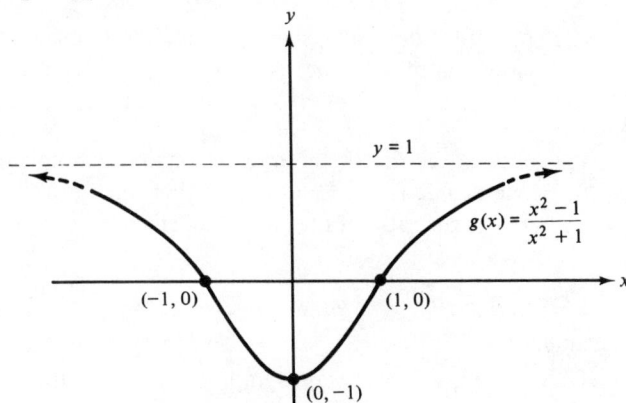

Figure 5-20

We said that $y = c$ is a horizontal asymptote of the graph of a function f if $f(x)$ approaches c as x becomes infinite. How can we determine whether there is such a number c? Let f be a rational function

$$f(x) = \frac{P(x)}{Q(x)} = \frac{A_n x^n + A_{n-1}x^{n-1} + \cdots + A_1 x + A_0}{B_m x^m + B_{m-1}x^{m-1} + \cdots + B_1 x + B_0} \qquad (A_n \neq 0, B_m \neq 0).$$

We consider three cases:

1. $n = m$
2. $n < m$
3. $n > m$.

In each of the first two cases we begin by dividing both numerator and denominator by x^m, the largest power of x that occurs in either place. (This is what we did in Example 3.) We then ask how the new numerator and denominator behave as x becomes infinite. Typically there will be terms of the form k/x^p, where p is a positive integer and k is constant. By Example 2, each such term approaches zero as x becomes infinite and any sum of such terms approaches zero.

Case 1. $n = m$

Example 4

Determine the horizontal asymptotes of the graph of

$$f(x) = \frac{2x^2 - 3x + 4}{4x^2 - 1}.$$

Solution. Dividing both numerator and denominator by x^2, we obtain

$$f(x) = \frac{2 - \dfrac{3}{x} + \dfrac{4}{x^2}}{4 + \dfrac{1}{x^2}}.$$

As $x \to +\infty$ and as $x \to -\infty$, the terms $-3/x$ and $4/x^2$ approach zero whereas the constant 2 does not change; hence the numerator approaches 2. Similarly, the denominator approaches 4; so $f(x)$ approaches $2/4 = 1/2$. Therefore the line $y = 1/2$ is a horizontal asymptote.

It is easy to see that if the numerator and denominator of a rational function have the same degree (as in Example 4), then the graph has a horizontal asymptote $y = c$, where c is the ratio of the leading coefficients of the numerator and denominator.

Case 2. $n < m$

Example 5

Determine any horizontal asymptotes of the graph of

$$g(x) = \frac{5x^2 - x + 1}{2x^3 + 4}.$$

Solution. Dividing numerator and denominator by x^3, we obtain

$$g(x) = \frac{\dfrac{5}{x} - \dfrac{1}{x^2} + \dfrac{1}{x^3}}{2 + \dfrac{4}{x^3}}.$$

The numerator is a sum of terms, all of which approach zero as x becomes infinite; so the numerator also approaches zero. The denominator approaches 2. Consequently the quotient $g(x)$ approaches $0/2 = 0$ and $y = 0$ (the x axis) is a horizontal asymptote.

A **proper** rational function is one in which the degree of the numerator is less than the degree of the denominator; the function g in Example 5 is a proper rational function. It is easy to see that any proper rational function has the x axis as a horizontal asymptote.

Case 3. $n > m$. Let us consider an example in which the degree of the numerator is greater than the degree of the denominator.

Example 6

Determine the limiting behavior of the function

$$f(x) = \frac{3x^3 + 4x}{x^2 + 1}$$

as $x \longrightarrow +\infty$, and sketch the graph.

Solution. By dividing $x^2 + 1$ into $3x^3 + 4x$ as in Section 5-1, we can express f as follows:

$$f(x) = 3x + \frac{x}{x^2 + 1}.$$

As $x \longrightarrow +\infty$, the second term approaches zero, for it is a proper rational function. For large x, therefore, $f(x)$ is approximately equal to $3x$, which approaches plus infinity as $x \longrightarrow +\infty$. So $f(x) \longrightarrow +\infty$ as $x \longrightarrow +\infty$. To complete the graph we note that f is odd and has a zero at 0. There are no vertical or horizontal asymptotes. The graph of f appears in Fig. 5-21.

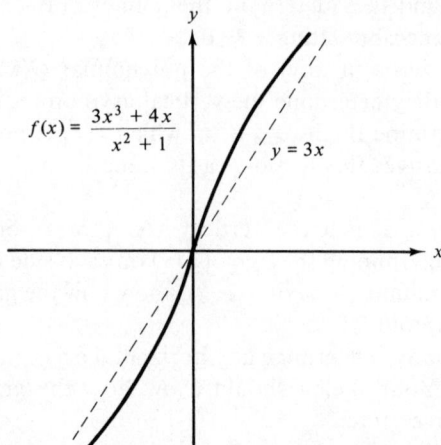

Figure 5-21

Example 6 suggests how to determine the behavior of a rational function as x becomes infinite when the degree of the numerator is greater than the degree of the denominator. First, use the division process to express the given function as a polynomial function plus a proper rational function. The proper rational function approaches zero and so the behavior of the given function for large x is the same as that of the polynomial function. Because the polynomial function becomes infinite as x does, no horizontal asymptote exists.

Summary: Determining Horizontal Asymptotes for a Rational Function.
Let f be a rational function

$$f(x) = \frac{P(x)}{Q(x)} = \frac{A_n x^n + A_{n-1} x^{n-1} + \cdots + A_1 x + A_0}{B_m x^m + B_{m-1} x^{m-1} + \cdots + B_1 x + B_0} \qquad (A_n \neq 0, B_m \neq 0),$$

where n is the degree of the numerator and m the degree of the denominator.
Case 1. If $n = m$, the line $y = c$ is a horizontal asymptote, where $c = A_m/B_m$ is the ratio of the leading coefficients of numerator and denominator.
Case 2. If $n < m$, the x axis is a horizontal asymptote of the graph of f.
Case 3. If $n > m$, there is no horizontal asymptote. The function f becomes infinite as x becomes infinite; the sign of $f(x)$ for large $|x|$ can be found by dividing and determining the sign of the polynomial part of the quotient.

Putting It All Together. The following steps indicate how we can use earlier tools, together with those just discussed, to draw the graph of a rational function f. We assume that $f(x) = P(x)/Q(x)$, where $P(x)$ and $Q(x)$ are polynomials without common factors. [If $P(x)$ and $Q(x)$ do have common factors, cancel them, but remember that any zeros of the canceled factors are excluded from the domain of f.]
1. *Symmetry.* Check to see whether $f(x)$ is even or odd by evaluating $f(-x)$.
2. *Intercepts.* Locate the zeros (if any) of the polynomial $P(x)$. They are also the zeros of $f(x)$ and they determine the x intercepts of the graph. The y intercept is obtained by setting $x = 0$.
3. *Poles.* Locate the zeros (if any) of the polynomial $Q(x)$. These are the poles of $f(x)$ and they determine the vertical asymptotes of the graph.
4. *Sign of $f(x)$.* Determine the intervals on which $f(x)$ is positive and those on which it is negative; this is most easily done by making a sign chart, as in Section 5-7.
5. *Vertical Asymptotes.* Sketch the vertical asymptotes—one at each pole of $f(x)$. Then by determining the sign of $f(x)$ on each side of a pole, decide how $f(x)$ becomes infinite ($+\infty$ or $-\infty$?). Show how the graph approaches each vertical asymptote.
6. *Horizontal Asymptotes.* Determine any horizontal asymptotes as indicated in the summary. Your sketch should show how the graph approaches each horizontal asymptote.

Example 7

Graph $f(x) = \dfrac{x^2 - 1}{x^2 - 4}$.

Solution.

1. First, we note that f is even, because

$$f(-x) = \frac{(-x)^2 - 1}{(-x)^2 - 4} = \frac{x^2 - 1}{x^2 - 4} = f(x).$$

The graph is therefore symmetric with respect to the y axis.

2. The zeros of f are -1 and 1; hence the x intercepts are $(-1, 0)$ and $(1, 0)$. The y intercept is $(0, 1/4)$.

3. The poles of f are -2 and 2 and so the lines $x = -2$ and $x = 2$ are vertical asymptotes.

4. The sign of $f(x)$ is the same as the sign of $(x^2 - 1)(x^2 - 4)$.

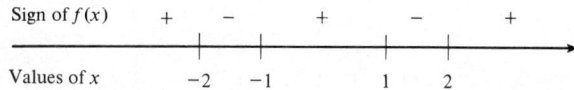

5. Because $f(x)$ is positive for $x > 2$, $f(x) \rightarrow +\infty$ as $x \rightarrow 2^+$. But if $1 < x < 2$, $f(x)$ is negative; so $f(x) \rightarrow -\infty$ as $x \rightarrow 2^-$. Figure 5-22(a) shows the graph of f on both sides of the asymptote $x = 2$. The asymptote $x = -2$ will be taken care of when we use symmetry.

6. Because the degrees of the numerator and denominator of f are equal and the ratio of the leading coefficients is 1, we see that the line $y = 1$ is a horizontal asymptote. In Fig. 5-22(b) we show how the graph approaches this asymptote.

The sketch now shows most features of the graph except the behavior when x is near zero. We plot the intercept $(0, 1/4)$ and connect it to the remainder of the graph, keeping in mind the symmetry; the result is shown in Fig. 5-22(c).

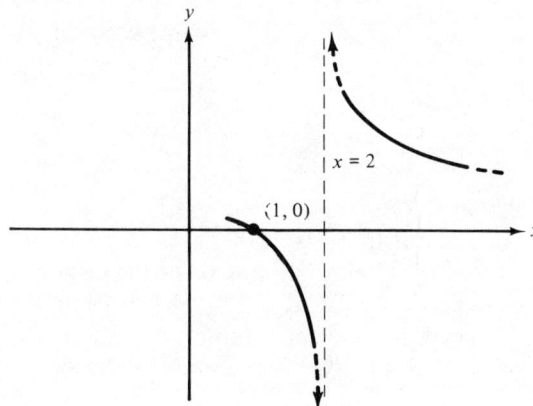

(a)

Figure 5-22(a)

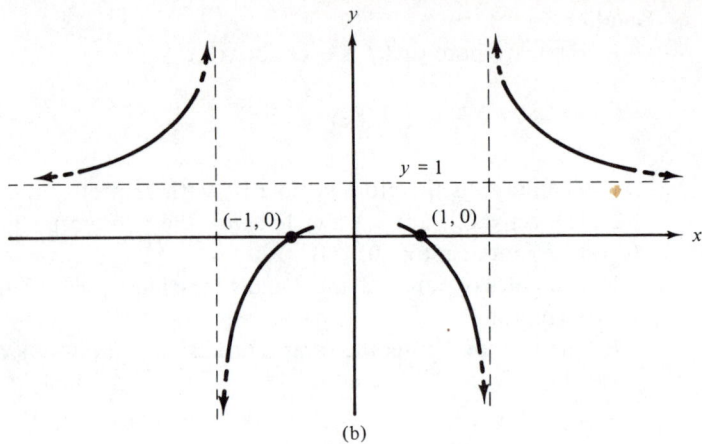

(b)

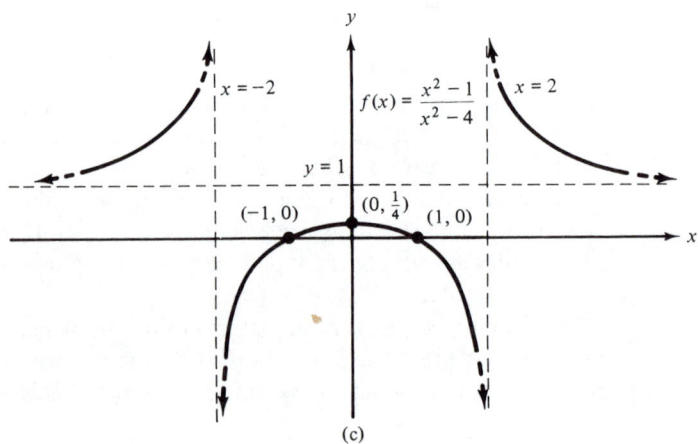

(c)

Figure 5-22(b), (c)

Example 8

Graph $g(x) = \dfrac{2x(x+1)}{(x-3)^2}$.

Solution.

1. The function g is neither even nor odd; so its graph is not symmetric with respect to either the y axis or the origin.
2. The zeros of g are $-1, 0$. Hence x intercepts are $(-1, 0)$ and $(0, 0)$.
3. There is a pole at 3 and so $x = 3$ is a vertical asymptote.
4. The sign of $g(x)$ is the same as the sign of $2x(x+1)(x-3)^2$.

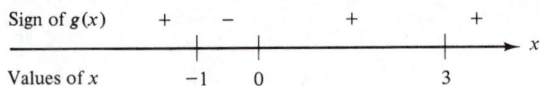

5. Figure 5-23(a) shows the graph of g on both sides of the pole at $x = 3$.

Polynomial and Rational Functions   Chap. 5

6. It is easy to see that line $y = 2$ is a horizontal asymptote. Does the graph cross this asymptote? We can find out by setting $g(x) = 2$.

$$\frac{2x(x+1)}{(x-3)^2} = 2$$

$$x^2 + x = x^2 - 6x + 9$$

$$7x = 9$$

$$x = \frac{9}{7}$$

Thus the graph intersects $y = 2$ at $(9/7, 2)$. It approaches its asymptote $y = 2$ from below on the left, from above on the right.

The information obtained is shown in Fig. 5-23(b). Our graph suggests that there is a minimum point corresponding to a value of x in the interval $0 < x < 1$, but this point can be determined only by methods of calculus.

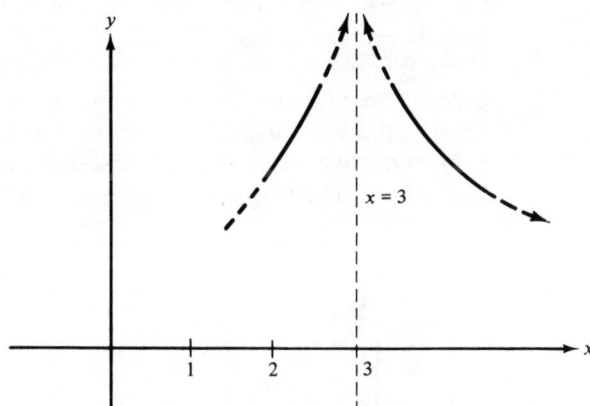

(a)

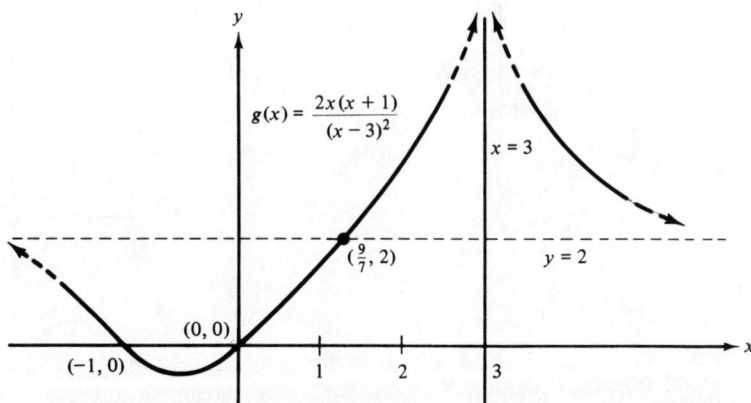

(b)

Figure 5-23

Example 9

Graph $f(x) = \dfrac{x^2 + 1}{x}$.

Solution.

1. By calculating $f(-x)$, it is easy to see that $f(-x) = -f(x)$. so f is odd and its graph is symmetric with respect to the origin.
2. This function nas no zeros (why?) and hence there are no x intercepts. It is undefined at 0 and so there are no y intercepts.
3. There is a pole at $x = 0$ and so the x axis is a vertical asymptote.
4. The sign of $f(x)$ is the same as the sign of $(x^2 + 1)x$.

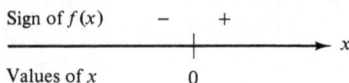

Sign of $f(x)$ $\quad - \quad +$

Values of x $\quad\quad 0$

5. On the basis of this information we can see how the graph approaches its vertical asymptote [Fig. 5-24(a)].
6. Using division, we have $f(x) = x + (1/x)$. So $f(x)$ becomes infinite as x becomes infinite and thus the graph has no horizontal asymptotes. For large x, the value of $f(x)$ is close to the value of x; the graph therefore approaches line $y = x$ as x becomes large. It does not, however, cross this line (why?). The graph is shown in Fig. 5-24(b).

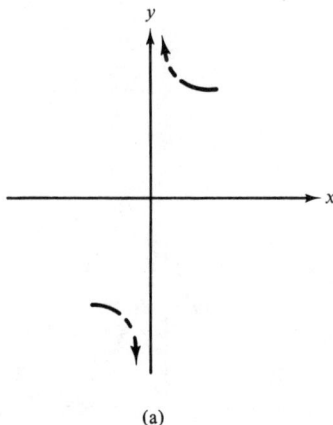

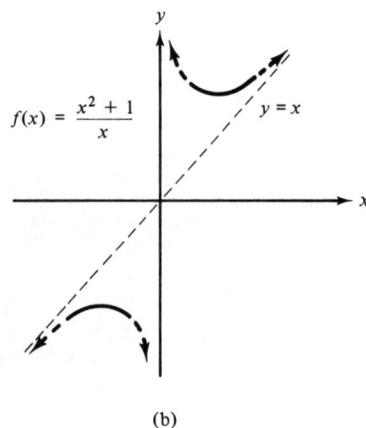

(a)

$f(x) = \dfrac{x^2 + 1}{x}$

$y = x$

(b)

Figure 5-24

Polynomial and Rational Functions   Chap. 5

EXERCISES 5-9

In Problems 1–11 determine any horizontal asymptotes.

1. $f(x) = \dfrac{x^2 + 4}{3x^2 + 2}$
2. $g(x) = \dfrac{1 - x}{1 + x}$
3. $h(x) = \dfrac{4x^3}{(1 + x)^3}$

4. $f(x) = \dfrac{2x - 1}{x^2 + 2x - 8}$
5. $h(x) = \dfrac{1}{x^2} - \dfrac{2}{x} + 1$
6. $f(x) = \dfrac{1}{x - 1} + \dfrac{1}{x + 1}$

7. $f(x) = \dfrac{4}{1 + x^3}$
8. $g(x) = 5 + \dfrac{1}{x}$
9. $f(x) = \dfrac{x^4 - 1}{5x^4 + x^2 + 7}$

10. $g(x) = \dfrac{1}{x} - \dfrac{3}{x^2}$
11. $g(x) = \dfrac{2}{x + 3} - \dfrac{3}{2x - 1}$

In Problems 12–20 sketch the graph.

12. Problem 1
13. Problem 2
14. Problem 3
15. Problem 4
16. Problem 5
17. Problem 6
18. Problem 7
19. Problem 8
20. Problem 11

In Problems 21–29 write $f(x)$ as the sum of a polynomial and a proper rational expression. What does $f(x)$ approach as $x \longrightarrow +\infty$? as $x \longrightarrow -\infty$? Sketch the graph.

21. $f(x) = \dfrac{x^2 + 4}{x}$
22. $f(x) = \dfrac{x^3 + 5x + 6}{x}$
23. $f(x) = \dfrac{x^2 + 3x + 1}{x - 2}$

24. $f(x) = \dfrac{x^2 + 2x + 5}{x}$
25. $f(x) = \dfrac{9 - x^2}{x}$
26. $f(x) = \dfrac{x^3 - 3x + 2}{x}$

27. $f(x) = \dfrac{3x^2 + 4x + 5}{x - 1}$
28. $f(x) = \dfrac{x^4}{x^2 + 1}$
29. $f(x) = \dfrac{2x^3 - 8x}{x^2 + 4}$

FOR THE ENTHUSIASTS

30. Show that $(0, 1)$ is a maximum point of the graph of f in Example 1.

31. Show that $(0, -1)$ is a minimum point of the graph of g in Example 3.

In Problems 32–36 write an expression for a rational function whose graph has the given properties.

32. horizontal asymptote: $y = 1$
    x intercepts: $(-\frac{1}{2}, 0), (2, 0)$

33. horizontal asymptote: $y = -\frac{3}{2}$
    vertical asymptotes: $x = -1, x = 1$
    x intercepts: $(0, 0)$

34. horizontal asymptote: $y = 0$
    vertical asymptotes: $x = -1, x = 1$
    x intercept: $(0, 0)$
    symmetry with respect to the origin.

35. horizontal asymptote: $y = 5$
    vertical asymptotes: $x = -1, x = 1$
    x intercepts: $(-2, 0), (0, 0), (2, 0)$
    symmetry with respect to the y axis

36. horizontal asymptote: $y = \frac{1}{2}$
    vertical asymptote: $x = 3$
    x intercepts: $(1, 0), (5, 0)$
    graph does not cross x axis

SUMMARY

In Chapter 5 we studied polynomial and rational functions. We were concerned with the algebra of these functions and with methods of graphing them.

1. You should know the definition of polynomial function and be able to apply it to determine whether a given function is of this type. You should know the meaning of the words degree, leading coefficient, and constant term.

2. You should be able to carry out the processes of addition, subtraction, multiplication, and division of polynomials and to simplify your results algebraically.

3. You should know what is meant by a factor of a polynomial and be able to state and prove the remainder and factor theorems. You should understand how the factor theorem can help in finding zeros of a polynomial function.

4. You should know what is meant by even/odd functions and be able to apply these properties (when they exist) to drawing graphs.

5. You should be able to determine the behavior of a polynomial function as its variable approaches $+\infty$ or $-\infty$.

6. You should be able to identify points (if any) at which a function has discontinuities.

7. You should understand Theorem 5-4 intuitively and know how it can be used to help locate or approximate zeros of a polynomial function.

8. You should know the meaning of the multiplicity of a zero and be able to draw a graph or chart showing the sign of a polynomial function once you know its zeros and their multiplicities.

9. You should know how to use the integral zeros theorem and the rational zeros theorem to limit the search for integral and rational zeros of a polynomial having integral coefficients.

10. You should know the definition of rational function and be able to determine whether a given function is of this type. You should know the meaning of the terms numerator and denominator as applied to rational functions.

11. You should be able to carry out the processes of addition, subtraction, multiplication, and division of rational functions and to simplify your results algebraically.

12. You should know how to find the zeros and the points of discontinuity of a rational function by solving certain polynomial equations.

13. You should know how to draw a chart showing the sign of a rational function.

14. You should be able to find the poles of a rational function and know how to obtain from them the vertical asymptotes of the graph. You should be able to use the sign chart in (13) to show how the graph approaches each vertical asymptote.

15. By comparing the degrees of numerator and denominator, you should be able to determine whether a rational function has a horizontal asymptote; if it has one, you should be able to find an equation of this asymptote and show in a sketch how the graph approaches it.

REVIEW EXERCISES

1. Which of the following expressions define polynomial functions? For those that do, give the degree, leading coefficient, and constant term.
   (a) $8 - 6x^2 + x^4$
   (b) $x^3 - x^2 + \sqrt{x} - 6$
   (c) $|x - 3| + 4$
   (d) $x^3 - \sqrt{2}x^2 + \sqrt{3}x$
   (e) $(t + 1)(t - 1)$
   (f) $(\sqrt{t} + 1)(\sqrt{t} - 1)$

2. Find the quotient $Q(x)$ and the remainder $R(x)$ when $P(x)$ is divided by $D(x)$.
   (a) $P(x) = x^4 + 3x^3 - x$
       $D(x) = x^2 - 2$
   (b) $P(x) = x^3 - 8$
       $D(x) = x^2 - 4$
   (c) $P(x) = x^3 - 8$
       $D(x) = x - 2$
   (d) $P(x) = x + 3$
       $D(x) = x - 3$

3. Use the remainder theorem to find the remainder when $P(x)$ is divided by $D(x)$.
   (a) $P(x) = x^4 - x + 7$
       $D(x) = x - 1$
   (b) $P(x) = x^3 - 8$
       $D(x) = x - 2$
   (c) $P(x) = x^3 - 3x + 1$
       $D(x) = x + 2$

4. Find all real zeros.
   (a) $9x - x^3$
   (b) $x^4 - 6x^2 + 8$
   (c) $x^3 - 3x^2 + 2x$
   (d) $x^3 - x^2 + x - 6$
   (e) $x^3 + 3x^2 - 2x - 2$
   (f) $3x^3 - x^2 + x + 2$
   (g) $16x^3 + 4x^2 - 36x - 9$

5. Find the intervals on which the expression is positive.
   (a) $160t - 16t^2$
   (b) $9x - x^3$
   (c) $(x + 2)^3(x - 2)^2(x - 4)$
   (d) $-2x^4(x - 2)^3(x - 4)^2$
   (e) $x^4 - 6x^2 + 8$
   (f) $\dfrac{x + 1}{(x - 2)^2(x - 3)}$
   (g) $\dfrac{(x - 1)^2}{x^2 + 4x + 3}$
   (h) $\dfrac{(x + 1)^2}{x^2 + 5x + 6}$

6. Are the following functions odd or even?
   (a) $f(x) = 9x - x^3$
   (b) $g(t) = 160t - 16t^2$
   (c) $F(x) = x^4 - 6x^2 + 8$
   (d) $g(x) = \dfrac{1}{1 + x^2}$
   (e) $h(x) = \dfrac{x}{x^2 + 9}$
   (f) $f(x) = \dfrac{x}{1 - x^2}$

7. Find any points of discontinuity.
   (a) $f(x) = \dfrac{x}{1 - x^2}$
   (b) $f(x) = 9x - x^3$
   (c) $g(x) = \dfrac{1}{1 + x^2}$
   (d) $h(x) = \dfrac{2x - 3}{x^2 - 5x + 6}$

8. For each of the functions in Problem 8, determine the behavior as $x \longrightarrow +\infty$ and as $x \longrightarrow -\infty$.

9. Find the zeros and the poles and make a sign chart.
   (a) $f(x) = \dfrac{x - 1}{(x + 2)^2}$
   (b) $g(x) = \dfrac{x(x - 1)}{x^2 - 3x + 2}$
   (c) $f(x) = \dfrac{x^2 - 4}{x^2 - 1}$
   (d) $f(x) = \dfrac{(x - 1)^2}{x^2 + 5x + 6}$
   (e) $1 - x^{-2}$
   (f) $(x^2 + 2)(1 + 2x)^{-2}$
   (g) $3 - x^{-1} - 2x^{-2}$
   (h) $(x + 3)^{-1}(x^{-1} + 3)$
   (i) $(4 - x^{-2})(1 + x + x^{-1})^{-1}$
   (j) $[x^2 + 1 - (x^2 + 1)^{-1}]^{-1}$

10. Find any vertical asymptotes and show in a sketch how the graph approaches these asymptotes.

(a) Problem 9(c)
(b) Problem 9(e)
(c) Problem 9(h)
(d) Problem 9(i)

11. Find any horizontal asymptotes and show in a sketch how the graph approaches its asymptote. If there is no horizontal asymptote, explain how the function behaves as $x \longrightarrow +\infty$ and as $x \longrightarrow -\infty$.

(a) $f(x) = \dfrac{x^2 - 4}{x^2 - 1}$

(b) $g(x) = \dfrac{1}{1 + x^2}$

(c) $f(x) = \dfrac{x - 1}{3x + 4}$

(d) $F(x) = \dfrac{x^2 + 2x}{x + 3}$

12. Sketch the graphs, noting symmetry, intercepts, sign of the function, vertical and horizontal asymptotes.

(a) $f(x) = \dfrac{2x^2 + 1}{x}$

(b) $f(x) = \dfrac{2x}{x^2 - 1}$

(c) $g(x) = \dfrac{3x + 2}{2x - 3}$

(d) $f(u) = \dfrac{u}{u^2 + 1}$

(e) $h(x) = x^2 + \dfrac{1}{x}$

(f) $f(t) = t - \dfrac{1}{t}$

(g) $g(x) = \dfrac{x^2 - 4}{x^2 + 1}$

(h) $f(x) = \dfrac{x(1 - x)}{x^2 + 2x + 1}$

(i) $f(x) = \dfrac{x^4 + 1}{x^2}$

Complex zeros of polynomial functions

6

6-1 THE ALGEBRA OF COMPLEX NUMBERS

In the real number system negative numbers do not have square roots. Consequently, many equations—for example, $x^2 + 1 = 0$—do not have real solutions. We are now going to study the system of complex numbers, a larger number system in which all polynomial equations of degree one or greater can be solved.

> **Definition.** A **complex number** is a number z of the form
>
> $$z = a + bi,$$
>
> where a and b are real numbers and i is a number with the property that $i^2 = -1$. The real number a is called the **real part** of z and the real number b is called the **imaginary part** of z.

Example 1

The following numbers are complex.

$z_1 = 2 + 3i$ (real part $= 2$, imaginary part $= 3$)

$z_2 = 4 + 0i$ (real part $= 4$, imaginary part $= 0$)

$z_3 = \dfrac{\sqrt{2}}{2} - \dfrac{\sqrt{2}}{2}i$ $\left(\text{real part} = \dfrac{\sqrt{2}}{2}, \text{imaginary part} = -\dfrac{\sqrt{2}}{2}\right)$

$z_4 = 0 - 3i$ (real part $= 0$, imaginary part $= -3$)

Notice that in z_3 and z_4 we wrote $a + (-b)i$ as $a - bi$.

213

The algebra of complex numbers is based on the following definitions.

---

1. Equality. The complex numbers $a + bi$ and $c + di$ (a, b, c, d real) are **equal** if and only if $a = c$ and $b = d$.

In other words, complex numbers are equal if and only if they have the same real parts and the same imaginary parts.

---

2. Addition. The **sum** of the complex numbers $a + bi$ and $c + di$ (a, b, c, d real) is the complex number $(a + c) + (b + d)i$.

To add two complex numbers, we add the real parts and add the imaginary parts.

Example 2

If $z_1 = 2 - 3i$ and $z_2 = 3 + 5i$, then
$$z_1 + z_2 = (2 + 3) + (-3 + 5)i = 5 + 2i.$$

---

3. Multiplication. The **product** of the complex numbers $a + bi$ and $c + di$ (a, b, c, d real) is the complex number $(ac - bd) + (ad + bc)i$.

Example 3

If $z_1 = 2 - 3i$ and $z_2 = 3 + 5i$, then
$$z_1 z_2 = [2 \cdot 3 - (-3) \cdot 5] + [2 \cdot 5 + (-3)(3)]i = 21 + i.$$

The definition of product appears complicated at first, but it amounts to the requirement that we can use the ordinary properties of real number algebra in conjunction with the additional stipulation that $i^2 = -1$:

$$(a + bi)(c + di) = ac + adi + bci + bdi^2$$
$$= ac + (ad + bc)i + bdi^2$$
$$= ac + (ad + bc)i + bd(-1)$$
$$= (ac - bd) + (ad + bc)i.$$

Multiplying $z_1 = 2 - 3i$ and $z_2 = 3 + 5i$ in this way, we get

$$(2 - 3i)(3 + 5i) = 6 + 10i - 9i - 15i^2 = 6 + i - 15(-1) = 21 + i.$$

It is evident from definitions (2) and (3) that the sum and the product of two complex numbers are both complex numbers so that the system of all complex numbers satisfies closure laws analogous to the closure laws of the real number system (see Section 1-1).

It is easy to show further that the laws of arithmetic hold for complex addition and multiplication.

> **Associative Laws.** If z_1, z_2, and z_3 are any complex numbers, then
> $$z_1 + (z_2 + z_3) = (z_1 + z_2) + z_3$$
> and
> $$z_1(z_2 z_3) = (z_1 z_2) z_3.$$

> **Commutative Laws.** If z_1 and z_2 are any complex numbers, then
> $$z_1 + z_2 = z_2 + z_1$$
> and
> $$z_1 z_2 = z_2 z_1.$$

> **Distributive Laws.** If z_1, z_2, and z_3 are any complex numbers, then
> $$z_1(z_2 + z_3) = z_1 z_2 + z_1 z_3$$
> and
> $$(z_1 + z_2)z_3 = z_1 z_3 + z_2 z_3.$$

As an illustration, let us verify the commutative laws. Suppose that $z_1 = a + bi$ and $z_2 = c + di$, where $a, b, c,$ and d are real numbers. Then by the definition of complex addition we have

$$z_1 + z_2 = (a + bi) + (c + di) = (a + c) + (b + d)i$$

whereas

$$z_2 + z_1 = (c + di) + (a + bi) = (c + a) + (d + b)i.$$

From the commutative law of the real number system $a + c$ and $c + a$ are equal and $b + d$ and $d + b$ are equal. Thus $z_1 + z_2$ and $z_2 + z_1$ have the same real parts and the same imaginary parts. Accordingly they are equal and we see that complex addition is commutative.

Looking at the product, we have from the definition

$$z_1 z_2 = (ac - bd) + (ad + bc)i$$
$$z_2 z_1 = (ca - db) + (da + cb)i.$$

Comparing the real and imaginary parts, we see that they are indeed the same; so $z_1 z_2 = z_2 z_1$ and complex multiplication is commutative.

The associative and distributive laws can be verified in much the same way.

Example 4

Simplify $(2 + 3i)(5i + 1) - 3i(i + 2)$.

Solution. Using the laws of arithmetic and the fact that $i^2 = -1$, we have

$$(2 + 3i)(5i + 1) - 3i(i + 2) = 10i + 2 + 15i^2 + 3i - 3i^2 - 6i$$
$$= 10i + 2 - 15 + 3i + 3 - 6i$$
$$= -10 + 7i.$$

The complex numbers $0 + 0i$ and $1 + 0i$ behave like 0 and 1. It is easy to show that they serve as identities for addition and multiplication, respectively. We will abbreviate $0 + 0i$ by 0 and $1 + 0i$ by 1.

Identity Laws. The complex number 0 has the property that

$$z + 0 = 0 + z = z$$

for every complex number z. The complex number 1 has the property that

$$z \cdot 1 = 1 \cdot z = z$$

for every complex number z.

The complex number 0 has two other properties like the real number 0:

1. $z \cdot 0 = 0$ for every complex number z and
2. if z_1 and z_2 are complex numbers such that $z_1 z_2 = 0$, then $z_1 = 0$ or $z_2 = 0$.

The complex number system satisfies inverse laws, too. If $z = a + bi$, it is clear that the complex number $(-a) + (-b)i$ is the opposite of z, for $(a + bi) + [(-a) + (-b)i] = 0$.

Looking for a reciprocal of $a + bi$, we are led to the equations

$$(a + bi)(u + vi) = 1 + 0i,$$

$$(au - bv) + (bu + av)i = 1 + 0i.$$

According to the definition of equality (1), the real and imaginary parts of both sides must be the same. Thus

$$au - bv = 1 \quad \text{and} \quad bu + av = 0.$$

These equations in the unknowns u and v have a solution, provided that a and b are not both zero; the solution is

$$u = \frac{a}{a^2 + b^2} \qquad v = \frac{-b}{a^2 + b^2}$$

(see Problem 44). Thus the reciprocal of $z = a + bi$ is

$$\frac{a}{a^2 + b^2} - \frac{b}{a^2 + b^2}i$$

provided that z is not the complex number 0.

Inverse Laws. For each complex number $z = a + bi$ (a, b real) the complex number $-z = (-a) + (-b)i$ is the **opposite** of z:

$$z + (-z) = 0.$$

If $z \neq 0$, then the complex number

$$\frac{1}{z} = \frac{a}{a^2 + b^2} + \frac{-b}{a^2 + b^2}i$$

is the **reciprocal** of z:

$$z \cdot \left(\frac{1}{z}\right) = 1.$$

Example 5

If $z = 3 - 4i$, then

$$-z = -3 + 4i \quad \text{and} \quad \frac{1}{z} = \frac{3}{25} + \frac{4}{25}i.$$

Example 6

If

$$z = \frac{1}{\sqrt{2}} + \frac{1}{\sqrt{2}}i$$

then

$$\frac{1}{z} = \frac{1}{\sqrt{2}} - \frac{1}{\sqrt{2}}i.$$

Example 7

If $z = i$, then $1/z = -i$.

We can solve linear equations in the complex number system by using inverses, just as we do for real number equations.

Example 8

Solve the following equation for z:

$$(1 + i)z + i = 1.$$

Solution. Add the opposite of i to both sides of the equation:

$$(1 + i)z = 1 - i.$$

Now multiply by the reciprocal of $1 + i$:

$$\frac{1}{1+i}(1 + i)z = \frac{1}{1+i}(1 - i)$$

$$z = \frac{1}{1+i}(1 - i) = \frac{1-i}{2} \cdot (1 - i)$$

$$= \frac{(1-i)^2}{2} = \frac{1 - 2i + i^2}{2} = -i.$$

So the solution is $z = -i$.

The operations of subtraction and division are defined for complex numbers just as they were for real numbers.

$$z_1 - z_2 \quad \text{means} \quad z_1 + (-z_2)$$

$$z_1 \div z_2 \quad \text{or} \quad \frac{z_1}{z_2} \quad \text{means} \quad z_1 \cdot \frac{1}{z_2} \quad (z_2 \neq 0)$$

If r is any real number, we can regard r as the complex number $r + 0i$. Under this interpretation the real number system becomes part of the complex number system, a complex number being real if and only if its imaginary part is zero.

Integral Powers of Complex Numbers

We use integral exponents with complex numbers in the usual way: z^2 means $z \cdot z$, z^3 means $z \cdot z \cdot z, \ldots, z^{-1}$ means $1/z$, and so on.

Example 9

The positive integral powers of i are

$$i^1 = i$$
$$i^2 = -1$$
$$i^3 = -i$$
$$i^4 = -i^2 = 1$$
$$i^5 = i, \text{ and so on.}$$

For negative powers of i we have

$$i^{-1} = -i$$
$$i^{-2} = -1$$
$$i^{-3} = i$$
$$i^{-4} = 1$$
$$i^{-5} = -i, \text{ and so on.}$$

It is easy to show that integral exponents for complex numbers satisfy the basic properties established in Theorem 1-1.

Properties of Integral Exponents. If z_1 and z_2 are any complex numbers different from zero and if m and n are any integers, then

$$z_1^{m+n} = z_1^m z_1^n$$

$$z_1^{m-n} = \frac{z_1^m}{z_1^n}$$

$$(z_1^m)^n = z_1^{mn}$$

$$(z_1 z_2)^n = z_1^n z_2^n$$

$$\left(\frac{z_1}{z_2}\right)^n = \frac{z_1^n}{z_2^n}.$$

The binomial expansion was discussed in Section 1-3. Its use with complex numbers is justified by the associative, commutative, and distributive laws and the properties of exponents.

Example 10

Show that

$$\left(\frac{1}{2} + \frac{\sqrt{3}}{2}i\right)^3 = -1.$$

Solution. The binomial expansion gives

$$\left(\frac{1}{2}\right)^3 + 3\left(\frac{1}{2}\right)^2 \frac{\sqrt{3}}{2}i + 3\left(\frac{1}{2}\right)\left(\frac{\sqrt{3}}{2}i\right)^2 + \left(\frac{\sqrt{3}}{2}i\right)^3$$

$$= \frac{1}{8} + \frac{3\sqrt{3}\,i}{8} - \frac{9}{8} - \frac{3\sqrt{3}\,i}{8} = -1.$$

Square Roots of Negative Real Numbers

If $n = -p$ is a negative real number ($p > 0$), then the two complex numbers $\pm\sqrt{p}\,i$ are square roots of n, for we have

$$(\sqrt{p}\,i)^2 = (\sqrt{p})^2 i^2 = -p = n$$

and

$$(-\sqrt{p}\,i)^2 = (-\sqrt{p})^2 i^2 = -p = n.$$

Example 11

The square roots of -25 are $\pm 5i$.

Example 12

The square roots of -32 are $\pm\sqrt{32}i$ or $\pm 4\sqrt{2}\,i$.

Example 13

Solve $z^2 + 9 = 0$.

Solution.

$$z^2 = -9$$

$$z = \pm 3i$$

Example 14

Solve $(z - 4)^2 + 12 = 0$.

Solution.

$$(z - 4)^2 = -12$$

$$z - 4 = \pm 2\sqrt{3}\,i$$

$$z = 4 \pm 2\sqrt{3}\,i$$

Complex Zeros of Quadratic Functions

In Chapter 4 we studied quadratic functions of the form

$$f(x) = Ax^2 + Bx + C,$$

where the coefficients A, B, and C are real numbers and $A \neq 0$. The domain of f was understood to be the set of real numbers. If the discriminant $B^2 - 4AC < 0$, we found that f has no real zeros.

Suppose now that we think of the domain of f as the set of complex numbers. Does f have any complex zeros? By completing the square as in Section 4-10, we obtain

$$\left(x + \frac{B}{2A}\right)^2 = \frac{B^2 - 4AC}{4A^2} = -\frac{4AC - B^2}{4A^2},$$

where the right member is a negative real number. Taking square roots gives

$$x + \frac{B}{2A} = \pm \frac{\sqrt{4AC - B^2}}{2A}i$$

$$x = -\frac{B}{2A} \pm \frac{\sqrt{4AC - B^2}}{2A}i.$$

So the function f has two complex zeros when $B^2 - 4AC < 0$.

Example 15

Find the zeros of $f(x) = x^2 + 2x + 3$.

Solution. The discriminant is $B^2 - 4AC = -8$ so that $\sqrt{4AC - B^2} = 2\sqrt{2}$ and the zeros are

$$x = \frac{-2 \pm 2\sqrt{2}\,i}{2} = -1 \pm \sqrt{2}\,i.$$

Example 16

Find the zeros of $f(x) = x^3 - 1$.

Solution. Clearly 1 is a zero and we have the factorization

$$f(x) = (x - 1)(x^2 + x + 1).$$

The quadratic factor gives us two additional zeros

$$-\frac{1}{2} \pm \frac{\sqrt{3}}{2}i.$$

EXERCISES 6-1

In Problems 1–20 perform the indicated operations and expresses your answers in the form $a + bi$ (a, b real).

1. $\dfrac{1+i}{2} - \dfrac{1-i}{2}$

2. $(i + 1)(i + 2) + (i + 2)(i + 3)$

3. $(2 - 3i)^2 - (2 + 3i)^2$

4. $i(i - 1)(i + 1)$

5. $\dfrac{1}{3 - 4i}$

6. $\dfrac{1}{3i + 1}$

7. $\dfrac{i + 1}{3i - 1}$

8. $\dfrac{2i - 1}{i(i + 2)}$

9. $\dfrac{1}{1 - i} + \dfrac{1}{1 + i}$

10. $\dfrac{5}{4 + 3i} + \dfrac{5}{4 - 3i}$

11. $\dfrac{1 + i}{1 - i} + \dfrac{1 - i}{1 + i}$

12. $i + i^{-1}$

13. $\dfrac{3}{i} + \dfrac{2}{1 + i}$

14. $(2 - 2i)^3$

15. $\left(\dfrac{1 + i}{\sqrt{2}}\right)^4$

16. $(1 - i)^4$

17. $(2 + i)^{-2}$

18. $(1 - i)^{-3}$

19. $(1 + i)^{-4}$

20. $\left(\dfrac{1 - i}{1 + i}\right)^2$

In Problems 21–25 solve the equation for z.

21. $\dfrac{z + i}{i} = 4$

22. $(1 + i)z + (1 - i) = 3$

23. $\dfrac{z}{3i} = 2 + 3i$ $\qquad\qquad\qquad\qquad$ **24.** $2iz + 3 = i - 1$

25. $(i - 2)z + 5 + 2i = 3 - i$

In Problems 26–33 find the complex square roots of the given numbers.

26. -36 $\qquad$ **27.** -100 $\qquad$ **28.** -160 $\qquad$ **29.** -18

30. $-\dfrac{5}{9}$ $\qquad$ **31.** $-\dfrac{36}{25}$ $\qquad$ **32.** $-\dfrac{49}{3}$ $\qquad$ **33.** $-\dfrac{64}{5}$

In Problems 34–43 find the complex zeros.

34. $f(x) = x^2 + 81$ $\qquad\qquad$ **35.** $f(x) = x^2 + 75$

36. $f(x) = x^2 + 3x + 6$ $\qquad\qquad$ **37.** $f(x) = 3x^2 + 2x + 4$

38. $g(x) = x(x + 1) + 1$ $\qquad\qquad$ **39.** $h(x) = -5 + 3x - 2x^2$

40. $f(x) = 4x^3 + x^2 + x$ $\qquad\qquad$ **41.** $f(x) = x^3 + 8$

42. $g(x) = x^4 - 16$ $\qquad\qquad$ **43.** $h(x) = x^4 - 9$

FOR THE ENTHUSIASTS

44. Show that if a and b are not both zero, the simultaneous equations

$$\left.\begin{array}{c} au - bv = 1 \\ bu + av = 0 \end{array}\right\}$$

have the solution

$$u = \frac{a}{a^2 + b^2} \qquad v = \frac{-b}{a^2 + b^2}.$$

45. Find two square roots of the complex number i. [*Hint:* Let $z = u + vi$ and solve the equations that result from $z^2 = (u + vi)^2 = i$.]

46. Show that every complex number except zero has two square roots.

47. Show that the complex number zero has the property stated in the text: $z \cdot 0 = 0$ for every complex number z.

48. Show that if z_1 and z_2 are complex numbers such that $z_1 z_2 = 0$, then at least one of z_1 and z_2 is zero.

6-2 THE GEOMETRY OF COMPLEX NUMBERS

In Chapter 1 we used the geometric representation of real numbers as points on a line. Each point on the line represents exactly one real number and each real number is represented by exactly one point. In order to accommodate the complex numbers, which are determined by ordered pairs of real numbers, we use the coordinate plane as a model. We represent the complex number $a + bi$ (a, b real) by the point (a, b) in the plane. Using this representation, each point in the plane determines exactly one complex number and each complex number determines exactly one point in the plane.

Example 1

The complex number $1 + 2i$ determines the point $(1, 2)$. The point $(-3, 4)$ determines the complex number $-3 + 4i$. (See Fig. 6-1.)

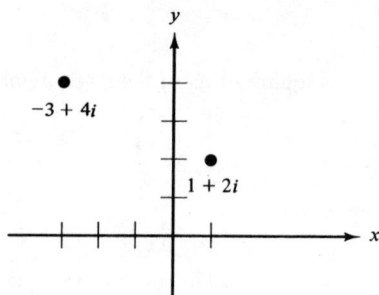

Figure 6-1

Those complex numbers that lie on the horizontal axis have imaginary part 0 and therefore are real numbers; for this reason, we call the horizontal axis the **real axis**. Complex numbers that are not on the real axis are called **imaginary**. Imaginary numbers whose real parts are zero are called **pure imaginary**; these numbers lie on the vertical axis.

Example 2

The number 2 lies on the real axis at $(2, 0)$. The pure imaginary number $3i$ is at $(0, 3)$. (See Fig. 6-2.)

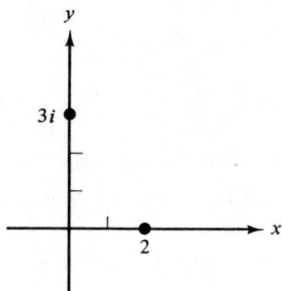

Figure 6-2

Vector Addition

Addition of complex numbers has a convenient geometric interpretation in terms of vectors.

Definition. A **vector** is a directed line segment.

Two vectors are **equivalent** if they have

1. the same length and
2. the same direction (i.e., are parallel and have the same sense).

We indicate the sense of a vector *from* one point (initial point) *to* another point (endpoint) by placing an arrowhead at the endpoint.

Example 3

In Fig. 6-3 we have drawn a vector $\mathbf{v}_1$ and several other vectors to compare with it: $\mathbf{v}_1$ is equivalent to $\mathbf{v}_2$ and to $\mathbf{v}_4$ as they do have the same length and direction. However, $\mathbf{v}_1$ is not equivalent to $\mathbf{v}_3$; although they are parallel and have the same length, $\mathbf{v}_1$ and $\mathbf{v}_3$ have opposite sense. Neither $\mathbf{v}_5$ nor $\mathbf{v}_6$ is equivalent to $\mathbf{v}_1$ because $\mathbf{v}_5$ does not have the same length as $\mathbf{v}_1$ and $\mathbf{v}_6$ is not parallel to $\mathbf{v}_1$.

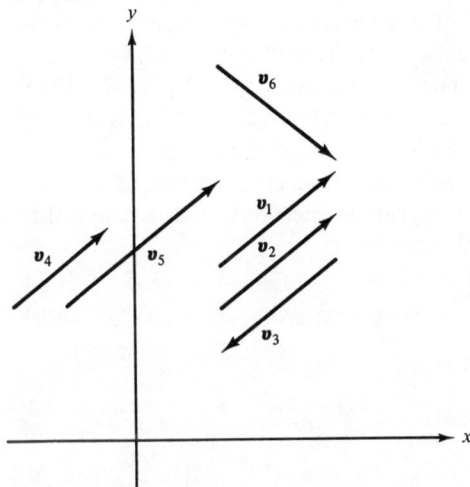

Figure 6-3

When the initial point and endpoint of a vector coincide, the vector is called a **zero vector**; you are asked to show [Problem 40(a)] that zero vectors behave like the number zero. The length of a zero vector is zero; it does not have any direction.

Equivalent vectors have the property that one can be moved parallel to itself (translated) so as to coincide with the other. Neither the length nor the direction is changed by a translation.

Definition
1. Let $\mathbf{v}_1$ and $\mathbf{v}_2$ be vectors such that the initial point of $\mathbf{v}_2$ coincides with the endpoint of $\mathbf{v}_1$. Then the **sum $\mathbf{v}_1 + \mathbf{v}_2$** is the vector from the initial point of $\mathbf{v}_1$ to the endpoint of $\mathbf{v}_2$ [see Fig. 6-4(a)].
2. Let $\mathbf{v}_1$ and $\mathbf{v}_2$ be vectors such that the initial point of $\mathbf{v}_2$ does not coincide with the endpoint of $\mathbf{v}_1$. Construct a vector $\mathbf{v}_2'$ equivalent to $\mathbf{v}_2$ such that the initial point of $\mathbf{v}_2$ does coincide with the endpoint of $\mathbf{v}_1$. Then the **sum $\mathbf{v}_1 + \mathbf{v}_2$** is defined to be $\mathbf{v}_1 + \mathbf{v}_2'$ [Fig. 6-4(b)].

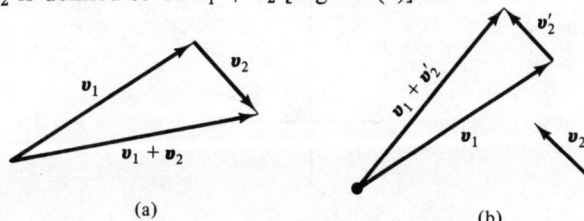

(a)   (b)

Figure 6-4

Note. Many physical quantities, such as forces, are represented mathematically as vectors. The sum of two forces is the force that results when one of the forces is applied after the other; for this reason, the vector sum is sometimes called the **resultant**.

To interpret complex number addition in terms of vectors, we will associate with each complex number $z = a + bi$ (a, b real) the vector from the origin to (a, b). Now consider the vectors associated with two complex numbers $z_1 = a_1 + b_1 i$ and $z_2 = a_2 + b_2 i$ [Fig. 6-5(a)]. Translate the vector associated with z_2 so that its initial point coincides with (a_1, b_1). The new position of the endpoint is $(a_1 + a_2, b_1 + b_2)$ [Fig. 6-5(b)]. By (2) of the definition, the vector from $(0, 0)$ to $(a_1 + a_2, b_1 + b_2)$ is the sum of the vectors associated with z_1 and z_2. But it is also the vector associated with the complex number $(a_1 + a_2) + (b_1 + b_2)i$, which is simply $z_1 + z_2$. So the vector associated with $z_1 + z_2$ is the sum of the vectors associated with z_1 and z_2; in other words, complex number addition corresponds to vector addition. Figure 6-5(c) shows that the sum $z_1 + z_2$ can be interpreted as the diagonal of a parallelogram determined by z_1 and z_2.

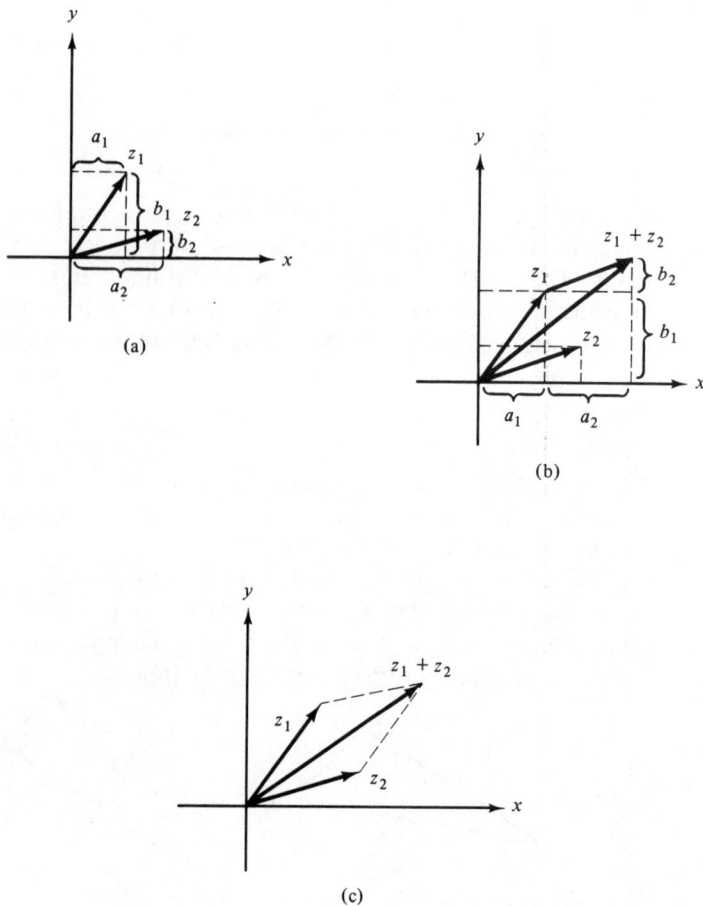

(a)

(b)

(c)

Figure 6-5

Absolute Value of a Complex Number

Definition. The **absolute value** of the complex number $z = a + bi$ (a, b real) is the nonnegative real number $\sqrt{a^2 + b^2}$; the absolute value of z is denoted by $|z|$.

Note. $|z|$ can be interpreted as the distance between the origin and (a, b), or as length of the vector associated with z.

Example 4

(a) $\qquad\qquad\qquad |1 + 2i| = \sqrt{1 + 4} = \sqrt{5}$

(b) $\qquad\qquad\qquad |-3 + 4i| = \sqrt{9 + 16} = 5$

(c) $\qquad\qquad\qquad |2| = \sqrt{4 + 0} = 2$

(d) $\qquad\qquad\qquad |3i| = \sqrt{0 + 9} = 3$

The absolute value as defined above is consistent with the absolute value as defined for real numbers; for if r is real, then $r = r + 0i$ and

$$|r| = \sqrt{r^2} = \begin{cases} r & \text{if } r \geq 0 \\ -r & \text{if } r < 0 \end{cases}.$$

Furthermore, it is easy to see that if $z_1 = a_1 + b_1 i$, $z_2 = a_2 + b_2 i$ are complex numbers, then $|z_1 - z_2|$ is the distance between z_1 and z_2 (Fig. 6-6):

$$|z_1 - z_2| = |(a_1 - a_2) + (b_1 - b_2)i| = \sqrt{(a_1 - a_2)^2 + (b_1 - b_2)^2}.$$

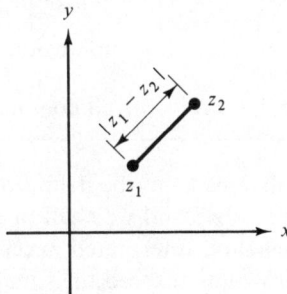

Figure 6-6

Example 5

A complex number z satisfies $|z - i| = 1$ if and only if its distance from i is 1. Accordingly, the set of all complex numbers z such that $|z - i| = 1$ is a circle of radius 1 centered at i (Fig. 6-7).

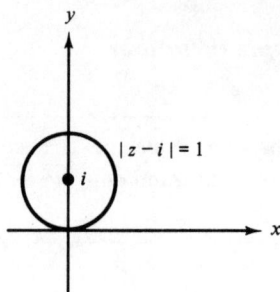

Figure 6-7

Example 6

The set of all complex numbers z such that $|z - 2| < 2$ is the interior of a circle of radius 2 centered at 2 (Fig. 6-8).

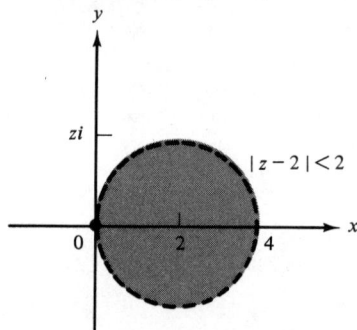

Figure 6-8

The complex absolute value has properties just like those studied in Section 1-6.

Properties of Absolute Value

$$|z| \geq 0 \quad \text{for every complex number } z \qquad (6\text{-}1)$$

$$|z_1 z_2| = |z_1||z_2| \quad \text{for all complex numbers } z_1, z_2 \qquad (6\text{-}2)$$

$$\left|\frac{z_1}{z_2}\right| = \frac{|z_1|}{|z_2|} \quad \text{for all complex numbers } z_1, z_2 \text{ if } z_2 \neq 0 \quad (6\text{-}3)$$

$$|z_1 + z_2| \leq |z_1| + |z_2| \quad \text{for all complex numbers } z_1, z_2 \qquad (6\text{-}4)$$

The inequality (6-1) is obvious from the definition of absolute value. Equations (6-2) and (6-3) are not hard to prove and we shall prove Eq. (6-2) below. Inequality (6-4) is called the **triangle inequality**. Interpreted in terms of vectors it states that the length of one side of a triangle cannot exceed the sum of the lengths of the other two sides (Fig. 6-9). We will not give an algebraic proof of this familiar fact.

Conjugate of a Complex Number

Definition. The **conjugate** of the complex number $z = a + bi$ (a, b real) is the complex number $a - bi$; the conjugate of z is denoted by $\bar{z}$.

Complex Zeros of Polynomial Functions   Chap. 6

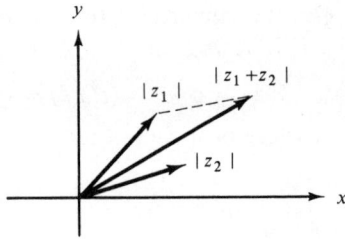

Figure 6-9

Example 7

(a)	$\overline{1 + 2i} = 1 - 2i$
(b)	$\overline{-3 + 4i} = -3 - 4i$
(c)	$\bar{2} = \overline{2 + 0i} = 2 - 0i = 2$
(d)	$\overline{3i} = \overline{0 + 3i} = 0 - 3i = -3i$

The geometric relationship between z and $\bar{z}$ is evident in Fig. 6-10: The two points are symmetric with respect to the real axis. It is easy to see that a number z is real (on the real axis) if and only if it is equal to its own conjugate $z = \bar{z}$.

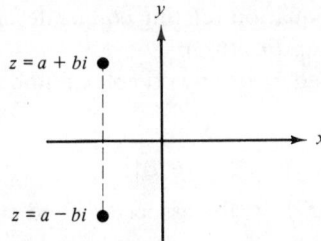

Figure 6-10

The conjugate has the following algebraic properties.

Properties of the Conjugate

$$z = \bar{z} \quad \text{if and only if } z \text{ is real} \tag{6-5}$$

$$\overline{z_1 + z_2} = \overline{z_1} + \overline{z_2} \quad \text{for all complex numbers } z_1, z_2 \tag{6-6}$$

$$\overline{z_1 z_2} = (\overline{z_1})(\overline{z_2}) \quad \text{for all complex numbers } z_1, z_2 \tag{6-7}$$

$$\overline{z^n} = (\bar{z})^n \quad \text{for every complex number } z \tag{6-8}$$

$$\overline{\bar{z}} = z \quad \text{for every complex number } z \tag{6-9}$$

We already noted that Eq. (6-5) holds. To verify Eq. (6-6), let $z_1 = a_1 + b_1 i$, $z_2 = a_2 + b_2 i$ (a_1, b_1, a_2, b_2 real) and calculate $\overline{z_1 + z_2}$ and $\overline{z_1} + \overline{z_2}$.

$$\overline{z_1 + z_2} = \overline{(a_1 + a_2) + (b_1 + b_2)i} = (a_1 + a_2) - (b_1 + b_2)i$$

$$\overline{z_1} + \overline{z_2} = (a_1 - b_1 i) + (a_2 - b_2 i) = (a_1 + a_2) - (b_1 + b_2)i$$

Because the two calculations give the same result, (6-6) holds. We can prove Eq. (6-7) in the same way (Problem 38). It follows from (6-7) that

$$\overline{z^2} = \overline{z \cdot z} = (\bar{z})^2, \qquad \overline{z^3} = \overline{z^2 \cdot z} = (\overline{z^2})(\bar{z}) = (\bar{z})^2\bar{z} = (\bar{z})^3,$$

and so on, so that $\overline{z^n} = (\bar{z})^n$ for every positive integer n. Furthermore, if $z \neq 0$,

$$\overline{z^n z^{-n}} = \bar{1} = 1$$

$$(\overline{z^n})(\overline{z^{-n}}) = 1$$

$$\overline{z^{-n}} = \frac{1}{\overline{z^n}} = \frac{1}{(\bar{z})^n} = (\bar{z})^{-n};$$

and so Eq. (6-8) holds for all integers, provided that $z \neq 0$. It is easy to verify Eq. (6-9) and we have left this as an exercise (Problem 37).

The sum of a complex number and its conjugate is real. If $z = a + bi$ (a, b real), then

$$z + \bar{z} = (a + bi) + (a - bi) = 2a.$$

Moreover, the product of a complex number and its conjugate is real and non-negative.

$$z\bar{z} = (a + bi)(a - bi) = a^2 + b^2 = |z|^2 \qquad (6\text{-}10)$$

This gives us an interesting equation relating conjugates and absolute values; we will make good use of it in proving Eq. (6-2).

To prove (6-2), let z_1 and z_2 be any complex numbers and consider $|z_1 z_2|$.

$$\begin{aligned}
|z_1 z_2|^2 &= (z_1 z_2)(\overline{z_1 z_2}) && \text{[by (6-10)]}\\
&= (z_1 z_2)(\bar{z}_1 \bar{z}_2) && \text{[by (6-7)]}\\
&= (z_1 \bar{z}_1)(z_2 \bar{z}_2) && \text{[by associative and commutative laws]}\\
|z_1 z_2|^2 &= |z_1|^2 |z_2|^2 && \text{[by (6-10)]}
\end{aligned}$$

Taking the positive square root of both sides, we get (6-2):

$$|z_1 z_2| = |z_1| |z_2|.$$

Conjugate Zeros of a Quadratic Function

Consider once again the zeros of the quadratic function $f(x) = Ax^2 + Bx + C$, where A, B, and C are real numbers for which $B^2 - 4AC < 0$. We saw in the last section that this function has a pair of imaginary zeros:

$$-\frac{B}{2A} \pm \frac{\sqrt{4AC - B^2}}{2A} i.$$

Notice that the complex numbers in this pair are conjugates of one another. It is a fact that if P is any polynomial function with *real* coefficients, then the imaginary zeros of P occur in conjugate pairs; that is, if z is a zero of P, then so is $\bar{z}$.

> **Theorem 6-1 (Conjugate Zeros).** If $P(x)$ is a polynomial with real coefficients and if z is a zero of P, then so is $\bar{z}$.

Proof. Let $P(x)$ be given by

$$A_n x^n + A_n x^{n-1} + \cdots + A_1 x + A_0.$$

By assumption, $P(z) = 0$ and it follows that $\overline{P(z)} = \bar{0} = 0$:

$$0 = \overline{P(z)} = \overline{A_n z^n + A_{n-1} z^{n-1} + \cdots + A_1 z + A_0}.$$

Repeated applications of (6-6) give

$$0 = \overline{A_n z^n} + \overline{A_{n-1} z^{n-1}} + \cdots + \overline{A_1 z} + \overline{A_0}.$$

Using (6-7) and (6-8), we have

$$0 = \overline{A_n}(\bar{z})^n + \overline{A_{n-1}}(\bar{z})^{n-1} + \cdots + \overline{A_1}\bar{z} + \overline{A_0}.$$

By hypothesis the coefficients $A_0, A_1, \ldots, A_{n-1}, A_n$ are real and so they are equal to their conjugates. Consequently,

$$0 = A_n(\bar{z})^n + A_{n-1}(\bar{z})^{n-1} + \cdots + A_1 \bar{z} + A_0 = P(\bar{z}).$$

This simply says that $\bar{z}$ is a zero of P.

Example 8

The zeros of $P(x) = x(x - 1)(x^2 + x + 1)$ are 0, 1, and $(-1 \pm \sqrt{3}\,i)/2$. (You should verify this fact.) The two imaginary zeros are conjugates.

Example 9

The zeros of $P(x) = x^4 + 3x^2 + 2$ are $x = \pm i$, $x = \pm\sqrt{2}\,i$. (Verify.) These four imaginary zeros occur in conjugate pairs.

EXERCISES 6-2

In Problems 1–8 plot the given complex number.

1. $1 - 2i$ **2.** $2 + i$ **3.** $-\dfrac{1}{2} + \dfrac{\sqrt{3}}{2}i$ **4.** $-\pi$

5. $\sqrt{2}(1 + i)$ **6.** $i(i - 2)$ **7.** $\dfrac{2}{3 + 4i}$ **8.** $\dfrac{1 + i}{1 - i}$

In Problems 9–18 find the absolute value and the conjugate.

9. $1 - 2i$ **10.** $3 + 4i$ **11.** $-\dfrac{1}{2} + \dfrac{\sqrt{3}}{2}i$

12. 2π **13.** $2\pi i$ **14.** $i(i - 2)$

15. $(i + 1)(3 - 2i)$ **16.** $\dfrac{1 + i}{1 - i}$ **17.** $\dfrac{1}{1 + i} + \dfrac{1}{1 - i}$

18. $\dfrac{3i + 2}{i + 1} + \dfrac{i - 1}{2i + 3}$

In Problems 19–28 identify or sketch the set of points z satisfying the equation or inequality.

19. $|z - 1| = 2$ **20.** $|z + 2| = 2$ **21.** $|z + 2| > 1$

22. $|z + 2| < 3$ **23.** $1 < |z + 2| < 3$ **24.** $|z - 2i| = 1$

25. $|z + 2i| = 1$ **26.** $|z - 1 - i| = \sqrt{2}$ **27.** $|z + 1 - i| > \sqrt{2}$

28. $|iz + 1| = 1$

In Problems 29–35 find any complex zeros of the given polynomial.

29. $x^2 + 3x + 4$ **30.** $2x^2 + x + 1$ **31.** $x^3 + 4x^2 + 8x$

32. $x^3 + x^2 + x + 1$ **33.** $x^4 - 16$ **34.** $x^3 - 27$

35. $x^4 + 9x^2 + 20$

36. Show that $1 + i$ is a zero of $x^2 - 2i$ but that $1 - i$ is not. Does this contradict Theorem 6-1?

37. Verify Eq. (6-9):

$$\bar{\bar{z}} = z \quad \text{for every complex number } z.$$

FOR THE ENTHUSIASTS

38. Verify Eq. (6-7):

$$\overline{z_1 z_2} = (\bar{z_1})(\bar{z_2}) \quad \text{for all complex numbers } z_1, z_2.$$

39. Show that if $|z| = 1$, then $\bar{z} = 1/z$.

40. More on Vector Algebra

(a) *Identity.* Show that the vector from a point P to P has the property of an identity for vector addition.

(b) *Inverse.* What is the additive inverse (opposite) of a vector?

(c) *Subtraction.* Define subtraction of vectors and show that $\mathbf{v}_1 - \mathbf{v}_2$ can be described as a vector $\mathbf{w}$ such that $\mathbf{v}_2 + \mathbf{w} = \mathbf{v}_1$.

6-3 FUNDAMENTAL THEOREM OF ALGEBRA

In this section we consider complex zeros of polynomial functions. First, however, we will generalize our notion of "polynomial" function to allow complex *coefficients* as well as complex values for the independent variable.

Definition. A function defined for complex numbers z by a rule of the type

$$P(z) = A_n z^n + A_{n-1} z^{n-1} + \cdots + A_1 z + A_0,$$

where the coefficients $A_0, A_1, \ldots, A_n$ are complex numbers, is called a **polynomial function over the complex numbers** (briefly, a "complex polynomial"). If $A_n \neq 0$, then n is the **degree** of P and A_n is its **leading coefficient**.

Example 1

The following are polynomial functions over the complex numbers.

$P_1(z) = z^2 + 1$ (degree 2, leading coefficient 1)

$P_2(z) = (4 + z)(4 - z) = 16 - z^2$ (degree 2, leading coefficient -1)

$P_3(z) = -2 - 9iz + 10z^2 + 3iz^3$ (degree 3, leading coefficient $3i$)

Notes.
1. Keep in mind that the complex numbers include the real numbers; so some or all of the coefficients of a complex polynomial can be real. By a "real polynomial" we mean a complex polynomial whose coefficients are all real, such as $P_1(z)$ and $P_2(z)$ in Example 1.
2. Throughout this section we assume that the domain of all polynomial functions is the set of complex numbers; it is to emphasize this fact that we denote the independent variable by z.

A complex polynomial function of degree zero is just a nonzero complex constant: $P(z) = C$. The zero function $P(z) = 0$ is a constant function and a polynomial function, but it has no degree.

A **zero** of a complex polynomial function P is a complex number u such that $P(u) = 0$. In Example 1, i and $-i$ are zeros of P_1; 4 and -4 are zeros of P_2. By computing $P_3(2i)$, you can verify that $2i$ is a zero of P_3.

The algebra of polynomials over the complex numbers is carried out by the same rules that govern real polynomials, the only change being that the calculation of coefficients involves complex numbers. In particular, the division process and its consequences, the remainder and factor theorems, are valid in the setting of complex numbers. Let us restate these theorems in this new setting and give some examples.

Division Process for Complex Polynomials. Let $P(z)$ and $D(z)$ be complex polynomials such that $D(z)$ is not the constant 0. Then there exist complex polynomials $Q(z)$ and $R(z)$ such that for all complex numbers z

$$P(z) = Q(z)D(z) + R(z) \tag{6-11}$$

and $R(z)$ either is the constant 0 or has degree less than the degree of $D(z)$.

Example 2

Divide $P(z) = z^2 + 6z$ by $D(z) = z + 2$.

Solution.

$$
\begin{array}{r}
z + 4 \\
z + 2 \overline{)\, z^2 + 6z} \\
\underline{z^2 + 2z} \\
4z \\
\underline{4z + 8} \\
-8
\end{array}
$$

So

$$P(z) = (z + 4)D(z) - 8;$$

$$Q(z) = z + 4 \quad \text{and} \quad R(z) = -8.$$

Theorem 6-2 (Remainder Theorem). If the complex polynomial $P(z)$ is divided by the complex polynomial $D(z) = z - u$, where u is a constant, then the remainder is the constant $P(u)$.

Example 3

Find the remainder when $P(z) = 3iz^2 + 2z + i$ is divided by $z - 2i$.

Solution. The remainder is $P(2i) = -12i + 4i + i = -7i$.

Theorem 6-3 (Factor Theorem). The complex number u is a zero of a complex polynomial $P(z)$ if and only if $z - u$ is a factor of $P(z)$.

Example 4

Let us consider the complex polynomial $P_3(z) = -2 - 9iz + 10z^2 + 3iz^3$ for which we saw above that $2i$ is a zero. By the factor theorem, $z - 2i$ is a factor of $P_3(z)$. Using the division process, we can find the other factor.

$$
\begin{array}{r}
3iz^2 + 4z - i \\
z - 2i \overline{)\ 3iz^3 + 10z^2 - 9iz - 2} \\
\underline{3iz^3 + 6z^2} \qquad\qquad \text{(Note: } -6i^2z^2 = 6z^2) \\
4z^2 - 9iz \\
\underline{4z^2 - 8iz} \\
- iz - 2 \\
\underline{- iz - 2} \\
0
\end{array}
$$

So $P_3(z) = (z - 2i)(3iz^2 + 4z - i)$. The quadratic factor can readily be factored.

$$3iz^2 + 4z - i = (3iz + 1)(z - i)$$

Corresponding to these two new factors of $P_3(z)$, we obtain two more zeros:

$$z = i \quad \text{and} \quad z = -\frac{1}{3i} = \frac{1}{3}i.$$

A landmark in mathematical history is the discovery by Gauss of the following theorem.

Theorem 6-4 (Fundamental Theorem of Algebra). Every complex polynomial function of degree $n \geq 1$ has a complex zero.

The proof of Gauss' theorem is extremely difficult and we will not attempt to prove it here. We will consider some of its important consequences, however.

Let $P(z)$ be a complex polynomial of degree $n \geq 1$; by the fundamental theorem, P has a zero, say u_1. By the factor theorem, $z - u_1$ is a factor of $P(z)$; so there is a complex polynomial $P_1(z)$ such that

$$P(z) = (z - u_1)P_1(z).$$

Clearly the degree of P_1 is $n - 1$. Now if $n > 1$, we can apply exactly the same reasoning to P_1: P_1 has a zero u_2 and by the factor theorem

$$P_1(z) = (z - u_2)P_2(z),$$

where the complex polynomial P_2 has degree $n - 2$. So for $P(z)$ we now have

$$P(z) = (z - u_1)(z - u_2)P_2(z).$$

Continuing in this way, we obtain zeros $u_1, u_2, \ldots, u_n$ and factors $z - u_1, z - u_2, \ldots, z - u_n$ so that

$$P(z) = (z - u_1)(z - u_2) \cdots (z - u_n)P_n(z),$$

where $P_n(z)$ is of degree zero—that is, a complex constant $c \neq 0$. So

$$P(z) = c(z - u_1)(z - u_2) \cdots (z - u_n). \tag{6-12}$$

Equation (6-12) displays the zeros of P. Of course, the complex numbers $u_1, u_2, \ldots, u_n$ need not be different from one another; a zero that occurs exactly m times in the list $u_1, u_2, \ldots, u_n$ is said to have **multiplicity** m.

Example 5

Suppose that we have factored a sixth-degree complex polynomial as follows:

$$P(z) = 3(z - i)(z - 2i)(z - 2i)(z + 3i)(z + 3i)(z + 3i).$$

Then i is a zero of multiplicity 1, $2i$ is a zero of multiplicity 2, $-3i$ is a zero of multiplicity 3. The sum of these multiplicities is $1 + 2 + 3 = 6$.

What we observe from Eq. (6-12) is that any complex polynomial $P(z)$ of degree $n \geq 1$ has a factorization consisting of a constant times n complex *linear* factors. Moreover, the number of zeros of P is at most n and, in any case, the sum of the multiplicities of the zeros of P is exactly n.

It should be emphasized that the fact that P *has* a factorization (6-12) does not necessarily make that factorization or the zeros of P easy to *find*. The search for general formulas giving the zeros of a polynomial in terms of the coefficients (as the quadratic formula does) has been fraught with difficulties. Formulas exist for degree $n = 1, 2, 3, 4$, but for $n = 3, 4$ they are extremely complicated. For $n \geq 5$ it has been shown that no such general formulas exist. So the problem of finding the zeros of a polynomial (and thereby factoring it) remains a difficult one.

The factorization (6-12), in conjunction with Theorem 6-1, nevertheless sheds valuable light on the factorization of real polynomial functions, as we shall see.

Real Polynomial Functions

For most purposes in calculus, we are interested in the *real* zeros and the *real* factors of polynomial functions with *real* coefficients. Remarkably, the theory of *complex* polynomials helps us in this search. First, we obtain a generalization of Theorem 6-1.

> **Theorem 6-5.** Let $P(z)$ be a real polynomial. If u is an imaginary zero of P having multiplicity m, then $\bar{u}$ is also an imaginary zero of P having the same multiplicity m.

Proof. Let $u = a + bi$ (a, b real) so that $\bar{u} = a - bi$. Now u and $\bar{u}$ are imaginary and thus $b \neq 0$ and $u \neq \bar{u}$. Because u is a zero of P, we know by the conjugate zeros theorem that $\bar{u}$ is also a zero of P. By the factor theorem, $z - u$ and $z - \bar{u}$ are both factors of $P(z)$. Let m be the multiplicity of u and n the multiplicity of $\bar{u}$. Then

$$P(z) = (z - u)^m (z - \bar{u})^n P_1(z), \tag{6-13}$$

where $P_1(u)$ and $P_1(\bar{u})$ are not zero. We want to show than $m = n$. Suppose that $m > n$; dividing both sides of (6-12) by $(z - u)^n (z - \bar{u})^n$ gives

$$\frac{P(z)}{(z - u)^n (z - \bar{u})^n} = (z - u)^{m-n} P_1(z) = Q(z).$$

Consider the quotient on the left: the numerator $P(z)$ has real coefficients by hypothesis, and the denominator has real coefficients because

$$[(z - u)(z - \bar{u})]^n = [z^2 - 2az + (a^2 + b^2)]^n.$$

It follows that the polynomial $Q(z)$ on the right also has real coefficients. Now clearly $Q(u) = 0$; so by Theorem 6-1 $Q(\bar{u}) = 0$. But then $(\bar{u} - u)^{m-n} P_1(\bar{u}) = 0$ and because $\bar{u} - u \neq 0$, we conclude that $P_1(\bar{u}) = 0$. This contradicts Eq. (6-13). A similar contradiction arises if we suppose that $m < n$. Therefore $m = n$, showing that u and $\bar{u}$ have the same multiplicity.

The upshot of Theorem 6-5 is that imaginary zeros of real polynomials occur in conjugate pairs—each such zero occurring exactly as often as its conjugate.

To see how our discussions reveal information about real zeros, consider first a cubic

$$f(x) = Ax^3 + Bx^2 + Cx + D,$$

where A, B, C, and D are real and $A \neq 0$. By Theorem 6-5, any imaginary zeros of f must occur in pairs. If f has any imaginary zeros at all, it must have an even number of them, and because it cannot have more than three zeros, it must have just two. Then Eq. (6-12) shows that there is a third zero, which must be real. If f does not have imaginary zeros, then, of course, all its zeros are real. Clearly the argument here can be adapted to apply to any real polynomial of odd degree.

> **Theorem 6-6.** If P is a real polynomial of odd degree, then P has at least one real zero.

Example 6

Given that $1 + 2i$ is a zero of $P(z) = 3z^3 - 7z^2 + 17z - 5$, factor $P(z)$ in the form (6-12) and find the other zeros.

Solution. Because $1 + 2i$ is a zero and the coefficients of $P(z)$ are real, $1 - 2i$ is also a zero. Hence

$$[z - (1 + 2i)][z - (1 - 2i)] = z^2 - 2z + 5$$

is a factor. Dividing $P(z)$ by this factor, we get

$$P(z) = (3z - 1)(z^2 - 2z + 5)$$

$$= 3\left(z - \frac{1}{3}\right)[z - (1 + 2i)][z - (1 - 2i)]$$

and we see that the third zero is $\frac{1}{3}$.

Example 7

Given that $1 + i$ is a zero of multiplicity 2 of $P(z) = z^5 - 3z^4 + 4z^3 - 4z + 4$, find the other zeros and factor $P(z)$ in the form (6-12).

Solution. Because $1 + i$ is a zero of multiplicity 2 and the coefficients of P are real, $1 - i$ is also a zero of multiplicity 2 and $P(z)$ is divisible by the polynomial

$$[z - (1 + i)]^2[z - (1 - i)]^2 = (z^2 - 2z + 2)^2 = z^4 - 4z^3 + 8z^2 - 8z + 4.$$

Carrying out the division, we have

$$P(z) = (z + 1)[z - (1 + i)]^2[z - (1 - i)]^2.$$

The conjugate zeros $1 + i$, $1 - i$ each have multiplicity 2 whereas -1 is a zero of multiplicity 1.

Example 8

Find the zeros and factors of

$$P(z) = z^6 + 10z^4 + 32z^2 + 32.$$

Solution. Clearly this polynomial has no real zeros. By trial and error, we discover that $2i$ is a zero; it follows that $-2i$ is also a zero and so we divide out the factor $z^2 + 4$:

$$P(z) = (z^4 + 6z^2 + 8)(z^2 + 4)$$

$$= (z^2 + 2)(z^2 + 4)(z^2 + 4)$$

$$= (z + \sqrt{2}\,i)(z - \sqrt{2}\,i)(z + 2i)^2(z - 2i)^2.$$

The conjugate zeros $2i$, $-2i$ each have multiplicity 2 whereas the conjugate zeros $\sqrt{2}\,i$, $-\sqrt{2}\,i$ have multiplicity 1.

Arranging Eq. (6-11) in a slightly different form, denoting real zeros by $r_1, \ldots,$ r_k and imaginary zeros by $u_1, \ldots, u_j, \bar{u}_1, \ldots, \bar{u}_j$, we have

$$P(z) = c(z - r_1)(z - r_2) \cdots (z - r_k)[(z - u_1)(z - \bar{u}_1)][(z - u_2)(z - \bar{u}_2)]$$

$$\cdots [(z - u_j)(z - \bar{u}_j)].$$

But as we saw in the proof of Theorem 6-5, each product $(z - u)(z - \bar{u})$ is equivalent to a real quadratic $z^2 - 2az + (a^2 + b^2)$. Thus

$$P(z) = c(z - r_1)(z - r_2) \cdots (z - r_k)[z^2 - 2a_1z + (a_1^2 + b_1^2)][z^2 - 2a_2z + (a_2^2 + b_2^2)]$$

$$\cdots [z^2 - 2a_jz + (a_j^2 + b_j^2)]. \qquad (6\text{-}14)$$

Each quadratic factor in (6-14) has only imaginary zeros and cannot be factored into real linear factors; such quadratics are called **irreducible**.

Theorem 6-7. Every real polynomial of degree $n \geq 1$ can be factored into a product of real polynomials of first and second degree, in which the second-degree factors (if any) are irreducible.

Example 9

Factor $P(z) = z^4 - 36$.

Solution. Start with the factorization for a difference of squares,

$$z^4 - 36 = (z^2 - 6)(z^2 + 6)$$
$$= (z + \sqrt{6})(z - \sqrt{6})(z^2 + 6).$$

The factor $z^2 + 6$ has only imaginary zeros and is therefore irreducible.

Example 10

Factor $P(z) = z^6 - 1$.

Solution. This is a difference of squares:

$$z^6 - 1 = (z^3 + 1)(z^3 - 1)$$
$$= (z + 1)(z^2 - z + 1)(z - 1)(z^2 + z + 1).$$

The quadratic factors are irreducible, as we can see by calculating their discriminants.

EXERCISES 6-3

In Problems 1–8 find the degree and the leading coefficient.

1. $3 - 2i + 2iz + (3 - 2i)z^2$ 2. $1 - z^2 + iz^4$

3. $i(z - i)(iz - 1)$ 4. $(z + i)(z - i)(4i - z)(4 + iz)$

5. $[(1 + i)z + 3 - 2i][(1 + i)z - 3 + 2i]$

6. $\dfrac{z + i}{3} + \dfrac{z - i}{4}$ 7. $\dfrac{z^2 + z}{2} + \dfrac{z - 2z^2}{4}$ 8. $(iz + 3)^3$

In Problems 9–16 you are given complex polynomials $P(z)$ and $D(z)$. Use the division process to find the complex polynomials $Q(z)$ and $R(z)$ as in (6-11).

9. $P(z) = 3z^3 + 2z + 5, \ D(z) = z + 2$

10. $P(z) = 3z^3 + 2z + 5, \ D(z) = z^2 - 2z + 1$

11. $P(z) = \frac{2}{3}z^4 + \frac{1}{3}z^2 + \frac{4}{3}$, $D(z) = z + 1$

12. $P(z) = \frac{2}{3}z^4 + \frac{1}{3}z^2 + \frac{4}{3}$, $D(z) = z^2 - z + 2$

13. $P(z) = 3iz^2 + 4z + 2i$, $D(z) = z - i$

14. $P(z) = 5z^3 - 3iz^2 + 2z - 3i$, $D(z) = z^2 + 2iz + 1$

15. $P(z) = 1 - 3iz + 5z^2 - 12iz^3$, $D(z) = z + 2i$

16. $P(z) = z + 4iz^3$; $D(z) = z^3 + iz^2$

17. Show that $1 - i$ is a zero of the complex polynomial $P(z) = 3z^2 - 2iz + 2 + 8i$. Find another zero of P.

18. Show that $1 + i$ is a zero of the complex polynomial $P(z) = 3iz^2 - 4z + 10 + 4i$. Find another zero of P.

19. Show that $(1 + i)/\sqrt{2}$ and $(-1 - i)/\sqrt{2}$ are zeros of the complex polynomial $P(z) = z^2 - i$.

20. Show that $1/2 + (\sqrt{3}/2)i$ and $1/2 - (\sqrt{3}/2)i$ are zeros of the real polynomial $P(z) = z^3 + 1$.

In Problems 21–24 find the remainder when $P(z)$ is divided by $z + i$.

21. $P(z) = 3iz^3 - 2z^2 + 4iz$ 22. $P(z) = z^2 - 2z + i$

23. $P(z) = z^3 + z^2 + z + 1$ 24. $P(z) = 4z^3 - 3iz^2 + 2z - i$

In Problems 25–38 find all complex zeros of the given complex polynomial. Then factor the polynomial into complex linear factors as in Eq. (6-12).

25. $z^3 - 8$ 26. $z^3 + 27$ 27. $z^4 + 3z^2 + 2$

28. $z^3 - 2iz^2 - z + 2i$ 29. $z^3 - iz^2 + 4z - 4i$ 30. $5iz^3 - z^2 + 45iz - 9$

31. $z^4 + z^3 - z - 1$

32. $z^4 - 5z^3 + 15z^2 - 23z + 20$, given that $1 + 2i$ is a zero

33. $z^4 - 6z^3 + 29z^2 - 24z + 100$, given that $3 - 4i$ is a zero

34. $z^4 - 4z^3 + 8z^2 - 8z + 4$, given that $1 + i$ is a zero

35. $z^6 + 19z^4 + 99z^2 + 81$, given that $3i$ is a zero of multiplicity 2

36. $z^6 + 3z^4 + z^5 + 3z^2 + 2z^3 + z + 1$, given that i is a zero of multiplicity 2

37. $2z^3 + 5z^2 + 5z + 3$ (*Hint:* Try for a rational zero.)

38. $3z^3 + 4z^2 + 7z + 2$ (*Hint:* Try for a rational zero.)

In Problems 39–46 factor the polynomial into real linear and irreducible quadratic factors as in Eq. (6-14).

39. Problem 31   40. Problem 32   41. Problem 33   42. Problem 34

43. Problem 35   44. Problem 36   45. Problem 37   46. Problem 38

In Problems 47–52 find a complex polynomial of degree as small as possible having the given zeros. Express your answer in the form (6-12).

47. $i, 2i$ 48. $1 - i, -2, 3i$ 49. $i + 1, i - 1$ 50. $3 + 2i, 3, 2$

51. $2i$ (multiplicity 2), $3i$ (multiplicity 3)

52. $1 - 2i$ (multiplicity 2), 3 (multiplicity 3)

In Problems 53–58 find a real polynomial of degree as small as possible having the given zeros. Express your answer in the form (6-14).

53. $i, 2i$ 54. $1 - i, -2, 3i$ 55. $i + 1, i - 1$ 56. $3 + 2i, 3, 2$

57. $2i$ (multiplicity 2), $3i$ (multiplicity 3)

58. $1 - 2i$ (multiplicity 2), 3 (multiplicity 3)

Sec. 6-3   Fundamental Theorem of Algebra   **237**

59. Consider a quadratic complex polynomial
$$P(z) = z^2 + Bz + C$$
with zeros u_1 and u_2 (not necessarily different). Show that $B = -(u_1 + u_2)$ and $C = u_1 u_2$.

60. Consider a cubic complex polynomial
$$P(z) = z^3 + Bz^2 + Cz + D$$
with zeros u_1, u_2, and u_3 (not necessarily different). Show that $B = -(u_1 + u_2 + u_3)$ and $D = -u_1 u_2 u_3$. Find a formula for C in terms of u_1, u_2, and u_3.

61. (a) Find six complex zeros of $z^6 - 1$.
(b) Find eight complex zeros of $z^8 - 1$.

SUMMARY

The purpose of this short chapter is to provide further insight into the processes of factoring polynomials and finding their zeros. In order to gain this insight, we need to look at the real numbers as part of a larger set of numbers—the complex numbers.

1. You should know the general form for complex numbers and be able to perform algebraic operations on these numbers.

2. You should be able to solve linear equations with complex coefficients, find square roots of negative real numbers, and find the zeros of quadratic polynomials with real coefficients.

3. You should know how to represent a complex number as a point in the plane and should recognize addition of complex numbers as vector addition.

4. You should know the definitions and properties of the complex conjugate and absolute value. You should be able to prove the theorem about conjugate zeros and use it in finding zeros of a polynomial function with real coefficients.

5. You should be able to perform algebraic operations on complex polynomials.

6. Given a zero of a complex polynomial, you should be able to use the factor theorem and the division process to find other zeros and factors. In the case of a real polynomial, you should know how to factor it into a product of real polynomials of first and second degree.

REVIEW EXERCISES

1. Simplify.
(a) $(1 + 2i)^4$ (b) $(1 + i)^{-3}$

2. Find the complex square roots of
(a) -49 (b) -288 (c) $-\dfrac{4}{5}$

3. Solve the equations:
   (a) $(2 - 3i)z + 1 - i = 2$
   (b) $3z + i = 3iz - 1$
   (c) $5z^2 + 2z + 1 = 0$
   (d) $6z^2 - iz + 1 = 0$

4. Find all complex numbers z having the property that the reciprocal of z is equal to the opposite of z.

5. Without looking in the text, show that for any complex number z both $z + \bar{z}$ and $z\bar{z}$ are real and $z\bar{z}$ is nonnegative.

6. Describe and sketch the set of points z for which $|z - 1 - i| < \sqrt{2}$.

7. (a) State the conjugate zeros theorem.
   (b) Find all the zeros of $12z^3 + 8z^2 + 3z + 2$, given that $i/2$ is a zero.

8. (a) State the fundamental theorem of algebra and explain how it leads to the factorization of complex polynomials into linear factors.
   (b) Factor $12z^3 + 8z^2 + 3z + 2$ [see Problem 7(b)] into linear factors.

9. Factor each of the following into a product of real polynomials of first and second degree.
   (a) $5z^6 + z^5 + 91z^4 + 18z^3 + 423z^2 + 81z + 81$, given that $3i$ is a zero of multiplicity 2
   (b) $25z^3 - 20z^2 + 5z - 4$, given that there is a rational zero.

Composition and inverses

7

7-1 INCREASING FUNCTIONS

If the values of a function f increase as the values of x increase on an interval, then the function itself is said to be increasing on that interval; if the values decrease as x increases, the function is decreasing.

> **Definition.**
> 1. A function f is **increasing** on an interval provided that, for all x_1, x_2 in the interval,
>
> $$x_1 < x_2 \quad \text{implies} \quad f(x_1) < f(x_2).$$
>
> 2. A function f is **decreasing** on an interval provided that, for all x_1, x_2 in the interval,
>
> $$x_1 < x_2 \quad \text{implies} \quad f(x_1) > f(x_2).$$

Figure 7-1 illustrates this definition.

A function that is increasing (decreasing) on the interval of all real numbers is said to be **increasing (decreasing) everywhere.**

Example 1

Show that the function $f(x) = 2x + 3$ is increasing everywhere.

Solution. Suppose that $x_1 < x_2$. Using the multiplication and addition prop-

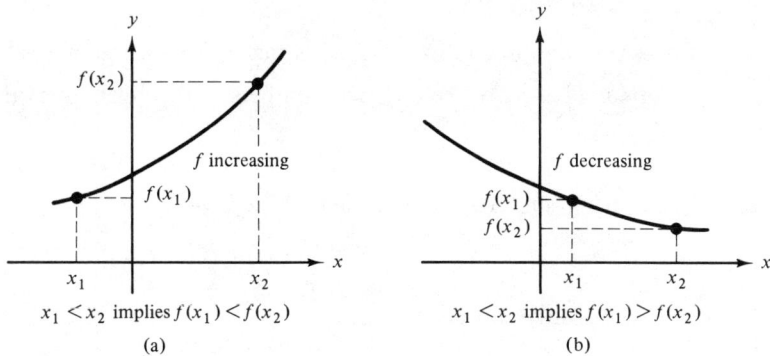

$x_1 < x_2$ implies $f(x_1) < f(x_2)$

(a)

$x_1 < x_2$ implies $f(x_1) > f(x_2)$

(b)

Figure 7-1

erties of Section 1-4, we get

$$2x_1 < 2x_2$$
$$2x_1 + 3 < 2x_2 + 3$$
$$f(x_1) < f(x_2).$$

So $x_1 < x_2$ implies $f(x_1) < f(x_2)$. Because x_1 and x_2 can be any numbers with $x_1 < x_2$, we have shown that f is increasing everywhere. [See Fig. 7-2(a).]

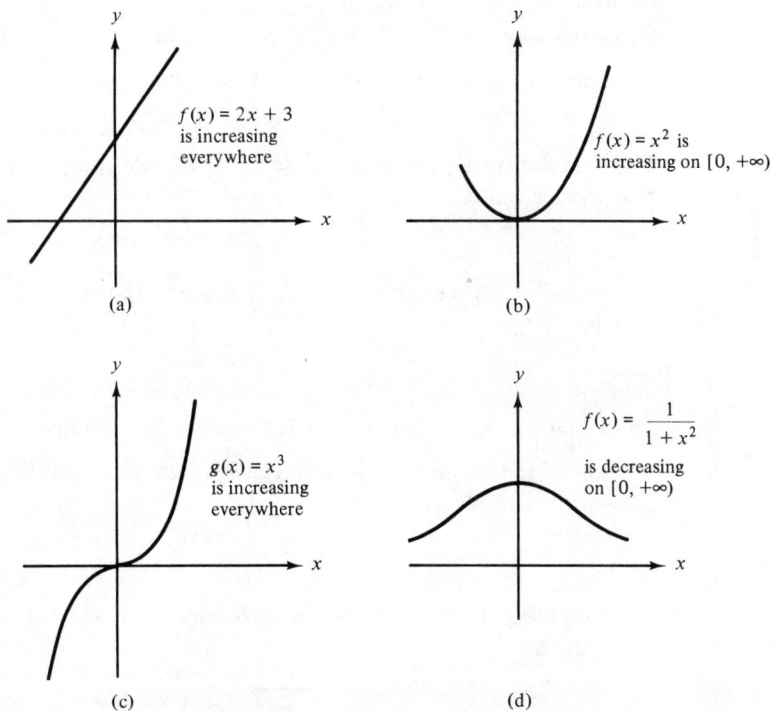

$f(x) = 2x + 3$ is increasing everywhere

(a)

$f(x) = x^2$ is increasing on $[0, +\infty)$

(b)

$g(x) = x^3$ is increasing everywhere

(c)

$f(x) = \dfrac{1}{1 + x^2}$ is decreasing on $[0, +\infty)$

(d)

Figure 7-2

Example 2

Show that the function $f(x) = x^2$ is increasing on the interval $[0, +\infty)$.

Solution. Let x_1, x_2 be any two numbers in the given interval with $x_1 < x_2$ so that

$$0 \leq x_1 < x_2.$$

Multiplying first by x_1 and then by x_2, we have

$$0 \leq x_1^2 \leq x_1 x_2$$
$$0 \leq x_1 x_2 < x_2^2.$$

By the transitive property,

$$0 \leq x_1^2 < x_1 x_2 < x_2^2.$$

But this shows that $x_1^2 < x_2^2$ and so $f(x_1) < f(x_2)$. The graph of f [Fig. 7-2(b)], is rising over the interval $[0, +\infty)$.

Another way to show that a function is increasing is to form the expression $f(x_2) - f(x_1)$ and show that $f(x_2) - f(x_1)$ is positive when $x_1 < x_2$. Applying this procedure in Example 2,

$$f(x_2) - f(x_1) = x_2^2 - x_1^2 = (x_2 - x_1)(x_2 + x_1).$$

Both factors are positive when $0 \leq x_1 < x_2$. So $f(x_2) - f(x_1) > 0$ and $f(x_1) < f(x_2)$.

Example 3

Show that the function $g(x) = x^3$ is increasing everywhere.

Solution. Let x_1 and x_2 be any real numbers with $x_1 < x_2$. Then

$$g(x_2) - g(x_1) = x_2^3 - x_1^3 = (x_2 - x_1)(x_2^2 + x_1 x_2 + x_1^2).$$

The first factor is positive because $x_1 < x_2$. By completing the square, we see that the second factor is

$$\left(x_2 + \frac{x_1}{2}\right)^2 + \frac{3}{4}x_1^2 > 0.$$

So $g(x_2) - g(x_1)$ is positive and $g(x_1) < g(x_2)$. Hence g is increasing everywhere. [See Fig. 7-2(c).]

Example 4

Show that the function $f(x) = 1/(1 + x^2)$ is decreasing on the interval $[0, +\infty)$.

Solution. Let x_1 and x_2 be nonnegative numbers such that $x_1 < x_2$. Then (as shown in Example 2)

$$x_1^2 < x_2^2.$$
$$1 + x_1^2 < 1 + x_2^2.$$

Now dividing both sides by the product $(1 + x_1^2)(1 + x_2^2)$ (this product is positive), we obtain

$$\frac{1}{1 + x_2^2} < \frac{1}{1 + x_1^2}.$$

This means that $f(x_2) < f(x_1)$, showing that f is decreasing on $[0, +\infty)$. [See Fig. 7-2(d).]

Determining whether a function is increasing (decreasing) on an interval is not always easy; in many cases, the methods of calculus are necessary. Nevertheless several useful rules can be stated at this point.

---

1. A linear function $f(x) = mx + b$ is increasing everywhere if $m > 0$; it is decreasing everywhere if $m < 0$.
2. If C is any constant and f is increasing on an interval, then the function g defined by

$$g(x) = f(x) + C$$

is also increasing on that interval. Similarly, if f is decreasing on the interval, so is g.
3. (a) If C is a *positive* constant and f is increasing on an interval, then the function g defined by

$$g(x) = Cf(x)$$

is also increasing on that interval; if f is decreasing, then g is decreasing.
   (b) If C is a *negative* constant and f is increasing, then the function g in part (a) is decreasing; if f is decreasing, then g is increasing.
4. Let f be a function whose values on an interval are either always positive or always negative. Let g be defined by

$$g(x) = \frac{1}{f(x)}.$$

If f is increasing on that interval, then g is decreasing there; if f is decreasing on the interval, g is increasing there.
5. If two or more functions are increasing on an interval, then their sum is increasing on that interval; if the functions are decreasing on an interval, then so is their sum.

For the most part, these rules are intuitively plausible and can be regarded as "common sense." They can, of course, be verified from our definitions. Let us verify (4).

Proof. Let f be positive valued and increasing on an interval and let $g(x) = 1/f(x)$. If x_1 and x_2 are points of the interval such that $x_1 < x_2$, then

$$f(x_1) < f(x_2)$$

because f is increasing. Dividing by the positive number $f(x_1)f(x_2)$, we get

$$\frac{f(x_1)}{f(x_1)f(x_2)} < \frac{f(x_2)}{f(x_1)f(x_2)}$$

$$\frac{1}{f(x_2)} < \frac{1}{f(x_1)}$$

$$g(x_2) < g(x_1).$$

Thus g is decreasing on the given interval. The same proof holds for a negative valued function f, because the product $f(x_1)f(x_2)$ is still positive.

Example 5

From earlier examples we know that the functions $f(x) = x^2$ and $g(x) = x^3$ are increasing on $[0, +\infty)$. Therefore on that interval
(a) $h(x) = 3x^2$ is increasing [by 3(a)].
(b) $k(x) = 3x^2 + 7$ is increasing (by 2).
(c) $p(x) = x^3 + (3x^2 + 7)$ is increasing (by 4).
(d) $q(x) = \dfrac{1}{x^3 + 3x^2 + 7}$ is decreasing (by 5).
In deriving (d), we used the fact that $x^3 + 3x^2 + 7$ does not change sign (it is always positive) on $[0, +\infty)$.

Example 6

The function $f(x) = x$ is increasing and always positive on the interval $(0, +\infty)$. Therefore, on that interval,
(a) $g(x) = 1/x$ is decreasing (by 4).
(b) $h(x) = 1 - (1/x)$ is increasing [by 3(b) and 2].

Example 7

Show that the function $h(x) = 3/(4 + x^2)$ is decreasing on the interval $[0, +\infty)$.

Solution. Because x^2 is increasing on $[0, +\infty)$ (by Example 2), we know that $4 + x^2$ is also increasing there (by 2). Now $4 + x^2$ is always positive; so we conclude that $1/(4 + x^2)$ and hence $3/(4 + x^2)$ are decreasing on $[0, +\infty)$ [by (4) and 3(a)].

EXERCISES 7-1

1. Show directly from the definition that the function $f(x) = x^2$ is decreasing on the interval $(-\infty, 0]$.
2. Show directly from the definition that the function $f(x) = 1/x$ is decreasing on the interval $(0, +\infty)$.
3. Is the function $f(x) = 1 + (1/x)$ increasing or decreasing on the interval $(0, +\infty)$? Justify your answer.
4. Is the function $f(x) = 1/(1 - x)$ increasing or decreasing
   (a) on the interval $(1, +\infty)$?
   (b) on the interval $(-\infty, 1)$?
5. Show that the function is neither increasing nor decreasing on the given interval.
   (a) $f(x) = 3; [-2, 1]$
   (b) $f(x) = |x|; (-\infty, +\infty)$
   (c) $f(x) = x^2 - x; [0, +\infty)$

7-2 COMPOSITION OF FUNCTIONS

It is often helpful to think of a function f as an "operator"—something that operates on x and produces $f(x)$. For instance, if f is the square function, $f(x) = x^2$ for real numbers x, then the operation that f performs on x is that of squaring it.

The function $g(x) = x + 1$ adds 1 to x. The function $h(x) = 2x$ doubles x and the function $k(x) = 2x + 1$ doubles x and adds 1. Notice that k has the same effect as first doing h and then doing g.

When we have two functions, h and g, we may wish to apply one and then the other. We can first apply h to x, getting $h(x)$, and then apply g to $h(x)$, getting $g(h(x))$. The operator that results by applying h and then g is called the **composition** of g on h and is denoted $g \circ h$. In order for the function $g \circ h$ to be defined for a particular real number x, not only must x be in the domain of h but also $h(x)$ must be in the domain of g.

Example 1

Let g and h be defined for all real numbers as follows:

$$g(x) = 2x + 1$$
$$h(x) = \sqrt{x^2 + 4}.$$

The composition of g on h is the function $g \circ h$ defined for all real numbers x by

$$(g \circ h)(x) = g(h(x)) = 2h(x) + 1 = 2\sqrt{x^2 + 4} + 1.$$

Example 2

With the same functions g and h as in Example 1, let us find the composition of h on g.

$$(h \circ g)(x) = h(g(x)) = \sqrt{g(x))^2 + 4}$$
$$= \sqrt{(2x + 1)^2 + 4}$$
$$= \sqrt{4x^2 + 4x + 5}$$

Notice particularly that $h(g(x))$ is not the same as $g(h(x))$, which we calculated in Example 1.

We now summarize the terminology and notation introduced in the preceding discussion.

Definition. Let g and h be functions. The **composition** of g on h is the function f defined by
$$f(x) = g(h(x))$$
for every x in the domain of h such that $h(x)$ is in the domain of g. The composition of g on h is denoted $g \circ h$.

Example 3

The expression $\sqrt{1 - x^2}$ can be understood as giving $g(h(x))$, where $h(x) = 1 - x^2$ and $g(x) = \sqrt{x}$. The domain of $g \circ h$ is $-1 \leq x \leq 1$. Although $h(x)$ is defined for real x, its values for $x > 1$ and for $x < -1$ are negative and therefore do not belong to the domain of g.

Example 4

Express $f(x) = \sqrt{x^2 + 1}$ as the composition of two functions.

Solution. One way to do so is to let $h(x) = x^2 + 1$ and $g(x) = \sqrt{x}$. Then $f(x) = g(h(x))$ for all real x. A different way is to let $H(x) = x^2$ and $G(x) = \sqrt{x + 1}$; again, $f(x) = G(H(x))$ for real x.

An especially important case of the composition of two functions occurs when one of the functions is the absolute value function. Let h be any function and define $f(x) = |h(x)|$ for all x in the domain of h. Then

$$f(x) = \begin{cases} h(x) & \text{for any } x \text{ such that } h(x) \geq 0 \\ -h(x) & \text{for any } x \text{ such that } h(x) < 0. \end{cases}$$

The graph of f can be obtained from the graph of h as follows. On any interval where h is nonnegative, the graph of f coincides with that of h. On any interval where h is negative, the graph of f is obtained by reflecting the graph of h through the x axis.

Example 5

Graph $f(x) = |x^2 - 1|$.

Solution. The graph of $y = h(x) = x^2 - 1$ is easy to draw [see Fig. 7-3(a)]. Because $h(x) < 0$ for $-1 < x < 1$, the graph of h on that interval is below the x axis; this portion of the graph is reflected through the x axis in the graph of f [Fig. 7-3(b)].

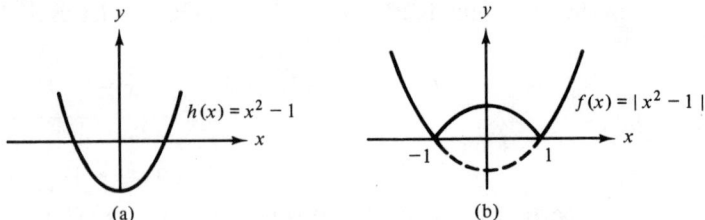

Figure 7-3

Composition and Inverses   Chap. 7

EXERCISES 7-2

In Problems 1–6 you are given functions g and h. Find $g(h(x))$ and $h(g(x))$ and specify their domains. (Assume that the domain of each function consists of all real numbers x for which the given expression makes sense.)

1. $g(x) = x^2$ **2.** $g(x) = \dfrac{1}{x}$ **3.** $g(x) = x - 1$

     $h(x) = x + 2$ $h(x) = 2x$ $h(x) = x^3$

4. $g(x) = \sqrt{4x - 2}$ **5.** $g(x) = x^3 - 1$ **6.** $g(x) = \sqrt{x + 2}$

     $h(x) = x^2$ $h(x) = \dfrac{1}{x}$ $h(x) = x^2 - 2$

In Problems 7–17 find functions g and h such that $g(h(x))$ is equal to the given expression.

7. $(x + 5)^2$ **8.** $\sqrt{3x - 2}$ **9.** $(x^2 + 3)^2$

10. $\sqrt{x(2x - 1)}$ **11.** $\dfrac{1}{1 + 1/x}$ **12.** $x^4 + 2x^2 + 1$

13. $\sqrt{1 + \sqrt{x}}$ **14.** $|x - 3|$ **15.** $|2x + 1|$

16. $2|x| - 1$ **17.** $|x^2 - 3x + 2|$

In Problems 18–27 graph the function.

18. $f(x) = |2x + 3|$ **19.** $f(x) = |7 - 3x|$ **20.** $g(x) = |x^2 - 4|$

21. $f(x) = |4x - x^2|$ **22.** $g(t) = |t^3 - 4t|$ **23.** $f(u) = |u^3 + 3u^2|$

24. $f(x) = |x(x + 1)|$ **25.** $g(x) = |x^2(x^2 - 1)|$

26. $f(t) = |2t^3 - t^2 - 8t + 4|$ **27.** $f(t) = |t^4 - 5t^2 + 4|$

28. Let $f(x) = 1/x$ for $x \neq 0$. Find $f \circ f$.

29. Let $f(x) = -x$ for real x. Find $f \circ f$.

30. Let $f(x) = (1 - x)/(1 + x)$ for $x \neq -1$. Find $f \circ f$.

FOR THE ENTHUSIASTS

31. Let h be a function and g be the square root function: $g(x) = \sqrt{x}$ for $x \geq 0$. Then the composition $f = g \circ h$ is defined by $f(x) = \sqrt{h(x)}$ for all x in the domain of h such that $h(x) \geq 0$. Explain how you would obtain the graph of f from the graph of h. Then apply your method to the following.

     (a) $f(x) = \sqrt{2 - 3x}$ **(b)** $f(x) = \sqrt{3x + 7}$

     (c) $f(x) = \sqrt{4 - x^2}$ **(d)** $f(x) = \sqrt{x^2 + 1}$

     (e) $f(x) = \sqrt{2x^2 - 5}$

32. **(a)** Show that if g and h are both linear, then the composition of g on h is also linear.

     (b) How does part (a) relate to Problem 11(a), Section 4-6?

33. **(a)** Show that if f and g are both increasing everywhere, then $f \circ g$ is increasing everywhere.

     (b) Suppose that f is increasing everywhere and g is decreasing everywhere. What about $f \circ g$? $g \circ f$? Justify your conclusions.

     (c) Suppose that f and g are both decreasing everywhere. What about $f \circ g$? Justify your conclusion.

7-3 INVERSE OF A FUNCTION

Suppose that f is a function with domain D and range R. The function f operates on the numbers in D and produces numbers in R. Let us try to find a function g that "undoes" the work of f. Such a function operates on the numbers in R and produces numbers in D in such a way that $g(f(x)) = x$.

Example 1

Let f be the function that adds 1 to every number: $f(x) = x + 1$. Clearly the function g that we want subtracts 1 from every number: $g(u) = u - 1$. Indeed, we see that

$$g(f(x)) = g(x + 1) = (x + 1) - 1 = x.$$

Moreover, we notice that

$$f(g(u)) = g(u) + 1 = (u - 1) + 1 = u.$$

Not only does g undo the work of f but f also undoes the work of g.

Example 2

Let f be the function that doubles every number in the interval $(0, 1)$:

$$f(x) = 2x \quad \text{for } 0 < x < 1.$$

The range R of f is the interval $(0, 2)$. Then g is the function that divides every number in R by 2:

$$g(u) = \frac{1}{2}u \qquad (0 < u < 2).$$

Again,

$$g(f(x)) = g(2x) = \frac{1}{2}(2x) = x$$

and

$$f(g(u)) = 2\left(\frac{1}{2}u\right) = u.$$

The function g described in these examples is the inverse of f. Here is a definition of inverse.

Definition. Let f be a function with domain D and range R. An **inverse** of f is a function g that satisfies the conditions
1. the domain of g is R.
2. the range of g is D.
3. $g(f(x)) = x$ for every x in D.
4. $f(g(u)) = u$ for every u in R.

Condition (3) says that g undoes the work of f. Furthermore, (4) says that f undoes the work of g. So f is an inverse of g whenever g is an inverse of f.

In Examples 1 and 2 we were successful in finding the desired inverse function. Let us consider another example.

Example 3

Let f be the function that squares every number: $f(x) = x^2$ for all real numbers x. What is g? If we try to undo the squaring operation, we encounter difficulty because we do not know, for instance, whether 25 was the result of squaring 5 or -5. We would have $g(25) = \pm 5$, which is unacceptable because we have not assigned a definite value to $g(25)$. The function f does not have an inverse. The trouble with f is that it assumes the same value (25) for different values of x (-5 and 5).

One-to-One Functions

A function f that takes the same value v for different values of x

$$v = f(x_1) = f(x_2) \qquad (x_1 \neq x_2)$$

cannot have an inverse. This was the difficulty we encountered in Example 3. However, if f never assumes the same value for different values of x, then f does have an inverse, as we will show in the proof of Theorem 7-1. First, we state a definition.

Definition. A function f is **one to one** provided that if x_1 and x_2 are different elements of the domain of f, then $f(x_1) \neq f(x_2)$.

Theorem 7-1. A function has an inverse if and only if it is one to one.

Proof. We prove the half of the theorem that states if f is one to one, then it has an inverse; the other half is left as an exercise (Problem 31).

Let f be one to one and let u be in the range R of f. We define $g(u)$ as follows. Because f is one to one, there is one and only one number x in the domain D of f such that $f(x) = u$. Define $g(u)$ to be that x.

We show that g is an inverse of f. Clearly the domain of g is R. Then if x is any number in D, let $u = f(x)$. Next, $g(u)$ is x by the way we defined g; hence $g(f(x)) = x$. This proves (3) and also shows that the range of g is D. Finally let u be any number in R. It follows from the way we defined g that if $u = f(x)$, then $x = g(u)$. Therefore $u = f(g(u))$, proving (4).

Theorem 7-1 gives us a criterion for determining whether a given function has an inverse: it does if and only if it is one to one. The next theorems give conditions under which we can be sure a function is one to one.

Theorem 7-2. If the domain of f is an interval and f is increasing on that interval, then f is one to one.

Proof. Suppose that x_1 and x_2 belong to the domain of f. If $x_1 \neq x_2$, then either $x_1 < x_2$ or $x_2 < x_1$. In the first case, because f is increasing, $f(x_1) < f(x_2)$. In the second case, $f(x_1) > f(x_2)$. In either case, $f(x_1) \neq f(x_2)$ and so f is one to one.

Example 4

The function $f(x) = 3x - 2$ is increasing everywhere. By Theorem 7-2, it is one to one; by Theorem 7-1, it has an inverse g. We can find g by using Eq. (4) in the definition:

$$f(g(u)) = u$$
$$3g(u) - 2 = u$$
$$3g(u) = u + 2$$
$$g(u) = \frac{1}{3}u + \frac{2}{3}.$$

Example 5

We saw that the function defined for all real x by $f(x) = x^2$ does not have an inverse. But if we restrict the domain to nonnegative numbers, defining $h(x) = x^2$ for $x \geq 0$, then h is increasing for $x \geq 0$ and therefore has an inverse by Theorem 7-2. The inverse of h is easily found to be the function $g(u) = \sqrt{u}$ $(u \geq 0)$. We can represent g just as well by the notation $g(x) = \sqrt{x}$ $(x \geq 0)$ because the name of the independent variable does not matter.

Theorem 7-3.   If the domain of f is an interval and f is decreasing on that interval, then f is one to one.

The proof of Theorem 7-3 is entirely similiar to the proof of Theorem 7-2 and is left as an exercise (Problem 29).

Example 6

The function f defined for x in the interval $[0, +\infty)$ by $f(x) = 1/(1 + x^2)$ is decreasing on that interval. Its range is the interval $(0, 1]$. By Theorems 7-3 and 7-1, f has an inverse g. We obtain $g(u)$ for $0 < u \leq 1$ by using (4):

$$f(g(u)) = u$$
$$\frac{1}{1 + [g(u)]^2} = u$$
$$1 + [g(u)]^2 = \frac{1}{u}$$
$$[g(u)]^2 = \frac{1}{u} - 1.$$

Because the range of g is the domain of f, $g(u) \geq 0$. Consequently,

$$g(u) = \sqrt{\frac{1}{u} - 1} \qquad (0 < u \leq 1).$$

If we wish to write the independent variable as x, we can do so:

$$g(x) = \sqrt{\frac{1}{x} - 1} \qquad (0 < x \leq 1).$$

Notation. When a function has an inverse, it is common practice to denote this inverse by f^{-1}. Thus for the function f in Example 6 we could use this notation to express the inverse of f by the equation

$$f^{-1}(x) = \sqrt{\frac{1}{x} - 1} \quad (0 < x \le 1).$$

[Be careful not to misinterpret $f^{-1}(x)$ as $1/f(x)$.]

Graph of the Inverse

Suppose that f is a function with an inverse f^{-1}. Let us investigate the relation between the graphs of $y = f(x)$ and $y = f^{-1}(x)$.

The equation $y = f^{-1}(x)$ is equivalent to $x = f(y)$. Therefore a point (a, b) lies on the graph of f^{-1} if and only if (b, a) lies on the graph of f. So the points that make up the graph of f^{-1} are obtained from those on the graph of f by reversing the coordinates.

Example 7

Consider $h(x) = x^2$ for $x \ge 0$. As shown in Example 5, the inverse of h is given by $h^{-1}(x) = \sqrt{x}$ $(x \ge 0)$. The following points are on the graph of h:

$$(0, 0), \quad \left(\frac{1}{2}, \frac{1}{4}\right), \quad (1, 1), \quad (2, 4), \quad (\sqrt{5}, 5), \quad \left(\frac{8}{3}, \frac{64}{9}\right).$$

By reversing the coordinates, we get points on the graph of h^{-1}:

$$(0, 0), \quad \left(\frac{1}{4}, \frac{1}{2}\right), \quad (1, 1), \quad (4, 2), \quad (5, \sqrt{5}), \quad \left(\frac{64}{9}, \frac{8}{3}\right).$$

These points are plotted on the graphs shown in Fig. 7-4.

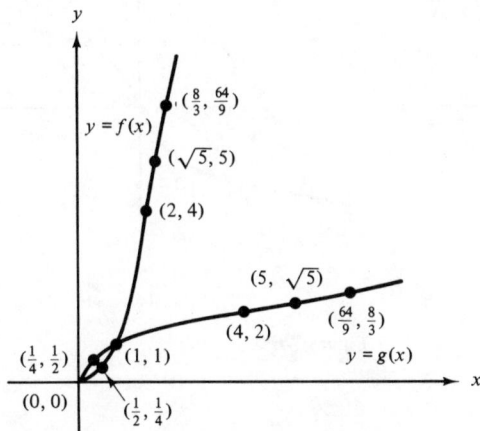

Figure 7-4

If we have a coordinate system with equal scales on the x and y axes, then (a, b) and (b, a) are symmetric with respect to the line $y = x$ because this line is the perpendicular bisector of the segment connecting the two points (see Fig. 7-5). You are asked to prove this fact in Problem 32.

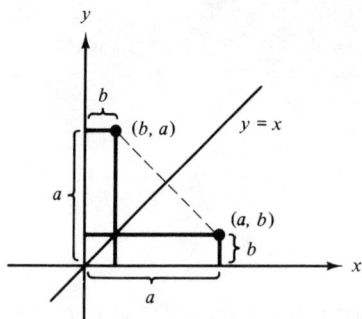

Figure 7-5

Because of this symmetry, we can obtain the graph of the inverse by reflecting the graph of the original function through the line $y = x$.

If f has an inverse, the graph of $y = f(x)$ and the graph of $y = f^{-1}(x)$ are symmetric with respect to line $y = x$.

Example 8

Graph the cube root function $g(x) = \sqrt[3]{x}$.

Solution. The function g is the inverse of $f(x) = x^3$, whose graph is shown in Fig. 7-6(a). We reflect the graph of f through the line $y = x$ and thereby obtain the graph of g, as in Fig. 7-6(b).

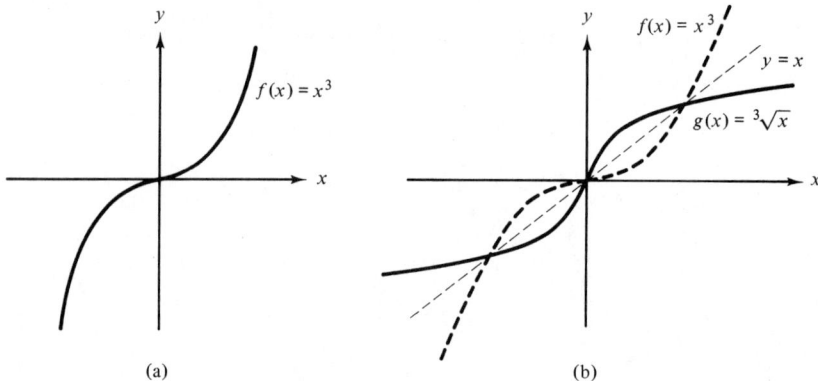

(a)

(b)

Figure 7-6

EXERCISES 7-3

In Problems 1–10 verify that the function has an inverse and find the inverse.

1. $f(x) = 3x$

2. $h(x) = -\frac{1}{2}x$

3. $f(t) = 3t + 4$

4. $h(v) = \frac{1}{3}v - 2$

5. $f(x) = 2x + 1$ for $0 \le x \le 2$ **6.** $h(x) = \dfrac{1}{1 + x^2}$ for $x \le 0$

7. $f(t) = t^2 + 1$ for $t \ge 0$ **8.** $f(x) = \dfrac{1}{1 + x}$ for $x \ge 0$

9. $f(x) = (2 + x)^3$ **10.** $h(x) = \sqrt{x}$ for $x \ge 0$

In Problems 11–19 determine whether the function has an inverse. Justify your conclusion.

11. $f(x) = \dfrac{1}{1 + x^2}$ $(x \ge 0)$ **12.** $g(x) = \dfrac{1}{1 + x^2}$ $(-4 \le x \le 4)$

13. $F(x) = x^2 + 6x + 8$ **14.** $h(x) = \dfrac{1}{1 - x}$ $(x > 1)$

15. $f(x) = x^2 + 6x + 8$ $(x \ge 0)$ **16.** $G(x) = x + \dfrac{1}{x}$ $(x > 0)$

17. $f(x) = |x|$ $(-1 \le x \le 1)$ **18.** $g(x) = |x|$ $(x \le 0)$

19. $g(x) = x + \dfrac{1}{x}$ $(x \ge 1)$

20. Let $f(x) = (1 - x)/(1 + x)$ for $x \ne -1$. Show that f is its own inverse.

In Problems 21–28 graph the function and its inverse on the same set of coordinate axes.

21. $f(x) = 2\sqrt{x}$ for $x \ge 0$ **22.** $f(x) = 3x^2$ for $x \ge 0$

23. $f(x) = \dfrac{1}{x^3}$ for $x \ne 0$ **24.** $h(x) = (x + 1)^3$

25. $h(x) = 2x + 3$ **26.** $f(x) = \dfrac{2 - x}{3}$

27. $f(u) = \sqrt{u^2 + 1}$ for $u \ge 0$ **28.** $g(x) = \sqrt{9 - x^2}$ for $0 \le x \le 3$

FOR THE ENTHUSIASTS

29. Prove Theorem 7-3.

30. (a) Without using Theorems 7-2 and 7-3, show that the linear function $f(x) = mx + b$ is one to one whenever $m \ne 0$. (Is this function one to one if $m = 0$?)
(b) Show that the inverse of a nonconstant linear function is linear.

31. Prove the second part of Theorem 7-1: If f has an inverse, then f is one to one.

32. Show that if the coordinate axes have equal scales and if $a \ne b$, then $y = x$ is the perpendicular bisector of the segment connecting (a, b) and (b, a).

SUMMARY

The purpose of this short chapter has been to introduce you to two operations involving functions that will be used repeatedly throughout the remainder of this book as well as in your calculus course: composition and inverses. The preliminary section on increasing functions is necessary background for the theorems in Section 7-3, but it is also important in its own right.

Composition occurs when two (or more) functions are combined by applying one after another.

1. You should be able to form the composition $f = g \circ h$ of two given functions g and h. Specifically you should be able both to determine the rule for f and to find its domain.

2. You should be able to "decompose" a composite function f and express it as the composition of two or more functions; often there is more than one way to do this.

The inverse of a function arises when we try to find a function that "undoes" the operation that the original function does.

3. You should know the four properties required by the definition of inverse.

4. You should know that not every function has an inverse and be able to give examples of functions that do not have inverses.

5. You should understand what is meant by a one to one function and should know that a function has an inverse if and only if it is one to one.

6. You should be able to determine whether a function is increasing (decreasing) on an interval.

7. You should know that if the domain of a function is an interval and if the function is increasing (decreasing) there, then it is one to one and hence has an inverse.

8. You should be able to find the inverse when it exists.

9. You should know how the graphs of $y = g(x)$ and $y = f(x)$ are related geometrically when g and f are inverses. You should therefore be able to graph one of the two functions, given the graph of the other.

REVIEW EXERCISES

1. Show that $f(x) = 1/(x^2 - 1)$ is
   (a) decreasing on $(0, 1)$ and $(1, +\infty)$.
   (b) increasing on $(-\infty, -1)$ and $(-1, 0)$.

2. Is the given function one to one? Justify your conclusion.
   (a) $f(x) = x^2 + 1$ (b) $g(x) = \sqrt{x^2 - 1}$ for $x \geq 1$
   (c) $h(x) = 1 - \dfrac{1}{x}$ for $x > 0$ (d) $H(x) = 1 - \dfrac{1}{x}$ for $x \neq 0$

3. Find $g(h(x))$ and $h(g(x))$ if
   (a) $g(x) = 1 - 3x$ (b) $g(x) = \sqrt{x}$
        $h(x) = 2x^2$; $h(x) = x^2$;
   (c) $g(x) = \dfrac{1}{x^2 + 1}$
        $h(x) = \sqrt{x}$.

4. Find functions g and h such that $f(x) = g(h(x))$.
   (a) $f(x) = |3x - 2|$

(b) $f(x) = \dfrac{1}{\sqrt{2x+3}}$

(c) $f(x) = (x - 2x^2)^2$

5. For each of the following functions find the inverse f^{-1} (if it exists) and draw the graph of $y = f^{-1}(x)$.

(a) $f(x) = \dfrac{1}{x^2}$ for $x > 0$ **(b)** $f(x) = x^3 - 3x$

(c) $f(x) = x(1 + x)$ for $x > 0$ **(d)** $f(x) = \dfrac{1}{x^2 - 1}$ for $x \geq 0$, $x \neq 1$

6. Let h be any function and g be the square function $g(x) = x^2$ for x real. Find $g \circ h$ and explain how you could obtain the graph of $g \circ h$ from the graph of h. Use your method to graph $g \circ h$, where

    (a) $h(x) = x$.   **(b)** $h(x) = 1 - x$.

    (c) $h(x) = \sqrt{x}$.   **(d)** $h(x) = x - x^2$.

7. The function h triples every real number; the function g subtracts 4 from every real number. Describe in words the operation performed on x by each of the following functions.

    (a) h^{-1} **(b)** g^{-1}

    (c) $h \circ g$ **(d)** $g^{-1} \circ h^{-1}$

Write an algebraic formula for the value at x of h, g, and each of the preceding functions. Verify that the functions in parts (c) and (d) are inverses.

Exponential and logarithmic functions

8

8-1 ROOTS, RADICALS, AND RATIONAL EXPONENTS

The nth Root Function

Let us consider the inverse of the nth power function f defined for all real numbers by $f(x) = x^n$.

Case 1. If n is *odd* [Fig. 8-1(a)], f is increasing everywhere. Hence f is one to one and has an inverse g. In this case, the range of f is all real numbers and the following facts about g follow directly from the definition of inverse in Section 7-3.

1. The domain of g is the set of all real numbers.
2. The range of g is the set of all real numbers.
3. $g(x^n) = x$ for every real number x.
4. $[g(x)]^n = x$ for every real number x.

Case 2. If n is *even* [Fig. 8-1(b)], the function defined for all real numbers by $f(x) = x^n$ is not one to one and does not have an inverse. But the restriction of f to the interval $[0, +\infty)$ is increasing on that interval, is one to one, and does have an inverse g. Keeping in mind that the range of the restricted function is $[0, +\infty)$, we see that the following facts about g follow from Section 7-3.

1. The domain of g is the interval $[0, +\infty)$.
2. The range of g is the interval $[0, +\infty)$.
3. $g(x^n) = x$ for $x \geq 0$.
4. $[g(x)]^n = x$ for $x \geq 0$.

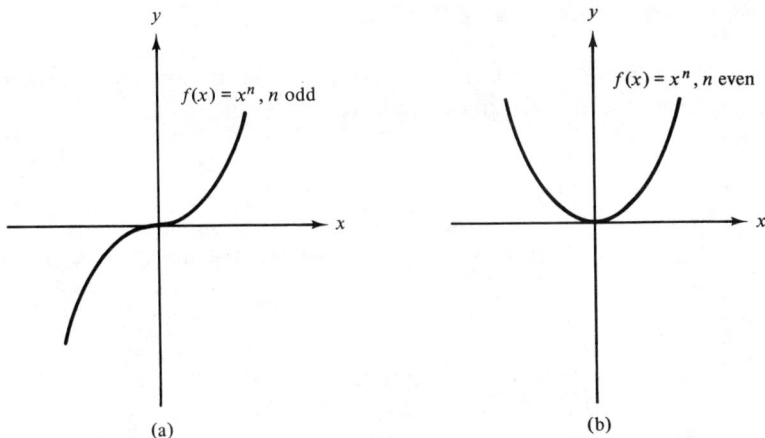

(a)

(b)

Figure 8-1

Looking at (4) in both cases, it is natural to write

$$g(x) = \sqrt[n]{x},$$

for doing so generalizes our notation is Section 1-1 concerning $\sqrt[2]{x}$ and $\sqrt[3]{x}$. We call $g(x)$ the **nth root** of x. It is important to remember that the domain of g is all real numbers when n is odd but only the interval $[0, +\infty)$ when n is even. The symbol $\sqrt[n]{}$ is called a **radical**. In the notation of radicals (3) and (4) become

3. $\sqrt[n]{x^n} = x$
4. $(\sqrt[n]{x})^n = x.$

These equations hold for all real x if n is odd, for $x \geq 0$ if n is even.

The graph of the inverse g can be obtained in each case by reflecting the graph of the original function through line $y = x$, as in Fig. 8-2.

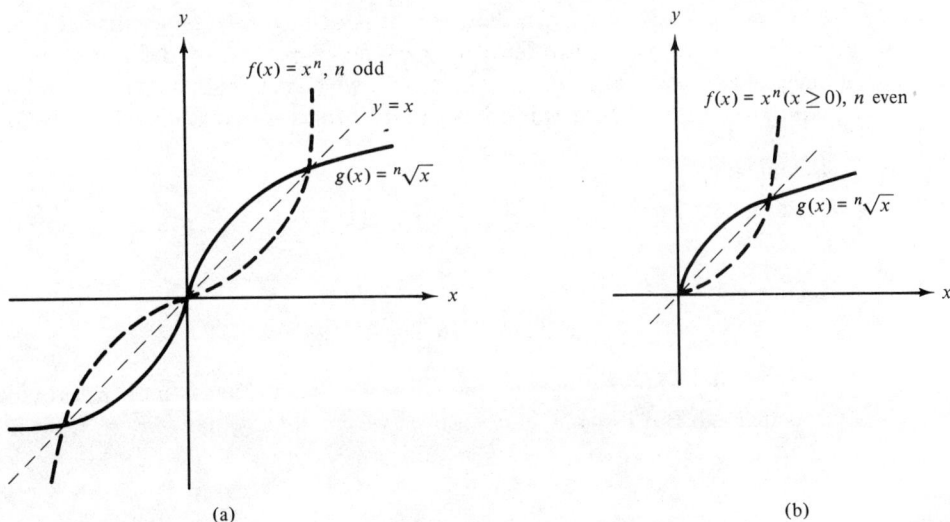

(a)

(b)

Figure 8-2

Algebraic Properties of Radicals

The nth root of a fixed real number a can be interpreted in terms of solutions of equations. Because of Eq. (4) we have:

If n is odd, then $\sqrt[n]{a}$ is the solution of the equation $u^n = a$.

If n is even and $a \geq 0$, then $\sqrt[n]{a}$ is the nonnegative solution of $u^n = a$. If n is even and $a < 0$, this equation has no solutions and $\sqrt[n]{a}$ is not defined.

Example 1

For the cube root we know that $\sqrt[3]{8} = 2$ because $u = 2$ is the solution of $u^3 = 8$. Also, $\sqrt[3]{-8} = -2$, $\sqrt[3]{1/27} = 1/3$.

Example 2

For the fourth root we have $\sqrt[4]{16} = 2$ because $u = 2$ is the nonnegative solution of $u^4 = 16$. Also, $\sqrt[4]{5} \approx 1.495$, whereas $\sqrt[4]{-7}$ is undefined.

The next two theorems are helpful in simplifying nth roots and radicals.

Theorem 8-1. If n is a positive integer and a and b are real numbers for which $\sqrt[n]{a}$ and $\sqrt[n]{b}$ are defined, then $\sqrt[n]{ab}$ is also defined and

$$\sqrt[n]{ab} = \sqrt[n]{a}\, \sqrt[n]{b}.$$

Proof. *Case 1.* If n is odd, then the nth root is defined for all real numbers. We can let $c = \sqrt[n]{a}$ and $d = \sqrt[n]{b}$. By definition, c satisfies the equation $u^n = a$; so $c^n = a$. Similarly, $d^n = b$. It follows that $c^n d^n = ab$. But $c^n d^n = (cd)^n$ [Theorem 1-1(4)] and it follows that cd is the solution of the equation $u^n = ab$. So $\sqrt[n]{ab} = cd = \sqrt[n]{a}\, \sqrt[n]{b}$.

Case 2. If n is even, then the nth root is defined for nonnegative real numbers. For $a \geq 0$ and $b \geq 0$ we can let $c = \sqrt[n]{a}$ and $d = \sqrt[n]{b}$. By definition, c and d are nonnegative. Exactly as in the previous case, we can show that cd is a solution of the equation $u^n = ab$. Because cd is nonnegative, it follows that $\sqrt[n]{ab} = cd = \sqrt[n]{a}\, \sqrt[n]{b}$.

Example 3

$$\sqrt[3]{-\frac{8}{27}} = \sqrt[3]{(-8) \cdot \frac{1}{27}} = \sqrt[3]{-8} \cdot \sqrt[3]{\frac{1}{27}} = (-2) \cdot \frac{1}{3} = -\frac{2}{3}$$

Example 4

$$\sqrt[4]{32} = \sqrt[4]{16 \cdot 2} = \sqrt[4]{16} \cdot \sqrt[4]{2} = 2\sqrt[4]{2}$$

Note. We observed earlier that $(\sqrt[n]{a})^n = a$ and that $\sqrt[n]{a^n} = a$, provided that $\sqrt[n]{a}$ is defined. But if $a < 0$ and n is even we cannot say that $\sqrt[n]{a^n} = a$; for instance, $\sqrt{(-5)^2} = \sqrt{25} = 5$; so $\sqrt{(-5)^2} \neq -5$.

Exponential and Logarithmic Functions Chap. 8

Theorem 8-2. If m and n are integers such that $n > 0$ and a is a nonzero real number for which $\sqrt[n]{a}$ is defined, then $\sqrt[n]{a^m}$ is also defined and

$$\sqrt[n]{a^m} = (\sqrt[n]{a})^m.$$

The proof of Theorem 8-2 is similar to the proof of Theorem 8-1 and is left as an exercise (Problem 80).

Note. Clearly if m is a positive integer, the equation in Theorem 8-2 also holds when $a = 0$.

Example 5

$$\sqrt[3]{2^{18}} = (\sqrt[3]{2})^{18} = [(\sqrt[3]{2})^3]^6 = 2^6 = 64$$

Rationalizing a Denominator

Often it is convenient to remove square roots from a denominator by a process called **rationalizing the denominator**. This process is illustrated in the next examples.

Example 6

Suppose that we wish to evaluate $6/\sqrt{3}$. We could divide a decimal approximation for $\sqrt{3}$ into 6:

$$1.732\,\overline{)\,6.000.}$$

It is easier, however, to multiply both numerator and denominator by $\sqrt{3}$:

$$\frac{6}{\sqrt{3}} = \frac{6\sqrt{3}}{\sqrt{3}\,\sqrt{3}} = \frac{6}{3}\sqrt{3} = 2\sqrt{3}.$$

The answer is obtained by multiplying 1.732 by 2:

$$\frac{6}{\sqrt{3}} = 2\sqrt{3} \approx 2(1.732) = 3.464.$$

When the denominator of a fraction is of the form $a + b\sqrt{r}$, we can rationalize it by multiplying numerator and denominator by $a - b\sqrt{r}$ (called the **conjugate** of $a + b\sqrt{r}$). The new denominator is $(a + b\sqrt{r})(a - b\sqrt{r}) = a^2 - b^2 r$.

Example 7

To evaluate $2/(3 + 2\sqrt{2})$, we multiply numerator and denominator by $3 - 2\sqrt{2}$, obtaining

$$\frac{2}{3 + 2\sqrt{2}} = \frac{2}{3 + 2\sqrt{2}} \cdot \frac{3 - 2\sqrt{2}}{3 - 2\sqrt{2}} = \frac{6 - 4\sqrt{2}}{9 - 8}$$

$$\approx 6 - 4 \cdot (1.414)$$

$$= 6 - 5.656$$

$$= .344.$$

When the denominator of a fraction is of the form $a\sqrt{r} + b\sqrt{s}$, we can rationalize it by multiplying by $a\sqrt{r} - b\sqrt{s}$. Then the new denominator is $a^2r - b^2s$.

Example 8

To evaluate $5/(2\sqrt{3} - \sqrt{2})$, we multiply numerator and denominator by $2\sqrt{3} + \sqrt{2}$, obtaining

$$\frac{5}{2\sqrt{3} - \sqrt{2}} = \frac{5}{2\sqrt{3} - \sqrt{2}} \cdot \frac{2\sqrt{3} + \sqrt{2}}{2\sqrt{3} + \sqrt{2}} = \frac{10\sqrt{3} + 5\sqrt{2}}{4 \cdot 3 - 2}$$

$$= \frac{10\sqrt{3} + 5\sqrt{2}}{10}$$

$$= \sqrt{3} + \frac{1}{2}\sqrt{2}$$

$$\approx 1.732 + .707$$

$$= 2.439.$$

Example 9

Simplify the expression $\dfrac{1}{2 - \sqrt{x}}$.

Solution. By rationalizing the denominator, we get

$$\frac{1}{2 - \sqrt{x}} \cdot \frac{2 + \sqrt{x}}{2 + \sqrt{x}} = \frac{2 + \sqrt{x}}{4 - x}.$$

This process is valid for $x \geq 0$, $x \neq 4$.

Rational Exponents

Let us return to the discussion of exponents begun in Section 1-2, where we discussed the use of positive and negative integral exponents. Here our purpose is to define a^r, where r is a rational number other than an integer. We can assume that $r = p/q$, where p and q are integers without common factors such that $q > 1$. We want our definition of $a^r = a^{p/q}$ to be consistent with the laws of exponents established for integers in Theorem 1-1. For convenience, we restate these laws, in which a and b are nonzero.

1. $a^{m+n} = a^m a^n$

2. $a^{m-n} = \dfrac{a^m}{a^n}$

3. $(a^m)^n = a^{mn}$

4. $(ab)^n = a^n b^n$

5. $\left(\dfrac{a}{b}\right)^n = \dfrac{a^n}{b^n}$

Suppose Eq. (3) were to hold when m and n are permitted to have rational values; we could then take $m = 1/q$ and $n = q$, obtaining

$$(a^{1/q})^q = a^{(1/q)q} = a^1 = a.$$

This means that $a^{1/q}$ would be a qth root of a and suggests that we define $a^{1/q} = \sqrt[q]{a}$. If q is odd, this is meaningful for all a; if q is even, it is meaningful for $a > 0$. Returning to Eq. (3), we now let $m = 1/q$ and $n = p$:

$$(a^{1/q})^p = a^{p/q}.$$

Accordingly, we define $a^{p/q}$ to be $(\sqrt[q]{a})^p$.

Definition.   Let r be a nonintegral rational number expressed in the form $r = p/q$, where p and q are integers without common factors and $q > 1$. We define

$$a^r = a^{p/q} = (\sqrt[q]{a})^p$$

for all values of a for which the right-hand member is defined.

Note 1.   Under the terms of this definition $a^{p/q}$ is undefined when

1. $a < 0$ and q is even (for then $\sqrt[q]{a}$ is not defined).
2. $a = 0$ and p is negative (for then $(\sqrt[q]{a})^p$ involves division by 0).

Note 2.   In determining a to a rational exponent, that exponent is expressed as a *reduced* fraction with a positive denominator. If, for example, we should encounter $a^{2/6}$, we will interpret it as $a^{1/3}$.

Example 10

Here are some examples of the use of fractional exponents.

(a) $2^{1/2} = \sqrt{2} \approx 1.414$

(b) $2^{3/2} = (\sqrt{2})^3 = 2\sqrt{2} \approx 2.828$

(c) $2^{-3/2} = (\sqrt{2})^{-3} = \dfrac{1}{(\sqrt{2})^3} = \dfrac{1}{2\sqrt{2}} \approx .354$

(d) $8^{2/3} = (\sqrt[3]{8})^2 = 2^2 = 4$

(e) $(-27)^{1/3} = \sqrt[3]{-27} = -3$

(f) $(-27)^{-4/3} = (\sqrt[3]{-27})^{-4} = \dfrac{1}{(-3)^4} = \dfrac{1}{81}$

(g) $(-27)^{1/4}$ is undefined

With the definition just adopted for rational exponents, it can be proved that if $a > 0$ and $b > 0$, then the laws of exponents [Eqs. (1) through (5) in Theorem 1-1] hold for *all rational* numbers m and n. The equation $a^{-n} = 1/a^n$ also holds for all rational numbers n, because it is a special case of Eq. (2).

Example 11

Write $\sqrt{x} + (1/\sqrt{x})$ in a form that has fractional and negative exponents.

Solution.   For $x > 0$ we have

$$\sqrt{x} + \frac{1}{\sqrt{x}} = x^{1/2} + \frac{1}{x^{1/2}} = x^{1/2} + x^{-1/2}.$$

Example 12

Write $\sqrt{xy^3}$ in a form using fractional exponents.

Solution. If $xy^3 \geq 0$, we certainly have

$$\sqrt{xy^3} = (xy^3)^{1/2}.$$

If x and y are both nonnegative, we can go further:

$$(xy^3)^{1/2} = x^{1/2}(y^3)^{1/2} = x^{1/2}y^{3/2}.$$

Example 13

Write $(x + 1)^{1/2}(x^2 + 1)^{-1/2}$ in a form without fractional or negative exponents.

Solution. For $x \geq -1$

$$(x + 1)^{1/2}(x^2 + 1)^{-1/2} = \frac{(x + 1)^{1/2}}{(x^2 + 1)^{1/2}} = \frac{\sqrt{x + 1}}{\sqrt{x^2 + 1}} = \sqrt{\frac{x + 1}{x^2 + 1}}.$$

Example 14

Find the zeros of the function

$$f(x) = x(x - 1)^{1/2}(x - 2)^{-1/3}.$$

Solution. This function can be expressed as a quotient:

$$f(x) = \frac{x\sqrt{x - 1}}{\sqrt[3]{x - 2}},$$

defined for real x such that $x \geq 1$ and $x \neq 2$. There is only one number in the domain of f at which the numerator is zero—namely, $x = 1$. Thus $x = 1$ is the only zero of f.

Example 15

Properties of rational exponents are often used in conjunction with the distributive law to simplify expressions. For instance,

$$x^{5/2} - 4x^{3/2} + 3x^{1/2} = x^{1/2}[x^{5/2-(1/2)} - 4x^{3/2-(1/2)} + 3x^{1/2-(1/2)}]$$

$$= x^{1/2}(x^2 - 4x + 3x^0)$$

$$= \sqrt{x}\,(x^2 - 4x + 3)$$

By factoring out $x^{1/2}$ in this way, we can see that the zeros of the original expression are 0, 1, and 3.

Example 16

Show that

$$x^{1/2}(2 + x)^2 + 2x^{3/2}(2 + x) = \sqrt{x}\,(2 + x)(2 + 3x).$$

Solution. Both terms on the left contain powers of x and $2 + x$. We factor out the smallest powers of these factors that occur in the two parts of the expression. The smallest power of x is $x^{1/2}$ and the smallest power of $2 + x$ is $(2 + x)^1$. The result is

$$x^{1/2}(2+x)[x^{1/2-(1/2)}(2+x)^{2-1} + 2x^{3/2-(1/2)}(2+x)^{1-1}]$$
$$= x^{1/2}(2+x)[x^0(2+x)^1 + 2x^1(2+x)^0]$$
$$= \sqrt{x}\,(2+x)[(2+x)+2x] = \sqrt{x}\,(2+x)(2+3x).$$

This calculation is valid for $x \geq 0$.

Example 17

Show that

$$x^{-1/2}(3x+1)^3 + x^{1/2}(3x+1)^2 = \frac{(3x+1)^2(4x+1)}{\sqrt{x}}.$$

Solution. Again, we factor out the smallest powers of x and $3x+1$ occurring in the two terms. The smallest power of x is $x^{-1/2}$ and the smallest power of $3x+1$ is $(3x+1)^2$. We obtain

$$x^{-1/2}(3x+1)^2[x^{-1/2-(-1/2)}(3x+1)^{3-2} + x^{1/2-(-1/2)}(3x+1)^{2-2}]$$
$$= x^{-1/2}(3x+1)^2[x^0(3x+1)^1 + x^1(3x+1)^0]$$
$$= \frac{1}{\sqrt{x}}(3x+1)^2[(3x+1)+x] = \frac{(3x+1)^2(4x+1)}{\sqrt{x}}.$$

Example 18

Show that

$$\frac{2}{3}(x+4)^{1/3}(x-2)^{-1/3} + \frac{1}{3}(x+4)^{-2/3}(x-2)^{2/3} = \frac{x+2}{(\sqrt[3]{x+4})^2\,\sqrt[3]{x-2}}.$$

Solution. We factor out $(1/3)(x+4)^{-2/3}(x-2)^{-1/3}$, obtaining

$$\frac{1}{3}(x+4)^{-2/3}(x-2)^{-1/3}[2(x+4)^{1/3-(-2/3)} + (x-2)^{2/3-(-1/3)}]$$

$$= \frac{1}{3}(x+4)^{-2/3}(x-2)^{-1/3}[2(x+4)^1 + (x-2)^1]$$

$$= \frac{3x+6}{3(x+4)^{2/3}(x-2)^{1/3}}$$

$$= \frac{x+2}{(\sqrt[3]{x+4})^2\,\sqrt[3]{x-2}}.$$

EXERCISES 8-1

In Problems 1–17 rationalize the denominator.

1. $\dfrac{5}{\sqrt{2}}$

2. $\dfrac{7}{\sqrt{5}}$

3. $1 - \dfrac{1}{\sqrt{5}}$

4. $3 + \dfrac{2}{\sqrt{3}}$

5. $\dfrac{1+\sqrt{2}}{3-2\sqrt{2}}$

6. $\dfrac{2-\sqrt{3}}{4+\sqrt{3}}$

7. $\dfrac{\sqrt{5}}{3 + 2\sqrt{5}}$ **8.** $\dfrac{\sqrt{3}}{2\sqrt{3} - 3}$ **9.** $\dfrac{\sqrt{5}}{2\sqrt{5} + 2}$

10. $\dfrac{1 + \sqrt{x}}{1 - \sqrt{x}}$ **11.** $\dfrac{2 - \sqrt{x}}{3 + \sqrt{x}}$ **12.** $\dfrac{2}{4 + 2\sqrt{x}}$

13. $\dfrac{1}{\sqrt{x} - \sqrt{y}}$ **14.** $\dfrac{x - y}{\sqrt{x} - \sqrt{y}}$ **15.** $\dfrac{x}{\sqrt{x + 4} - 2}$

16. $\dfrac{x^2}{\sqrt{x^2 + 1} - 1}$ **17.** $\dfrac{x^2 - y^2}{\sqrt{x} + \sqrt{y}}$

In Problems 18–31 simplify, leaving your answer in terms of $\sqrt{2}$.

18. $2^{1/2} + 2^{3/2}$ **19.** $2^{-1/2} - 2^{-3/2}$

20. $2^{1/2} - 2^{3/2} + 2^{5/2}$ **21.** $2^{-1/2} + 2^{-3/2} + 2^{-5/2}$

22. $2^{1/2}(2^{1/2} + 2^{-1/2})$ **23.** $(2^{1/2} + 2^{-1/2})/2^{1/2}$

24. $(2^{3/2} - 2^{-1/2})/2^{1/2}$ **25.** $(2^{1/2} + 2^{-1/2})^2$

26. $\left(\dfrac{2^{1/2} - 2^{-1/2}}{2^{1/2} + 2^{-1/2}}\right)^2$ **27.** $(2^{3/2} - 2^{1/2})^2$

28. $(3 + 2^{3/2})^2$ **29.** $(1 + 3\cdot 2^{1/2})^2$

30. $(3 + 2^{3/2})^{-1}$ **31.** $(3 + 2^{3/2})^{-2}$

In Problems 32–41 write the given expression in an equivalent form, using fractional and negative exponents instead of radicals and quotients.

32. $x\sqrt{x}$ **33.** $\sqrt{\dfrac{a^2 + 1}{b^2 + 1}}$

34. $4\sqrt[3]{x + y} - 5\sqrt[3]{3x - y}$ **35.** $\dfrac{8ab}{\sqrt[5]{(a + 4b)^3}}$

36. $\sqrt[3]{xy^2z^5}$ **37.** $(3x - 2)\sqrt{3x - 2}$

38. $\sqrt{\dfrac{1 - (1/a^2)}{3 + b^2}}$ **39.** $\sqrt[2]{x^2 + y^2} - \sqrt[3]{x^3 + 8y^3}$

40. $\sqrt[5]{\dfrac{(3x - 2)^2}{(5x^2 + 4y^2)^3}}$ **41.** $xy^2z^5\sqrt[3]{xy^2z^5}$

In Problems 42–51 eliminate fractional and negative exponents, writing the given expression in terms of radicals and/or quotients.

42. $(2x^2 + 3)^{1/2}$ **43.** $(1 + a^{-2})^{3/2}$

44. $(b^2 + 4)^{-5/2}$ **45.** $(4 - x)^{-2/3}$

46. $(1 + 3x^2)^{-1/2}$ **47.** $6x(1 + 3x^2)^{-1/2}$

48. $2u(u^2 + 4)^{3/2}$ **49.** $\dfrac{1}{2}x^2(1 + x)^{-1/2} + 2x(1 + x)^{1/2}$

50. $x(3x + 1)^{-1/2} + (3x + 1)^{1/2}$ **51.** $4t^2(t^2 + 1)^{-1/3} + 6(t^2 + 1)^{2/3}$

In Problems 52–61 find any zeros of the given expression:

52. $(x - 3)(x - 2)^{1/2}$ **53.** $(x - 2)(x - 3)^{1/2}$

54. $(x - 3)(x - 2)^{-1/2}$ **55.** $x(x + 1)^{-1/2}$

56. $(x - 3)(x - 2)^{1/3}$ **57.** $(x - 3)(x - 2)^{-1/3}$

58. $(u^2 - 1)(u^2 - 4)^{1/2}$ **59.** $(t^2 - 9)^{1/3}(t^2 - 4)^{-2/3}$

60. $(x^2 + x - 1)^{2/3}(x^2 + 3x + 2)^{-2/3}$ **61.** $(x^2 + x + 1)^{1/3}(x^2 - 8)^{-2/3}$

In Problems 62–68 show that the given expressions are equal.

62. $x^{4/3}(2x - 1) + x^{1/3}(2x - 1)^2 = \sqrt[3]{x}\,(2x - 1)(3x - 1)$

63. $(x + 1)^{1/2}(x + 3)^3 - 3(x + 1)^{3/2}(x + 3)^2 = -2x\sqrt{x + 1}(x + 3)^2$

64. $(2x + 1)^{1/2}(x + 3) + 5(2x + 1)^{3/2} = \sqrt{2x + 1}(11x + 8)$

65. $y^{1/2}(y^2 + 4)^2 + 3y^{3/2}(y^2 + 4) = \sqrt{y}\,(y^2 + 4)(y^2 + 3y + 4)$

66. $x^{4/3}(2x - 1) + x^{1/3}(2x - 1)^2 = \sqrt[3]{x}\,(2x - 1)(3x - 1)$

67. $-\dfrac{2}{3}x^{1/3}(1 - x)^{-1/3} + \dfrac{1}{3}x^{-2/3}(1 - x)^{2/3} = \dfrac{1 - 3x}{3\sqrt[3]{x^2}\sqrt[3]{1 - x}}$

68. $x^{3/5}(1 - x)^{-1/2} - x^{-2/5}(1 - x)^{1/2} = \dfrac{2x - 1}{\sqrt[5]{x^2}\sqrt{1 - x}}$

In Problems 69–74 solve for the rational number r.

69. $2^r = 8\sqrt{2}$ **70.** $3^r = \dfrac{1}{3\sqrt{3}}$ **71.** $3^r = 1$

72. $4^r = 8\sqrt{2}$ **73.** $6^r = 2^r \cdot 3\sqrt[3]{3}$ **74.** $8^r = \dfrac{1}{16\sqrt{2}}$

FOR THE ENTHUSIASTS

In Problems 75–77 solve for the rational number r.

75. $3^{3r} - 3^{2r} - 3^{r+1} + 3 = 0$ *Hint:* Let $u = 3^r$ and solve a cubic.

76. $2^{4r} - 10 \cdot 2^{2r} + 16 = 0$ **77.** $2^{3r+1} - 2^{2r+1} - 2^r + 1 = 0$

†78. Kepler's Third Law states that the square of the time required for a planet to orbit the sun is proportional to the cube of the distance of the planet from the sun. If the distance of Mars from the sun is 1.41×10^8 miles whereas that of the earth from the sun is 9.29×10^7 miles, find the number of days required for Mars to complete one orbit.

†79. The distance of the planet Jupiter from the sun is about 4.832×10^8 miles. How many years does it take for Jupiter to complete one orbit? (See Problem 78.)

80. Prove Theorem 8-2.

8-2 EXPONENTIAL FUNCTIONS

We saw in Section 8-1 how a^x is defined in terms of powers and roots when the exponent x is a rational number and the base a is positive. Because an irrational number cannot be expressed exactly as a quotient of integers, the definition given in Section 8-1 cannot be applied when x is irrational.

The Base 3

For the moment, let us take the base $a = 3$ and suppose that $x = \sqrt{2} = 1.41421\ldots$. We want to define $3^{\sqrt{2}}$. The irrational number $\sqrt{2}$ is approximated more and more closely by the successive terms of the sequence

$$1,\ 1.4,\ 1.41,\ 1.414,\ .1.4142,\ 1.41421,\ \ldots \longrightarrow \sqrt{2}. \qquad (8\text{-}1)$$

These are rational numbers. The terms of the corresponding sequence

$$3^1,\ 3^{1.4},\ 3^{1.41},\ 3^{1.414},\ 3^{1.4142},\ 3^{1.41421},\ \ldots \qquad (8\text{-}2)$$

all involve rational powers of 3. For instance, $3^{1.4} = 3^{14/10} = 3^{7/5} = (\sqrt[5]{3})^7$; and using a calculator, we find $3^{1.4} \approx 4.65554$.

Approximate values for some other terms in the sequence (8-2) are shown in tabular form.

r	3^r
1.	3.00000
1.4	4.65554
1.41	4.70696
1.414	4.72770
1.4142	4.72873
1.41421	4.72879

It can be shown that the values of the terms in (8-2) approach a definite real number. We define $3^{\sqrt{2}}$ to be this number. Its value to four decimal places is 4.7288.

More generally, if x is any irrational number, then x can be approximated by a sequence of rational numbers:

$$r_1, r_2, r_3, \ldots \longrightarrow x.$$

It can then be proved that the corresponding sequence

$$3^{r_1}, 3^{r_2}, 3^{r_3}, \ldots$$

approaches a definite real number. The value of this number is taken as the definition of 3^x. Thus we have

$$3^{r_1}, 3^{r_2}, 3^{r_3}, \ldots \longrightarrow 3^x.$$

With 3^x defined for every real number x, we can consider the function whose value for each real number x is 3^x. It is called the **exponential function with base 3** and its graph appears in Fig. 8-3. It can be shown that this function is increasing and continuous

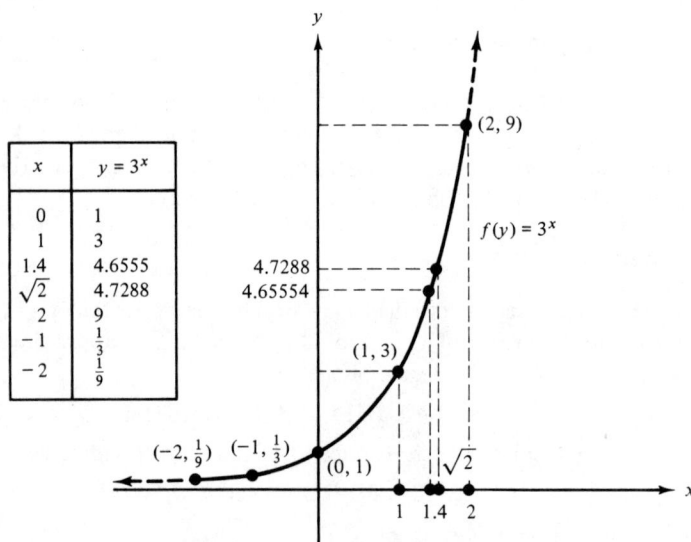

x	$y = 3^x$
0	1
1	3
1.4	4.6555
$\sqrt{2}$	4.7288
2	9
-1	$\frac{1}{3}$
-2	$\frac{1}{9}$

Figure 8-3

Exponential and Logarithmic Functions   Chap. 8

on the interval of all real numbers. As $x \rightarrow +\infty$, we find that $3^x \rightarrow +\infty$ also. You will see how fast 3^x increases if you calculate $3^3, 3^5, 3^{10}, \ldots$. As $x \rightarrow -\infty$, the values of 3^x become very small, approaching zero. You can convince yourself of this fact by calculating $3^{-3}, 3^{-5}, 3^{-10}, \ldots$. Notice that the values of 3^x are always positive and so the graph approaches the x axis as a horizontal asymptote from above.

The Base a

Let us consider the possibility of using other numbers as a base. If $a > 0$, we can define a^x in much the same way as we defined 3^x. If x is the real number approached by a sequence of rational numbers

$$r_1, r_2, r_3, \ldots \longrightarrow x,$$

then a^x is the number approached by the corresponding sequence

$$a^{r_1}, a^{r_2}, a^{r_3}, \ldots \longrightarrow a^x.$$

The function f defined for all real numbers by $f(x) = a^x$ is called the **exponential function with base a**.

Some important properties of the exponential functions are contained in the following theorem, which we state without proof.

Theorem 8-3. If a is any positive real number other than one, then the exponential function with base a

$$f(x) = a^x,$$

is defined and continuous for all real numbers. Its range is the set of positive numbers. If $a > 1$, it is increasing whereas if $0 < a < 1$, it is decreasing. In either case, f is one to one.

Note 1. The exponential function with base $a = 1$ is of little interest, for it is just the constant 1; clearly this function is not one to one.

Note 2. It is not possible to use a base a that is either zero or negative. If $a = 0$, we cannot define a^x for $x < 0$. If $a < 0$, a^x is undefined for certain rational values of x, such as 1/2 and 1/4.

Exponential Graphs

Example 1

Graph the exponential function with base $a = 2$ and compare this graph with the graph of $y = 3^x$.

Solution. The graph of $y = 2^x$ appears with that of $y = 3^x$ in Fig. 8-4. For $x > 0$ we see that $2^x < 3^x$, but for $x < 0$ we have $2^x > 3^x$. The two graphs intersect at $(0, 1)$.

Example 2

Graph the function $g(x) = 2 + 3^x$.

x	2^x	3^x
0	1	1
1	2	3
2	4	9
−1	$\frac{1}{2}$	$\frac{1}{3}$

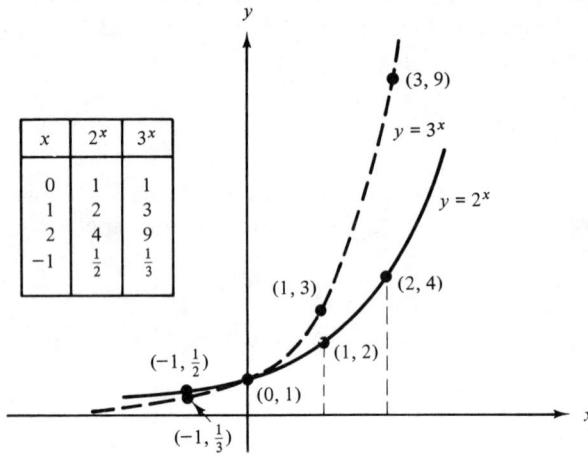

Figure 8-4

Solution. This graph is a vertical translation of the graph of the exponential function with base 3. The line $y = 2$ is a horizontal asymptote, and the graph of $g(x)$ is always above this line (see Fig. 8-5).

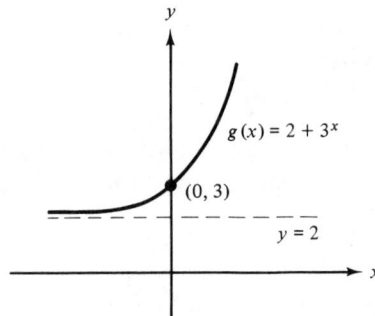

Figure 8-5

Example 3

Draw the graph of the exponential function with base $a = \frac{1}{3}$ and compare it with that of $y = 3^x$.

Solution. Since $\left(\frac{1}{3}\right)^x = \frac{1}{3^x} = 3^{-x}$ the graph of $y = \left(\frac{1}{3}\right)^x$ is the same as the graph of $y = 3^{-x}$. For each point (x, y) on this graph the point $(-x, y)$ is on the graph of $y = 3^x$. Therefore we can obtain the graph of $y = 3^{-x}$ by reflecting the graph of $y = 3^x$ through the y axis, as in Fig. 8-6.

Example 4

Find the intervals on which the function $F(x) = (x^2 - 1)3^x$ is positive.

Solution. Since 3^x is always positive, $F(x) > 0$ if and only if $x^2 - 1$ is positive. So $F(x) > 0$ on the intervals $(-\infty, -1)$, $(1, +\infty)$.

Exponential and Logarithmic Functions   Chap. 8

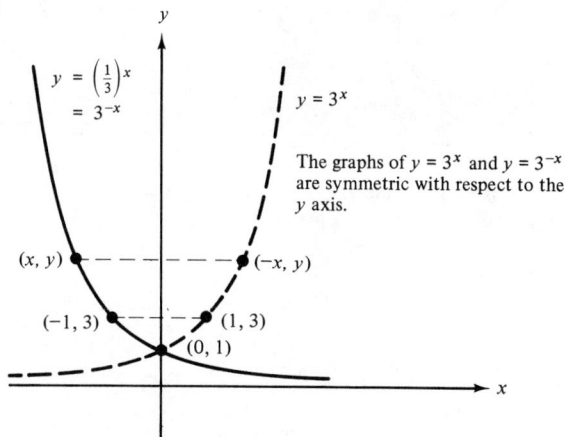

$y = \left(\dfrac{1}{3}\right)^x$
$= 3^{-x}$

$y = 3^x$

The graphs of $y = 3^x$ and $y = 3^{-x}$ are symmetric with respect to the y axis.

(x, y) $(-x, y)$

$(-1, 3)$ $(1, 3)$

$(0, 1)$

Figure 8-6

Example 5

Find the zeros of $G(x) = (3^x - 1)(x + 1)$.

Solution. Since $G(x) = 0$ if and only if $3^x - 1 = 0$ or $x + 1 = 0$, it suffices to solve $3^x = 1$ and $x + 1 = 0$. Now $x = 0$ is a solution of $3^x = 1$ and furthermore it is the *only* solution because the function $f(x) = 3^x$ is one to one. The solution of $x + 1 = 0$ is $x = -1$. So the function G has only the two zeros, $x = 0, -1$.

Algebra of Real Exponents

The algebraic properties of exponents discussed in Sections 1-2 and 8-1 can be shown to hold for all real exponents.

> **Theorem 8-4.** If a and b are positive real numbers and x and y are any real numbers, then
> 1. $a^{x+y} = a^x a^y$
> 2. $a^{x-y} = \dfrac{a^x}{a^y}$
> 3. $a^{xy} = (a^x)^y$
> 4. $(ab)^x = a^x b^x$
> 5. $\left(\dfrac{a}{b}\right)^x = \dfrac{a^x}{b^x}$
> 6. $a^0 = 1$
> 7. $a^{-x} = \dfrac{1}{a^x} = \left(\dfrac{1}{a}\right)^x$

Example 6

Find all zeros of the function $g(x) = 3^x - 3^{-x}$.

Solution. Set $3^x - 3^{-x} = 0$, multiply by 3^x, and use (1) and (6) of Theorem 8-4 to obtain

$$3^x 3^x - 3^x 3^{-x} = 0$$
$$3^{2x} - 3^0 = 0$$
$$3^{2x} - 1 = 0$$
$$3^{2x} = 1$$

Clearly $x = 0$ is a solution and because the function $F(x) = 3^{2x} = 9^x$ is one to one, this is the *only* zero of g.

Example 7

Solve $3^x + 1 = 0$.

Solution. This equation has no solutions because 3^x is positive for all x by Theorem 8-3.

Example 8

Solve $3^{3x} - 3 \cdot 3^{2x} - 3 \cdot 3^x + 9 = 0$.

Solution. Substitute $u = 3^x$. Then $3^{3x} = u^3$ and $3^{2x} = u^2$ so that the equation becomes
$$u^3 - 3u^2 - 3u + 9 = 0.$$
The left side factors:
$$(u - 3)(u - \sqrt{3})(u + \sqrt{3}) = 0.$$
Solutions are $u = 3, \sqrt{3}, -\sqrt{3}$. But $u = 3^x > 0$ and so we reject $u = -\sqrt{3}$. The other two solutions give $u = 3^x = 3$ or $x = 1$; and $u = 3^x = \sqrt{3}$ or $x = 1/2$.

Inequalities

If $a > 1$, the function $f(x) = a^x$ is increasing everywhere so that $x_1 < x_2$ implies $f(x_1) < f(x_2)$. Conversely, if $f(x_1) < f(x_2)$, then $x_1 < x_2$, for otherwise $x_1 \geq x_2$, which would imply $f(x_1) \geq f(x_2)$. Thus when $a > 1$, the inequality $a^{x_1} < a^{x_2}$ is equivalent to $x_1 < x_2$.

Example 9

Solve $2^x > 8$.

Solution. Because $8 = 2^3$, the given inequality is equivalent to $2^x > 2^3$, which is, in turn, equivalent to $x > 3$.

Example 10

Find the intervals on which $g(x) = 5^{x^2} - 25$ is positive.

Solution. The inequality $g(x) > 0$ is equivalent to $5^{x^2} > 5^2$, which is, in turn, equivalent to $x^2 > 2$. The function g is positive on the intervals $(-\infty, -\sqrt{2})$, $(\sqrt{2}, +\infty)$.

EXERCISES 8-2

†1. Graph the exponential function with base 5, $f(x) = 5^x$. Label on your graph the points corresponding to $x = 0, 1, 2, 3, 1/2, 3/2, -1/2, -1, -2$. Use a calculator to obtain the values of $5^1, 5^{1.4}, 5^{1.414}, 5^{1.4142}$, and use the last of these values as an approximation for the y coordinate of point $(\sqrt{2}, 5^{\sqrt{2}})$.

†2. Graph the exponential function with base 1/5, $g(x) = 1/5^x$. Label on your graph the points corresponding to $x = 0, 1, 2, -1, -2, -3, -1/2, 1/2, 3/2$. How is the graph of g related geometrically to the graph of f in Problem 1?

In Problems 3–10 find all zeros; if none exists, explain.

3. $f(x) = 5^x - 25$
4. $g(x) = 2^x + 4$
5. $f(t) = 3^t - 1$
6. $g(x) = 2^x - 2^{-x}$
7. $f(y) = 2^y(y + 3)$
8. $g(x) = 3^{-x}(x^2 - 18)$
9. $h(x) = 2^x + 8 \cdot 2^{-x} - 6$
10. $g(x) = 2^x - 8 \cdot 2^{-x} - 2$

In Problems 11–29 find the intervals on which the function is positive.

11. $f(t) = 3^t$
12. $f(x) = x2^x$
13. $g(x) = \left(\frac{1}{2}\right)^x + \left(\frac{1}{2}\right)^{-x}$
14. $g(x) = 3^{-x^2}$
15. $f(y) = 2^y(y + 3)$
16. $g(x) = \left(\frac{2}{3}\right)^{-x}(x^2 - 18)$
17. $f(x) = 2^x - 4$
18. $f(x) = 2^x + 4$
19. $g(t) = 3^t - 1$
20. $h(t) = \left(\frac{1}{3}\right)^t - 9$
21. $f(x) = 2^{x^2} - 4$
22. $g(x) = 3^x - 3^{-x}$
23. $h(u) = 3^{-u^2} - \frac{1}{9}$
24. $f(x) = x(2^x - 4)$
25. $f(x) = (x - 1)(3^{2x} - 27)$
26. $g(x) = (x^2 - 1)(2^x - 4)$
27. $G(x) = (x^2 - 1)(3^{x^2} - 81)$
28. $g(x) = 10^x(x^2 + 1)$
29. $F(x) = 10^x(x^2 - 5x + 6)$

In Problems 30–37 solve the inequality.

30. $3^x > 27$
31. $2^x \le 16$
32. $3^{-x} > \frac{1}{27}$
33. $9 - 3^x > 6$
34. $3^{x^2} > 27$
35. $2^{x^2} \le 16$
36. $3^{-x^2} > \frac{1}{27}$
37. $3^{-x^2} > 27$

38. Use a translation of axes to graph.
(a) $f(x) = 3^{x-2} + 4$
(b) $g(x) = \frac{1}{3^{x+1}} + 2$
(c) $h(x) = 3^{x+2} - 4$

FOR THE ENTHUSIASTS

39. Graph.
(a) $f(x) = 3^{x^2}$
(b) $g(x) = 3^{-x^2}$
(c) $h(x) = x3^x$
(d) $f(x) = x3^{-x}$

40. (a) Show that the function

$$g(x) = \frac{3^x + 3^{-x}}{2}$$

is even and draw its graph.

(b) Show that the function

$$h(x) = \frac{3^x - 3^{-x}}{2}$$

is odd and draw its graph.

41. Show that if $0 < a < b$, then $a^x < b^x$ for all $x > 0$ but $a^x > b^x$ for all $x < 0$.

42. Show that if $0 < a < 1$, the inequality $a^{x_1} < a^{x_2}$ is equivalent to $x_1 > x_2$.

8-3 LOGARITHMIC FUNCTIONS

Consider the exponential function with base a where $a > 0$ and $a \neq 1$. The domain of f is the set of real numbers; its range is the set of positive real numbers. If $a > 1$, f is increasing everywhere, whereas if $0 < a < 1$, f is decreasing everywhere. In either case, f is one to one and therefore has an inverse g, called the **logarithmic function with base** a and denoted by $\log_a$. The value $g(x)$ is often written $\log_a x$ rather than $\log_a(x)$.

Definition.   Let a be a positive number different from one. The **logarithmic function with base a**

$$g(x) = \log_a x$$

is the inverse of the exponential function with base a.

From the definition of the inverse of a function in Section 7-3 we immediately have the following properties of $g(x) = \log_a x$.

---

1. The domain of the logarithmic function with base a is the set of positive real numbers.
2. The range of the logarithmic function with base a is the set of all real numbers.
3. $\log_a (a^x) = x$ for every real number x.
4. $a^{\log_a x} = x$ for every $x > 0$.

Equation (4) says that $\log_a x$ is the exponent we raise a to in order to get x.

Because f and g are inverses, the equations $y = g(x)$ and $x = f(y)$ are equivalent. Therefore

$$\log_a x = y \text{ if and only if } a^y = x.$$

Example 1

Consider the exponential and logarithmic functions with base 3. In Table 8-1 we list some equations in exponential form and beside each one its equivalent in logarithmic form.

TABLE 8-1

Exponential form	Logarithmic form
$3^1 = 3$	$\log_3 3 = 1$
$3^2 = 9$	$\log_3 9 = 2$
$3^3 = 27$	$\log_3 27 = 3$
$3^{1/2} = \sqrt{3}$	$\log_3 \sqrt{3} = 1/2$
$3^{1/3} = \sqrt[3]{3}$	$\log_3 \sqrt[3]{3} = 1/3$
$3^{-2} = 1/9$	$\log_3 (1/9) = -2$
$3^{7/3} = 9\sqrt[3]{3}$	$\log_3 (9\sqrt[3]{3}) = 7/3$
$3^0 = 1$	$\log_3 1 = 0$
$3^v = u$	$\log_3 u = v$

Example 2

Table 8-2 shows some additional pairs of equivalent statements, using various bases.

TABLE 8-2

Exponential form	Logarithmic form
$2^3 = 8$	$\log_2 8 = 3$
$5^{1/2} = \sqrt{5}$	$\log_5 \sqrt{5} = 1/2$
$10^{-5} = .00001$	$\log_{10} (.00001) = -5$
$4^0 = 1$	$\log_4 1 = 0$
$a^v = u$	$\log_a u = v$

Example 3

Solve for x.

$$3^{2x} = 27$$

Solution. Converting to logarithmic form:

$$2x = \log_3 27.$$

The right-hand side is 3. So

$$2x = 3$$

$$x = \frac{3}{2}.$$

Example 4

Solve for x.

$$3^{-x^2} = \frac{1}{9}$$

Solution. Converting to logarithmic form:

$$-x^2 = \log_3\left(\frac{1}{9}\right)$$
$$-x^2 = -2$$
$$x = \pm\sqrt{2}$$

Example 5

Solve for x.

$$\log_3 x = 4$$

Solution. Converting to exponential form:

$$x = 3^4 = 81$$

Example 6

Solve for a.

$$\log_a 16 = 2$$

Solution. Converting to exponential form:

$$a^2 = 16$$

and because the base must be positive, we have

$$a = 4.$$

Graphs of Logarithmic Functions

The graph of $g(x) = \log_a x$ can be obtained from the graph of $f(x) = a^x$ by reflection through the line $y = x$, as in Fig. 8-7. Notice that if $a > 1$, both f and its inverse are increasing; if $0 < a < 1$, both are decreasing.

For most practical purposes, we need only work with logarithms to bases greater than 1 and so we will concentrate in examples and exercises on the case $a > 1$, shown in Fig. 8-7(a). It can be shown for this case that $\log_a x \longrightarrow +\infty$ as $x \longrightarrow +\infty$ and that $\log_a x \longrightarrow -\infty$ as $x \longrightarrow 0^+$. The logarithmic function is undefined for $x \leq 0$.

Algebraic Properties of Logarithms

Some years ago logarithms were very important for the purpose of carrying out complicated numerical calculations. Their usefulness in this connection stems from the algebraic properties stated in the following theorem.

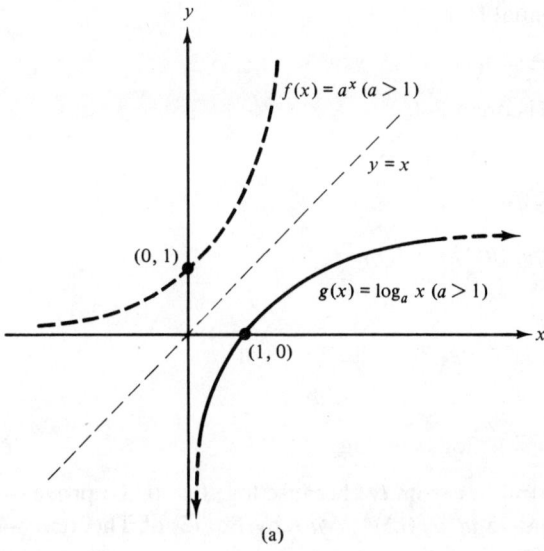

(a)

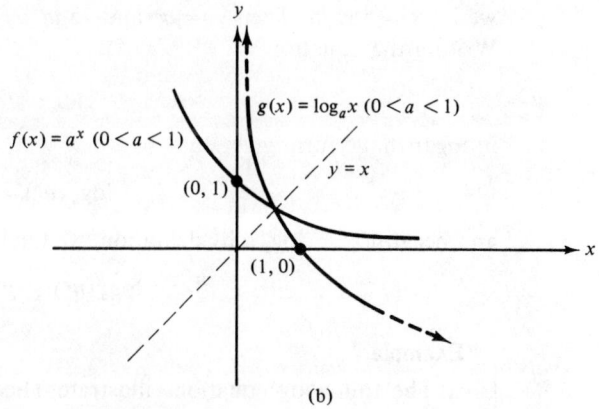

(b)

Figure 8-7

Theorem 8-5. Let a be a positive number different from 1. If u and v are positive numbers, then

1. $\log_a (uv) = \log_a u + \log_a v$

2. $\log_a \left(\dfrac{u}{v}\right) = \log_a u - \log_a v$

3. $\log_a \left(\dfrac{1}{v}\right) = -\log_a v$

4. $\log_a (u^p) = p \log_a u$ for any real number p.

Proof. Each part of this theorem is proved by referring to a corresponding property of the exponential function. For instance, to prove (1), let

$$x = \log_a u \qquad y = \log_a v.$$

Writing these equations in exponential form,

$$a^x = u \qquad a^y = v.$$

Multiplying and using Eq. (1) of Theorem 8-4,

$$uv = a^x a^y = a^{x+y}.$$

Returning to logarithmic form, we get

$$\log_a (uv) = x + y.$$

But by definition of x and y,

$$x + y = \log_a u + \log_a v$$

and so

$$\log_a (uv) = \log_a u + \log_a v.$$

The proof of (2) is similar and (3) follows from (2) because $\log_a 1 = 0$. To prove (4), we let $x = \log_a u$. Then $u = a^x$ and so $u^p = (a^x)^p = a^{xp}$ by Eq. (3) of Theorem 8-4. Writing the equation

$$u^p = a^{xp}$$

in logarithmic form gives

$$\log_a (u^p) = xp$$

and because $x = \log_a u$ by definition, we finally obtain

$$\log_a (u^p) = p \log_a u.$$

Example 7

The following equations illustrate Theorem 8-5. (See Table 8-1 for values of $\log_3$.)

$$\log_3 (9 \cdot 27) = \log_3 9 + \log_3 27 = 2 + 3 = 5 \qquad \text{[Theorem 8-5(1)]}$$

$$\log_3 \left(\frac{\sqrt{3}}{3} \right) = \log_3 \sqrt{3} - \log_3 3 = \frac{1}{2} - 1 = -\frac{1}{2} \qquad \text{[Theorem 8-5(2)]}$$

$$\log_3 \left(\frac{1}{\sqrt[3]{3}} \right) = -\log_3 \sqrt[3]{3} = -\frac{1}{3} \qquad \text{[Theorem 8-5(3)]}$$

$$\log_3 [(\sqrt{3})^3] = 3 \log_3 \sqrt{3} = 3 \cdot \frac{1}{2} = \frac{3}{2} \qquad \text{[Theorem 8-5(4)]}$$

Common Logarithms

Logarithms with base 10 are called **common** logarithms. Approximations of the common logarithms of some particular numbers are shown in Table 8-3. Much more extensive tables of logarithms can be found in books on trigonometry, algebra, or calculus.

TABLE 8-3 A BRIEF TABLE
OF COMMON LOGARITHMS

x	$\log_{10} x$
0.01	−2.0000
0.1	−1.0000
1.0	0.0000
2	0.3010
3	0.4771
4	0.6021
5	0.6990
10	1.0000
15	1.1761
20	1.3010
50	1.6990
100	2.0000

Example 8

Use Table 8-3 and Theorem 8-5 to find

$$\log_{10} 6, \quad \log_{10}\left(\frac{3}{2}\right), \quad \log_{10}\left(\frac{1}{2}\right), \quad \log_{10} 27, \quad \log_{10} 24.$$

Solution. Because $6 = 3 \cdot 2$, we know from (1) of Theorem 8-5 that

$$\log_{10} 6 = \log_{10} 3 + \log_{10} 2 \approx .4771 + .3010 = .7781.$$

Next, to obtain $\log_{10} (3/2)$, we use (2):

$$\log_{10}\left(\frac{3}{2}\right) = \log_{10} 3 - \log_{10} 2$$

$$\approx .4771 - .3010 = .1761.$$

Property (3) enables us to get the logarithm of the reciprocal of a number.

$$\log_{10}\left(\frac{1}{2}\right) = -\log_{10} 2$$

$$\approx -.3010.$$

Logarithms of powers are obtained by (4).

$$\log_{10} 27 = \log_{10} (3^3) = 3 \log_{10} 3$$

$$\approx 3(.4771)$$

$$= 1.4313$$

Finally, we can calculate the logarithm of 24 by using several properties in the theorem.

$$\log_{10} 24 = \log_{10} (3 \cdot 8) = \log_{10} 3 + \log_{10} 8$$

$$= \log_{10} 3 + \log_{10} (2^3)$$

$$= \log_{10} 3 + 3 \log_{10} 2$$

$$\approx .4771 + .9030 = 1.3801$$

Clearly Table 8-3 is inadequate for obtaining the logarithms of most numbers. The same applies even to more extensive log tables. To find the logarithms of some numbers, it may be necessary to approximate them by linear interpolation.

Example 9

Find $\log_{10} 26$.

Solution. In Example 8 we found $\log_{10} 24 \approx 1.3801$ and $\log_{10} 27 \approx 1.4313$. Linear interpolation gives

$$\frac{\log_{10} 26 - \log_{10} 24}{26 - 24} \approx \frac{\log_{10} 27 - \log_{10} 24}{27 - 24}$$

$$\frac{\log_{10} 26 - 1.3801}{2} \approx \frac{1.4313 - 1.3801}{3}$$

$$\log_{10} 26 \approx 1.3801 + \frac{2}{3}(0.0512)$$

$$\approx 1.3801 + .0341 = 1.4142.$$

(The correct value of $\log_{10} 26$ to four decimal places is 1.4150.)

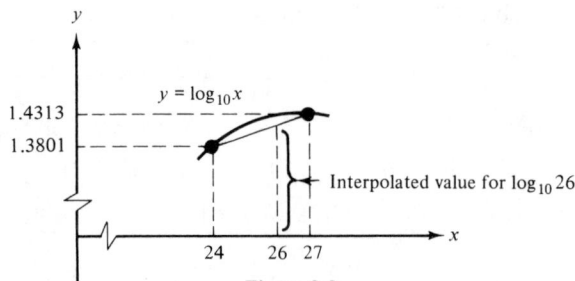

Figure 8-8

Example 10

Use logarithms to calculate the value of

$$N = \frac{(5.96)^4}{(2.46)^5}.$$

Solution. We take the common logarithm of both sides, applying Eqs. (2) and (4) of Theorem 8-5.

$$\log_{10} N = 4 \log_{10} 5.96 - 5 \log_{10} 2.46$$

Using a table to approximate $\log_{10} 5.96$ and $\log_{10} 2.46$, we get

$$\log_{10} N \approx 4(.7752) - 5(.3909)$$

$$= 3.1008 - 1.9545$$

$$= 1.1463.$$

Notice that in dealing with the logarithms, we subtract (rather than divide) and multiply (rather than raise to a power). Now that we know the logarithm of N we can find N, approximately, by referring to a table and interpolating. We get

$$N \approx 14.006.$$

Since the development of electronic computers and hand calculators, logarithms no longer play an important role in arithmetic computation. Nevertheless, logarithmic functions and their algebraic properties formulated in Theorem 8-5 continue to be extremely important for other purposes. The next few examples demonstrate some simplifications typically used in calculus.

Example 11

Simplify $\log_a [x^3(1 - x)^4]$, assuming $0 < x < 1$.

Solution. Using (1) and (4) of Theorem 8-5, we have

$$\log_a [x^3(1 - x)^4] = \log_a (x^3) + \log_a [(1 - x)^4]$$
$$= 3 \log_a x + 4 \log_a (1 - x).$$

Example 12

Simplify $\log_a \sqrt{x^2 + y^2}$.

Solution. Using (4) of Theorem 8-5, we get

$$\log_a \sqrt{x^2 + y^2} = \log_a [(x^2 + y^2)^{1/2}] = \frac{1}{2} \log_a (x^2 + y^2).$$

[Caution: The logarithm of a sum is not the sum of the logarithms; hence $\log_a (x^2 + y^2)$ is *not* equal to $\log_a (x^2) + \log_a (y^2)$.]

Example 13

Simplify $\log_a (x^2 y^3 / z^5)$, assuming x, y, and z are positive.

Solution. Using (1), (2), and (4) of Theorem 8-5, we obtain

$$\log_a \frac{x^2 y^3}{z^5} = [\log_a (x^2) + \log_a (y^3)] - \log_a (z^5)$$
$$= 2 \log_a x + 3 \log_a y - 5 \log_a z.$$

Example 14

Simplify

$$\log_a \sqrt[3]{\frac{(x^2 + 1)^4(x^2 - 1)}{2x^3}}$$

assuming $x > 1$.

Solution.

$$\log_a \sqrt[3]{\frac{(x^2 + 1)^4(x^2 - 1)}{2x^3}} = \frac{1}{3} \log_a \frac{(x^2 + 1)^4(x^2 - 1)}{2x^3}$$

$$= \frac{1}{3} [\log_a (x^2 + 1)^4 + \log_a (x^2 - 1) - \log_a (2x^3)]$$

$$= \frac{1}{3} [4 \log_a (x^2 + 1) + \log_a (x^2 - 1) - \log_a 2 - 3 \log_a x]$$

$$= \frac{4}{3} \log_a (x^2 + 1) + \frac{1}{3} \log_a (x^2 - 1) - \frac{1}{3} \log_a 2 - \log_a x.$$

Change of Base

Suppose that we are working with a logarithmic function with base a and wish to convert to another base b. For $x > 0$, we would like to express $\log_b x$ in terms of $\log_a x$. We can see how to do so as follows. Let $y = \log_b x$ so that

$$x = b^y.$$

Next, write $b = a^{\log_a b}$ and thus

$$b^y = (a^{\log_a b})^y = a^{y \log_a b}.$$

Then

$$x = a^{y \log_a b}.$$

Writing this last equation in logarithmic form, we have

$$\log_a x = y \log_a b = \log_b x \log_a b$$

Solving for $\log_b x$ gives

$$\log_b x = \frac{\log_a x}{\log_a b}$$

Thus if we have a table of logarithms, base a, we can convert to base b by dividing each entry by $\log_a b$.

Theorem 8-6. If a and b are positive numbers both different from 1, and if $x > 0$, then

$$\log_b x = \frac{\log_a x}{\log_a b}.$$

Example 15

Convert the common logarithms in Table 8-3 to logarithms to the base 2.

Solution. For each x in the table we divide the corresponding common logarithm by $\log_{10} 2$, which is approximately .3010. For instance,

$$\log_2 3 = \frac{\log_{10} 3}{\log_{10} 2} \approx \frac{.4771}{.3010} \approx 1.585.$$

TABLE 8-4

x	$\log_{10} x$	$\log_2 x$
0.01	−2.0000	−6.645
0.1	−1.0000	−3.322
1.0	0.0000	0.000
2.0	0.3010	1.000
3.0	0.4771	1.585
4.0	0.6021	2.000
5	0.6990	2.322
10	1.0000	3.322
15	1.1761	3.907
20	1.3010	4.322
50	1.6990	5.644
100	2.0000	6.645

Exponential and Logarithmic Functions   Chap. 8

EXERCISES 8-3

In Problems 1–6 write each equation in an equivalent logarithmic from with base 3 (as in Table 8-1).

1. $3^5 = 243$

2. $3^{-1} = \dfrac{1}{3}$

3. $3^{3/2} = 3\sqrt{3}$

4. $3^{5/2} = 9\sqrt{3}$

5. $3^{-2/3} = \dfrac{1}{\sqrt[3]{9}}$

6. $3^{-4/3} = \dfrac{1}{\sqrt[3]{81}}$

In Problems 7–12 write each equation in an equivalent exponential form with a suitable base (as in Table 8-2).

7. $\log_3 81 = 4$

8. $\log_{10}(1000) = 3$

9. $\log_4\left(\dfrac{1}{16}\right) = -2$

10. $\log_2\left(\dfrac{1}{16}\right) = -4$

11. $\log_{10}(10\sqrt{10}) = \dfrac{3}{2}$

12. $\log_{10}\left(\dfrac{1000}{\sqrt{10}}\right) = \dfrac{5}{2}$

In Problems 13–18 write each equation in an equivalent logarithmic form with a suitable base (as in Table 8-2).

13. $3^4 = 81$

14. $3^{7/3} = 9\sqrt[3]{3}$

15. $2^{-10} = \dfrac{1}{1024}$

16. $5^{-2} = .04$

17. $5^{2x} = y$

18. $5^{2x} = y^3$

19. Sketch the graph of the exponential function $f(x) = 3^x$ and on the same coordinate system sketch the inverse function, $g(x) = \log_3 x$.

In Problems 20–34 evaluate the given expression.

20. $\log_3 9$

21. $\log_3\left(\dfrac{1}{27}\right)$

22. $\log_3 3(\sqrt[3]{3})$

23. $\log_3\left(\dfrac{1}{3^{2/3}}\right)$

24. $\log_3 1$

25. $\log_3(3^5)$

26. $\log_3(3^x)$

27. $\log_3(9^{5/2})$

28. $3^{\log_3 9}$

29. $3^{\log_3 2}$

30. $3^{\log_3(1/3)}$

31. $3^{\log_3 u}$

32. $3^{\log_3 \sqrt{2}}$

33. $3^{2\log_3 5}$

34. $3^{2\log_3 2} + 1$

In Problems 35–51 solve for x.

35. $2^x = 2048$

36. $3^x = \dfrac{1}{81}$

37. $5^x = 1$

38. $2^{5x} = 8$

39. $2^{1/x} = .125$

40. $2^{-x^2} = .125$

41. $\log_2 x = 5$

42. $\log_3 x = 0$

43. $\log_{10} x = 3$

44. $\log_5 x = 1$

45. $\log_2 x = 1.5$

46. $\log_x 8 = 3$

47. $\log_x 81 = 4$

48. $\log_x 2 = 3$

49. $\log_x 3 = 2$

50. $\log_x 3 = \dfrac{1}{2}$

51. $\log_x 8 = \dfrac{1}{3}$

52. $\log_x 8 = -3$

In Problems 53–60 use Table 8-3 and logarithm properties stated in Theorem 8-5 to find an approximation. Interpolation may be necessary.

53. $\log_{10} 12$

54. $\log_{10}(.75)$

55. $\log_{10} 400$

56. $\log_{10} .004$

57. $\log_{10} 7$

58. $\log_{10} 80$

59. $\log_{10} 75$

60. $\log_{10} 48$

†61. Given $\log_{10} 71 \approx 1.8513$ and $\log_{10} 45.6 \approx 1.6590$, use Theorem 8-5 to find approximate values of
  (a) $\log_{10} \sqrt[3]{71}$.
  (b) $\log_{10}[71^2 \cdot (45.6)^4]$.
  (c) $\log_{10} \sqrt[3]{\dfrac{71^8}{(45.6)^4}}$.

Compare your results with those given by a calculator.

In Problems 62–79, simplify the expression and indicate the values of the variable x for which the simplified expression is meaningful.

62. $\log_{10} [(3x - 4)^5]$

63. $\log_{10} [5(x - 3)^3]$

64. $\log_{10} [(x + 1)(x^2 + 1)]$

65. $\log_{10} [(x^2 + 4)(x^2 + 1)^3]$

66. $\log_{10} \dfrac{1}{x^2 + 1}$

67. $\log_5 \dfrac{1}{(3x - 1)^3}$

68. $\log_{10} \sqrt{x^2 + 25}$

69. $\log_2 \sqrt{x^2 + 25}$

70. $\log_{10} \sqrt[3]{x^3 + 27}$

71. $\log_{10} \dfrac{x^2 - 4}{x^2 + 4}$

72. $\log_{10} \dfrac{x - 1}{\sqrt{x^2 + 1}}$

73. $\log_2 \dfrac{1}{\sqrt[4]{x^2 + 2x + 4}}$

74. $\log_{10} [x(x + 1)^2(x + 2)^{-2}]$

75. $\log_{10} [x(3x + 4)^2\sqrt{3x - 2}]$

76. $\log_3 \dfrac{\sqrt{x - 3}}{x\sqrt{x + 3}}$

77. $\log_5 \dfrac{(2x^2 + 3)^2}{\sqrt{x + 5}(x^2 + 1)^3}$

78. $\log_5 \dfrac{(x^2 + 5)^3(x - 1)}{5x}$

79. $\log_{10} \dfrac{(3x^2 + 4)^2\sqrt{3x - 2}}{x^2 + 9}$

In Problems 80–85, show that the equation holds for all values of x indicated and for any base a that is positive and different from one.

80. $\log_a [a^2(a + x)^2] = 2 + 2 \log_a (a + x) \qquad (x > -a)$

81. $\log_a [a^4(1 + x)^3] = 4 + 3 \log_a (1 + x) \qquad (x > -1)$

82. $\log_a \sqrt{(a + x)^2 - x^2} = \dfrac{1}{2} + \dfrac{1}{2} \log_a (a + 2x) \qquad \left(x > -\dfrac{a}{2}\right)$

83. $\log_a \sqrt{\left(a^2 - \dfrac{x^2}{a^2}\right)^2 + 4x^2} = \log_a (a^4 + x^2) - 2 \qquad$ (all real x)

84. $\log_a (x + \sqrt{x^2 - 1}) = -\log_a (x - \sqrt{x^2 - 1}) \qquad (x > 1)$

85. $\log_a \dfrac{2x - \sqrt{4x^2 - 1}}{2x + \sqrt{4x^2 - 1}} = 2 \log_a (2x - \sqrt{4x^2 - 1}) \qquad \left(x > \dfrac{1}{2}\right)$

In Problems 86–91 solve for x.

86. $\log_{10} (x - 1) + \log_{10} 5x = 1$

87. $\log_{10} (x^2 + 16) - \log_{10} x = 1$

88. $\log_3 (9x + 54) - \log_3 x = 3$

89. $\log_5 (x^2 + 4) - \log_5 x = 1$

90. $\log_6 (5x - 7) + \log_6 (3x + 6) = 2$

91. $\log_{12} (2x - 20) + \log_{12} (x - 4) = 2$

†92. Using Table 8-4 and Theorem 8-6 find $\log_3 5$ and $\log_3 20$.

FOR THE ENTHUSIASTS

93. Show that the function

$$h(x) = \frac{3^x - 3^{-x}}{2}$$

has an inverse and express this inverse in terms of logarithms. Graph h^{-1}.

94. Show that if a and b are positive numbers different from one, then $\log_a b$ and $\log_b a$ are reciprocals.

95. Prove Eq. (2) in Theorem 8-5.

8-4 EXPONENTIAL AND LOGARITHMIC FUNCTIONS WITH BASE e

The Number e

Let $f(u) = (1 + 1/u)^u$ for $u > 0$ and consider the behavior of $f(u)$ as $u \to +\infty$. The base $1 + 1/u$ approaches one whereas the exponent u approaches plus infinity. It is not immediately clear exactly how $f(u)$ behaves, but the values in Table 8-5 can be

TABLE 8-5

u	$f(u)$
1	$(1 + 1)^1 = 2$
2	$\left(1 + \dfrac{1}{2}\right)^2 = 2.25$
3	$\left(1 + \dfrac{1}{3}\right)^3 \approx 2.3704$
4	$\left(1 + \dfrac{1}{4}\right)^4 \approx 2.4414$
5	$\left(1 + \dfrac{1}{5}\right)^5 \approx 2.4883$
10	$\left(1 + \dfrac{1}{10}\right)^{10} \approx 2.5937$
100	$\left(1 + \dfrac{1}{100}\right)^{100} \approx 2.7048$

obtained by using a calculator. It can be shown that f is increasing on $(0, +\infty)$ and that the values of $f(u)$ remain below 3; in fact, there exists a real number, denoted by e, such that $f(u)$ approaches e as $u \to +\infty$. To five decimal places

$$e \approx 2.71828$$

but, like the number π in geometry, e is irrational.

Soon we shall see that the number e arises naturally in connection with certain applications.

Compound Interest

In Section 1-2 we discussed interest. We showed that if an amount P of money is invested at an interest rate r compounded annually, then after n years it will be worth $P(1 + r)^n$, provided that all interest is left on deposit. Usually interest is compounded more frequently than once a year. Let us see how this works.

In one type of passbook account interest is compounded quarterly. The year is divided into four quarters, beginning January 1, April 1, July 1, and October 1. Interest is paid four times, once at the end of each quarter, at a rate of $r/4$; this rate is applied to the minimum balance during the quarter. If the resulting interest is left in the account, then it becomes part of the amount to which interest is applied at the end of the next quarter. Consequently, the compounding of interest is beneficial to the investor.

Example 1

On January 1 you invest $1000 in a savings account to be compounded quarterly at $5\frac{1}{2}\%$ per year. You leave all interest on deposit for the year. At the end of the first quarter your interest is

$$\$1000(.055/4) = \$13.75$$

and the amount in your account becomes

$$\$1013.75 = \$1000\left(1 + \frac{.055}{4}\right).$$

At the end of the second quarter your interest is

$$\$1013.75(.055/4) = \$13.94$$

and the amount in your account becomes

$$\$1027.69 = \$1013.75\left(1 + \frac{.055}{4}\right)$$

$$= \$1000\left(1 + \frac{.055}{4}\right)\left(1 + \frac{.055}{4}\right)$$

$$= \$1000\left(1 + \frac{.055}{4}\right)^2.$$

At the end of the third quarter your interest is

$$\$1027.69(.055/4) = \$14.13$$

and your account totals

$$\$1000\left(1 + \frac{.055}{4}\right)^3 = \$1041.82.$$

At the end of the last quarter your account totals $1056.14.

Notice that if the annual interest rate, $5\frac{1}{2}\%$, had been applied just once at the end of the year, your account would have totaled $1055.00.

If you leave all interest on deposit for q quarters, your account will then total $(\$1000)[1 + (.055/4)]^q$.

Interest is not always compounded annually or quarterly. Sometimes it is compounded monthly or even daily. To formulate the idea of compound interest generally, suppose that interest is paid at an annual rate r and applied at n equal intervals of time during the year. Then at the end of each interval interest is credited in the amount of r/n times the minimum balance during that interval. If all interest is left on deposit, then after one interval a deposit of P dollars will be worth $P[1 + (r/n)]$, after two intervals it will be worth $P[1 + (r/n)]^2$, and so on; after m intervals it will be worth $P[1 + (r/n)]^m$. Because t years represents nt intervals, we can express the total amount as follows.

Compound Interest. If n is a positive integer, the formula

$$A = P\left(1 + \frac{r}{n}\right)^{nt}$$

---

gives the amount A in a savings account after t full years, where

$$P = \text{principal}$$

$$r = \text{annual interest rate}$$

assuming no withdrawals and that interest is compounded n times each year.

Example 2

On January 1 you invest $1000 to be compounded monthly $(n = 12)$ at $5\frac{1}{2}\%$ per year, leaving all interest on deposit. Table 8-6 shows the number of dollars in your account at the end of each of the first 12 months.

TABLE 8-6

January	$1000\left(1 + \frac{.055}{12}\right)$	$= 1004.58$
February	$1000\left(1 + \frac{.055}{12}\right)^2$	$= 1009.19$
March	$1000\left(1 + \frac{.055}{12}\right)^3$	$= 1013.81$
April	$1000\left(1 + \frac{.055}{12}\right)^4$	$= 1018.46$
May	$1000\left(1 + \frac{.055}{12}\right)^5$	$= 1023.13$
June	$1000\left(1 + \frac{.055}{12}\right)^6$	$= 1027.82$
July	$1000\left(1 + \frac{.055}{12}\right)^7$	$= 1032.53$
August	$1000\left(1 + \frac{.055}{12}\right)^8$	$= 1037.26$
September	$1000\left(1 + \frac{.055}{12}\right)^9$	$= 1042.01$
October	$1000\left(1 + \frac{.055}{12}\right)^{10}$	$= 1046.79$
November	$1000\left(1 + \frac{.055}{12}\right)^{11}$	$= 1051.58$
December	$1000\left(1 + \frac{.055}{12}\right)^{12}$	$= 1056.40$

Continuously Compounded Interest

You may wonder what the result will be if interest is compounded daily, hourly, every minute, and so forth. In other words, what happens as $n \longrightarrow +\infty$? We examine the expression $[1 + (r/n)]^{nt}$. Letting u denote n/r, we have $n = ur$, $nt = urt$, and $r/n = 1/u$ so that

$$\left(1 + \frac{r}{n}\right)^{nt} = \left(1 + \frac{1}{u}\right)^{urt} = \left[\left(1 + \frac{1}{u}\right)^u\right]^{rt}.$$

It is clear that as $n \longrightarrow +\infty$, then $u \longrightarrow +\infty$ also, for $u = n/r$. The expression $[1 + (1/u)]^u$ approaches e and it follows that

$$\left[\left(1 + \frac{1}{u}\right)^u\right]^{rt} \longrightarrow e^{rt}$$

Therefore the effect of letting $n \longrightarrow +\infty$ is that

$$A = P\left(1 + \frac{r}{n}\right)^{nt} \longrightarrow Pe^{rt}$$

Interest determined in this way is said to be **compounded continuously** at the annual rate r.

Continuous Interest. The formula

$$A = Pe^{rt}$$

gives the amount in a savings account after t years, where

$$P = \text{principal}$$

$$r = \text{annual interest rate}$$

assuming no withdrawals and that interest is compounded continuously.

Although in previous examples we considered t to be a positive integer, the formula $A = Pe^{rt}$ can be shown to be valid for other values of t as well.

Example 3

You deposit $1000 in a new savings account with interest compounded continuously at an annual rate of 8%. If you do not withdraw any money from your account, what will it be worth after $2\frac{1}{2}$ years?

Solution. In the formula

$$A = Pe^{rt}$$

we let $t = 2.5$, $r = .08$, and $P = 1000$, obtaining

$$A = 1000e^{(.08)(2.5)}.$$

Using Table 8-7, we find $e^{.2} \approx 1.2214$ and

$$A \approx 1000(1.2214).$$

So the value of your account in $2\frac{1}{2}$ years will be $1221.40.

Exponential Growth

Let Q be any quantity that is growing continuously in such a way that its rate of growth at a given time is proportional to the value of Q at that time. An example, as we have just seen, is the money in a savings account compounded continuously.

Such a quantity Q can be expressed as a function of time t by an equation that resembles the formula for continuously compounded interest:

$$Q = Q_0 e^{rt}, \tag{8-3}$$

where Q_0 is the amount of the quantity present when $t = 0$ (the initial value) and where $r > 0$ is the proportionality constant associated with the growth rate. Q is said to satisfy the **law of exponential growth**. A graph illustrating exponential growth appears in Fig. 8-9(a).

TABLE 8-7 VALUES OF NATURAL EXPONENTIAL AND LOGARITHMIC FUNCTIONS

x	e^x	e^{-x}	$\log_e x$	x	e^x	e^{-x}	$\log_e x$
0.05	1.0513	0.9512	−2.9957	2.10	8.1662	0.1225	0.7419
0.10	1.1052	0.9048	−2.3026	2.20	9.0250	0.1108	0.7885
0.15	1.1618	0.8607	−1.8971	2.30	9.9742	0.1003	0.8329
0.20	1.2214	0.8187	−1.6094	2.40	11.0232	0.0907	0.8755
0.25	1.2840	0.7788	−1.3863	2.50	12.1825	0.0821	0.9163
0.30	1.3499	0.7408	−1.2040	2.60	13.4637	0.0743	0.9555
0.35	1.4191	0.7047	−1.0498	2.70	14.8797	0.0672	0.9933
0.40	1.4918	0.6703	−0.9163	2.80	16.4446	0.0608	1.0300
0.45	1.5683	0.6376	−0.7985	2.90	18.1741	0.0550	1.0647
0.50	1.6487	0.6065	−0.6931	3.00	20.0855	0.0498	1.0986
0.55	1.7333	0.5770	−0.5978	3.10	22.1979	0.0450	1.1314
0.60	1.8221	0.5488	−0.5108	3.20	24.5325	0.0408	1.1632
0.65	1.9155	0.5220	−0.4308	3.30	27.1126	0.0369	1.1939
0.70	2.0138	0.4966	−0.3567	3.40	29.9641	0.0334	1.2238
0.75	2.1170	0.4724	−0.2877	3.50	33.1154	0.0302	1.2528
0.80	2.2255	0.4493	−0.2231	3.60	36.5982	0.0273	1.2809
0.85	2.3396	0.4274	−0.1625	3.70	40.4473	0.0247	1.3083
0.90	2.4596	0.4066	−0.1054	3.80	44.7012	0.0224	1.3350
0.95	2.5857	0.3867	−0.0513	3.90	49.4024	0.0202	1.3610
1.00	2.7183	0.3679	0	4.00	54.5981	0.0183	1.3863
1.05	2.8577	0.3499	0.0488	4.10	60.3403	0.0166	1.4110
1.10	3.0042	0.3329	0.0953	4.20	66.6863	0.0150	1.4351
1.15	3.1582	0.3166	0.1398	4.30	73.6998	0.0136	1.4586
1.20	2.3201	0.3012	0.1823	4.40	81.4509	0.0123	1.4816
1.25	3.4903	0.2865	0.2231	4.50	90.0171	0.0111	1.5041
1.30	3.6693	0.2725	0.2624	4.60	99.4843	0.0101	1.5261
1.35	3.8574	0.2592	0.3001	4.70	109.947	0.0091	1.5476
1.40	4.0552	0.2466	0.3365	4.80	121.510	0.0082	1.5686
1.45	4.2631	0.2346	0.3716	4.90	134.290	0.0074	1.5892
1.50	4.4817	0.2231	0.4055	5.00	148.413	0.0067	1.6094
1.55	4.7115	0.2122	0.4383	5.10	164.022	0.0061	1.6292
1.60	4.9530	0.2019	0.4700	5.20	181.272	0.0055	1.6487
1.65	5.2070	0.1921	0.5008	5.30	200.337	0.0050	1.6677
1.70	5.4740	0.1827	0.5306	5.40	221.406	0.0045	1.6864
1.75	5.7546	0.1738	0.5596	5.50	244.692	0.0041	1.7048
1.80	6.0497	0.1653	0.5878	5.60	270.426	0.0037	1.7228
1.85	6.3598	0.1572	0.6152	5.70	298.867	0.0033	1.7405
1.90	6.6859	0.1496	0.6419	5.80	330.000	0.0030	1.7579
1.95	7.0287	0.1423	0.6678	5.90	365.038	0.0027	1.7750
2.00	7.3891	0.1353	0.6931	6.00	403.429	0.0025	1.7918

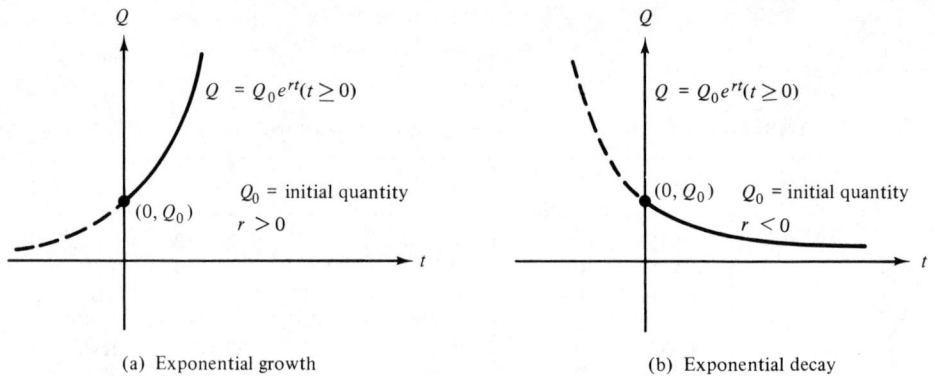

(a) Exponential growth

(b) Exponential decay

Figure 8-9

Example 4

The number Q of cells in a growing organism is sometimes assumed to obey our assumptions about exponential growth[1] and so Q is given by an equation of the form (8-3). If the number of cells at noon is Q_0 and the number doubles after an hour, at what time will the number have tripled?

Solution. We know that

$$Q = Q_0 e^{rt}$$

for some constant r, where t is time in hours after noon. We also know that $Q = 2Q_0$ when $t = 1$ and so

$$2Q_0 = Q_0 e^r$$
$$2 = e^r.$$

This shows that r is $\log_e 2$ and thus the formula for Q is

$$Q = Q_0 e^{t \log_e 2}.$$

To determine the time at which the number will have tripled, we set $Q = 3Q_0$ and solve for t.

$$3Q_0 = Q_0 e^{t \log_e 2}$$
$$3 = e^{t \log_e 2}$$
$$\log_e 3 = t \log_e 2$$
$$t = \frac{\log_e 3}{\log_e 2}$$

The value of t can be approximated by using Table 8-7.

$$t \approx \frac{1.0986}{.6931} \approx 1.585$$

The solution of Example 4 can be simplified by substituting 2 for e^r as follows:

---

[1]Of course, this is an idealization because the number of cells is always an integer and cannot increase continuously. This idealization gives a good approximation, however.

Exponential and Logarithmic Functions   Chap. 8

$$Q = Q_0 e^{rt}$$
$$= Q_0 (e^r)^t$$
$$= Q_0 2^t.$$

The time at which the number will have tripled is determined by setting $Q = 3Q_0$:

$$3Q_0 = Q_0 2^t$$
$$3 = 2^t$$
$$t = \log_2 3.$$

To find $\log_2 3$, we use the change of base formula in Theorem 8-6, obtaining

$$\log_2 3 = \frac{\log_e 3}{\log_e 2}.$$

The point to be noted here is that it may be simpler to express Eq. (8-3) in terms of a base other than e.

Example 5

The 1980 census showed the population of a village to be 3500. If the 1970 population was 3000 and the population satisfied the law of exponential growth over the decade from 1970 to 1980, what was the population in 1975?

Solution. Let us measure time t in decades after 1970. The population Q satisfies

$$Q = 3000 e^{rt}$$

and we are given $Q = 3500$ when $t = 1$. Thus

$$3500 = 3000 e^{r \cdot 1}$$
$$\frac{7}{6} = e^r.$$

So we have

$$Q = 3000 (e^r)^t = 3000 \left(\frac{7}{6}\right)^t.$$

The year 1975 corresponds to $t = 1/2$. Therefore the 1975 population was

$$3000 \left(\frac{7}{6}\right)^{1/2} = 3000 \sqrt{\frac{7}{6}}$$
$$\approx 3000 \cdot (1.080) = 3240$$

Exponential Decay

If the constant r in Eq. (8-3) is negative, the function $Q = Q_0 e^{rt}$ is decreasing and instead of growth we have decay. In this case, the quantity is decreasing at a rate proportional at any time to the amount present at that time. A quantity satisfying (8-3) with $r < 0$ is said to satisfy the **law of exponential decay**. A graph illustrating exponential decay appears in Fig. 8-9(b).

Example 6

A radioactive substance can be assumed to decay exponentially. If 10 grams are on hand initially and 8 grams remain after 100 days, express the quantity Q (grams) remaining after t days as a function of t.

Solution. We know that $Q = 10e^{rt}$, where $Q = 8$ when $t = 100$. Thus

$$8 = 10e^{100r}$$

$$e^{100r} = \frac{4}{5}$$

$$Q = 10(e^{100r})^{t/100}$$

$$= 10\left(\frac{4}{5}\right)^{t/100}$$

Example 7

In Example 6 find the amount of the substance remaining after 300 days.

Solution. At $t = 300$

$$Q = 10\left(\frac{4}{5}\right)^{300/100} = 10\left(\frac{4}{5}\right)^3 = 5.12 \text{ grams.}$$

A quantity that is subject to the law of exponential decay is often described in terms of a **half-life**. A half-life is the length of time during which half the quantity decomposes. Radium (Ra^{226}), for instance, has a half-life of about 1600 years. This means that it would take 1600 years for one gram of radium to be reduced to 1/2 gram. It would then take another 1600 years for the 1/2 gram to be reduced to 1/4 gram and so on. Whatever the amount at a given time, half that amount will remain after 1600 years. You are asked to verify the half-life property for exponential decay in Problem 37.

Example 8

A substance is decaying exponentially with a half-life of 30 years. If the present amount is Q_0, express the amount t years from now as a function of t.

Solution. In the equation $Q = Q_0e^{rt}$ we want to find r. Because $Q = \frac{1}{2}Q_0$ when $t = 30$, we know that

$$\frac{1}{2}Q_0 = Q_0e^{30r}$$

$$\frac{1}{2} = e^{30r}.$$

Thus

$$Q = Q_0(e^{30r})^{t/30}$$

$$= Q_0\left(\frac{1}{2}\right)^{t/30}$$

$$= Q_0 2^{-t/30}.$$

Example 9

For the substance in Example 8 find the proportion of the present amount remaining
  (a) 60 years from now.
  (b) 120 years from now.
  (c) 15 years from now.
How long will it take for the amount of substance to be reduced to 20% of the initial amount Q_0?

Solution. Let Q_0 be the present amount.
  (a) After 60 years we have

$$Q = Q_0 \cdot 2^{-2} = \frac{Q_0}{4}.$$

So one-quarter remains.
  (b) After 120 years

$$Q = Q_0 \cdot 2^{-4} = \frac{Q_0}{16}.$$

So one-sixteenth remains.
  (c) After 15 years

$$Q = Q_0 \cdot 2^{-1/2} = \frac{Q_0}{\sqrt{2}} \approx .707 Q_0.$$

So about 71% remains.

To determine how long until the substance is reduced to 20% of the original amount, we set $Q_0 2^{-t/30}$ equal to $.2Q_0$ and solve for t.

$$.2Q_0 = Q_0 2^{-t/30}$$
$$.2 = 2^{-t/30}$$

Taking logarithms on both sides gives

$$\log_e (.2) = \left(\frac{-t}{30}\right) \log_e 2.$$

Using (4) of Theorem 8-5 and looking up $\log_e (.2)$ and $\log_e 2$ in Table 8-7, we get

$$-1.6094 \approx \frac{-t}{30}(.6931)$$

and with a calculator

$$t \approx 69.7 \text{ years.}$$

Natural Logarithms

In several of the preceding examples logarithms to the base e occurred. The logarithmic function with base e is the one most widely used in calculus. It is called the **natural logarithm** and is abbreviated by the special symbol **ln**; thus $\ln x = \log_e x$. A graph of $y = \ln x$ appears in Fig. 8-10.

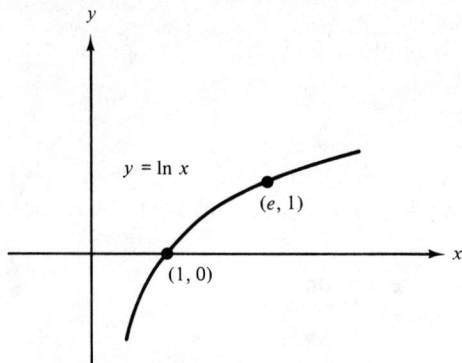

Figure 8-10

Example 10

Graph the function $f(x) = \ln(x - 2) + 1$.

Solution. The domain of f is the interval $(2, +\infty)$. Letting $y = f(x)$, we get

$$y - 1 = \ln(x - 2)$$

and we can make a translation of axes

$$u = x - 2$$
$$v = y - 1$$

to the point $(2, 1)$. The new equation is $v = \ln u$; its graph in Fig. 8-11 is simply a translation of the graph of $y = \ln x$. Notice that $f(x) \to -\infty$ as $x \to 2^+$. The intercept on the x axis is obtained by solving $\ln(x - 2) + 1 = 0$.

$$\ln(x - 2) = -1$$
$$x - 2 = e^{-1}$$
$$x = 2 + \frac{1}{e}$$

So the intercept is $(2 + 1/e, 0)$.

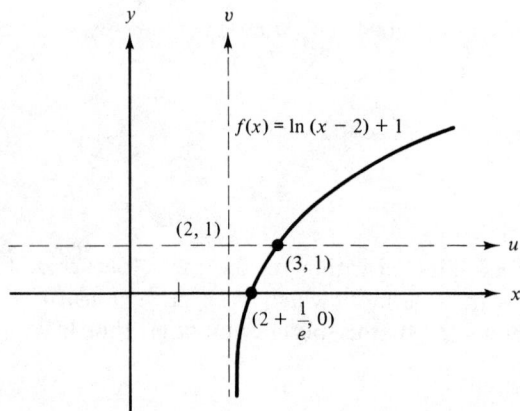

Figure 8-11

Example 11

Graph $g(x) = e^{x-1} + 3$.

Solution.   Letting $y = g(x)$, we have

$$y - 3 = e^{x-1},$$

which we recognize as a translation of the graph of the exponential function with base e. The equations of this translation are

$$u = x - 1$$
$$v = y - 3$$

and so the new center is $(1, 3)$. (See Fig. 8-12.) As $x \to -\infty$, we see that $f(x)$ approaches zero and so the line $y = 3$ is a horizontal asymptote. The y intercept is $(0, 3 + e^{-1})$.

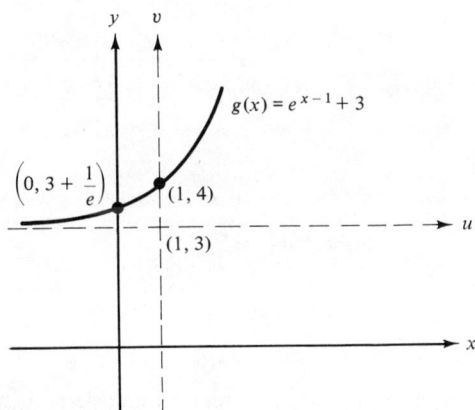

Figure 8-12

Example 12

Graph the function $f(x) = \ln |x|$.

Solution.   This is an even function, defined for all real x except zero. For $x > 0$, $f(x) = \ln x$; the remainder of the graph is obtained by reflecting through the y axis. (See Fig. 8-13.)

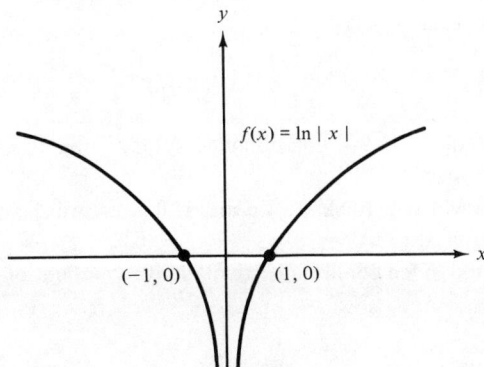

Figure 8-13

Example 13

Graph the function $h(x) = e^{-x^2}$.

Solution. This function is defined for all real x. There is an intercept at $(0, 1)$. For $x \geq 0$ we know that x^2 is increasing and so $-x^2$ is decreasing Because $e > 1$, we see that e^{-x^2} is decreasing. As $x \longrightarrow +\infty$, $-x^2 \longrightarrow -\infty$ and thus e^{-x^2} approaches zero. This gives us the graph in Fig. 8-14(a). Using the fact that h is even, and reflecting through the y axis, we get the complete graph in Fig. 8-14(b).

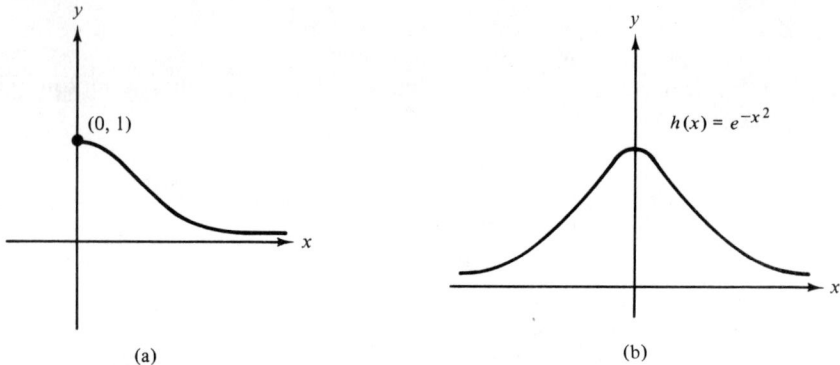

(a)   (b)

Figure 8-14

EXERCISES 8-4

†1. You invest $500 in a savings account at 5% compounded quarterly, leaving all interest on deposit.
   (a) What is the amount of money in your account at the end of 2 years?
   (b) How much interest is paid to you during the first year?
   (c) How much interest is paid to you during the second year?

†2. Mr. Johnson saves $6000 for his son's education. If he puts this money into an account at 8% compounded quarterly, how much will it be worth after 4 years?

†3. Find the amount in a savings account if $1000 is invested at 8% and interest is compounded
   (a) semiannually (twice a year).
   (b) quarterly.
   (c) monthly.
   (d) continuously.

†4. Mrs. Sullivan invests $10,000 at 15% compounded monthly. How much is in her account at the end of one year?

†5. An investment of $1000 is worth $1500 after 4 years. If interest was compounded continuously, what was the interest rate?

†6. At what interest rate, compounded continuously, will a given amount of money double
   (a) in 12 years?
   (b) in 8 years?
   (c) in 6 years?

†7. If the population of a country, initially 25 million, increases exponentially and reaches 30 million when $t = 10$ years, express the population as a function of t. What will the population be when $t = 20$ years?

†8. A biologist is studying the growth of a population of insects in a controlled environment. Assuming that the population grows exponentially and that it was 100,000 one day and 250,000 4 days later, when would the population reach a million?

†9. If the population of a village grows exponentially and increases by 10% in 4 years, how many more years will it take for the original population to double?

10. A radioactive substance is decaying according to the exponential law. If 1 gram is present initially and .9 gram is present after 50 years, express the amount remaining after t years as a function of t.

†11. Of the initial amount of a radioactive substance 30% remains after a period of 50 years. What percentage remains after 75 years? after 100 years?

†12. The quantity of a drug in the bloodstream diminishes according to the law of exponential decay. If half the initial amount is present after one week, how many weeks will it take before less than .1% of the initial amount remains?

In Problems 13–22 sketch a graph of $Q = f(t)$ for t in the given interval. Does the function represent exponential growth or decay?

13. $Q = 3e^{.2t}$ $[0, 10]$
14. $Q = \frac{1}{2}e^{.5t}$ $[0, 10]$
15. $Q = \frac{1}{3}e^{.05t}$ $[0, 10]$
16. $Q = 4e^{.01t}$ $[0, 10]$
17. $Q = 5e^{3t}$ $[0, 2]$
18. $Q = 300e^{6t}$ $[0, 1]$
19. $Q = \frac{1}{2}e^{-.05t}$ $[0, 20]$
20. $Q = 2e^{-.1t}$ $[0, 20]$
21. $Q = (18)2^{-t/20}$ $[0, 60]$
22. $Q = (56)2^{-t/10}$ $[0, 4]$

In Problems 23–36 sketch the graph.

23. $y = e^{x-1}$
24. $y + 2 = e^{x-3}$
25. $y = 3e^{x-2}$
26. $y = 3e^x - 2$
27. $y = e^{|x|}$
28. $y = e^{x^2}$
29. $y = e^{x^2+1}$
30. $y = \ln(x - 3)$
31. $y = \ln(x + 1) + 2$
32. $y = \ln(x^2 + e)$
33. $y = (\ln x)^2$
34. $y = |\ln x|$
35. $y = \sqrt{\ln x}$
36. $y = \ln|x| + 3$

FOR THE ENTHUSIASTS

37. Show that the law of exponential decay

$$f(t) = Q_0 e^{rt}$$

has the half-life property: there is a fixed number $H > 0$ such that for any time t,

$$f(t + H) = \frac{1}{2}f(t).$$

In other words, for any interval of the form $[t, t + H]$ there is half as much of the quantity at the end of the interval as there was at the beginning.

38. What property of exponential growth is analogous to the half-life property in Problem 37? Verify your conclusion.

39. Two functions C and S are defined for real numbers x as follows:

$$C(x) = \frac{e^x + e^{-x}}{2} \qquad S(x) = \frac{e^x - e^{-x}}{2}.$$

(a) Show that C is even and S is odd.

(b) Graph the two functions.

(c) Prove that for all real x,

    (i) $[C(x)]^2 - [S(x)]^2 = 1$.

    (ii) $[C(x)]^2 + [S(x)]^2 = C(2x)$.

    (iii) $2C(x)S(x) = S(2x)$.

(d) Solve the equations $C(x) = 2$ and $S(x) = 2$.

(e) Show that S has an inverse and express S^{-1} in terms of the natural logarithm function.

Note. The functions C and S are called the **hyperbolic cosine** and the **hyperbolic sine**, respectively. They arise in applications of calculus to various engineering problems.

SUMMARY

This chapter has developed the exponential and logarithmic functions.

The use of exponents was initially discussed in Chapter 1 where we considered the case in which the exponent is an integer. In this chapter we saw how the basic exponential properties of Theorem 1-1 can be extended to apply to rational and irrational exponents, provided that we adopt suitable definitions.

1. You should know the definition of a^r, where the exponent r is rational. You should be able to convert expressions involving radicals into a form with fractional exponents (and vice versa).

2. You should know the algebraic properties of radicals and rational exponents as stated in Theorems 8-1 and 8-2 and properties (1) through (5) on page 260. You should be able to use these properties to simplify algebraic expressions.

3. You should know the characteristics of graphs of exponential functions (Theorem 8-3) and the algebraic properties of these functions (Theorem 8-4).

4. You should know that a logarithmic function is the inverse of an exponential function and be able to convert equations from exponential to logarithmic form.

5. You should understand how the algebraic properties of logarithms (Theorem 8-5) arise from corresponding properties of exponentials.

6. You should be able to convert from one base of logarithms to another, using Theorem 8-6.

7. You should know how the special number e occurs in connection with continuously compounded interest and should be familiar with the graphs of the exponential and logarithmic functions to base e.

8. You should be able to use the exponential and logarithmic functions in applications to continuously compounded interest, radioactive decay, and other instances of exponential growth and decay.

REVIEW EXERCISES

1. Express in terms of rational exponents. Assume that the variables are nonnegative.
   (a) $\sqrt[3]{a^2bc^4}$

   (b) $\sqrt{\dfrac{x^2+y^2}{z}}$

   (c) $5\sqrt[3]{x} - \dfrac{4}{\sqrt[3]{x^2}} + \dfrac{7}{x}$

   (d) $(2\sqrt{x}\,\sqrt[3]{y})^6$

   (e) $\sqrt{\sqrt{a}\,\sqrt[3]{b}}$

2. Express in terms of radicals without using fractional or negative exponents. Assume that the variables are nonnegative.
   (a) $u^{3/4}v^{1/3}w^{-1/2}$

   (b) $(x^2+y^2)^{-1/2}z^{1/3}$

   (c) $3x^{2/3} + 3x^{-2/3}$

   (d) $(4 + x^{1/2})^{1/2}$

3. Find the zeros.
   (a) $(x^2 - 1)(x^2 + x + 1)^{-1/2}$

   (b) $(x^2 - 1)(x^2 + 2x)^{1/2}$

   (c) $(x^2 - 1)(x^2 - 2x)^{-1/2}$

   (d) $(x^2 - 1)(x^2 + 2x)^{-1/2}$

   (e) $(x^2 - 1)(x^2 + 2x)^{-2/3}$

4. Graph.
   (a) $f(x) = 3^x + 4$

   (b) $g(x) = \dfrac{1}{3^x + 4}$

   (c) $f(x) = \dfrac{3^x - 3^{-x}}{2}$

   (d) $g(x) = 1 - 3^{1-x}$

   (e) $h(x) = e^{1-x^2}$

   (f) $f(x) = e^{1/x}$

5. Solve.
   (a) $2^x + 2^{x-1} - 12 = 0$

   (b) $2^x + 2^{-x} = \dfrac{3\sqrt{2}}{2}$

   (c) $3 \cdot 3^{2x} - 10 \cdot 3^x + 3 = 0$

   (d) $3^{3x+2} - 3^{2x+1} - 3^{x+1} + 1 = 0$

   (e) $2^x + 2^{-x} > 0$

   (f) $2^x - 2^{-x} \le 0$

   (g) $2^{-x^2} < \dfrac{1}{16}$

   (h) $2^{x^2} \ge 32$

6. Rewrite the equation in terms of logarithms to the base 5.
   (a) $5^x = u$

   (b) $5^{1-x^2} = v$

   (c) $\dfrac{1}{5^{1-x^2}} = w$

7. Rewrite the equation in terms of exponentials to the base 5.
   (a) $\log_5 u = \dfrac{1}{x}$

   (b) $\log_5 v + 1 = 3x$

   (c) $\log_5 w - 1 = x^2$

8. If $\log_a 2 = 1.386$ and $\log_a 3 = 2.197$, find the following logarithms.
   (a) $\log_a 4$
   (b) $\log_a 6$
   (c) $\log_a 27$
   (d) $\log_a (1/\sqrt{2})$

9. Simplify.
   (a) $\log_5 [25\sqrt{x+3}\,(x+3)^2]$

   (b) $\log_5 \dfrac{\sqrt[3]{(x+5)^2}(x+3)^2}{5(x^2+1)^4}$

   (c) $\ln \dfrac{\sqrt[3]{x^3+1}}{\sqrt[3]{e^3+1}}$

   (d) $\ln \dfrac{x + \sqrt{x^2-1}}{x - \sqrt{x^2-1}}$

10. Graph.
    (a) $f(x) = \log_{10}(x^2 - 1)$

    (b) $g(x) = 3 + \log_3(x - 1)$

    (c) $f(x) = \log_{10}\sqrt{x}$

    (d) $f(x) = \log_{10}(x^2 + 100)$

    (e) $g(x) = \ln\left(e + \dfrac{1}{x}\right)$

    (f) $h(x) = \ln(\ln x)$

11. Solve.
    (a) $\log_3 (x^2 - 2) + \log_3 (x^2 + 3) = \log_3 14$
    (b) $\log_5 (x^2 + 3) - \log_5 (x^2 - 3) = \log_5 2$

12. If a population grows exponentially and doubles after 5 years, after how long will it triple?

13. What proportion of an initial amount of radium Ra^{226} (half-life 1600 years) remains after 800 years?

The trigonometric functions

9-1 DEFINITIONS OF SINE AND COSINE

Periodic Functions

Many natural phenomena are periodic in the sense that they exhibit repetitions of behavior after a fixed period of time. The displacement (up-and-down motion) of a weight hanging on a spring, the swinging of a pendulum, or the vibration of a plucked string, for instance, are periodic phenomena. The changing of weather patterns throughout the seasons of the year can also be regarded as being approximately periodic. Underlying the mathematical study of periodicity are the trigonometric functions, the subject of this chapter. Before defining these particular functions, we consider the general concept of periodicity.

Definition.　A function f having the property that there is a number $p > 0$ such that

$$f(x) = f(x + p)$$

for every x in the domain of f is said to be **periodic** with **period** p. If there is a least positive number p satisfying this equation, it is called the **principal period** of f or just **the period** of f.

Example 1

In Example 5, Section 4-4, we used the symbol $[x]$ to denote the greatest integer less than or equal to x. Thus $[1.5] = 1$, $[\pi] = 3$, $[-3/2] = -2$ and so on. Now let us define a function D for real numbers x by the equation

$$D(x) = x - [x].$$

For positive numbers, $D(x)$ represents the "decimal" part of x because we have subtracted the integral part of x from x. We certainly should expect D to have period 1. To verify this property we use the result of Problem 8(a), Section 4-4: $[x + 1] = [x] + 1$. For every x we have

$$\begin{aligned} D(x + 1) &= (x + 1) - [x + 1] \\ &= x + 1 - ([x] + 1) \qquad \text{[Problem 8(a), Section 4-4]} \\ &= x - [x] \\ &= D(x) \end{aligned}$$

and so D does indeed have period 1. Because D is increasing on the interval $[0, 1)$, it is clear that there is no positive number p less than 1 such that D has period p. So 1 is the principal period of D. The graph appears in Fig. 9-1.

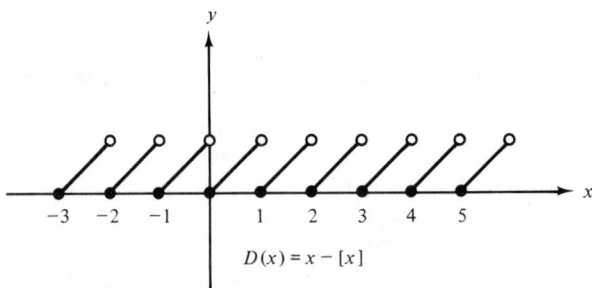

$$D(x) = x - [x]$$

Figure 9-1

If f is periodic with period p, it is easy to show that $f(x) = f(x \pm p) = f(x \pm 2p) = \cdots$ and, in general, $f(x + np) = f(x)$ for every integer n and every x in the domain of f.

Consequently, if the values of f are known on an interval of length p, then *all* its values can be determined. To graph a function with period p, we first draw that part of the graph over some interval of length p; the remainder of the graph consists of repetitions of this part.

Example 2

A periodic function with period 4 is defined for $0 \le x < 4$ by the equations

$$f(x) = \begin{cases} x & \text{for } 0 \le x < 3 \\ 12 - 3x & \text{for } 3 \le x < 4 \end{cases}.$$

Draw a graph of f and find $f(x)$ for $x = 1, 3, 4, 5, 8, 9, 12, -1,$ and -4.

Solution. The graph of f over the interval $[0, 4)$ is shown in Fig. 9-2(a). Because of the periodicity, it can be extended as in Fig. 9-2(b). The values of f at 5, 9, and -3 are the same as the value at 1 because 5, 9, and -3 differ from 1 by integral multiples of 4. Thus

$$f(5) = f(9) = f(-3) = f(1) = 1.$$

For similar reasons, the values of f at 4, 8,12, and -4 are the same as the value at zero. Thus

$$f(4) = f(8) = f(12) = f(-4) = f(0) = 0.$$

If we should wish to find $f(39.5)$, we observe that $39.5 = 3.5 + 9 \cdot 4$. So $f(39.5) = f(3.5) = 12 - 10.5 = 1.5$.

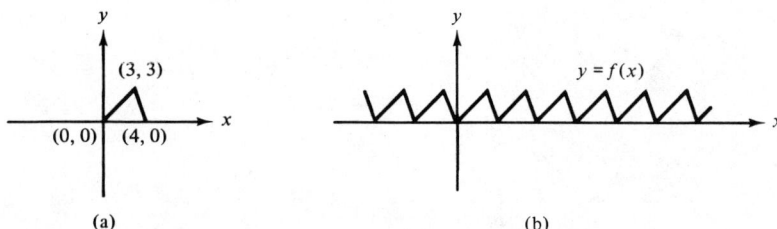

Figure 9-2

The Point P_x

Consider the **unit circle** C with radius 1 and center at the origin of the u, v plane. This circle is the graph (Fig. 9-3) of the equation

$$u^2 + v^2 = 1. \tag{9-1}$$

We are going to describe a correspondence that associates with each real number x a point P_x on C. Here is the rule of correspondence. Beginning at the point $A(1, 0)$ on C, we construct a path along C of length $|x|$; this path is taken in the counterclockwise direction around C if $x > 0$, in the clockwise direction if $x < 0$. The endpoint of this path is defined to be P_x. (See Fig. 9-4.) If $x = 0$, let us agree that P_x is just $A(1, 0)$.

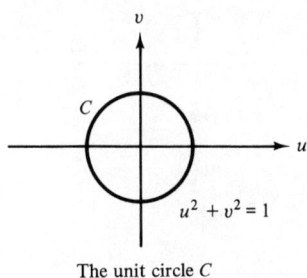

The unit circle C

Figure 9-3

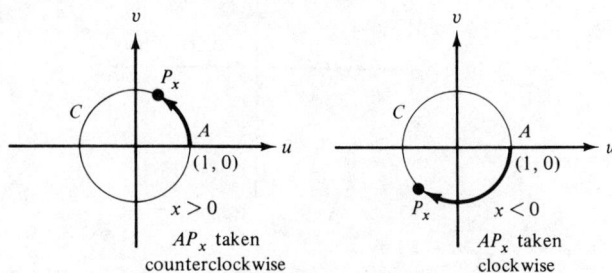

Figure 9-4

We have listed P_x for various values of x in Table 9-1. Let us see how some of the entries were found. Because the circumference of the unit circle C is 2π, the path determining the point $P_{2\pi}$ requires exactly one full trip around C so that $P_{2\pi} = A(1, 0)$ [Fig. 9-5(a)]. The path determining P_π covers just one-half the unit circle, ending at $(-1, 0)$ [Fig. 9-5(b)], and the path determining $P_{\pi/2}$ ends one-quarter of the way

TABLE 9-1

x	P_x
0	$(1, 0)$
$\pi/2$	$(0, 1)$
π	$(-1, 0)$
$3\pi/2$	$(0, -1)$
2π	$(1, 0)$
$5\pi/2$	$(0, 1)$
3π	$(-1, 0)$
$-\pi/2$	$(0, -1)$
$-\pi$	$(-1, 0)$
-2π	$(1, 0)$

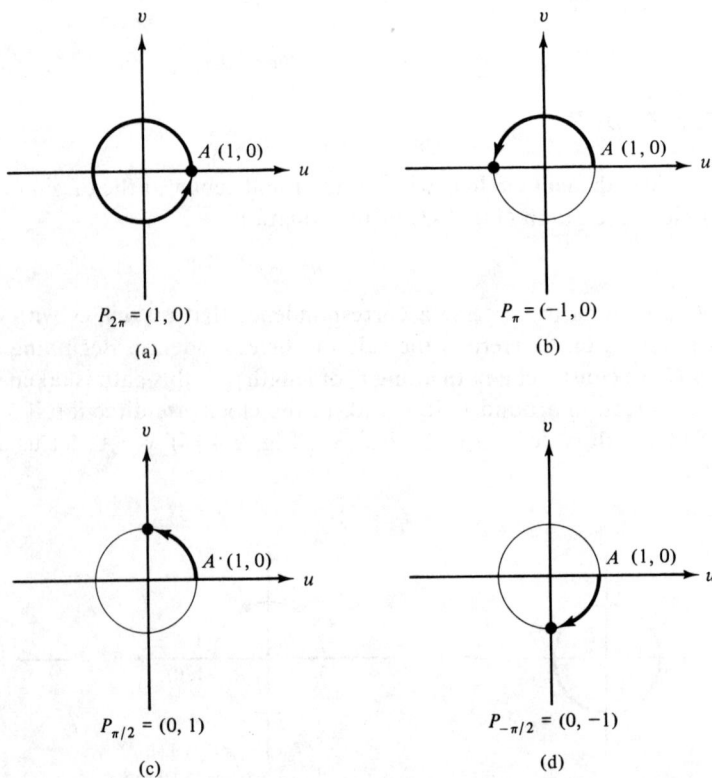

Figure 9-5

around C at $(0, 1)$ [Fig. 9-5(c)]. Because the number $-\pi/2$ is negative, $P_{-\pi/2}$ is obtained by traveling clockwise one-quarter of the way around C to $(0, -1)$ [Fig. 9-5(d)].

When $|x| > 2\pi$, the path determining P_x overlaps itself; the points $P_{3\pi}$ and $P_{-5\pi/2}$ are shown in Fig. 9-6.

The Trigonometric Functions   Chap. 9

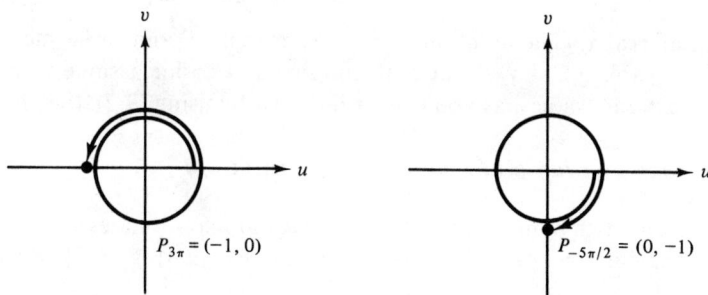

Figure 9-6

Sine and Cosine Functions

We are now ready to define two of the six trigonometric functions.

Definition.

1. The **sine** function is defined for all real numbers by

$$\text{sine } (x) = \text{the ordinate } (v \text{ coordinate) of } P_x.$$

2. The **cosine** function is defined for all real numbers by

$$\text{cosine } (x) = \text{the abscissa } (u \text{ coordinate) of } P_x.$$

It is customary to write $\sin x$ for the value sine (x) and $\cos x$ for cosine (x). Referring to Table 9-1, we easily get the values for the sine and cosine functions shown in Table 9-2.

TABLE 9-2

x	P_x	$\sin x$	$\cos x$
0	$(1, 0)$	0	1
$\pi/2$	$(0, 1)$	1	0
π	$(-1, 0)$	0	-1
$3\pi/2$	$(0, -1)$	-1	0
2π	$(1, 0)$	0	1
$5\pi/2$	$(0, 1)$	1	0
3π	$(-1, 0)$	0	-1
$-\pi/2$	$(0, -1)$	-1	0
$-\pi$	$(-1, 0)$	0	-1
-2π	$(1, 0)$	0	1

Because the point P_x always lies on the unit circle in the u, v plane, its coordinates must satisfy Eq. (9-1) and so

$$\cos^2 x + \sin^2 x = 1 \qquad \text{(9-2)}$$

for all real x. [The notation $\cos^2 x$ means $(\cos x)^2$ and $\sin^2 x$ means $(\sin x)^2$.]

Table 9-2 shows that both the sine and cosine assume values 1 and -1. These are extreme values, as you can easily show by using (9-2). (See Problems 50 and 51.)

Values of Sine and Cosine at $\pi/4$ and $5\pi/4$

There are two points on C at which the u and v coordinates are equal. They correspond to the intersections of C with the line $v = u$ [Fig. 9-7]. Solving the system of simultaneous equations

$$u^2 + v^2 = 1$$

$$v = u,$$

we obtain $2u^2 = 1$, $u = \pm\sqrt{2}/2$. The two points of intersection are $(\sqrt{2}/2, \sqrt{2}/2)$ and $(-\sqrt{2}/2, -\sqrt{2}/2)$. It is easy to see that the first point is P_x for $x = \pi/4$ whereas the second is P_x for $x = 5\pi/4$. We therefore have

$$\sin\frac{\pi}{4} = \cos\frac{\pi}{4} = \frac{\sqrt{2}}{2}$$

$$\sin\frac{5\pi}{4} = \cos\frac{5\pi}{4} = -\frac{\sqrt{2}}{2}.$$

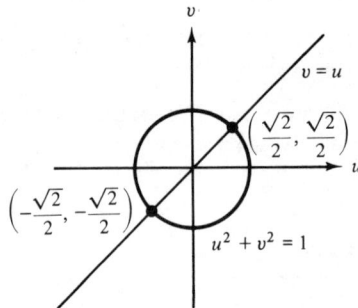

Figure 9-7

Values of Sine and Cosine at $\pi/3$

It is possible to inscribe within any circle of radius r a regular hexagon H whose sides are each of length r. By connecting the six vertices of H with the center of the circle, six equilateral triangles are formed. We have carried out this construction in Fig. 9-8 with the unit circle C, choosing one vertex of the hexagon to be at $A(1, 0)$. Now the vertices of H divide C into arcs of length $(\frac{1}{6})(2\pi) = \pi/3$; so the first of these vertices following A (counterclockwise) is simply $P_{\pi/3}$. If we can determine the coordinates of this vertex, we will know the values of cosine and sine at $\pi/3$.

Because the triangle with vertices 0, A, and $P_{\pi/3}$ is equilateral the altitude from $P_{\pi/3}$ bisects the base OA and so the abscissa of $P_{\pi/3}$ is $1/2$. Applying the Pythagorean theorem to the right triangle $OMP_{\pi/3}$, we then find the ordinate of $P_{\pi/3}$ to be $\sqrt{1 - (\frac{1}{2})^2} = \sqrt{3}/2$. Therefore

The Trigonometric Functions   Chap. 9

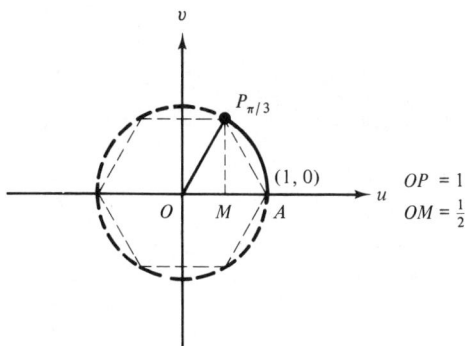

Figure 9-8

$$\sin \frac{\pi}{3} = \frac{\sqrt{3}}{2}$$

$$\cos \frac{\pi}{3} = \frac{1}{2}.$$

Periodicity of Sine and Cosine

It is obvious that the points P_x and $P_{x+2\pi}$ coincide, for $P_{x+2\pi}$ lies exactly one circumference around C from P_x. Because P_x and $P_{x+2\pi}$ are the same point, their respective coordinates are equal and we have

$$\left.\begin{array}{l} \sin x = \sin (x + 2\pi) \\ \cos x = \cos(x + 2\pi) \end{array}\right\} \tag{9-3}$$

for all real numbers. These equations express the periodicity of the sine and cosine functions. Clearly 2π is the first value of x greater than zero for which cosine has the value 1; thus 2π is the principal period of cosine. Similarly, 2π is also the principal period of sine.

Theorem 9-1. The functions sine and cosine are both periodic with principal period 2π.

Example 3

Evaluate.

(a) $\sin \dfrac{9\pi}{2}$

(b) $\sin \dfrac{21\pi}{4}$

Solution. We use the periodicity.

(a) The value of sine at $9\pi/2$ is the same as its value at $\pi/2$ because $9\pi/2$ and $\pi/2$ differ by 4π, an integral multiple of the period 2π. Thus $\sin 9\pi/2 = \sin \pi/2 = 1$.

(b) Similarly, $\sin 21\pi/4 = \sin 5\pi/4 = -\sqrt{2}/2$ because $21\pi/4$ and $5\pi/4$ differ by an integral multiple of 2π.

Example 4

Evaluate $\cos\left(\dfrac{-5\pi}{3}\right)$.

Solution. Because of the periodicity,

$$\cos\left(\frac{-5\pi}{3}\right) = \cos\left(\frac{-5\pi}{3} + 2\pi\right) = \cos\frac{\pi}{3} = \frac{1}{2}.$$

EXERCISES 9-1

In Problems 1–4 a periodic function with given period p is defined over an interval of length p. Graph the function over the interval and then extend the graph so as to show the periodicity. Evaluate the function at the given values of x and show the corresponding points on your graph.

1. $f(x) = \begin{cases} x & \text{for } 0 \le x < \dfrac{1}{2} \\ 1 - x & \text{for } \dfrac{1}{2} \le x < 1 \end{cases}$ $(p = 1)$

$x = 0, \dfrac{1}{4}, \dfrac{1}{2}, \dfrac{3}{4}, 1, \dfrac{3}{2}, 2, -\dfrac{1}{4}, -\dfrac{1}{2}, -\dfrac{3}{4}, -1$

2. $g(x) = \sqrt{4 - x^2}$ for $-2 \le x < 2$ $(p = 4)$

$x = -2, -1, 0, 1, 2, 3, 4, 6, 8$

3. $h(x) = x$ for $-1 \le x < 1$ $(p = 2)$

$x = .1, .5, .9, 1.1, 1.5, 1.9, -.1, -.5, -.9, -1.1, -1.5, -1.9$

4. $F(x) = \dfrac{1}{x}$ for $0 < x < 1$ $(p = 1)$

$x = .1, .5, .9, 1.1, 1.5, 1.9, -.1, -.5, -.9$

In Problems 5–18 find the coordinates of P_x for the specified value of x.

5. $x = 3\pi$

6. $x = \dfrac{7\pi}{2}$

7. $x = -5\pi$

8. $x = -2\pi$

9. $x = 5\pi$

10. $x = \dfrac{9\pi}{4}$

11. $x = \dfrac{13\pi}{4}$

12. $x = \dfrac{-11\pi}{4}$

13. $x = \dfrac{-3\pi}{4}$

14. $x = \dfrac{-7\pi}{4}$

15. $x = \dfrac{7\pi}{3}$

16. $x = \dfrac{-5\pi}{3}$

17. $x = \dfrac{-11\pi}{3}$

18. $x = \dfrac{19\pi}{3}$

In Problems 19–36 find the indicated function value.

19. $\sin 3\pi$

20. $\sin \dfrac{7\pi}{2}$

21. $\sin (-5\pi)$

22. $\cos (-2\pi)$

23. $\cos \dfrac{7\pi}{2}$

24. $\cos 5\pi$

25. $\sin \dfrac{9\pi}{4}$

26. $\cos \dfrac{13\pi}{4}$

27. $\cos \dfrac{-11\pi}{4}$ **28.** $\sin \dfrac{-3\pi}{4}$ **29.** $\sin \dfrac{13\pi}{4}$ **30.** $\cos \dfrac{-7\pi}{4}$

31. $\sin \dfrac{7\pi}{3}$ **32.** $\cos \dfrac{7\pi}{3}$ **33.** $\sin \dfrac{-5\pi}{3}$ **34.** $\sin \dfrac{19\pi}{3}$

35. $\cos \dfrac{-11\pi}{3}$ **36.** $\cos \dfrac{-5\pi}{3}$

37. By finding the points of intersection of the unit circle C with line $v = -u$, show that

$$\sin \frac{3\pi}{4} = \frac{\sqrt{2}}{2}, \quad \cos \frac{3\pi}{4} = -\frac{\sqrt{2}}{2},$$

$$\sin \frac{7\pi}{4} = -\frac{\sqrt{2}}{2}, \quad \cos \frac{7\pi}{4} = \frac{\sqrt{2}}{2}.$$

In Problems 38–45 use the results of Problem 37 to find the indicated function value.

38. $\sin \dfrac{15\pi}{4}$ **39.** $\sin \dfrac{-\pi}{4}$ **40.** $\cos \dfrac{-9\pi}{4}$ **41.** $\cos \dfrac{11\pi}{4}$

42. $\sin \dfrac{-13\pi}{4}$ **43.** $\cos \dfrac{15\pi}{4}$ **44.** $\sin \dfrac{11\pi}{4}$ **45.** $\cos \dfrac{-13\pi}{4}$

FOR THE ENTHUSIASTS

46. Given that $\sin x = 1/3$, use Eq. (9-2) to find two possible values for $\cos x$. If it is also given that $0 \le x \le \pi/2$, what is $\cos x$?

47. Given that $\cos x = -1/3$, use (9-2) to find two possible values for $\sin x$. If it is also given that $\pi \le x \le 2\pi$, what is $\sin x$?

48. Let α denote that real number between 0 and $\pi/2$ such that $\cos \alpha = 3/5$.
  (a) Show geometrically that there is another real number β between 0 and 2π that $\cos \beta = 3/5$.
  (b) Express β in terms of α.
  (c) Show that $\sin \alpha = 4/5$ and $\sin \beta = -4/5$.

49. Let α denote that real number between 0 and $\pi/2$ such that $\sin \alpha = 12/13$.
  (a) Show geometrically that there is another real number β between 0 and 2π such that $\sin \beta = 12/13$.
  (b) Express β in terms of α.
  (c) Show that $\cos \alpha = 5/13$ and $\cos \beta = -5/13$.

50. Use (9-2) to show that the sine function assumes a maximum value of 1 and a minimum value of -1. At what values of x are these extrema assumed?

51. Use (9-2) to show that the cosine function assumes a maximum value of 1 and a minimum value of -1. At what values of x are these extrema assumed?

52. Let f be defined for all real numbers as follows.

$$f(x) = \begin{cases} 1 & \text{if } x \text{ is rational} \\ 0 & \text{if } x \text{ is irrational} \end{cases}$$

Show that f has period p for every positive rational number p and, consequently, f does not have a principal period.

9-2 SOME FUNDAMENTAL PROPERTIES OF SINE AND COSINE

In this section we develop additional properties involving the sine and cosine functions. First, let us exhibit what we already know about these functions by sketching their graphs.

Graphs of Sine and Cosine

Because of periodicity, we can restrict our attention to the interval $0 \leq x \leq 2\pi$. Consider first the sine function. As x increases from 0 to $\pi/2$, point P_x moves counterclockwise around unit circle C from $(1, 0)$ to $(0, 1)$. The ordinate of P_x, $\sin x$, increases from 0 to 1 (its maximum value). In fact, from the last section we know several values along the way.

x	0	$\dfrac{\pi}{4}$	$\dfrac{\pi}{3}$	$\dfrac{\pi}{2}$
$\sin x$	0	$\dfrac{\sqrt{2}}{2}$	$\dfrac{\sqrt{3}}{2}$	1

As x increases from $\pi/2$ to π, P_x continues along C from $(0, 1)$ to $(-1, 0)$ and $\sin x$ decreases from 1 to 0. As x increases from π to $3\pi/2$ and then to 2π, P_x moves from $(-1, 0)$ to $(0, -1)$ to $(1, 0)$; $\sin x$ decreases from 0 to -1 (its minimum value) and then increases to 0. The graph in Fig. 9-9(a) exhibits these features. Using the periodicity, we now extend this graph, obtaining the "sine wave" in Fig. 9-9(b). The x intercepts are located at the numbers $x = n\pi$ where n is an integer.

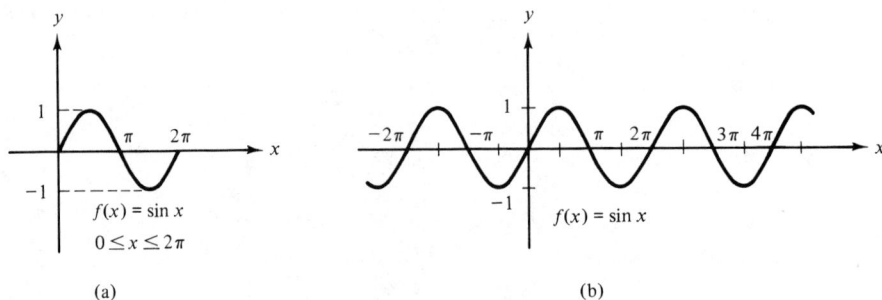

(a) (b)

Figure 9-9

A graph of the cosine function can be obtained in the same way, noting the behavior of the abscissa of P_x as x increases through interval $0 \leq x \leq 2\pi$. The graph for $0 \leq x \leq 2\pi$ is shown in Fig. 9-10(a) and its periodic extension in Fig. 9-10(b). Its x intercepts are at the points $x = \pi/2 + n\pi$ where n is an integer.

The Trigonometric Functions   Chap. 9

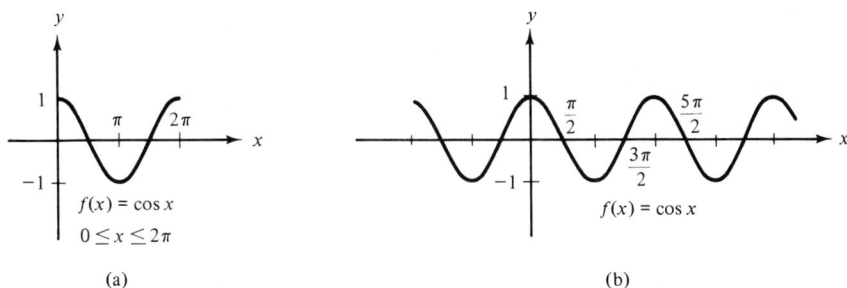

$f(x) = \cos x$

$0 \le x \le 2\pi$

(a)

$f(x) = \cos x$

(b)

Figure 9-10

There is a clear resemblance between the graphs of sine and cosine. In fact, each is a horizontal translation of the other (see Problem 35).

In the last section we saw that $\sin \pi/4 = \cos \pi/4$ and $\sin 5\pi/4 = \cos 5\pi/4$. This means that the graphs of the sine and cosine intersect where $x = \pi/4$ or $5\pi/4$ (Fig. 9-11). Because both functions have period 2π, their graphs also intersect when $x = \pi/4 + 2n\pi$ or $5\pi/4 + 2n\pi$, for some integer n.

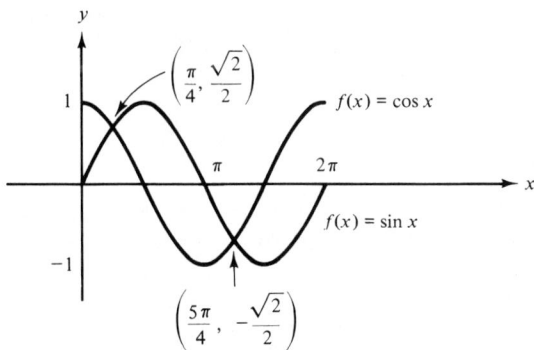

$\left(\dfrac{\pi}{4}, \dfrac{\sqrt{2}}{2} \right)$

$f(x) = \cos x$

$f(x) = \sin x$

$\left(\dfrac{5\pi}{4}, -\dfrac{\sqrt{2}}{2} \right)$

Figure 9-11

Symmetry of P_x and P_{-x}

It is clear geometrically (see Fig. 9-12) that P_x and P_{-x} are symmetric with respect to the u axis. Although the situation depicted in Fig. 9-12 shows a value of x between 0 and $\pi/2$, the symmetry of P_x and P_{-x} holds for all x and we have

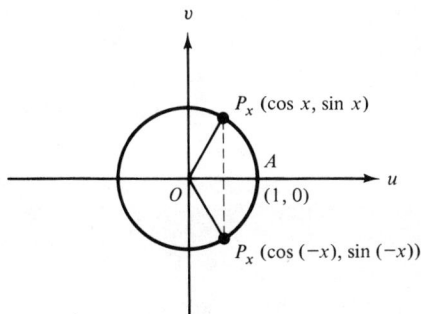

$P_x (\cos x, \sin x)$

A

$(1, 0)$

$P_x (\cos (-x), \sin (-x))$

Figure 9-12

$$\sin(-x) = \text{ordinate of } P_{-x} = -(\text{ordinate of } P_x) = -\sin x$$
$$\cos(-x) = \text{abscissa of } P_{-x} = \text{abscissa of } P_x = \cos x.$$

The results

$$\left.\begin{array}{r} \sin(-x) = -\sin x \\ \cos(-x) = \cos x \end{array}\right\} \qquad (9\text{-}4)$$

can be formulated in terms of odd and even functions:

Theorem 9-2. The sine function is odd and the cosine function is even.

Example 1

Find $\sin\left(\dfrac{-\pi}{3}\right)$ and $\cos\left(\dfrac{-\pi}{3}\right)$.

Solution. Because sine is odd,

$$\sin\left(\frac{-\pi}{3}\right) = -\sin\frac{\pi}{3} = -\frac{\sqrt{3}}{2}.$$

Because cosine is even,

$$\cos\left(\frac{-\pi}{3}\right) = \cos\frac{\pi}{3} = \frac{1}{2}.$$

Symmetry of P_x and $P_{\pi-x}$

The points P_x and $P_{\pi-x}$ are symmetric with respect to the v axis (Fig. 9-13). The ordinates of $P_{\pi-x}$ and P_x are the same whereas the abscissas are opposite in sign. Consequently

$$\left.\begin{array}{r} \sin(\pi - x) = \sin x \\ \cos(\pi - x) = -\cos x \end{array}\right\} . \qquad (9\text{-}5)$$

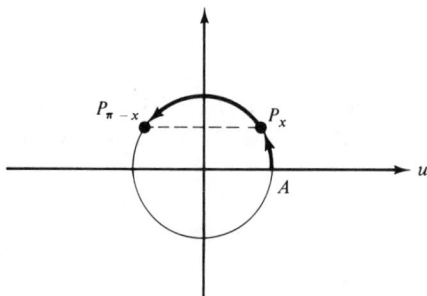

Figure 9-13

Example 2

Find $\sin\dfrac{3\pi}{4}$ and $\cos\dfrac{3\pi}{4}$.

Solution. By (9-5),

$$\sin \frac{3\pi}{4} = \sin\left(\pi - \frac{\pi}{4}\right) = \sin \frac{\pi}{4} = \frac{\sqrt{2}}{2}$$

and

$$\cos \frac{3\pi}{4} = \cos\left(\pi - \frac{\pi}{4}\right) = -\cos \frac{\pi}{4} = -\frac{\sqrt{2}}{2}.$$

Symmetry of P_x and $P_{x+\pi}$

The points P_x and $P_{x+\pi}$ are symmetric with respect to the origin (Fig. 9-14). Consequently,

$$\left.\begin{aligned}\sin(x+\pi) &= -\sin x\\ \cos(x+\pi) &= -\cos x\end{aligned}\right\}. \tag{9-6}$$

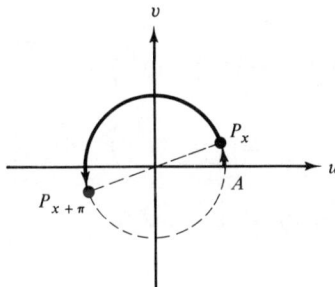

Figure 9-14

Example 3

Find $\sin \dfrac{4\pi}{3}$ and $\cos \dfrac{4\pi}{3}$.

Solution. By (9-6),

$$\sin \frac{4\pi}{3} = \sin\left(\frac{\pi}{3} + \pi\right) = -\sin \frac{\pi}{3} = -\frac{\sqrt{3}}{2}$$

$$\cos \frac{4\pi}{3} = \cos\left(\frac{\pi}{3} + \pi\right) = -\cos \frac{\pi}{3} = -\frac{1}{2}.$$

Symmetry of P_x and $P_{\pi/2-x}$

The points P_x and $P_{\pi/2-x}$ are symmetric with respect to the line $v = u$ (Fig. 9-15). According to Section 7-3, therefore, the coordinates of $P_{\pi/2-x}$ can be obtained by interchanging the coordinates of P_x. Thus

$$\left.\begin{aligned}\sin\left(\frac{\pi}{2} - x\right) &= \cos x\\ \cos\left(\frac{\pi}{2} - x\right) &= \sin x\end{aligned}\right\} \tag{9-7}$$

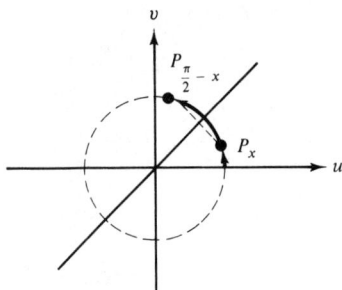

Figure 9-15

Example 4

Find $\sin \dfrac{\pi}{6}$ and $\cos \dfrac{\pi}{6}$.

Solution. By (9-7),

$$\sin \frac{\pi}{6} = \sin\left(\frac{\pi}{2} - \frac{\pi}{3}\right) = \cos \frac{\pi}{3} = \frac{1}{2},$$

$$\cos \frac{\pi}{6} = \cos\left(\frac{\pi}{2} - \frac{\pi}{3}\right) = \sin \frac{\pi}{3} = \frac{\sqrt{3}}{2}.$$

Solving Equations Involving Sines and Cosines

Example 5

Find all values of x in the interval $[0, 2\pi]$ that satisfy

$$\sin^2 x - 1 = 0.$$

Solution. The given equation is equivalent to $\sin x = 1$ or $\sin x = -1$. So either $x = \pi/2$ or $x = 3\pi/2$.

Example 6

Find all values of x in the interval $[0, 2\pi]$ that satisfy

$$2 \cos x - \sqrt{3} = 0.$$

Solution. The given equation is equivalent to $\cos x = \sqrt{3}/2$. Recalling that $\cos x$ is the abscissa of P_x we expect two solutions (see Fig. 9-16), one in the interval $[0, \pi/2]$ and one in the interval $[3\pi/2, 2\pi]$. The first solution is $x = \pi/6$ (Example 4) and the second is $x = 11\pi/6$.

More on Graphs

Example 7

Graph the function $f(x) = 2 \cos x$ for $-\pi \leq x \leq 3\pi$.

Solution. The graph resembles the graph of the cosine function over that interval except that each function value is multipled by 2. The result is shown in Fig. 9-17.

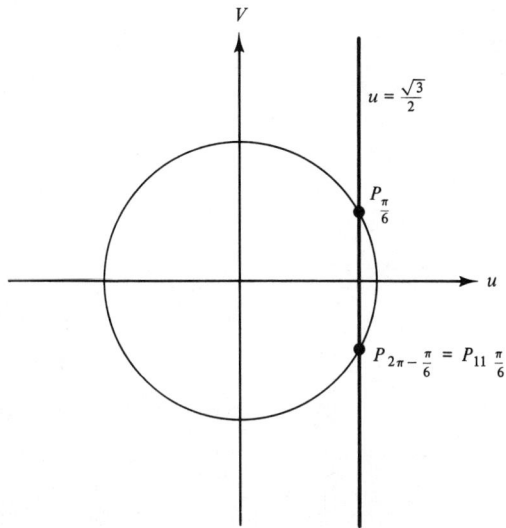

Figure 9-16

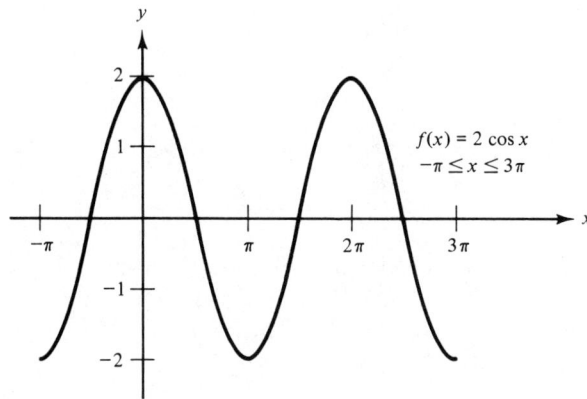

Figure 9-17

Example 8

Graph the function $g(x) = 2 \cos x + 1$ for $-\pi \leq x \leq 3\pi$.

Solution. The graph (Fig. 9-18) is a vertical translation of the graph of $2 \cos x$. Its x intercepts correspond to the solutions of $2 \cos x + 1 = 0$. There are four such solutions in the given interval:

$$x = \frac{-2\pi}{3}, \quad x = \frac{2\pi}{3}, \quad x = \frac{4\pi}{3}, \quad \frac{8\pi}{3}.$$

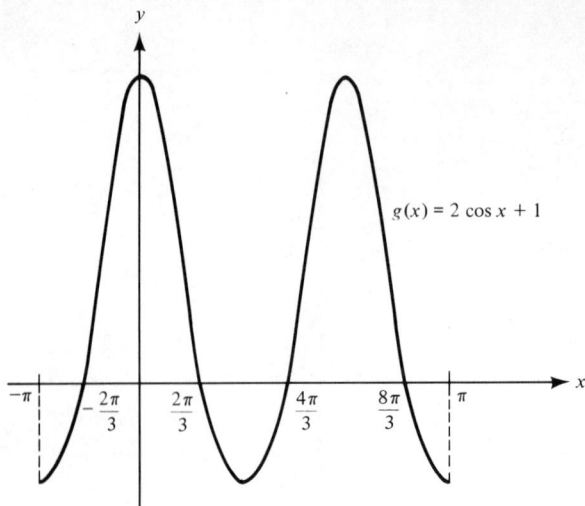

$g(x) = 2 \cos x + 1$

Figure 9-18

EXERCISES 9-2

1. **(a)** Fill in Table 9-3. Do as much as you can from memory. Check your answers against the results obtained in Sections 9-1 and 9-2. Learn the correct values if you do not already know them.

TABLE 9-3

x	$\sin x$	$\cos x$
0		
$\pi/6$		
$\pi/4$		
$\pi/3$		
$\pi/2$		

(b) Using the equations derived from the symmetries in this section, extend Table 9-3 in part (a) to include the following additional values of x.

$$\frac{2\pi}{3}, \quad \frac{3\pi}{4}, \quad \frac{5\pi}{6}, \quad \pi, \quad \frac{7\pi}{6}, \quad \frac{5\pi}{4}, \quad \frac{4\pi}{3}, \quad \frac{3\pi}{2}, \quad \frac{5\pi}{3}, \quad \frac{7\pi}{4}, \quad \frac{11\pi}{6}, \quad 2\pi$$

(c) Using Eqs. (9-4), together with the results already obtained, extend your table in (b) to include the following negative values of x.

$$\frac{-\pi}{6}, \quad \frac{-\pi}{4}, \quad \frac{-\pi}{3}, \quad \frac{-\pi}{2}, \quad \frac{-2\pi}{3}, \quad \frac{-3\pi}{4}, \quad \frac{-5\pi}{6}, \quad -\pi$$

2. Draw a graph of the sine function on the interval $[-\pi, 2\pi]$ by plotting the points $(x, \sin x)$ obtained in Problem 1. Draw a graph of the cosine function on the same interval.

In Problems 3-22 find all values of x in the indicated interval that satisfy the given equation. (Use the tables in Problem 1 and, if necesary, the periodicity of the sine and cosine functions.)

3. $2 \sin x - 1 = 0$ $[0, 2\pi]$ 4. $\sin x + \frac{1}{2} = 0$ $[0, 2\pi]$

5. $2 \cos x - 1 = 0$ $[-2\pi, 2\pi]$ 6. $2 \cos x + \sqrt{3} = 0$ $[-2\pi, 2\pi]$

The Trigonometric Functions   Chap. 9

7. $\sin^2 x = 1$ $[0, 4\pi]$
8. $\cos^2 x = 1$ $[0, 4\pi]$
9. $2 \sin^2 x = 1$ $[-\pi, \pi]$
10. $4 \sin^2 x = 1$ $[0, 2\pi]$
11. $2 \cos^2 x = 1$ $[-\pi, 3\pi]$
12. $4 \cos^2 x = 3$ $[-\pi, 3\pi]$
13. $2 \sin^2 x - 3 \sin x + 1 = 0$ $[0, 4\pi]$
14. $2 \cos^2 x - \cos x - 1 = 0$ $[-2\pi, 2\pi]$
15. $2 \sin x \cos x = \sin x$ $[-2\pi, 2\pi]$
16. $2 \cos^2 x + \sqrt{3} \cos x = 0$ $[0, 4\pi]$
17. $2 \sin^2 x - \sin x - 1 = 0$ $[-2\pi, \pi]$
18. $2 \cos^2 x = 2 - \sin x$ $[0, 2\pi]$
19. $2 \sin^2 x - \cos x + 1 = 0$ $[0, 2\pi]$
20. $\cos^2 x + 2 \sin x - 1 = 0$ $[0, 2\pi]$
21. $2 \cos^2 x + 3 \cos x - 2 = 0$ $[-\pi, \pi]$
22. $2 \cos^2 x + \sin x - 2 = 0$ $[-\pi, \pi]$

In Problems 23–35 graph the function on the indicated interval.

23. $f(x) = 3 \sin x$ $[0, 2\pi]$
24. $f(x) = -\sin x$ $[0, 4\pi]$
25. $g(x) = 2 \cos x$ $[0, 4\pi]$
26. $h(x) = -\frac{1}{2} \cos x$ $[-\pi, \pi]$
27. $h(t) = \frac{1}{3} \sin t$ $[-\pi, 3\pi]$
28. $f(x) = 1 + \cos x$ $[0, 4\pi]$
29. $g(x) = 1 - \sin x$ $[-\pi, \pi]$
30. $f(x) = 2 + 2 \sin x$ $[0, 2\pi]$
31. $g(x) = \frac{1}{2}(1 - \cos x)$ $[-\pi, \pi]$
32. $F(x) = |\sin x|$ $[0, 2\pi]$
33. $g(x) = |\cos x|$ $[-\pi, \pi]$
34. $h(t) = \sin^2 t + \cos^2 t$ $[0, 2\pi]$
35. Use Eqs. (9-7) to show that the graph of $y = \sin x$ is a translation of the graph of $y = \cos x$.

9-3 DEFINITIONS OF THE OTHER TRIGONOMETRIC FUNCTIONS

Altogether there are six trigonometric functions: sine, cosine, tangent, secant, cotangent, and cosecant. Each of the latter four is defined in terms of the sine and cosine.

Tangent

> **Definition.** The **tangent** function is the quotient of sine divided by cosine.

The value of the tangent at x is denoted by $\tan x$ and so

$$\tan x = \frac{\sin x}{\cos x} \tag{9-8}$$

provided that $\cos x \neq 0$. The tangent is undefined at $x = \pi/2 + n\pi$, for every integer n.

Example 1

By definition,

$$\tan \frac{\pi}{4} = \frac{\sin \pi/4}{\cos \pi/4} = \frac{\sqrt{2}/2}{\sqrt{2}/2} = 1$$

$$\tan \frac{2\pi}{3} = \frac{\sin 2\pi/3}{\cos 2\pi/3} = \frac{\sqrt{3}/2}{-1/2} = -\sqrt{3}$$

$$\tan \pi = \frac{\sin \pi}{\cos \pi} = \frac{0}{-1} = 0$$

$$\tan \frac{3\pi}{2} \text{ is undefined.}$$

The number tan x can be interpreted geometrically as the slope of the segment connecting the origin O and P_x (Fig. 9-19). This segment is vertical if $x = \pi/2 + n\pi$ where n is an integer, and in such a case, neither its slope nor tan x is defined.

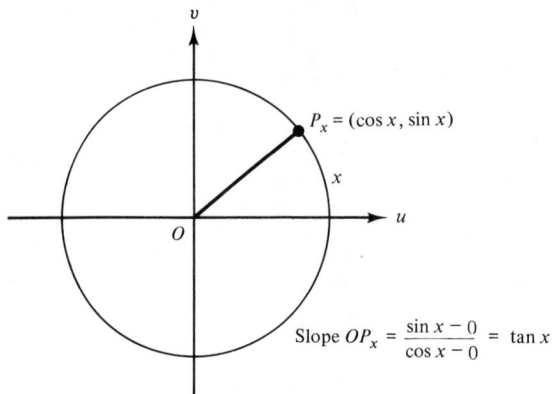

Slope $OP_x = \dfrac{\sin x - 0}{\cos x - 0} = \tan x$

Figure 9-19

The points P_x and $P_{x+\pi}$ lie on the same line through the origin (Fig. 9-14) so that OP_x and $OP_{x+\pi}$ have the same slope. Therefore

$$\tan x = \tan (x + \pi) \tag{9-9}$$

for all x in the domain of the tangent function, showing that this function has period π. It is clear that π is the principal period.

> **Theorem 9-3.** The tangent function is periodic with principal period π.

The graph of $y = \tan x$ in Fig. 9-20 shows its periodicity. There are intercepts

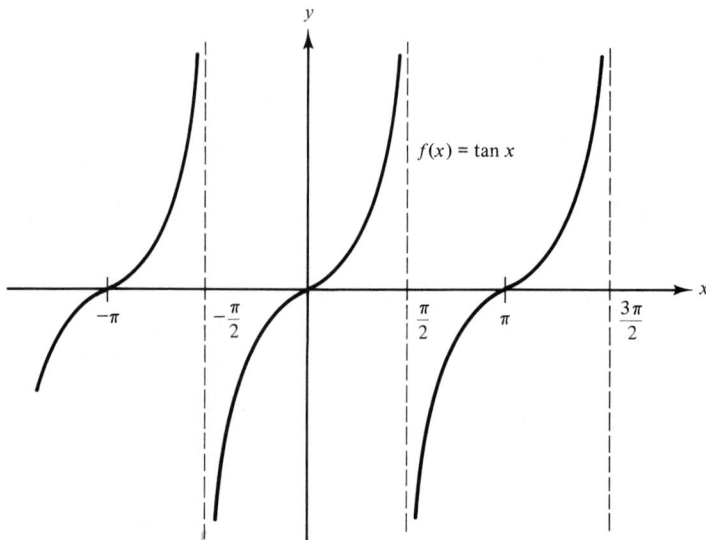

$f(x) = \tan x$

Figure 9-20

at the origin and $(n\pi, 0)$ for every integer n. As x increases through the interval $-\pi/2 < x < \pi/2$, $\tan x$ also increases. The graph has vertical asymptotes at $x = -\pi/2$ and $x = \pi/2$ because $\tan x \longrightarrow -\infty$ as x approaches $-\pi/2$ from the right and $\tan x \longrightarrow +\infty$ as x approaches $\pi/2$ from the left. Because of periodicity, there is a vertical asymptote at $x = \pi/2 + n\pi$ for every integer n; these numbers are points of discontinuity of the tangent function. The graph is symmetric with respect to the origin because tangent is an odd function:

$$\tan(-x) = \frac{\sin(-x)}{\cos(-x)} = \frac{-\sin x}{\cos x} = -\tan x. \tag{9-10}$$

Secant

> **Definition.** The **secant** function is the reciprocal of the cosine.

The value of the secant at x is denoted by $\sec x$ and so

$$\sec x = \frac{1}{\cos x} \tag{9-11}$$

provided that $\cos x \neq 0$. The secant is undefined at $x = \pi/2 + n\pi$, for every integer n.

Example 2

By definition,

$$\sec \frac{\pi}{3} = \frac{1}{\cos \pi/3} = \frac{1}{1/2} = 2$$

$$\sec \frac{5\pi}{6} = \frac{1}{\cos 5\pi/6} = \frac{1}{-\sqrt{3}/2} = \frac{-2\sqrt{3}}{3}$$

$\sec \dfrac{5\pi}{2}$ is undefined.

Like the cosine, the secant is an even function and has period 2π [Problem 2(a)]. Because the values of cosine are restricted to the interval $[-1, 1]$, the values of the secant always satisfy $|\sec x| \geq 1$.

The graph of the secant (Fig. 9-21) has vertical asymptotes at $x = \pi/2 + n\pi$, for every integer n; these numbers are points of discontinuity of the function.

If $\cos x \neq 0$, we can divide both sides of Eq. (9-2) by $\cos^2 x$.

$$\frac{\cos^2 x}{\cos^2 x} + \frac{\sin^2 x}{\cos^2 x} = \frac{1}{\cos^2 x}$$

$$1 + \tan^2 x = \sec^2 x \tag{9-12}$$

This result holds for every x except $x = \pi/2 + n\pi$, where n is an integer.

Cotangent

> **Definition.** The **cotangent** function is the quotient of cosine divided by sine.

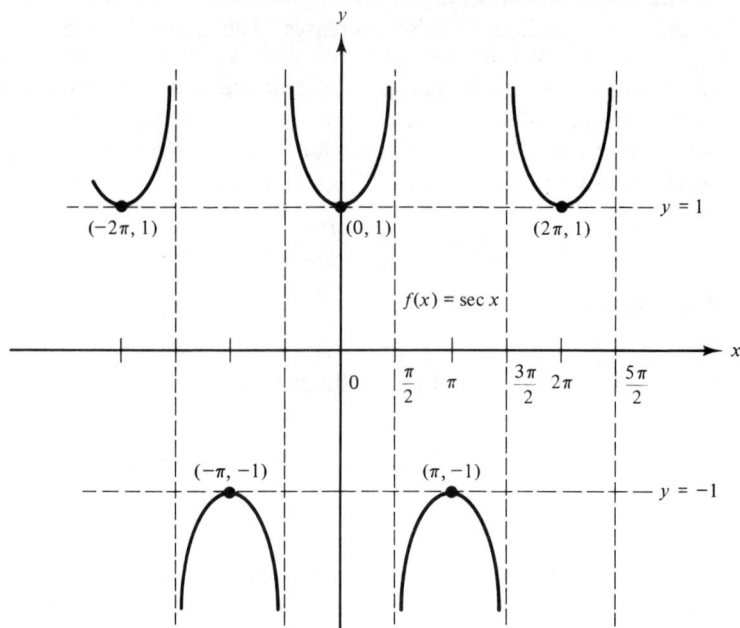

Figure 9-21

The value of the cotangent at x is denoted $\cot x$ so that

$$\cot x = \frac{\cos x}{\sin x} \tag{9-13}$$

provided that $\sin x \neq 0$. The cotangent is undefined at $x = n\pi$, for every integer n.

Example 3

By definition,

$$\cot \frac{\pi}{6} = \frac{\cos \pi/6}{\sin \pi/6} = \frac{\sqrt{3}/2}{1/2} = \sqrt{3}$$

$$\cot \frac{4\pi}{3} = \frac{\cos 4\pi/3}{\sin 4\pi/3} = \frac{-1/2}{-\sqrt{3}/2} = \frac{\sqrt{3}}{3}$$

$\cot 0$ is undefined.

Like the sine and tangent, the cotangent function is odd [Problem 2(b)]. Whenever $\tan x$ is different from zero, we have

$$\cot x = \frac{1}{\tan x} \tag{9-14}$$

and so the cotangent, like the tangent, has period π. The graph of $y = \cot x$ (Fig. 9-22) has vertical asymptotes at $x = n\pi$ for every integer n; these numbers are points of discontinuity of the cotangent function.

The Trigonometric Functions   Chap. 9

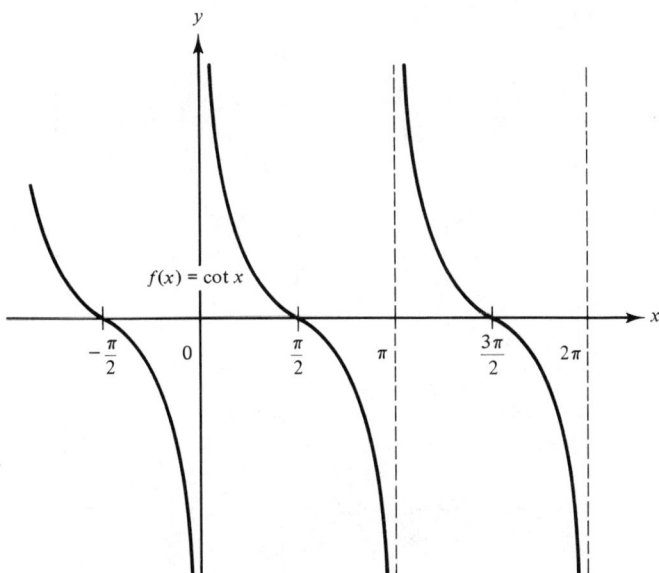

$f(x) = \cot x$

Figure 9-22

Cosecant

Definition.   The **cosecant** function is the reciprocal of the sine.

The value of the cosecant at x is denoted by csc x and so

$$\csc x = \frac{1}{\sin x} \qquad (9\text{-}15)$$

provided that $\sin x \neq 0$. The cosecant is undefined at $x = n\pi$, for every integer n.

Example 4

By definition,

$$\csc \frac{3\pi}{2} = \frac{1}{\sin 3\pi/2} = \frac{1}{-1} = -1$$

$$\csc \frac{-\pi}{3} = \frac{1}{\sin(-\pi/3)} = \frac{1}{-\sqrt{3}/2} = \frac{-2\sqrt{3}}{3}.$$

Like the sine, the cosecant is an odd function and has period 2π (Problem 2(c)). The values of the cosecant satisfy

$$|\csc x| \geq 1.$$

The graph (Fig. 9-23) has vertical asymptotes at $x = n\pi$ for every integer n; these numbers are points of discontinuity of the cosecant function.

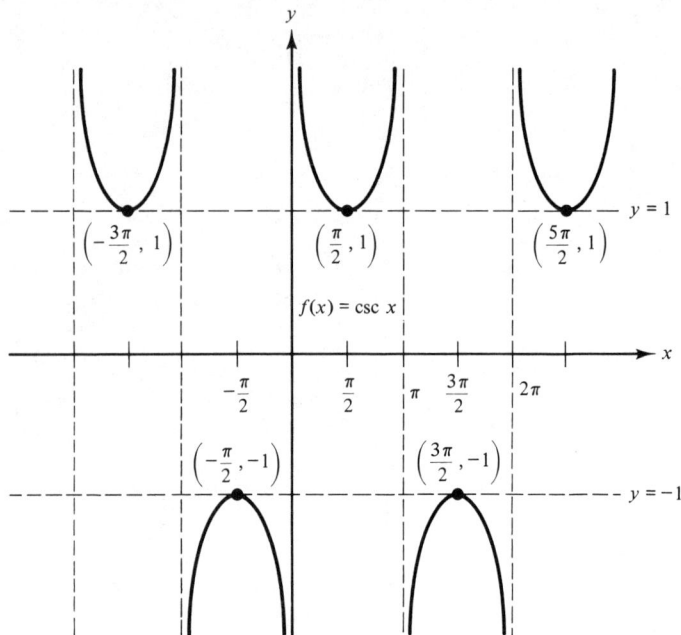

Figure 9-23

Starting with the basic equation

$$\cos^2 x + \sin^2 x = 1,$$

we can divide both sides by $\sin^2 x$, obtaining

$$\frac{\cos^2 x}{\sin^2 x} + \frac{\sin^2 x}{\sin^2 x} = \frac{1}{\sin^2 x}$$

$$\cot^2 x + 1 = \csc^2 x. \qquad (9\text{-}16)$$

This result holds whenever $\sin x \neq 0$—that is, for every x except $x = n\pi$, where n is an integer.

Example 5

Graph the function defined for real x by $f(x) = 2\sec x - 1$.

Solution. This graph can be obtained by multiplying each ordinate on the graph of the secant function by 2 and then translating downward by one unit. The graph appears in Fig. 9-24.

Example 6

Graph the function defined for real x by $g(x) = \tan^2 x$.

Solution. This graph can be obtained by squaring each ordinate on the graph of the tangent function. The graph appears in Fig. 9-25.

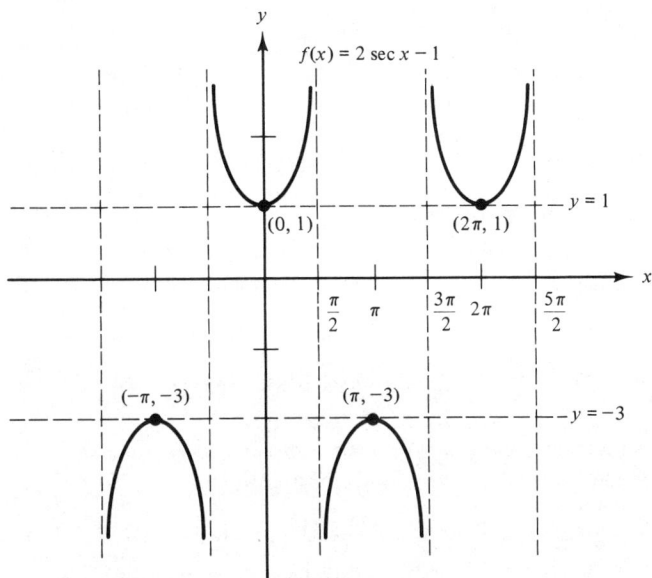

Figure 9-24

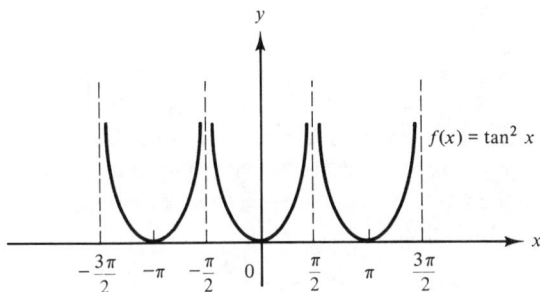

Figure 9-25

EXERCISES 9-3

1. Extend Table 9-3 to include columns for tangent, secant, cotangent, and cosecant. Fill in the values of all six trigonometric functions for all values of x indicated in Problem 1, parts (a) to (c) (Section 9-2).

2. Show that
   (a) secant is an even function.
   (b) cotangent is an odd function.
   (c) cosecant is an odd function.

In Problems 3–16 graph the function on the indicated interval. Identify any vertical asymptotes.

3. $f(x) = 2 \tan x \quad \left(-\dfrac{\pi}{2}, \dfrac{3\pi}{2}\right)$ **4.** $g(x) = \tan x + 1 \quad \left(-\dfrac{\pi}{2}, \dfrac{3\pi}{2}\right)$

5. $f(x) = 3 \sec x \quad [0, 4\pi]$ **6.** $f(x) = \dfrac{1}{2} \cot x \quad (-2\pi, 2\pi)$

7. $g(x) = \csc x - 1 \quad (-2\pi, 2\pi)$ **8.** $h(u) = 3(\sec u + 1) \quad [0, 2\pi]$

9. $f(t) = 1 - \tan t \quad [-\pi, \pi]$ **10.** $g(z) = \sec^2 z \quad [0, 2\pi]$

11. $h(z) = \tan^2 z - 2 \quad [0, 2\pi]$ **12.** $f(x) = \csc^2 x \quad (0, 2\pi)$

13. $g(u) = \cot^2 u + 1 \quad (0, 2\pi)$ **14.** $h(x) = |\sec x| \quad [0, 2\pi]$

15. $f(u) = |\csc u - 1| \quad (0, 2\pi)$ **16.** $g(z) = |\cot z| \quad (0, 2\pi)$

17. Graph $f(x) = \sec^2 x - 1$ and then compare your graph with Fig. 9-25.

FOR THE ENTHUSIASTS

18. Show that each of the following functions has period π.

(a) $\sec^2 x$ (b) $\csc^2 x$ (c) $\cos^2 x$

(d) $\sin^2 x$ (e) $|\cos x|$ (f) $|\sin x|$

19. Let α and β be the real numbers described in Problem 48 of Section 9-1. Evaluate the functions tangent, secant, cotangent, and cosecant at α and β.

20. Let α and β be the real numbers described in Problem 49 of Section 9-1. Evaluate the functions tangent, secant, cotangent, and cosecant at α and β.

9-4 TRIGONOMETRIC IDENTITIES

An equation involving one or more functions is called an **identity** if it holds for all values of the variables for which the functions are defined. Equations (9-2) through (9-16) established in Sections 9-1 through 9-3 are all identities. To avoid having this list seem too formidable a challenge to the memory, let us reorganize it somewhat.

Definitions

Several of the equations just mentioned are either definitions or immediate consequences of definitions.

$$\tan x = \frac{\sin x}{\cos x} \tag{9-8}$$

$$\sec x = \frac{1}{\cos x} \tag{9-11}$$

$$\cot x = \frac{\cos x}{\sin x} \tag{9-13}$$

$$\cot x = \frac{1}{\tan x} \tag{9-14}$$

$$\csc x = \frac{1}{\sin x} \tag{9-15}$$

Circular Properties

The identity

$$\cos^2 x + \sin^2 x = 1 \qquad (9\text{-}2)$$

can be remembered simply as saying that the point $P_x(\cos x, \sin x)$ lies on the unit circle. *This is the most fundamental relationship underlying the trigonometric functions.* Two additional identities are obtained directly from (9-2) by dividing by $\cos^2 x$ and $\sin^2 x$, respectively.

$$1 + \tan^2 x = \sec^2 x \qquad (9\text{-}12)$$

$$\cot^2 x + 1 = \csc^2 x \qquad (9\text{-}16)$$

These identities sometimes appear in the form $1 - \sin^2 x = \cos^2 x$, $\sec^2 x - 1 = \tan^2 x$, and so on.

Periodicity Properties

You should remember that the sine and cosine functions have period 2π, while the tangent function has period π.

$$\left.\begin{array}{l} \sin (x + 2\pi) = \sin x \\ \cos (x + 2\pi) = \cos x \end{array}\right\} \qquad (9\text{-}3)$$

$$\tan (x + \pi) = \tan x \qquad (9\text{-}9)$$

(What periods do secant, cotangent, and cosecant have?)

Symmetry Properties

The equations

$$\left.\begin{array}{l} \sin (-x) = -\sin x \\ \cos (-x) = \cos x \end{array}\right\} \qquad (9\text{-}4)$$

$$\tan (-x) = -\tan x \qquad (9\text{-}10)$$

were derived from the fact that P_x and P_{-x} are symmetric with respect to the origin. These identities express the fact that sine and tangent are odd whereas cosine is even. (What can you say in this regard concerning the other three functions?)

Several other identities were also derived from symmetry properties.

$$\left.\begin{array}{l} \sin (\pi - x) = \sin x \\ \cos (\pi - x) = -\cos x \end{array}\right\} \qquad (9\text{-}5)$$

$$\left.\begin{array}{l} \sin (x + \pi) = -\sin x \\ \cos (x + \pi) = -\cos x \end{array}\right\} \qquad (9\text{-}6)$$

$$\left.\begin{array}{l} \sin \left(\dfrac{\pi}{2} - x\right) = \cos x \\[2mm] \cos \left(\dfrac{\pi}{2} - x\right) = \sin x \end{array}\right\} \qquad (9\text{-}7)$$

As an aid in remembering them, you should reconstruct for yourself the diagrams in Figs. 9-12 through 9-15, exhibiting the symmetries on which these identities are based.

Proving Identities

Occasionally it is desirable to represent one trigonometric expression by another that is identical to it.

Example 1

Verify the identity

$$\frac{\sec x}{\csc x} = \tan x.$$

Solution. If x is any number for which $\sec x/\csc x$ and $\tan x$ are defined, we have

$$\frac{\sec x}{\csc x} = \frac{1/\cos x}{1/\sin x} \text{ by (9-11) and (9-15)}$$

$$= \frac{\sin x}{\cos x}$$

$$= \tan x \text{ by (9-8)}$$

Example 2

Verify the identity

$$\frac{1 - \sin x}{\cos x} = \frac{\cos x}{1 + \sin x}.$$

Solution. We first multiply the numerator and denominator on the left by $1 + \sin x$, and proceed as follows.

$$\frac{1 - \sin x}{\cos x} = \frac{(1 - \sin x)(1 + \sin x)}{\cos x \, (1 + \sin x)}$$

$$= \frac{1 - \sin^2 x}{\cos x \, (1 + \sin x)}$$

$$= \frac{\cos^2 x}{\cos x \, (1 + \sin x)} \text{ by (9-2)}$$

$$= \frac{\cos x}{1 + \sin x}$$

Each step is valid for all x such that $1 + \sin x$ and $\cos x$ are not zero.

Note. It is worth remembering the first step in the solution of Example 2. Multiplying $1 - \sin x$ by $1 + \sin x$ permits us to reduce the numerator to a single term, because $1 - \sin^2 x = \cos^2 x$. In the same way multiplying $1 - \cos x$ by $1 + \cos x$ is often helpful.

Example 3

Verify the identity

$$\frac{1}{1 - \sin x} + \frac{1}{1 + \sin x} = 2 \sec^2 x.$$

Solution. Combining the terms on the left, and noting that the common denominator $(1 + \sin x)(1 - \sin x)$ is $1 - \sin^2 x$ or $\cos^2 x$ we have

$$\frac{1}{1-\sin x} + \frac{1}{1+\sin x} = \frac{(1+\sin x) + (1-\sin x)}{1-\sin^2 x}$$

$$= \frac{2}{1-\sin^2 x}$$

$$= \frac{2}{\cos^2 x} \text{ by (9-2)}$$

$$= 2\sec^2 x \text{ by (9-11)}.$$

The following example makes use of (9-12) in the form $\sec^2 x - 1 = \tan^2 x$.

Example 4

Verify the identity

$$\tan^2 x - \sin^2 x = \sin^2 x \tan^2 x.$$

Solution. Using the definition of $\tan x$ we have

$$\tan^2 x - \sin^2 x = \frac{\sin^2 x}{\cos^2 x} - \sin^2 x \text{ by (9-8)}$$

$$= \sin^2 x \left(\frac{1}{\cos^2 x} - 1\right)$$

$$= \sin^2 x (\sec^2 x - 1) \text{ by (9-11)}$$

$$= \sin^2 x \tan^2 x \text{ by (9-12)}$$

Example 5

Verify the identity

$$\frac{\cos u}{1-\tan u} + \frac{\sin u}{1-\cot u} = \cos u + \sin u.$$

Solution. Using (9-8) and (9-13) we express $\tan u$ and $\cot u$ in terms of the basic functions $\sin u$ and $\cos u$. The result is simplified as follows.

$$\frac{\cos u}{1-\tan u} + \frac{\sin u}{1-\cot u} = \frac{\cos u}{1-\dfrac{\sin u}{\cos u}} + \frac{\sin u}{1+\dfrac{\cos u}{\sin u}} \text{ by (9-8), (9-13)}$$

$$= \frac{\cos^2 u}{\cos u - \sin u} + \frac{\sin^2 u}{\sin u - \cos u}$$

$$= \frac{\cos^2 u - \sin^2 u}{\cos u - \sin u}$$

$$= \cos u + \sin u$$

Sum and Difference Formulas

None of the identities (9-2) through (9-16) gives us any way of simplifying such expressions as $\cos(x - y)$ or $\sin(x + y)$. We begin by deriving an identity for $\cos(x - y)$. This identity, in combination with earlier identities, will enable us to express $\cos(x - y)$, $\sin(x - y)$, $\cos(x + y)$, and $\sin(x + y)$ in terms of $\cos x$, $\sin x$, $\cos y$, and $\sin y$.

Our derivation is based on the geometric fact that equal arcs on a circle subtend equal chords. Let x and y be two real numbers and consider the points P_x, P_y, P_{x-y} on the unit circle shown in Fig. 9-26. (For simplicity, we have drawn the diagrams with x and y positive and $x > y$, but the argument can be adapted to apply in any case.)

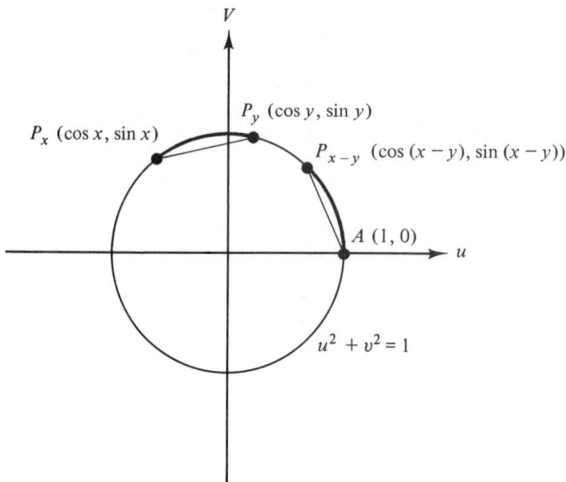

Figure 9-26

The chord from P_y to P_x has the same length as the chord from $A(1, 0)$ to P_{x-y} because these chords are subtended by arcs of equal length. Using the distance formula, we get

$$\sqrt{(\cos x - \cos y)^2 + (\sin x - \sin y)^2} = \sqrt{[\cos(x-y) - 1]^2 + [\sin(x-y) - 0]^2}.$$

Squaring both sides and simplifying,

$$(\cos x - \cos y)^2 + (\sin x - \sin y)^2$$
$$= [\cos(x-y) - 1]^2 + \sin^2(x-y)$$
$$(\cos^2 x - 2\cos x \cos y + \cos^2 y) + (\sin^2 x - 2\sin x \sin y + \sin^2 y)$$
$$= [\cos^2(x-y) - 2\cos(x-y) + 1] + \sin^2(x-y).$$

Using (9-2) three times, we obtain

$$2 - 2\cos x \cos y - 2\sin x \sin y = 2 - 2\cos(x-y).$$

Solving for $\cos(x-y)$, we get the identity

$$\cos(x-y) = \cos x \cos y + \sin x \sin y. \qquad (9\text{-}17)$$

Example 6

Using (9-17) with $x = \pi/3$ and $y = \pi/4$, evaluate $\cos \pi/12$.

Solution.

$$\cos \frac{\pi}{12} = \cos\left(\frac{\pi}{3} - \frac{\pi}{4}\right) = \cos \frac{\pi}{3} \cos \frac{\pi}{4} + \sin \frac{\pi}{3} \sin \frac{\pi}{4}$$
$$= \frac{1}{2} \cdot \frac{\sqrt{2}}{2} + \frac{\sqrt{3}}{2} \cdot \frac{\sqrt{2}}{2} = \frac{\sqrt{2} + \sqrt{6}}{4}$$
$$\approx .9659$$

Replacing y by $-y$ in (9-17) and remembering that cosine is even and sine is odd, we obtain an identity for $\cos(x + y)$.

$$\cos(x + y) = \cos[x - (-y)] = \cos x \cos(-y) + \sin x \sin(-y)$$

$$\cos(x + y) = \cos x \cos y - \sin x \sin y \qquad (9\text{-}18)$$

Example 7

Using (9-18) with $x = 2\pi/3$ and $y = \pi/4$, evaluate $\cos 11\pi/12$.

Solution.

$$\cos \frac{11\pi}{12} = \cos\left(\frac{2\pi}{3} + \frac{\pi}{4}\right) = \cos \frac{2\pi}{3} \cos \frac{\pi}{4} - \sin \frac{2\pi}{3} \sin \frac{\pi}{4}$$

$$= \left(-\frac{1}{2}\right) \cdot \frac{\sqrt{2}}{2} - \frac{\sqrt{3}}{2} \cdot \frac{\sqrt{2}}{2} = \frac{-\sqrt{2} - \sqrt{6}}{4}$$

$$\approx -.9659$$

To obtain formulas for $\sin(x - y)$ and $\sin(x + y)$, we make use of (9-7) and (9-18).

$$\sin(x - y) = \cos\left[\frac{\pi}{2} - (x - y)\right] = \cos\left[\left(\frac{\pi}{2} - x\right) + y\right]$$

$$= \cos\left(\frac{\pi}{2} - x\right) \cos y - \sin\left(\frac{\pi}{2} - x\right) \sin y$$

$$\sin(x - y) = \sin x \cos y - \cos x \sin y \qquad (9\text{-}19)$$

Finally, we derive

$$\sin(x + y) = \sin x \cos(-y) - \cos x \sin(-y)$$

$$\sin(x + y) = \sin x \cos y + \cos x \sin y. \qquad (9\text{-}20)$$

If we set $y = x$ in (9-18) and (9-20), we get special cases of these equations that are frequently used in calculus.

$$\cos 2x = \cos(x + x) = \cos x \cos x - \sin x \sin x$$

$$\cos 2x = \cos^2 x - \sin^2 x \qquad (9\text{-}21)$$

$$\sin 2x = \sin(x + x) = \sin x \cos x + \cos x \sin x$$

$$\sin 2x = 2 \sin x \cos x \qquad (9\text{-}22)$$

Example 8

Using (9-21) with $x = \pi/3$, evaluate $\cos 2\pi/3$.

Solution.

$$\cos \frac{2\pi}{3} = \cos^2 \frac{\pi}{3} - \sin^2 \frac{\pi}{3} = \left(\frac{1}{2}\right)^2 - \left(\frac{\sqrt{3}}{2}\right)^2 = \frac{1}{4} - \frac{3}{4} = -\frac{1}{2}$$

Example 9

Using (9-21) and (9-22) with $x = \pi/8$, find $\sin \pi/8$ and $\cos \pi/8$.

Solution. We know that the values of $\sin \pi/4$ and $\cos \pi/4$ are both $\sqrt{2}/2$. So

$$\cos^2 \frac{\pi}{8} - \sin^2 \frac{\pi}{8} = \cos \frac{\pi}{4} = \frac{\sqrt{2}}{2}$$

$$2 \sin \frac{\pi}{8} \cos \frac{\pi}{8} = \sin \frac{\pi}{4} = \frac{\sqrt{2}}{2}.$$

If we substitute $u = \cos \pi/8$, $v = \sin \pi/8$, our equations become

$$u^2 - v^2 = \frac{\sqrt{2}}{2}$$

$$2uv = \frac{\sqrt{2}}{2}.$$

Using the second equation to replace v by $\sqrt{2}/(4u)$ in the first, we have

$$u^2 - \frac{1}{8u^2} = \frac{\sqrt{2}}{2}$$

or

$$u^4 - \frac{\sqrt{2}}{2}u^2 - \frac{1}{8} = 0.$$

This quadratic in u^2 can be solved by setting $q = u^2$. We obtain $q = \sqrt{2}/4 \pm 1/2$. We reject the negative value of q (why?) and so $q = (\sqrt{2} + 2)/4$. Because u is positive (why?), we get $u = \sqrt{2 + \sqrt{2}}/2$, which leads to $v = \sqrt{2 - \sqrt{2}}/2$. Therefore

$$\cos \frac{\pi}{8} = u = \frac{\sqrt{2 + \sqrt{2}}}{2} \qquad \sin \frac{\pi}{8} = v = \frac{\sqrt{2 - \sqrt{2}}}{2}.$$

In the following box we summarize for later reference the various sum and difference formulas derived earlier.

Sum and Difference Formulas

$$\cos (x - y) = \cos x \cos y + \sin x \sin y \qquad \text{(9-17)}$$

$$\cos (x + y) = \cos x \cos y - \sin x \sin y \qquad \text{(9-18)}$$

$$\sin (x - y) = \sin x \cos y - \cos x \sin y \qquad \text{(9-19)}$$

$$\sin (x + y) = \sin x \cos y + \cos x \sin y \qquad \text{(9-20)}$$

$$\cos 2x = \cos^2 x - \sin^2 x \qquad \text{(9-21)}$$

$$\sin 2x = 2 \sin x \cos x \qquad \text{(9-22)}$$

We conclude this section with two identities that enable us to express the *squares* of $\sin x$ and $\cos x$ in terms of $\cos 2x$. They are obtained by solving the identities

$$1 = \cos^2 x + \sin^2 x$$

$$\cos 2x = \cos^2 x - \sin^2 x$$

simultaneously for $\cos^2 x$ and $\sin^2 x$. These results, too, will be of considerable importance in calculus.

$$\cos^2 x = \frac{1}{2}(1 + \cos 2x)$$

$$\sin^2 x = \frac{1}{2}(1 - \cos 2x) \qquad\qquad\qquad (9\text{-}24)$$

EXERCISES 9-4

In Problems 1–20 use (9-2) through (9-16) to verify the identities.

1. $\cot x \sin x = \cos x$

2. $\sin^2 x + \sin^2 x \cot^2 x = 1$

3. $\cos^4 x - \sin^4 x = \cos^2 x - \sin^2 x$

4. $(\sin y + \cos y)(\sin y - \cos y) = 2 \sin^2 y - 1$

5. $(\sec x + \tan x)(1 - \sin x) = \cos x$

6. $\dfrac{\sin x + \tan x}{1 + \cos x} = \tan x$

7. $\dfrac{1}{1 + \cos x} - \dfrac{1}{1 - \cos x} = -2 \cot x \csc x$

8. $\dfrac{1}{1 + \cos x} + \dfrac{1}{1 - \cos x} = 2 \csc^2 x$

9. $\cot^2 y - \cos^2 y = \cot^2 y \cos^2 y$

10. $\dfrac{\cos y}{1 - \tan y} - \dfrac{\sin y}{1 - \cot y} = \dfrac{\cos y + \sin y}{1 - 2 \sin^2 y}$

11. $\dfrac{1}{\sec x - 1} + \dfrac{1}{\sec x + 1} = 2 \csc^2 x \cos x$

12. $\dfrac{1 - \cos z}{z} = \dfrac{\sin^2 z}{z + z \cos z}$

13. $\tan(-x) \cos x = \sin(-x)$

14. $\dfrac{\tan(-x)}{1 - \tan^2 x} = \dfrac{1}{\cot(-x) - \tan(-x)}$

15. $\dfrac{\cos(\pi/2 - z)}{z} = \dfrac{\sin z}{z}$

16. $\dfrac{\cos(z - \pi/2)}{z^2} = \dfrac{\sin z}{z^2}$

17. $[\sin(-x) + \cos(-x)](\sin x + \cos x) = 1 - 2 \sin^2 x$

18. $[\sec(-x) + \tan(-x)](1 + \sin x) = \cos x$

19. $\sin\left(\dfrac{\pi}{2} - x\right) \tan x = \cos\left(\dfrac{\pi}{2} - x\right)$

20. $\tan^2(-x) + \cot^2(-x) = \sec^2\left(\dfrac{\pi}{2} - x\right) \csc^2\left(\dfrac{\pi}{2} - x\right) - 2$

21. Verify the following identities by means of the sum and difference formulas.

(a) $\sin(\pi - x) = \sin x$ (b) $\cos(\pi - x) = -\cos x$

(c) $\sin(x + \pi) = -\sin x$ (d) $\cos(x + \pi) = -\cos x$

22. Use parts (c) and (d) of Problem 21 to show that the tangent function has period π.

In Problems 23–36 use the sum and difference formulas (9-17) through (9-20) to evaluate.

23. $\sin \dfrac{\pi}{12}$ **24.** $\sin \dfrac{7\pi}{12}$ **25.** $\cos \dfrac{5\pi}{12}$ **26.** $\cos \dfrac{7\pi}{6}$

27. $\tan \dfrac{\pi}{12}$ **28.** $\cot \dfrac{7\pi}{12}$ **29.** $\csc \dfrac{5\pi}{12}$ **30.** $\sec \dfrac{\pi}{12}$

31. $\sin \left(x + \dfrac{\pi}{6} \right)$, where $\sin x = \dfrac{4}{5}$ and $\cos x = -\dfrac{3}{5}$

32. $\cos \left(x + \dfrac{\pi}{3} \right)$, where $\sin x = \dfrac{4}{5}$ and $\cos x = \dfrac{3}{5}$

33. $\sin \left(y - \dfrac{\pi}{4} \right)$, where $\sin y = \dfrac{3}{5}$ and $\cos y = -\dfrac{4}{5}$

34. $\cos \left(\dfrac{\pi}{4} - y \right)$, where $\sin y = -\dfrac{3}{5}$ and $\cos y = -\dfrac{4}{5}$

35. $\sin \left(x + \dfrac{2\pi}{3} \right)$, where $\sin x = \dfrac{5}{13}$ and $\cos x = -\dfrac{12}{13}$

36. $\cos \left(x - \dfrac{2\pi}{3} \right)$, where $\sin x = -\dfrac{5}{13}$ and $\cos x = \dfrac{12}{13}$

In Problems 37–42 use (9-21) and (9-22) to evaluate.

37. $\cos 2x$, where $0 < x < \dfrac{\pi}{2}$ and $\sin x = \dfrac{3}{5}$

38. $\sin 2y$, where $0 < y < \dfrac{\pi}{2}$ and $\cos y = \dfrac{3}{5}$

39. $\cos 2x$, where $0 < x < \pi$ and $\cos x = -\dfrac{5}{13}$

40. $\cos 2y$, where $\pi < y < 2\pi$ and $\cos y = \dfrac{12}{13}$

41. $\tan 2z$, where $0 < z < \dfrac{\pi}{2}$ and $\sin z = \dfrac{4}{5}$

42. $\tan 2w$, where $0 < w < \pi$ and $\cos w = -\dfrac{3}{5}$

In Problems 43–49 verify the identities.

43. $\cot 2x = \dfrac{\cos^2 x - \sin^2 x}{\sin 2x}$

44. $\dfrac{\sin x \cos x}{\sin^2 x - \cos^2 x} = -\dfrac{1}{2} \tan 2x$

45. $\tan 2x = \dfrac{2 \tan x}{1 - \tan^2 x}$

46. $\cos 2x + 2 \sin^2 x = 1$

47. $\cos^4 y - \sin^4 y = \cos 2y$

48. $\dfrac{\sin (x + y) - \sin x}{y} = \sin x \cdot \dfrac{\cos y - 1}{y} + \cos x \cdot \dfrac{\sin y}{y}$

49. $\tan \left(x + \dfrac{\pi}{4} \right) = \dfrac{\cos x + \sin x}{\cos x - \sin x}$

FOR THE ENTHUSIASTS

50. By starting with $\cos 3x = \cos (x + 2x)$, verify the identity
(a) $\cos 3x = 4 \cos^3 x - 3 \cos x$.
Similarly, verify the identity
(b) $\sin 3x = 3 \sin x - 4 \sin^3 x$.

51. Verify the identity
$$\tan (x + y) = \frac{\tan x + \tan y}{1 - \tan x \tan y}.$$
52. Verify the identities.
  (a) $\sin (x + y) + \sin (x - y) = 2 \sin x \cos y$
  (b) $\cos (x + y) + \cos (x - y) = 2 \cos x \cos y$
53. Show that
$$\sin \tfrac{1}{2} z = \pm \sqrt{\frac{1 - \cos z}{2}}$$
and
$$\cos \tfrac{1}{2} z = \pm \sqrt{\frac{1 + \cos z}{2}}.$$

(*Hint:* Let $\alpha = \tfrac{1}{2} z$ so that $z = 2\alpha$.)

9-5 TRIGONOMETRIC FUNCTIONS AND HARMONIC MOTION

As indicated in the introduction to this chapter, the trigonometric functions play a central role in the mathematical representation of periodic phenomena. One of the most important of these phenomena is **harmonic motion**—that is, motion that can be described as vibration. Examples of harmonic motion include

1. the motion of a weight that has been attached to a spring, pulled down from its equilibrium position and released.
2. the motion of a swinging pendulum.
3. the motion of a buoy bobbing up and down in a channel of water.
4. the motion of a point on the plucked string of a musical instrument.

Theoretical analysis of harmonic motion is normally studied in a course in calculus or differential equations and so we will not undertake it here. An interesting result, from our point of view, is that descriptions of these phenomena all involve functions of the form

$$f(t) = A \sin (kt + c), \tag{9-25}$$

where A, k, and c are constants; the independent variable t usually denotes time. The wavelike character of the sine graph makes it plausible that these functions should appear in the mathematics of harmonic motion.

Our present objective is to learn to graph functions like (9-25) and the closely related form

$$f(t) = A \cos (kt + c). \tag{9-26}$$

We begin with the simplest case and build up from there.

Graph of f(t) = A sin kt

Let us assume fitst that $A = 1$, $c = 0$, and $k > 0$. We expect the function $f(t) = \sin kt$ to be periodic, of course, but what is its period? We know that $\sin kt$ begins to repeat its values as soon as kt increases by 2π. Now kt increases by 2π whenever t increases by $2\pi/k$ and so we would guess that the period of $\sin kt$ is $2\pi/k$. This fact is easily verified from the definition of periodicity; with $f(t) = \sin kt$ we have, for every real number t,

$$f\left(t + \frac{2\pi}{k}\right) = \sin k\left(t + \frac{2\pi}{k}\right) = \sin(kt + 2\pi) = \sin kt = f(t).$$

Example 1

Graph $f(t) = \sin 2t$.

Solution. The period of f is $2\pi/2 = \pi$. Let us divide the interval $[0, \pi]$ into quarters so as to locate the points in that interval where f has its zeros or assumes its extreme values: $0, \pi/4, \pi/2, 3\pi/4, \pi$. The graph over $[0, \pi]$ appears in Fig. 9-27(a) and its periodic extension in Fig. 9-27(b).

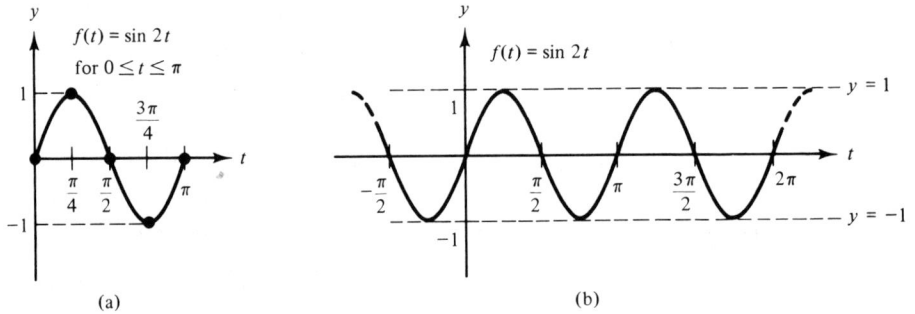

(a)   (b)

Figure 9-27

Example 2

Graph $g(t) = \sin \dfrac{t}{3}$.

Solution. The period of g is $2\pi/(1/3) = 6\pi$. Dividing the interval $[0, 6\pi]$ into quarters, we obtain the points $0, 3\pi/2, 3\pi, 9\pi/2, 6\pi$. The graph over $[0, 6\pi]$ appears in Fig. 9-28. We leave it for the reader to draw the periodic extension.

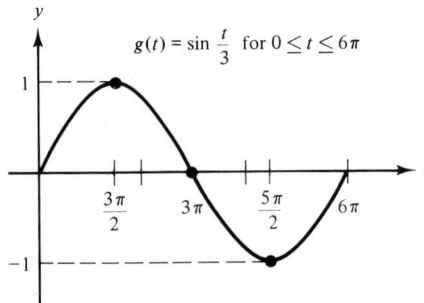

Figure 9-28

If the constant A is not equal to one, we must consider the fact that every function value is multiplied by A.

The Trigonometric Functions   Chap. 9

Example 3

Graph
  (a) $y = 2 \sin 3t$
  (b) $y = -\frac{1}{2} \sin 3t$
over one period.

Solution. The period of $\sin 3t$ is $2\pi/3$ and so we can do our work on the interval $[0, 2\pi/3]$.
  (a) We begin by sketching the graph of $y = \sin 3t$ and then double each ordinate to get $y = 2 \sin 3t$ [see Fig. 9-29(a)].
  (b) To obtain the graph of $y = -(1/2) \sin 3t$, we take half of each ordinate and reflect the resulting graph through the t axis. The completed graph appears in Fig. 9-29(b).

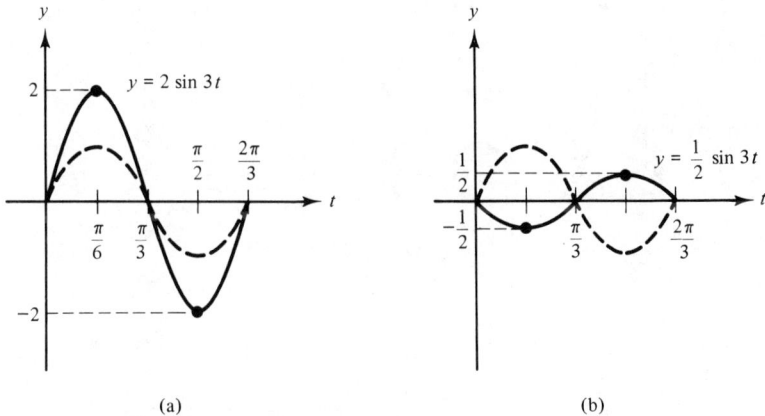

(a)   (b)

Figure 9-29

If the constant k is negative no difficulty arises. To graph $y = 4 \sin(-3t)$, for instance, we recall that sine is odd and write the equation in an equivalent form $y = -4 \sin 3t$.

Graph of f(t) = A sin (kt + c)

To deal with the constant c, we use a translation of axes. First, we note that if $y = f(t)$, we can write

$$y = A \sin k\left(t + \frac{c}{k}\right)$$

and the substitution

$$u = t + \frac{c}{k}$$

$$v = y$$

reduces the given equation to $y = A \sin ku$, the type dealt with earlier. The substitution amounts to a horizontal translation in which the new origin is $(-c/k, 0)$.

Example 4

Graph $y = \frac{1}{2} \sin (3t - \pi)$.

Solution. Because $\sin (3t - \pi) = \sin 3(t - \pi/3)$, the translation is

$$u = t - \frac{\pi}{3}$$

$$v = y.$$

We begin by sketching the graph of $y = (1/2) \sin 3u$ in a u, v coordinate system with origin at $(\pi/3, 0)$. The period is $2\pi/3$ and so we draw the graph (Fig. 9-30) on the interval $[\pi/3, \pi]$ and extend it periodically.

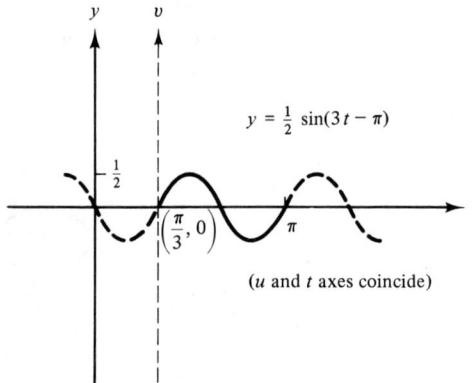

$$y = \tfrac{1}{2} \sin(3t - \pi)$$

$\left(\frac{\pi}{3}, 0\right)$ π

(*u* and *t* axes coincide)

Figure 9-30

To graph an equation of the form (9-26), in which sine is replaced by cosine, we use the identity (9-7), which converts the cosine to a sine, and then proceed as before.

Example 5

Graph $f(t) = 2 \cos (6t - \pi)$.

Solution. We rewrite f by means of (9-7) and (9-4).

$$f(t) = 2 \cos (6t - \pi) = 2 \sin \left[\frac{\pi}{2} - (6t - \pi) \right]$$

$$= 2 \sin \left(\frac{3\pi}{2} - 6t \right)$$

$$= -2 \sin \left(6t - \frac{3\pi}{2} \right)$$

$$= -2 \sin 6 \left(t - \frac{\pi}{4} \right)$$

The period is $2\pi/6 = \pi/3$ and we sketch the graph over the interval $[\pi/4, 7\pi/12]$ (see Fig. 9-31).

     The Trigonometric Functions   Chap. 9

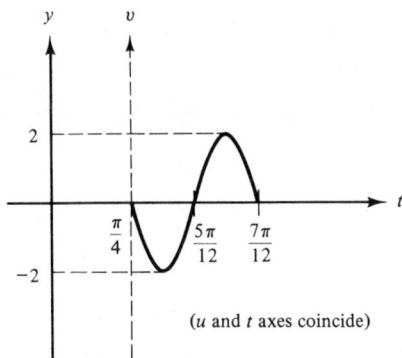

(u and t axes coincide)

Figure 9-31

Several significant features of the graph of (9-25) have become apparent in our discussions and are summarized below.

If $f(t) = A \sin(kt + c)$, then
1. the **period** of f is $2\pi/|k|$.
2. the maximum value of f is $|A|$ and the minimum value of f is $-|A|$. ($|A|$ is called the **amplitude** of the harmonic motion described by the equation.)
3. the number $-c/k$ is a **zero** of f.
4. the graph of f is similar to a sine wave.

EXERCISES 9-5

In Problems 1–12 determine the period of the function and draw its graph over an interval of length equal to one period. Label maximum and minimum points.

1. $f(t) = 4 \sin 3t$

2. $f(t) = \dfrac{3}{2} \sin \dfrac{1}{2} t$

3. $g(t) = \dfrac{1}{2} \sin 3t$

4. $g(t) = -5 \sin 4t$

5. $F(x) = -\dfrac{1}{3} \sin \dfrac{1}{2} x$

6. $f(t) = -\dfrac{1}{2} \cos \dfrac{3}{2} t$

7. $g(t) = 3 \cos 2t$

8. $f(t) = 2 \sin \left(2t - \dfrac{\pi}{3} \right)$

9. $f(u) = -3 \sin(6u + \pi)$

10. $g(t) = -\dfrac{1}{2} \sin \left(t - \dfrac{3\pi}{2} \right)$

11. $F(t) = \dfrac{2}{3} \cos \left(\dfrac{1}{2} t - \dfrac{\pi}{3} \right)$

12. $f(t) = \dfrac{1}{4} \cos(4t + \pi)$

FOR THE ENTHUSIASTS

In Problems 13–16 draw the graph by means of the methods of this section in combination with a vertical translation. Determine the period, and label maximum and minimum points.

13. $f(t) = -2 + 3 \sin 2t$

14. $h(x) = 2 + 3 \cos 2x$

15. $F(t) = 2 \cos 2t - 1$

16. $f(x) = \dfrac{1}{2} \sin \left(2x - \dfrac{\pi}{3} \right) - \dfrac{1}{2}$

Sometimes a vibration is described by a function of the form

$$f(t) = a \sin kt + b \cos kt \tag{9-27}$$

where a, b, and k are constants. The purpose of Problems 17–20 is to show how this equation can be expressed in the form (9-25).

17. Show that the point $(a/\sqrt{a^2 + b^2}, b/\sqrt{a^2 + b^2})$ is on the unit circle. Conclude that, if a and b are real numbers, then there exists a real number c such that $0 \leq c < 2\pi$ and

$$\frac{a}{\sqrt{a^2 + b^2}} = \cos c \quad \text{and} \quad \frac{b}{\sqrt{a^2 + b^2}} = \sin c.$$

18. Show that (9-27) can be written in the form

$$f(t) = A \sin(kt + c)$$

where $A = \sqrt{a^2 + b^2}$ and c is the number determined in Problem 17.
19. Show that the maximum value assumed by f in (9-27) is $\sqrt{a^2 + b^2}$.
20. Graph the function over an interval of length equal to one period. Label the maximum and minimum points and the zeros.
    (a) $f(t) = \sin t + \cos t$ (b) $f(t) = 3 \sin 2t + 4 \cos 2t$
    (c) $g(t) = 4 \sin 3t - 3 \cos 3t$ (d) $g(u) = 5 \cos 2u - 12 \sin 2u$

DAMPED VIBRATION

Friction, or perhaps some other resistance, may slow down a vibrating object; then the vibration is said to be **damped**. In Problems 21–24 you are given functions that describe damped vibrations. Draw the graphs, using the hint for Problem 21 as a guide.

21. $f(t) = \dfrac{1}{t} \sin t$ for $t > 0$ (*Hint:* First notice that $|f(t)| \leq 1/t$ so that the graph of f lies between $y = -1/t$ and $y = 1/t$; it touches one of these graphs when $\sin t = -1$ or $\sin t = 1$.)

22. $g(t) = -\dfrac{1}{t} \sin 2t$ for $t > 0$

23. $f(t) = e^{-t} \sin t$ for $t > 0$
24. $F(t) = e^{-t} \cos 2t$ for $t > 0$

SUMMARY

Chapter 9 introduced the basic trigonometric functions, their graphs, and the important identities that relate them to one another.

1. You should know what is meant by a periodic function and how to show that a function is periodic.
2. You should know the unit circle definitions of P_x, $\sin x$, and $\cos x$; that these functions have period 2π, and that they satisfy the fundamental identity (9-2).
3. By drawing a sketch, you should be able to recall the symmetry properties that involve P_x, P_{-x}, $P_{\pi-x}$, $P_{x+\pi}$, $P_{\pi/2-x}$, and the resulting identities relating sine and cosine.

4. You should be able to evaluate the sine and cosine functions at integral multiples of $\pi/4$ and $\pi/6$, using the periodicity and the various identities in (3).

5. You should be able to define tangent, cotangent, secant, and cosecant in terms of sine and cosine. You should know how to derive dentities (9-12) and (9-16) as variants of (9-2).

6. You should be able to verify trigonometric identities, including those based on the sum and difference identities.

7. You should be familiar with the graphs of all six of the trigonometric functions and functions obtained from them by simple translations or multiplication by constants.

8. You should recognize the form of a function that describes harmonic motion and be able to identify its period, maximum and minimum points, and zeros. You should be able to graph such a function.

REVIEW EXERCISES

1. Draw the graph of a function f that is linear over the interval $[0, 1)$ such that f has period 1, $f(0) = 2$, and $f(1/2) = 0$. Give a rule for $f(x)$.

2. Find the coordinates of P_x and the values of $\sin x$ and $\cos x$ if

   (a) $x = -\dfrac{7\pi}{4}$.

   (b) $x = \dfrac{22\pi}{3}$.

   (c) $x = \dfrac{99\pi}{4}$.

   (d) $x = \dfrac{23\pi}{6}$.

   (e) $x = \dfrac{37\pi}{2}$.

   (f) $x = -36\pi$.

3. Given that $\sin x = 2/3$ and P_x is in the second quadrant, find $\cos x$, $\tan x$, and $\sec x$.

4. Given that $\sin x = -4/7$ and P_x is in the third quadrant, find $\cos x$, $\tan x$, and $\csc x$.

5. Determine whether the function is odd or even.

   (a) $f(x) = x \sin x$

   (b) $f(x) = x \cos x$

   (c) $g(x) = \sin \dfrac{1}{x} (x \neq 0)$

   (d) $h(x) = \cos \dfrac{1}{x} (x \neq 0)$

   (e) $h(x) = \sin (\pi + x)$

   (f) $f(x) = \cos (\pi + x)$

   (g) $f(x) = \sin \left(\dfrac{\pi}{2} - x\right)$

   (h) $g(x) = \cos \left(\dfrac{\pi}{2} - x\right)$

   (i) $f(x) = \sin (\pi - x)$

   (j) $h(x) = \cos (\pi - x)$

6. (a) Give the definitions of tangent and secant. Without looking in the text, derive the identity relating $\tan^2 x$ and $\sec^2 x$ from the identity $\cos^2 x + \sin^2 x = 1$.

   (b) Do the same for the functions cotangent and cosecant.

7. Show that

   (a) $\tan \left(\dfrac{\pi}{2} - x\right) = \cot x$.

   (b) $\sec (x + \pi) = -\sec x$.

   (c) $\csc (\pi - x) = \csc x$.

8. Establish the following identities.

   (a) $(\cos x + \sin x)^2 = 1 + 2 \cos x \sin x$

   (b) $1 - 2 \sin^2 x = \cos^4 x - \sin^4 x$

   (c) $\sec^2 x - \csc^2 x = (\tan x + \cot x)(\tan x - \cot x)$

(d) $\dfrac{1 - \tan^2 x}{1 - \cot^2 x} = -\tan^2 x$

(e) $\cos^4 x = \dfrac{3}{8} + \dfrac{1}{2}\cos 2x + \dfrac{1}{8}\cos 4x$

(f) $\sin^2 x \cos^2 x = \dfrac{1}{8} - \dfrac{1}{8}\cos 4x$

9. (a) Sketch the graph of $f(t) = 2 \sin 3\,(t - \pi/3)$. Find the period and draw the graph over an interval of length equal to one period. Label maximum and minimum points.

   (b) In an alternating current line the voltage V is given by

$$V = 110 \sin 120\pi t,$$

   where t is time in seconds. What is the period? What is the maximum voltage? At what times t is the voltage maximal?

10. Graph.

   (a) $f(x) = \sin^2 x - 1$
   (c) $g(x) = \tan^2 x - 1$

   (b) $f(x) = (\cos x - 1)^2$
   (d) $f(x) = \sec x - \tan x$

Applications of trigonometric functions

10-1 TRIGONOMETRY OF ANGLES

The word trigonometry literally means "triangle measurement." Indeed, the trigonometric functions defined in the last chapter originally arose in connection with angles and triangles. In this section we explain how the trigonometric functions can be interpreted in terms of angles and indicate some applications that stem from this interpretation.

Angles

When a half line rotates in the plane around its endpoint O from a position OM to a position ON, an **angle** is generated. The amount of rotation is called the **measure** of the angle. In Figure 10-1 OM is the **initial** side of the angle, ON its **terminal** side, O its **vertex**. The measure of an angle is taken to be positive if the rotation is counterclockwise, negative if the rotation is clockwise. There is no limit on how large the measure of an angle can be because the rotation can include any number of complete revolutions.

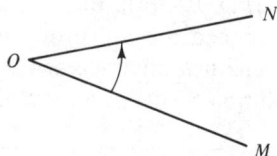

Figure 10-1

In high school geometry angles are usually measured in degrees (°). A degree is 1/360 of a full revolution. A right angle measures 90°, for there are four right angles in a full revolution. A straight angle measures 180°.

An angle is said to be in **standard position** if it is placed in a coordinate plane so that its initial side lies along the positive side of the horizontal axis and its vertex is at the origin. Figure 10-2 shows several angles in standard position and indicates their measures in degrees.

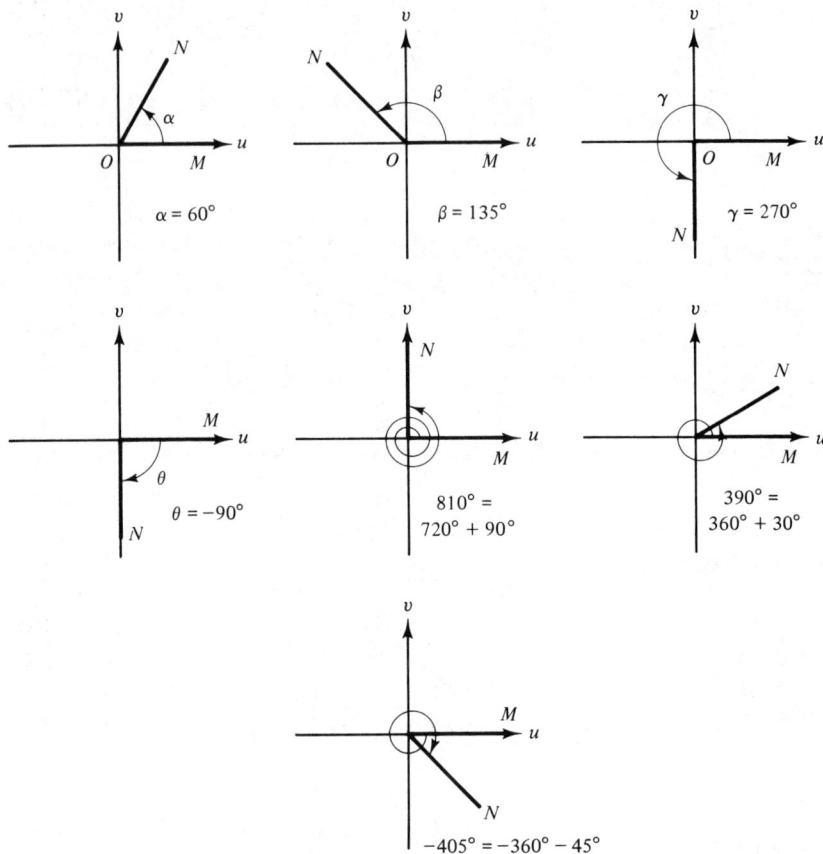

Figure 10-2

Radian Measure

For many purposes, it is desirable to measure angles in radians rather than degrees. An angle of 1 **radian** is one that if placed in standard position, subtends an arc of length 1 on the unit circle C [Fig. 10-3(a)]. Because the angle at the center of the circle and the corresponding subtended arc are proportional, we see that an angle of 2 radians subtends an arc of length 2, an angle of $1/2$ radians subtends an arc of length $1/2$, and so on. A full revolution subtends an arc of length 2π.

It follows from the definitions of sine and cosine that if x is the radian measure of an angle in standard position, then the terminal side of this angle intersects the unit circle C at the point P_x ($\cos x$, $\sin x$) [Fig. 10-3(b)].

Applications of Trigonometric Functions   Chap. 10

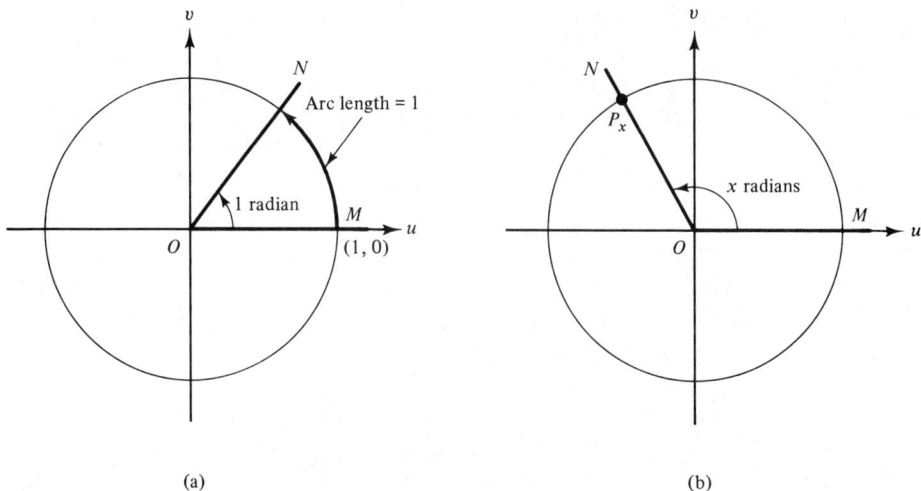

Figure 10-3

The degree measure of an angle can readily be converted to radians by observing that a full revolution is 2π radians or $360°$.

$$2\pi \text{ radians} = 360°$$
$$\pi \text{ radians} = 180° \qquad (10\text{-}1)$$

Example 1

Many familiar angles can be recognized easily as rational multiples of $180°$. Thus $60° = (1/3)(180°) = \pi/3$ radians; $45° = (1/4)(180°) = \pi/4$ radians; $540° = 3(180°) = 3\pi$ radians. An angle whose radian measure is $3\pi/2$ is three times a right angle so that $3\pi/2$ radians $= 270°$. Table 10-1 shows some familiar angles measured in degrees and radians.

TABLE 10-1

Degrees	Radians
0	0
30	$\pi/6$
45	$\pi/4$
60	$\pi/3$
90	$\pi/2$
120	$2\pi/3$
135	$3\pi/4$
150	$5\pi/6$
180	π

Of course, we can use (10-1), obtaining the conversion formulas

$$1 \text{ radian} = \left(\frac{180}{\pi}\right)^{\circ} \approx 57.2958^{\circ},$$

$$1^{\circ} = \frac{\pi}{180} \text{ radians} \approx .01745 \text{ radians.}$$

(10-2)

Example 2

(a) Convert 2.5 radians to degrees.
(b) Convert 17° to radians.

Solution.

(a) 2.5 radians $\approx$ (2.5)(57.2958)° $\approx$ 143.24°
(b) 17° $\approx$ 17 (.01745) radians
$\approx$ .297 radians

Example 3

A windmill is making 10 revolutions per minute. Express in degrees and radians the angle generated by one of its sails (blades) in one second.

Solution. Because 10 revolutions are made each minute, each sail turns through $10/60 = 1/6$ of a revolution in one second. In degrees the measure of this angle[1] is $(1/6)(360°) = 60°$; in radians it is $(1/6) (2\pi \text{ radians}) = \pi/3$ radians.

Example 4

A girl is riding on a merry-go-round, 12 feet from its center. How fast is she moving if the merry-go-round makes 8 revolutions per minute?

Solution. For each revolution, she travels $2\pi \cdot 12 = 24\pi$ feet. So her speed is $24\pi \cdot 8 = 192\pi$ feet per minute.

Trigonometric Ratios

The rotation of the terminal side of an angle in standard position is clearly periodic. Therefore it is not surprising that the trigonometric functions can be used to describe this rotation.

Consider an angle in standard position and let $R(u, v)$ be any point other than $(0, 0)$ on the terminal side of this angle (Fig. 10-4). The signs of u and v depend on the quadrant in which the terminal side lies, but, in any case, the lengths of the sides of the right triangle ROQ are $OQ = |u|$, $QR = |v|$. The length of the hypotenuse OR is

$$r = \sqrt{u^2 + v^2}.$$

---

[1] In many situations, the sign of an angle is irrelevant; in such cases, we will express the measure of the angle as a positive number.

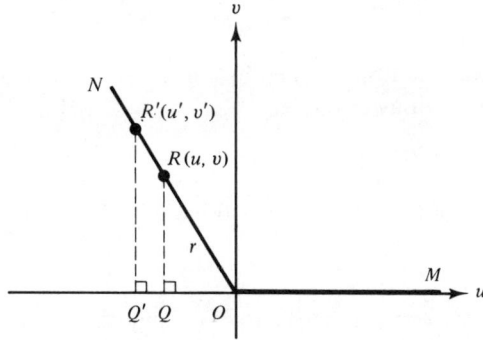

Figure 10-4

We form six ratios from the three numbers u, v, and r:

$$\frac{u}{r}, \quad \frac{v}{r}, \quad \frac{v}{u}, \quad \frac{r}{u}, \quad \frac{u}{v}, \quad \frac{r}{v}.$$

These ratios are independent of the position of R in the following sense. If $R' = (u', v')$ is any other point on the terminal side of the given angle and if $r' = \sqrt{(u')^2 + (v')^2}$, then the corresponding ratios are equal. That is,

$$\frac{u}{r} = \frac{u'}{r'}, \quad \frac{v}{r} = \frac{v'}{r'}, \quad \frac{v}{u} = \frac{v'}{u'}, \quad \frac{r}{u} = \frac{r'}{u'}, \quad \frac{u}{v} = \frac{u'}{v'}, \quad \frac{r}{v} = \frac{r'}{v'} \quad (10\text{-}3)$$

It is easy to verify these equations by using the similarity of the triangles ROQ and $R'OQ'$.

Suppose that the radian measure of the given angle is x, and that the terminal side is not on either of the coordinate axes. Then the terminal side intersects the unit circle C at P_x $(\cos x, \sin x)$ and so that side certainly contains P_x. Applying (10-3) with $R' = P_x$, $u' = \cos x$, $v' = \sin x$, and noting that $r' = \sqrt{\cos^2 x + \sin^2 x} = 1$, we have

$$\left.\begin{array}{ccc}
\dfrac{u}{r} = \dfrac{\cos x}{1} = \cos x, & \dfrac{v}{r} = \dfrac{\sin x}{1} = \sin x, & \dfrac{v}{u} = \dfrac{\sin x}{\cos x} = \tan x \\[2ex]
\dfrac{r}{u} = \dfrac{1}{\cos x} = \sec x, & \dfrac{u}{v} = \dfrac{\cos x}{\sin x} = \cot x, & \dfrac{r}{v} = \dfrac{1}{\sin x} = \csc x.
\end{array}\right\} \quad (10\text{-}4)$$

Thus the values of the trignometric functions can be interpreted as ratios of the three numbers u, v, r appearing in Fig. 10-4.

Example 5

Consider the angle in standard position that measures $135°$. Find the ratios u/r, v/r, and v/u, where $R(u, v)$ is any point other than $(0, 0)$ on the terminal side of this angle and $r = \sqrt{u^2 + v^2}$.

Solution. The radian measure of the given angle is $3\pi/4$. So

$$\frac{u}{r} = \frac{u}{\sqrt{u^2 + v^2}} = \cos\frac{3\pi}{4} = -\frac{\sqrt{2}}{2}$$

$$\frac{v}{r} = \frac{v}{\sqrt{u^2 + v^2}} = \sin\frac{3\pi}{4} = \frac{\sqrt{2}}{2}$$

$$\frac{v}{u} = \tan\frac{3\pi}{4} = -1.$$

Example 6

Find an equation satisfied by u and v if (u, v) is on the terminal side of the angle in the previous example.

Solution. We have seen that (u, v) satisfies $v/u = -1$, which leads to the linear equation $u + v = 0$. (However, the terminal side consists only of that portion of the line $u + v = 0$ lying in the second quadrant.)

Example 7

Let x be the radian measure of an angle in standard position whose terminal side is the half of the line $v = 5u$ that lies in the third quadrant. Find $\sin x$, $\cot x$, $\sec x$.

Solution. We can choose for (u, v) any point on the terminal side of the angle, say $(-1, -5)$. Then $r = \sqrt{1 + 25} = \sqrt{26}$ and so

$$\sin x = \frac{v}{r} = \frac{-5}{\sqrt{26}} = \frac{-5\sqrt{26}}{26}$$

$$\cot x = \frac{u}{v} = \frac{-1}{-5} = \frac{1}{5}$$

$$\sec x = \frac{r}{u} = \frac{\sqrt{26}}{-1} = -\sqrt{26}.$$

Notation for Angles

Angles are frequently denoted by Greek letters, such as $\alpha, \beta, \gamma, \theta, \phi, \ldots$ (respectively, alpha, beta, gamma, theta, phi, ...). When an angle α measures $30°$, we write "$\alpha = 30°$" even though an angle and its measure are not really the same thing. It is similar to saying "$PQ = 5$," where PQ is a line segment and 5 is its length. If we were using radians as a unit of measure, then instead of "$\alpha = 30°$," we would write "$\alpha = \pi/6$ radians." When radian measure is used, however, the word "radians" is often not written. That is, when units are not specified, the measure of an angle will be understood to be in radians.

Example 8

The statement $\alpha = 2$ means that the measure of α is 2 radians. The statement $\alpha = 2°$ means that the measure of α is 2 degrees.

It is often useful to view the trigonometric functions as being defined for *angles*. We do so by considering the sine of an angle whose radian measure is α to be the sine of the real number α (and similarly with the other trigonometric functions).

Example 9

Let α denote the angle described in Example 5. Find the values of $\sin \alpha$, $\cos \alpha$, $\tan \alpha$.

Solution. Because $\alpha = 135°$ and $135°$ corresponds to $3\pi/4$ radians, we have

$$\sin \alpha = \sin \frac{3\pi}{4} = \frac{\sqrt{2}}{2}$$

$$\cos \alpha = \cos \frac{3\pi}{4} = -\frac{\sqrt{2}}{2}$$

$$\tan \alpha = \tan \frac{3\alpha}{4} = -1.$$

Example 10

Evaluate.
  (a) $\sin 30°$
  (b) $\tan 60°$
  (c) $\sec(-450°)$

Solution.

  (a) $\sin 30° = \sin \dfrac{\pi}{6} = \dfrac{1}{2}$

  (b) $\tan 60° = \tan \dfrac{\pi}{3} = \sqrt{3}$

  (c) For the angle $-450°$, we have $\cos(-450°) = \cos(-90°) = \cos(-\pi/2) = 0$; hence $\sec(-450°)$ is undefined.

Right Triangles

Let OQR be a right triangle with a right angle at Q; let α denote the acute[2] angle at O (Fig. 10-5). If we choose a coordinate system with the positive side of the horizontal axis along OQ and vertex at O, then α is in standard position and its terminal side lies in the first quadrant. Letting R have coordinates (u, v), we see that u, v, and $r = \sqrt{u^2 + v^2}$ are the lengths of the sides of OQR.

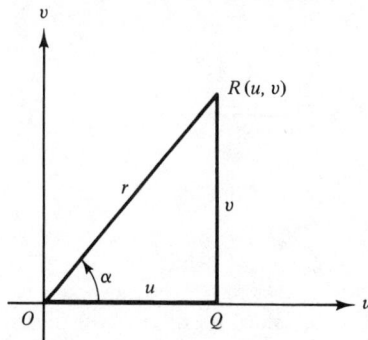

Figure 10-5

[2] An acute angle is one whose radian measure is between 0 and $\pi/2$.

$$u = OQ = \text{side adjacent to } \alpha$$

$$v = QR = \text{side opposite } \alpha$$

$$r = OR = \text{hypotenuse}$$

Consequently, by (10-4) the values of the trigonometric functions at α are ratios of sides of OQR.

$$\cos \alpha = \frac{u}{r} = \frac{\text{adjacent}}{\text{hypotenuse}}, \qquad \sin \alpha = \frac{v}{r} = \frac{\text{opposite}}{\text{hypotenuse}},$$

$$\tan \alpha = \frac{v}{u} = \frac{\text{opposite}}{\text{adjacent}}, \qquad \sec \alpha = \frac{r}{u} = \frac{\text{hypotenuse}}{\text{adjacent}}, \qquad (10\text{-}5)$$

$$\cot \alpha = \frac{u}{v} = \frac{\text{adjacent}}{\text{opposite}}, \qquad \csc \alpha = \frac{r}{v} = \frac{\text{hypotenuse}}{\text{opposite}}.$$

Example 11

Use geometric methods to obtain the trigonometric ratios for an isosceles right triangle.

Solution. Let OQR be an isosceles right triangle, as in Fig. 10-6, where $v = RQ = OQ = u$. Then $r = \sqrt{u^2 + v^2} = \sqrt{2}\,u$ and the acute angles are both $\pi/4$ radians. We obtain

$$\cos \frac{\pi}{4} = \frac{u}{\sqrt{2}\,u} = \frac{\sqrt{2}}{2}, \qquad \sin \frac{\pi}{4} = \frac{u}{\sqrt{2}\,u} = \frac{\sqrt{2}}{2}, \qquad \tan \frac{\pi}{4} = \frac{u}{u} = 1$$

and so on. These results agree with our work in Sections 9-1 and 9-3.

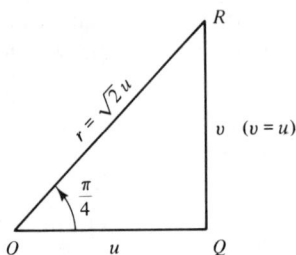

Figure 10-6

EXERCISES 10-1

1. Sketch an angle in standard position whose measure is as shown.
   (a) $75°$ (b) $-50°$ (c) $450°$ (d) $-270°$ (e) $900°$

2. Sketch an angle in standard position whose measure is between 0 and $360°$, given that its terminal side
   (a) is the half of $v = 2u$ in the first quadrant.
   (b) is the half of $v + 3u = 0$ in the second quadrant.
   (c) contains $(3, -8)$.

(d) contains $(-2, 0)$.

(e) has slope $1/2$ and lies in the upper half plane.

(f) has slope -2 and lies in the lower half plane.

3. Sketch an angle in standard position whose measure is between $-90°$ and $90°$, given that its terminal side

(a) is part of the line $u + 2v = 0$. (b) contains the point $(1, -1)$.

(c) has slope $2/3$. (d) is perpendicular to $v = 5u$.

4. Sketch an angle in standard position whose radian measure is as shown.

(a) $\dfrac{2\pi}{3}$ (b) $-\dfrac{\pi}{3}$ (c) 7π (d) $\dfrac{9\pi}{4}$ (e) 3

In Problems 5–12 convert the angle from degrees to radians.

5. $150°$ 6. $-210°$ 7. $750°$ 8. $75°$

9. $-50°$ 10. $405°$ 11. $-900°$ 12. $113°$

In Problems 13–20 convert the angle from radians to degrees.

13. $\dfrac{4\pi}{3}$ 14. $-\dfrac{3\pi}{2}$ 15. 7π 16. $\dfrac{11\pi}{4}$

17. 1.5 18. -3 19. 8 20. $2/3$

†21. A record is turning at 45 revolutions per minute (clockwise). Express in degrees and in radians the measure of the angle generated at the center by a point on the rim while playing a song lasting 2 minutes and 18 seconds.

22. Express in degrees and in radians the angle generated between 11:00 A.M. and 3:00 P.M.

(a) by the hour hand of a clock.

(b) by the minute hand.

†23. The propeller of an airplane makes 3500 revolutions per minute. What angle is generated by one of the blades of the propeller in one second?

†24. If the wheel of an automobile has a radius of 1.25 feet, how many revolutions per minute does it turn when traveling at 55 miles per hour?

25. For each angle described in Problem 2 find $\sin \alpha$, $\cos \alpha$, $\tan \alpha$, $\sec \alpha$.

26. For each angle described in Problem 3 find $\sin \alpha$, $\cos \alpha$, $\cot \alpha$, $\csc \alpha$.

27. Find the angles and trigonometric ratios associated with a right triangle whose sides are $1, \sqrt{3}$, and 2 units.

10-2 EVALUATION OF TRIGONOMETRIC FUNCTIONS FOR ANGLES

Quadrantal Angles

An angle in standard position whose terminal side coincides with half of one of the coordinate axes is called a **quadrantal** angle because such a terminal side is a boundary of two quadrants of the u, v plane. Examples of quadrantal angles are 0, $\pm\pi/2$, $\pm\pi$, $\pm 3\pi/2$, and so on. You should remember that not every trigonometric function is defined for all quadrantal angles; for instance, the tangent function is undefined at $\pm\pi/2$, $\pm 3\pi/2$, ... When they exist, however, the values of the trigonometric functions at quadrantal angles are very easy to obtain. We have done so in Sections 9-1 through 9-3.

The evaluation of sin α, cos α, tan α, and so on for a nonquadrantal angle α can be carried out by using tables in conjunction with linear interpolation. Although in practice this method is rapidly giving way to the easier and quicker process of pressing a button on a calculator, it is valuable for any mathematics student to learn and understand the techniques presented here. For this reason, we strongly suggest that calculators not be used for direct evaluation of these functions during the work of this section.

Acute Angles

> **Definition.** Two acute angles whose sum is a right angle are called **complementary** angles.

For example, $\pi/3$ and $\pi/6$ are complementary angles, as are 40° and 50°.

Approximate values of the trigonometric functions for angles measuring 0 to 90° are listed in Table 10-2. The table actually has only 46 lines, an abbreviation made possible by the following relationships. If α and β are complementary angles, then it follows from Eq. (9-7) that sin α = cos β, tan α = cot β, sec α = csc β. Thus the cosine of an angle greater than 45° is entered in the same position as the sine of its complementary angle; similarly for cotangent and tangent, cosecant and secant.

Consider, for instance, the angle 30°. For this angle the entries under the headings sin, cos, tan, cot, sec, csc are

$$.5000 \quad .8660 \quad .5774 \quad 1.732 \quad 1.155 \quad 2.000.$$

That is,

$$\sin 30° \approx .5000 \qquad \cot 30° \approx 1.732$$
$$\cos 30° \approx .8660 \qquad \sec 30° \approx 1.155$$
$$\tan 30° \approx .5774 \qquad \csc 30° \approx 2.000.$$

This row of entries also gives the values of the functions for the complementary angle 60°, but in the order cos, sin, cot, tan, csc, sec.

$$\cos 60° \approx .5000 \qquad \tan 60° \approx 1.732$$
$$\sin 60° \approx .8660 \qquad \csc 60° \approx 1.155$$
$$\cot 60° \approx .5774 \qquad \sec 60° \approx 2.000$$

To use Table 10-2, we read angles from 0 to 45° on the *left* column and use the headings at the *top*; we read angles from 46 to 90° on the *right* column and use the titles at the *bottom*. For convenience, the radian measure of each angle is also listed.

Example 1

Evaluate.

(a) tan 32°

(b) csc 75°

(c) cot 47°

(d) sec 45°

TABLE 10-2 VALUES OF TRIGONOMETRIC FUNCTIONS

Degrees	Radians	sin	cos	tan	cot	sec	csc		
0	.0000	.0000	1.0000	.0000	—	1.000	—	1.571	90
1	.0175	.0175	.9998	.0175	57.290	1.000	57.299	1.553	89
2	.0349	.0349	.9994	.0349	28.636	1.001	28.654	1.536	88
3	.0524	.0523	.9986	.0524	19.081	1.001	19.107	1.518	87
4	.0698	.0698	.9976	.0699	14.301	1.002	14.336	1.501	86
5	.0873	.0872	.9962	.0875	11.430	1.004	11.474	1.484	85
6	.1047	.1045	.9945	.1051	9.514	1.006	9.567	1.466	84
7	.1222	.1219	.9925	.1228	8.144	1.008	8.206	1.449	83
8	.1396	.1392	.9903	.1405	7.115	1.010	7.185	1.431	82
9	.1571	.1564	.9877	.1584	6.314	1.012	6.392	1.414	81
10	.1745	.1736	.9848	.1763	5.671	1.015	5.759	1.396	80
11	.1920	.1908	.9816	.1944	5.145	1.019	5.241	1.379	79
12	.2094	.2079	.9781	.2126	4.705	1.022	4.810	1.361	78
13	.2269	.2249	.9744	.2309	4.331	1.026	4.445	1.344	77
14	.2443	.2419	.9703	.2493	4.011	1.031	4.134	1.326	76
15	.2618	.2588	.9659	.2679	3.732	1.035	3.864	1.309	75
16	.2793	.2756	.9613	.2867	3.487	1.040	3.628	1.292	74
17	.2967	.2924	.9563	.3057	3.271	1.046	3.420	1.274	73
18	.3142	.3090	.9511	.3249	3.078	1.051	3.236	1.257	72
19	.3316	.3256	.9455	.3443	2.904	1.058	3.072	1.239	71
20	.3491	.3420	.9397	.3640	2.747	1.064	2.924	1.222	70
21	.3665	.3584	.9336	.3839	2.605	1.071	2.790	1.204	69
22	.3840	.3746	.9272	.4040	2.475	1.079	2.669	1.187	68
23	.4014	.3907	.9205	.4245	2.356	1.086	2.559	1.169	67
24	.4189	.4067	.9135	.4452	2.246	1.095	2.459	1.152	66
25	.4363	.4226	.9063	.4663	2.145	1.103	2.366	1.134	65
26	.4538	.4384	.8988	.4877	2.050	1.113	2.281	1.117	64
27	.4712	.4540	.8910	.5095	1.963	1.122	2.203	1.100	63
28	.4887	.4695	.8829	.5317	1.881	1.133	2.130	1.082	62
29	.5061	.4848	.8746	.5543	1.804	1.143	2.063	1.065	61
30	.5236	.5000	.8660	.5774	1.732	1.155	2.000	1.047	60
31	.5411	.5150	.8572	.6009	1.664	1.167	1.942	1.030	59
32	.5585	.5299	.8480	.6249	1.600	1.179	1.887	1.012	58
33	.5760	.5446	.8387	.6494	1.540	1.192	1.836	.9948	57
34	.5934	.5592	.8290	.6745	1.483	1.206	1.788	.9774	56
35	.6109	.5736	.8192	.7002	1.428	1.221	1.743	.9599	55
36	.6283	.5878	.8090	.7265	1.376	1.236	1.701	.9425	54
37	.6458	.6018	.7986	.7536	1.327	1.252	1.662	.9250	53
38	.6632	.6157	.7880	.7813	1.280	1.269	1.624	.9076	52
39	.6807	.6293	.7771	.8098	1.235	1.287	1.589	.8901	51
40	.6981	.6428	.7660	.8391	1.192	1.305	1.556	.8727	50
41	.7156	.6561	.7547	.8693	1.150	1.325	1.524	.8552	49
42	.7330	.6691	.7431	.9004	1.111	1.346	1.494	.8378	48
43	.7505	.6820	.7314	.9325	1.072	1.367	1.466	.8203	47
44	.7679	.6947	.7193	.9657	1.036	1.390	1.440	.8029	46
45	.7854	.7071	.7071	1.0000	1.000	1.414	1.414	.7854	45
		cos	sin	cot	tan	csc	sec	Radians	Degrees

Solution. (a) Using the left column to find the row for 32°, we locate tan 32° in that row under the heading tan at the top; the value is .6249, approximately.

(b) Using the right column to find the row for 75°, we locate csc 75° ≈ 1.035 in that row above the title csc at the bottom.

(c) cot 47° ≈ .9325.

(d) sec 45° ≈ 1.414.

If the acute angle whose trigonometric function values are to be found is not listed in the table, we can use linear interpolation.

Example 2

Evaluate.

(a) sin 37.34°

(b) cos 1.042

Solution. (a) The table gives sin 37° ≈ .6018 and sin 38° ≈ .6157.

$$\frac{\sin 37.34° - \sin 37°}{37.34 - 37} \approx \frac{\sin 38° - \sin 37°}{38 - 37}$$

$$\sin 37.34° \approx \sin 37° + (.34)(.6157 - .6018)$$

$$\approx .6018 + .0047$$

$$= .6065$$

(b) Here the angle is measured in radians. (How can you tell?) Using the column for radians on the right, 1.042 is between 1.030 and 1.047. We have

$$\frac{\cos 1.042 - \cos 1.030}{1.042 - 1.030} \approx \frac{\cos 1.047 - \cos 1.030}{1.047 - 1.030}$$

$$\cos 1.042 \approx .5150 + (.012)\frac{.5000 - .5150}{.017}$$

$$\approx .5150 + (.012)(-.8823)$$

$$\approx .5150 - .0106 = .5044.$$

Finding the Angle

We can use Table 10-2 and interpolation to find an acute angle if one of its trigonometric function values is given.

Example 3

Find acute angles α, β, γ, given that

(a) sin α = .5592.

(b) tan β = 1.165.

(c) cos γ = −2.4637.

Solution. (a) Looking down the column of sine values, we find the value .5592 in the row corresponding to 34°. We also see from the table that the radian measure of α is .5934, approximately.

(b) The entry 1.165 does not occur in the tangent column; so we need to interpolate. We see that $\tan 49° \approx 1.150$ and $\tan 50° \approx 1.192$. Therefore β is between 49 and 50°.

$$\frac{\tan \beta - \tan 49°}{\beta - 49} \approx \frac{\tan 50° - \tan 49°}{50 - 49}$$

$$\frac{1.165 - 1.150}{\beta - 49} \approx \frac{1.192 - 1.150}{50 - 49}$$

$$\frac{.015}{\beta - 49} \approx \frac{.042}{1}$$

$$\beta \approx 49 + (.015)\left(\frac{1}{.042}\right)$$

$$\approx 49.36°$$

If we wish to have β in radians, we use Eq. (10-2) and obtain

$$\beta \approx (49.36)\frac{\pi}{180} \approx .861.$$

(c) The range of the cosine function is the interval $[-1, 1]$ and so there is no γ such that $\cos \gamma = -2.4637$.

Other Angles

We have already seen how to evaluate the trigonometric functions at α when α is an acute or quadrantal angle. We now consider other possible values of α. Suppose first that $\pi/2 < \alpha < \pi$; then the terminal side of α lies in the second quadrant. The point P_α on the unit circle has coordinates $(\cos \alpha, \sin \alpha)$. The reflection of this point through the v axis [Fig. 10-7(a)] is $P_{\hat{\alpha}}$, where $\hat{\alpha} = \pi - \alpha$ is an acute angle with coordinates $(\cos \hat{\alpha}, \sin \hat{\alpha})$ such that

$$\cos \alpha = -\cos \hat{\alpha},$$

$$\sin \alpha = \sin \hat{\alpha}.$$

Thus the cosine of α can be obtained from the cosine of $\hat{\alpha}$ by changing its sign whereas $\sin \alpha$ is equal to $\sin \hat{\alpha}$. The other function values at α can also be obtained from those for $\hat{\alpha}$.

$$\tan \alpha = \frac{\sin \alpha}{\cos \alpha} = \frac{\sin \hat{\alpha}}{-\cos \hat{\alpha}} = -\tan \hat{\alpha}$$

$$\cot \alpha = \frac{\cos \alpha}{\sin \alpha} = \frac{-\cos \hat{\alpha}}{\sin \hat{\alpha}} = -\cot \hat{\alpha}$$

$$\sec \alpha = \frac{1}{\cos \alpha} = \frac{1}{\cos \hat{\alpha}} = -\sec \hat{\alpha}$$

$$\csc \alpha = \frac{1}{\sin \alpha} = \frac{1}{\sin \hat{\alpha}} = \csc \hat{\alpha}$$

For $\pi < \alpha < 3\pi/2$, the terminal side of α lies in the third quadrant. The reflection of $P_\alpha(\cos \alpha, \sin \alpha)$ through the origin is the point $P_{\hat{\alpha}}$, where $\hat{\alpha} =$

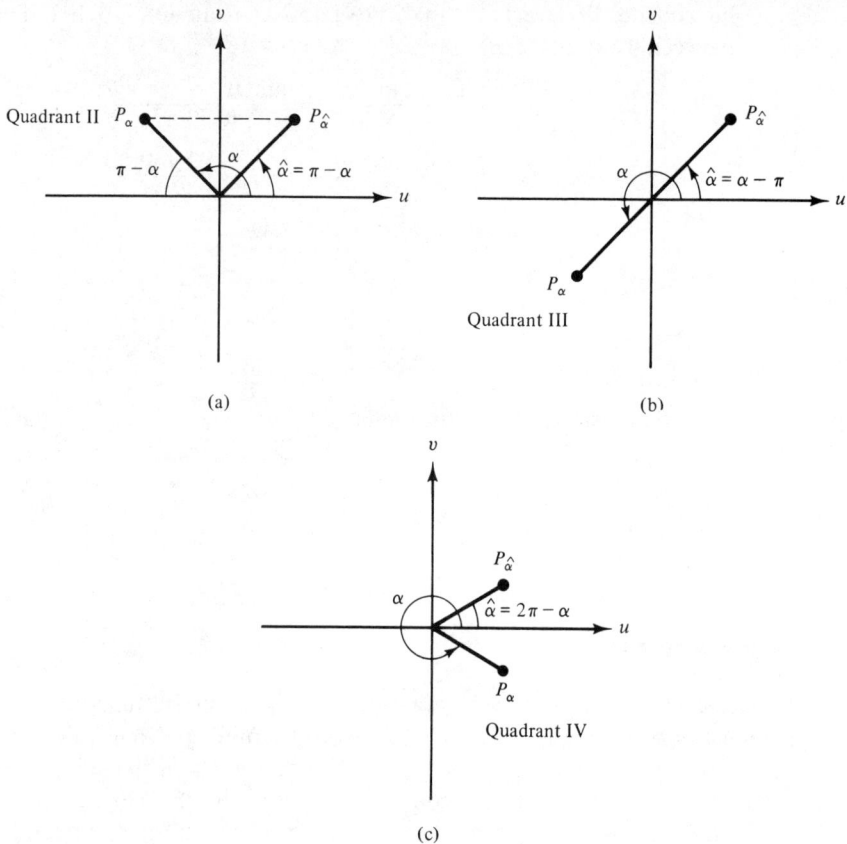

Quadrant II P_α $P_{\hat{\alpha}}$

$\pi - \alpha$ α $\hat{\alpha} = \pi - \alpha$ u

v

(a)

v

$P_{\hat{\alpha}}$

α $\hat{\alpha} = \alpha - \pi$ u

P_α

Quadrant III

(b)

v

$P_{\hat{\alpha}}$

α $\hat{\alpha} = 2\pi - \alpha$ u

P_α

Quadrant IV

(c)

Figure 10-7

$\alpha - \pi$ [Fig. 10-7(b)] is an acute angle for α such that

$$\cos \alpha = -\cos \hat{\alpha}, \qquad \sin \alpha = -\sin \hat{\alpha}, \qquad \tan \alpha = \tan \hat{\alpha},$$

$$\cot \alpha = \cot \hat{\alpha}, \qquad \sec \alpha = -\sec \hat{\alpha}, \qquad \csc \alpha = -\csc \hat{\alpha}.$$

For $3\pi/2 < \alpha < 2\pi$, the terminal side of α lies in the fourth quadrant. The reflection of $P_\alpha(\cos \alpha, \sin \alpha)$ through the u axis is the point $P_{\hat{\alpha}}(\cos \hat{\alpha}, \sin \hat{\alpha})$, where $\hat{\alpha} = 2\pi - \alpha$ [Fig. 10-7(c)] is an acute angle such that

$$\cos \alpha = \cos \hat{\alpha}, \qquad \sin \alpha = -\sin \hat{\alpha}, \qquad \tan \alpha = -\tan \hat{\alpha}$$

$$\cot \alpha = -\cot \hat{\alpha}, \qquad \sec \alpha = \sec \hat{\alpha}, \qquad \csc \alpha = -\csc \hat{\alpha}.$$

Our discussion shows that we can obtain the values of the trigonometric functions at any nonquadrantal angle α between 0 and 2π by making use of a certain angle $\hat{\alpha}$, called the reference angle for α, whose description differs from one quadrant to another.

If you examine the three preceding cases, you will see that the reference angle can be described as follows.

The **reference angle** for α is the smallest positive angle made by the terminal side of α and the horizontal axis.

Furthermore, the values of the trigonometric functions at α always have the same absolute values as those of $\hat{\alpha}$. Consequently, all we need to know to evaluate one of the functions at α is

1. the reference angle $\hat{\alpha}$.
2. whether the given function has positive or negative values in the quadrant containing the terminal side of α.

We have just indicated how to obtain (1); (2) is summarized in Fig. 10-8, in which the functions named in each quadrant have positive values there.

Signs of trigonometric functions

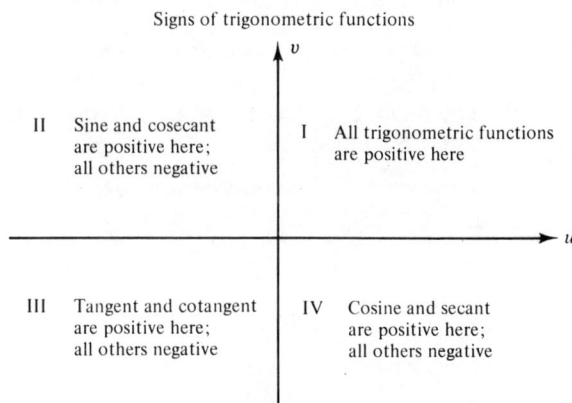

| II | Sine and cosecant are positive here; all others negative | I | All trigonometric functions are positive here |

| III | Tangent and cotangent are positive here; all others negative | IV | Cosine and secant are positive here; all others negative |

Figure 10-8

Negative angles and angles exceeding 2π radians are dealt with by using periodicity; each such angle is coterminal with an angle between 0 and 2π.

Example 4

Evaluate.

(a) $\sin \dfrac{2\pi}{3}$ (b) $\cos 150°$ (c) $\sin 102°$

(d) $\tan 212°$ (e) $\csc \dfrac{5\pi}{3}$ (f) $\sin 822°$

Solution. (a) For $\alpha = 2\pi/3$, $\hat{\alpha} = \pi - 2\pi/3 = \pi/3$ and we know from Section 9-1 that $\sin \pi/3 = \sqrt{3}/2$. Because $\alpha = 2\pi/3$ is in the second quadrant, its sine is positive. Therefore

$$\sin \frac{2\pi}{3} = \sin \frac{\pi}{3} = \frac{\sqrt{3}}{2}.$$

(b) For $\alpha = 150°$, $\hat{\alpha} = 180° - 150° = 30°$. Because the cosine is negative

in the second quadrant,
$$\cos 150° = -\cos 30° = -\cos\frac{\pi}{6} = -\frac{\sqrt{3}}{2}.$$

(c) For $\alpha = 102°$, $\hat{\alpha} = 180° - 102° = 78°$ and from the table we find $\sin 78° \approx .9781$. Because the sine is positive in the second quadrant,
$$\sin 102° \approx .9781.$$

(d) For $\alpha = 212°$, $\hat{\alpha} = 212° - 180° = 32°$. Because the tangent is positive in the third quadrant,
$$\tan 212° = \tan 32° \approx .6249.$$

(e) For $\alpha = 5\pi/3$, $\hat{\alpha} = 2\pi - 5\pi/3 = \pi/3$ and
$$\csc\frac{\pi}{3} = \frac{2}{\sqrt{3}} = \frac{2\sqrt{3}}{3}.$$

Because the cosecant is negative in the fourth quadrant,
$$\csc\frac{5\pi}{3} = -\frac{2\sqrt{3}}{3}.$$

(f) The angle $822°$ has the same terminal side as $102°$ because it differs from the latter by $720° = 2\cdot360°$. Therefore
$$\sin 822° = \sin 102° \approx .9781$$
by part (c).

Example 5

Find the angle β, given that

(a) $\cos\beta = -\dfrac{1}{2}$ and $0 < \beta < \pi$

(b) $\cos\beta = -.1823$ and $0 < \beta < 180°$.

Solution. (a) Because $\cos\beta < 0$, the terminal side of β is in the second quadrant. Then $\hat{\beta} = \pi - \beta$ is acute and $\cos\hat{\beta} = 1/2$. We recognize $\hat{\beta} = \pi/3$; so $\beta = 2\pi/3$.

(b) Again, $\cos\beta < 0$; so the terminal side of β is in the second quadrant and $\hat{\beta} = 180° - \beta$ is an acute angle with $\cos\hat{\beta} = .1823$. By interpolation, we find

$$\frac{\cos\hat{\beta} - \cos 79°}{\hat{\beta} - 79} \approx \frac{\cos 80° - \cos 79°}{80 - 79} \approx \frac{.1736 - .1908}{80 - 79} \approx -.0172.$$

Because $\cos\hat{\beta} = .1823$, we get

$$\frac{.1823 - .1908}{\hat{\beta} - 79} \approx -.0172$$

$$\hat{\beta} \approx 79° + \frac{.1823 - .1908}{-.0172}$$

$$\approx 79.49°.$$

Therefore
$$\beta \approx 180° - 79.49° = 100.51°.$$

EXERCISES 10-2

In Problems 1–16 evaluate, using Table 10-2 and linear interpolation if necessary.

1. $\sin 36°$ **2.** $\cos 42°$ **3.** $\cos 72°$ **4.** $\sin 68°$

5. $\tan 35°$ **6.** $\cot 56°$ **7.** $\sec 70°$ **8.** $\csc 70°$

9. $\sin 25.32°$ **10.** $\cos 32.25°$ **11.** $\tan \pi/5$ **12.** $\sin \pi/8$

13. $\sin 1$ **14.** $\cos 1$ **15.** $\tan .7893$ **16.** $\sec .1603$

In Problems 17–30 evaluate; state the reference angle in each case.

17. $\sin 128°$ **18.** $\cos 312°$ **19.** $\tan 243°$ **20.** $\sin (-48°)$

21. $\cos (-117°)$ **22.** $\sec 740°$ **23.** $\tan 582°$ **24.** $\cot (-412°)$

25. $\csc (-20°)$ **26.** $\tan (12\pi/5)$ **27.** $\sin 18\pi/5$ **28.** $\sin 8$

29. $\cos 4.132$ **30.** $\sec (-8)$

In Problems 31–46 find the angle α (in radians between 0 and 2π) satisfying the description.

31. $\cos \alpha = \sqrt{3}/2$ and $0 < \alpha < \pi/2$ **32.** $\sin \alpha = \sqrt{3}/2$ and $\pi/2 < \alpha < \pi$

33. $\sin \alpha = -1/2$ and $\dfrac{3\pi}{2} < \alpha < 2\pi$ **34.** $\tan \alpha = -1$ and $0 < \alpha < \pi$

35. $\sin \alpha = .5446$ and $0 < \alpha < \pi/2$ **36.** $\cos \alpha = .5446$ and $0 < \alpha < \pi$

37. $\tan \alpha = .7813$ and $\pi < \alpha < 2\pi$ **38.** $\tan \alpha = .7813$ and $\sin \alpha > 0$

39. $\cos \alpha = .5230$ and $\tan \alpha > 0$ **40.** $\cos \alpha = -.3082$ and $\sin \alpha < 0$

41. $\tan \alpha = -1.532$ and $\cos \alpha < 0$ **42.** $\sin \alpha = -.7880$ and $\cos \alpha > 0$

43. $\cos \alpha = -.7880$ and $\sin \alpha > 0$ **44.** $\cos \alpha = -.7880$ and $\sin \alpha < 0$

45. $\sin \alpha = .7880$ and $\tan \alpha < 0$ **46.** $\sin \alpha = -.7880$ and $\tan \alpha > 0$

47. A right triangle has sides 3, 4, and 5. Find its acute angles (in degrees).

48. A right triangle has sides 5, 12, and 13. Find its acute angles (in degrees).

10-3 TRIGONOMETRY OF TRIANGLES [3]

Solving Right Triangles

A right triangle ABC (Fig. 10-9) has two acute angles, say α and β. Because the sum of all three angles of the triangle must be a straight angle (180°), it is clear that α and β must add up to a right angle. Therefore α and β are complementary.

In Fig. 10-9 we have denoted the side opposite the vertex A by a and the angle at A by α; the side opposite B by b and the angle there by β. The side opposite the right angle at C is denoted by c; this is the hypotenuse.

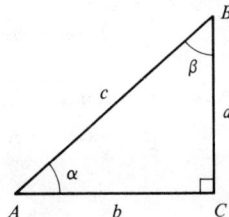

Figure 10-9

[3] In the remainder of this chapter your instructor may allow you to use a hand calculator instead of interpolation to approximate trigonometric function values and angles. This will occasionally lead to slight discrepancies between your results and those in the text or answer section.

If we know one of the acute angles and one of the sides, we can find the other acute angle and the two remaining sides. We call this process "solving" the right triangle ABC. In the following examples the notation indicated in Fig. 10-9 is used.

Example 1

Solve the right triangle ABC if $\alpha = 30°$ and $a = 1$.

Solution. First, $\beta = 90° - 30° = 60°$. Because $\sin \alpha = a/c = 1/c$, we get

$$\frac{1}{c} = \sin \alpha = \sin 30° = \frac{1}{2}.$$

Therefore $c = 2$. The third side b can be found from the relation

$$a^2 + b^2 = c^2$$

and we get $b = \sqrt{3}$. Thus the triangle ABC has these measurements:

$$\alpha = 30° \qquad a = 1$$
$$\beta = 60° \qquad b = \sqrt{3}$$
$$c = 2$$

Example 2

Solve the right triangle ABC if $\alpha = 60°$ and $b = 2$.

Solution. First, $\beta = 90° - 60° = 30°$. Now $\sin \beta = b/c = 2/c$ and because $\beta = 30°$, we get

$$\frac{2}{c} = \sin 30° = \frac{1}{2}.$$

Therefore $c = 4$. Because $a^2 + b^2 = c^2$, we have $a = \sqrt{12} = 2\sqrt{3}$.

$$\alpha = 60° \qquad a = 2\sqrt{3}$$
$$\beta = 30° \qquad b = 2$$
$$c = 4$$

Example 3

Solve the right triangle ABC if $\alpha = 20°$ and $c = 3$.

Solution. First, $\beta = 90° - 20° = 70°$. Then $\sin \alpha = a/c = a/3$ and we use Table 10-2 to obtain

$$a = 3 \sin 20° \approx 3(.3420) = 1.0260.$$

Similarly, $\sin \beta = b/3$ and we get

$$b = 3 \sin 70° \approx 3(.9397) = 2.8191.$$

Thus

$$\alpha = 20° \qquad a \approx 1.03$$
$$\beta = 70° \qquad b \approx 2.82$$
$$c = 3.$$

Example 4

Solve the right triangle ABC if $\beta = \pi/3$ and $c = 4$.

Solution. First,

$$\alpha = \frac{\pi}{2} - \frac{\pi}{3} = \frac{\pi}{6}.$$

Because $\sin \beta = b/c = b/4$, we have

$$b = 4 \sin \beta = 4 \sin\frac{\pi}{3} = 4\frac{\sqrt{3}}{2} = 2\sqrt{3}.$$

Similarly, $\cos \beta = a/4$. So

$$a = 4 \cos \frac{\pi}{3} = 4 \cdot \frac{1}{2} = 2.$$

Then

$$\alpha = \frac{\pi}{6} \qquad a = 2$$

$$\beta = \frac{\pi}{3} \qquad b = 2\sqrt{3}$$

$$c = 4.$$

Indirect Measurement

Many classical applications of trigonometry to surveying, navigating, engineering, and other fields arise from solutions of right triangles. The following examples show how certain distances can be obtained mathematically without measuring them directly. Other distances and angles are first measured and the procedures described earlier are applied to calculate the desired distance.

Example 5

A boy is flying a kite with 300 feet of string out. The angle made by the kite string with the horizontal direction is 50°. How high is the kite?

Solution. In Fig. 10-10 we are given $\alpha = 50°$ and $c = 300$ feet. We want to find a. Now

$$\frac{a}{c} = \sin 50° \approx .7660.$$

So

$$a = c \sin 50° \approx (300)(.7660)$$
$$= 229.8 \text{ feet.}$$

Example 6

Surveyors are trying to calculate the distance between two points A and B lying across an impenetrable swamp (Fig. 10-11). They choose a point C such that AC and BC are perpendicular and such that BC can be measured directly. They measure the angle $\beta = CBA$ and find $\beta = 75°$; they find $a = BC = 326$ meters.

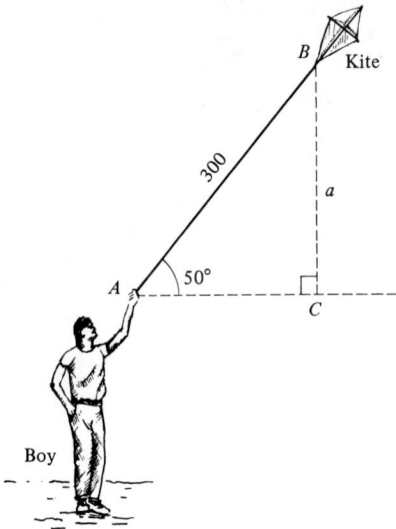

Figure 10-10

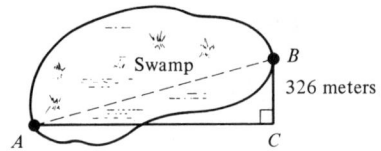

Figure 10-11

Then they calculate $c = AB$ as follows:

$$\frac{a}{c} = \cos 75°$$

$$c = \frac{a}{\cos 75°} \approx \frac{326}{.2588}$$

$$c \approx 1259.66 \text{ meters.}$$

If an object is above a viewer, the angle between the horizontal direction and the line of sight is called the **angle of elevation** of the object; if the object is below the viewer, this angle is called the **angle of depression**.

Example 7

The angle of elevation of a treetop from a point on the ground 20 feet from the base of the tree is 65°. How tall is the tree?

Solution. From Fig. 10-12 we have

$$\frac{a}{20} = \tan 65° \approx 2.145$$

$$a \approx (20)(2.145) \approx 42.9 \text{ feet.}$$

Example 8

A destroyer D detects a submarine S at a distance of 250 meters and at an angle of depression of 35°. How deep in the water is the submarine?

Solution. In Fig. 10-13 the depth is s.

$$\frac{s}{250} = \sin 35° \approx .5736$$

Applications of Trigonometric Functions   Chap. 10

so that

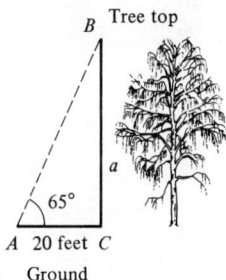

Figure 10-12

$$s \approx 143.4 \text{ meters.}$$

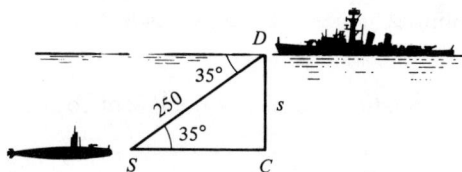

Figure 10-13

Solution of Oblique Triangles

An **oblique** triangle is one that does not have a right angle. The trigonometric functions can be used to solve oblique triangles as well as right triangles. Given the measurements of three of the six parts (angles and sides) of the triangle in Fig. 10-14, we are required to find the other three. Depending on which parts are given, there are a number of cases, of which we shall consider the following three.

Case 1. Two sides and the included angle are given.

Case 2. Three sides are given.

Case 3. Two angles and a side are given.

In order to deal with these cases, we need to develop two famous theorems from trigonometry, the law of cosines and the law of sines.

Suppose that ABC has an acute angle α at A. Let us choose an x, y coordinate system with origin at A and x axis along AC and such that $B(x, y)$ lies above the x axis (see Fig. 10-15). Let D be the foot of the perpendicular from B to the x axis.

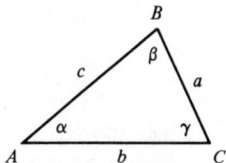

Figure 10-14

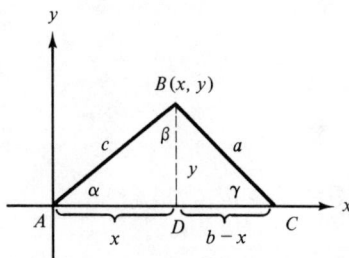

Figure 10-15

Applying the Pythagorean theorem to the right triangles ABD and DBC, we have

$$c^2 = x^2 + y^2$$
$$a^2 = y^2 + (b - x)^2.$$

Expanding the right side of the second equation and substituting for y^2 from the first, we get

$$a^2 = y^2 + (b - x)^2 = y^2 + b^2 - 2bx + x^2$$
$$= (c^2 - x^2) + b^2 - 2bx + x^2$$
$$= b^2 + c^2 - 2bx.$$

Noting that $x/c = \cos \alpha$, we have $x = c \cos \alpha$ and so

$$a^2 = b^2 + c^2 - 2bc \cos \alpha. \qquad (10\text{-}6a)$$

This equation, known as the law of cosines, can be stated in words as follows.

> **Theorem 10-1 (Law of Cosines).** The square of one side of a triangle is the sum of the squares of the other two sides, minus twice the product of these two sides times the cosine of the angle opposite the first side.

In our derivation of the law of cosines we assumed that angle α is acute. We leave it for the reader to verify that the proof also holds if α is greater than 90°. Of course, the law applies just as well to the sides opposite β and γ so that

$$b^2 = a^2 + c^2 - 2ac \cos \beta, \qquad (10\text{-}6b)$$
$$c^2 = a^2 + b^2 - 2ab \cos \gamma. \qquad (10\text{-}6c)$$

Example 1 (Case 1)

Using the notation in Fig. 10-14, solve the triangle ABC, where $b = 10$, $c = 7$, and $\alpha = 36°$.

Solution. We begin by sketching a triangle that approximates the given measurements (see Fig. 10-16). Next, we apply Eq. (10-6a):

$$a^2 = 10^2 + 7^2 - 2 \cdot 10 \cdot 7 \cdot \cos 36°.$$

Then because $\cos 36° \approx .8090$, we obtain

$$a^2 \approx 35.74,$$
$$a \approx 5.98.$$

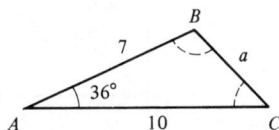

Figure 10-16

To obtain γ, we use Eq. (10-6c) and solve for $\cos \gamma$:

$$\cos \gamma = \frac{a^2 + b^2 - c^2}{2ab}.$$

Substituting for a, b, and c, we have

$$\cos \gamma \approx .7252.$$

From tables (using interpolation) we obtain

$$\gamma \approx 43.51°.$$

Finally, we apply the law of cosines in the form (10-6b), obtaining

$$\cos \beta = \frac{a^2 + c^2 - b^2}{2ac}$$

$$\approx -.1823.$$

Using the methods of Section 10-2 [in particular, the result of Example 5(b)], we find $\beta \approx 100.51°$. As a check on our work, we add α, β, and γ; the fact that the sum is 180.02° rather than 180° is attributable to the approximate values employed. (We could have calculated β by subtracting $\alpha + \gamma$ from 180°, but then we would not be able to check our work this way.)

Example 2 (Case 2)

Solve the triangle ABC, given its three sides $a = 9$, $b = 15$, $c = 8$.

Solution.   (See Fig. 10-17.) We use the law of cosines three times, obtaining the cosines of α, β, and γ and then the angles themselves.

$$\cos \alpha = \frac{b^2 + c^2 - a^2}{2bc} \approx .8667; \alpha \approx 29.92°$$

$$\cos \beta = \frac{a^2 + c^2 - b^2}{2ac} \approx -0.5556; \beta \approx 123.75°$$

$$\cos \gamma = \frac{a^2 + b^2 - c^2}{2ab} \approx .8963; \gamma \approx 26.32°$$

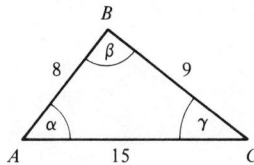

Figure 10-17

Law of Sines

We know from geometry that if α and β are angles of a triangle with $\alpha > \beta$, then the side opposite α is greater than the side opposite β. The following theorem shows that, in fact, the sides are proportional to the *sines* of the angles.

> **Theorem 10-2 (Law of Sines).**   If α, β, and γ are the angles of a triangle and a, b, and c are the lengths of the respective opposite sides, then
>
> $$\frac{a}{\sin \alpha} = \frac{b}{\sin \beta} = \frac{c}{\sin \gamma}. \qquad (10\text{-}7)$$

To prove Theorem 10-2, we let h be the length of the altitude BB' from B to the opposite side. Figure 10-18 shows three possible cases. In Fig. 10-18(a) α is acute and

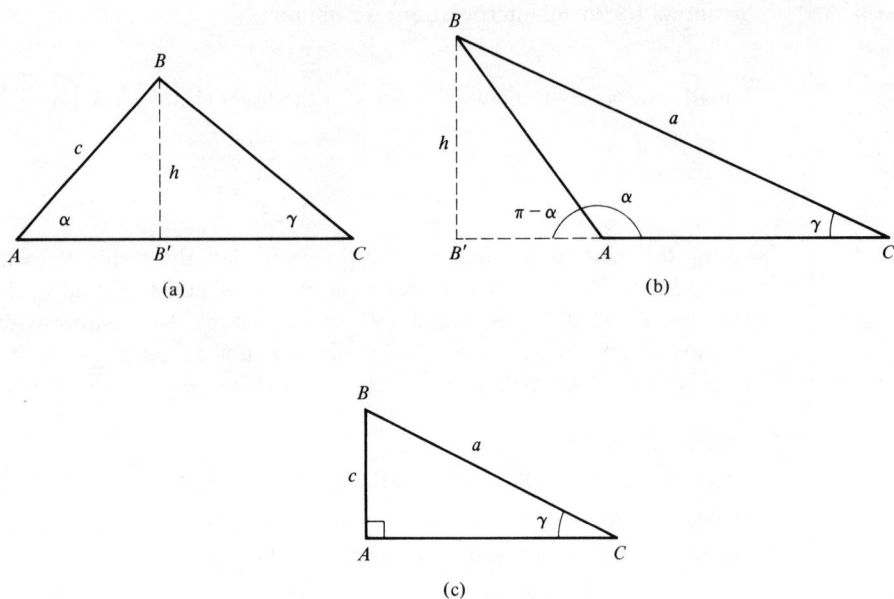

Figure 10-18

we can use the interpretation of $\sin \alpha$ as the ratio of two sides of the right triangle ABB' to obtain

$$\sin \alpha = \frac{h}{c}$$

$$h = c \sin \alpha.$$

Similarly,

$$\sin \gamma = \frac{h}{a}$$

$$h = a \sin \gamma.$$

Therefore

$$a \sin \gamma = c \sin \alpha$$

$$\frac{a}{\sin \alpha} = \frac{c}{\sin \gamma}.$$

In Fig. 10-18(b), where $\pi/2 < \alpha < \pi$, we see that B' meets the extension of side AC outside ABC. Considering the triangle $B'BA$, in which the angle at A is $\pi - \alpha$ radians, we have

$$\frac{h}{c} = \sin (\pi - \alpha) = \sin \alpha$$

$$h = c \sin \alpha$$

and as before

$$h = a \sin \gamma.$$

So again

$$\frac{a}{\sin \alpha} = \frac{c}{\sin \gamma}.$$

In Fig. 10-18(c), α is a right angle, $\sin \alpha = 1$ and $\sin \gamma = c/a$. So

$$a \sin \gamma = c = c \sin \alpha$$

and it follows once more that

$$\frac{a}{\sin \alpha} = \frac{c}{\sin \gamma}.$$

Because nothing special was assumed about the angle α, we could just as well have carried out the preceding argument with the angle β, obtaining in all possible cases

$$\frac{b}{\sin \beta} = \frac{c}{\sin \gamma}.$$

Therefore we have the double equation (10-7)

$$\frac{a}{\sin \alpha} = \frac{b}{\sin \beta} = \frac{c}{\sin \gamma}.$$

Example 3 (Case 3)

Solve triangle ABC, given $\alpha = 26°$, $\beta = 39°$, and $c = 24$.

Solution. First, we have $\gamma = 180° - 26° - 39° = 115°$. By the law of sines,

$$\frac{a}{\sin 26°} = \frac{b}{\sin 39°} = \frac{24}{\sin 115°}.$$

The last ratio is approximately 26.48 and so

$$b \approx (\sin 39°)(26.48)$$
$$\approx 16.66$$

and

$$a \approx (\sin 26°)(26.48)$$
$$\approx 11.61.$$

EXERCISES 10-3

Using the notation in Fig. 10-9, solve the right triangles whose parts are indicated in Problems 1–10.

1. $\alpha = 30°$, $a = 5$
2. $\alpha = 30°$, $b = 4$
3. $\alpha = 50°$, $c = 3$.
4. $\beta = 30°$, $a = 3$
5. $\beta = 60°$, $b = 4$
6. $\beta = 20°$, $c = 5$
7. $\alpha = \frac{\pi}{4}$, $a = 4$
8. $\beta = \frac{\pi}{3}$, $a = 6$
9. $\alpha = \frac{\pi}{6}$, $b = 4$
10. $\beta = \frac{\pi}{4}$, $c = 3$

11. A building casts a shadow 25 meters long. If the angle of elevation of the top of the building (measured from the end of the shadow) is $30°$, what is the height of the building?

12. A guy wire is fastened to the top of a telephone pole and anchored in the ground at a point 30 feet from the base of the pole. If the guy wire makes an angle of $40°$ with the ground, how high is the pole?

13. To calculate the width of a straight river, a man stands at a point A and locates point C directly across the river. He walks along the river's edge to a point B 100 yards from A and observes that the line BC makes an angle of $40°$ with the river. How wide is the river?

14. A plane traveling at an altitude of 10,000 feet is approaching a mountain 4000 feet high. The pilot observes the angle of depression of the mountaintop to be $10°$. What is the horizontal distance between the plane and the mountain?

15. A swimming pool is 20 meters long. The bottom of the pool makes a uniform angle of $10°$ with the horizontal (Fig. 10-19). If the depth at the deep end is 4 meters, what is the depth at the shallow end?

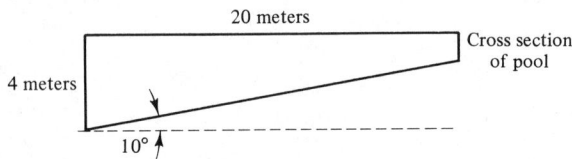

Figure 10-19

Using the notation in Fig. 10-14, solve the triangles whose parts are indicated in Problems 16–32.

16. $\beta = 53°$, $a = 8$, $c = 12$

17. $\alpha = 81°$, $b = 18$, $c = 16$

18. $\gamma = 35°$, $a = 7$, $b = 4$

19. $\beta = 40°$, $a = 14$, $c = 7$

20. $\alpha = .623$, $b = 3$, $c = 2$

21. $\gamma = .882$, $a = 4$, $b = 7$

22. $\alpha = .604$, $b = 19$, $c = 24$

23. $a = 3$, $b = 5$, $c = 7$

24. $a = 12$, $b = 5$, $c = 13$

25. $a = 6$, $b = 8$, $c = 10$

26. $a = 11$, $b = 20$, $c = 25$

27. $\beta = 36°$, $\gamma = 47°$, $a = 12$

28. $\alpha = 80°$, $\beta = 10°$, $a = 10$

29. $\alpha = 32.6°$, $\gamma = 28.4°$, $b = 12$

30. $\beta = .304$, $\gamma = .782$, $c = 15.3$

31. $\alpha = 1.2$, $\beta = .724$, $c = 18$

32. $\alpha = \frac{\pi}{5}$, $\beta = \frac{3\pi}{5}$, $a = 1$

10-4 INVERSE TRIGONOMETRIC FUNCTIONS

On numerous occasions we want to find an angle, given the value of one of the trigonometric functions at that angle. Accordingly, we investigate the existence of inverses of the trigonometric functions.

Arcsine

Consider first the sine function

$$f(x) = \sin x$$

where we can think of x as any real number or as an angle measured in radians. It is obvious that the sine function does not have an inverse because it is not one to one. For instance, there are many values of x for which $\sin x = 1/2$—$\pi/6$, $5\pi/6$, $13\pi/6$, and so on. There is no *one* angle having a sine equal to 1/2.

Applications of Trigonometric Functions   Chap. 10

We recall, however, that the sine function is increasing on the interval $[-\pi/2, \pi/2]$. In fact, on this interval the sine increases from its minimum value -1 to its maximum value 1, assuming each value in its range, $[-1, 1]$ once and only once. The only number x between $-\pi/2$ and $\pi/2$ satisfying $\sin x = 1/2$ is $x = \pi/6$. So although the sine function does not have an inverse itself, the function obtained by restricting it to the interval $[-\pi/2, \pi/2]$ [Fig. 10-20(a)] is one to one and does have an inverse. The inverse is called the **arcsine** function, denoted Arcsin. The value Arcsin (x) is often written without parentheses, Arcsin x.

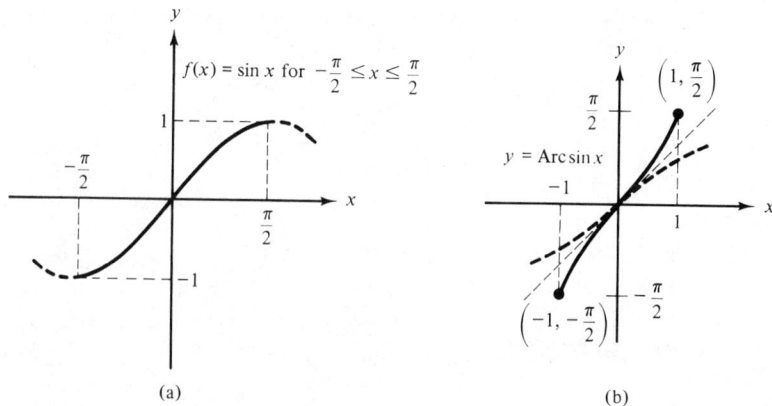

(a)   (b)

Figure 10-20

Properties of Arcsin.   Applying general properties of inverse functions (Sec. 7-3) we obtain the following facts about Arcsin.

1. The domain of Arcsin is the interval $[-1, 1]$.
2. The range of Arcsin is the interval $[-\pi/2, \pi/2]$.
3. Arcsin $(\sin x) = x$ for $-\pi/2 \le x \le \pi/2$.
4. $\sin$ (Arcsin $x) = x$ for $-1 \le x \le 1$.
5. $y =$ Arcsin x if and only if $x = \sin y$ and $-\pi/2 \le y \le \pi/2$.

Accordingly, we can describe the arcsine function in words as follows.

6. Arcsin x is the number (or angle) in the interval $[-\pi/2, \pi/2]$ whose sine is x.

When we say "the angle whose sine is x," it is always understood that this angle is measured in radians and is in the interval $[-\pi/2, \pi/2]$.

The graph of $y =$ Arcsin x is the reflection through line $y = x$ of the graph of $f(x) = \sin x$ for $-\pi/2 \le x \le \pi/2$. These graphs appear in Fig. 10-20(b).

Example 1

Evaluate.

(a) Arcsin $\dfrac{1}{2}$

(b) Arcsin 1

(c) Arcsin $\left(-\dfrac{\sqrt{3}}{2}\right)$

(d) Arcsin $\dfrac{3}{4}$

Solution. (a) We are looking for α in $[-\pi/2, \pi/2]$ such that $\sin \alpha = 1/2$; we should recognize that $\alpha = \pi/6$.

(b) The angle in $[-\pi/2, \pi/2]$ whose sine is 1 is $\pi/2$.

(c) The angle in $[-\pi/2, \pi/2]$ whose sine is $-\sqrt{3}/2$ is $-\pi/3$.

(d) To find the angle whose sine is 3/4, we use Table 10-2 and interpolate, obtaining .8482.

Example 2

Evaluate.

(a) $\sin\left(\text{Arcsin } \dfrac{2}{3}\right)$

(b) Arcsin $\sin\left(\dfrac{\pi}{3}\right)$

(c) Arcsin $\sin\left(\dfrac{2\pi}{3}\right)$

Solution.

(a) $\sin\left(\text{Arcsin } \dfrac{2}{3}\right) = \dfrac{2}{3}$ [by property (4)]

(b) Arcsin $\sin\left(\dfrac{\pi}{3}\right) = \dfrac{\pi}{3}$ [by property (3)]

(c) Arcsin $\sin\left(\dfrac{2\pi}{3}\right) = \text{Arcsin } \left(-\dfrac{\sqrt{3}}{2}\right)$

$= -\dfrac{\pi}{3}$

(Note that in (c) property (3) does not apply because $2\pi/3$ is not in the interval $[-\pi/2, \pi/2]$.)

Example 3

A boat is pulled in toward a dock 6 feet above the water by means of a rope (Fig. 10-21). Express the acute angle α made by the rope with the horizontal direction as a function of the length y of rope out.

Solution. Because $6/y = \sin \alpha$ and α is acute, we can solve for α:

$$\alpha = \text{Arcsin } \dfrac{6}{y} \text{ for } y \geq 6.$$

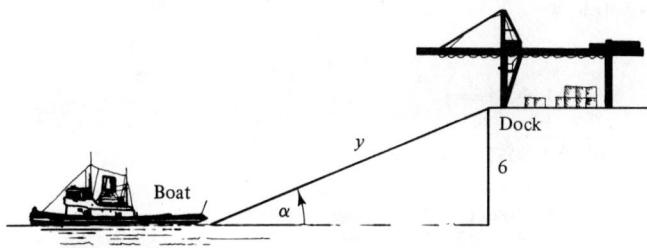

Figure 10-21

Arccosine

Like the sine, the cosine function is not one to one and hence does not have an inverse. But the function obtained by restricting the cosine to the interval $[0, \pi]$ is decreasing and therefore one to one on that interval. It has an inverse, the **arccosine** function, which will be denoted Arccos. The value Arccos (x) is often written without parentheses, Arccos x.

Properties of Arccos

1. The domain of Arccos is the interval $[-1, 1]$.
2. The range of Arccos is the interval $[0, \pi]$.
3. Arccos $(\cos x) = x$ for $0 \leq x \leq \pi$.
4. $\cos$ (Arccos x) $= x$ for $-1 \leq x \leq 1$.
5. $y = $ Arccos x if and only if $x = \cos y$ and $0 \leq y \leq \pi$.

Accordingly we can describe the arccosine function in words as follows.

6. Arccos x is the number (or angle) in the interval $[0, \pi]$ whose cosine is x.

A graph of $y = $ Arccos x appears in Fig. 10-22.

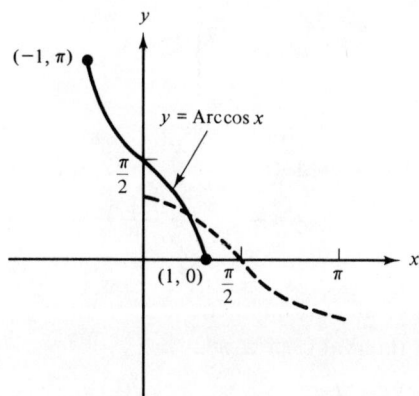

Figure 10-22

Example 4

Evalute.

(a) $\text{Arccos}\left(\dfrac{1}{2}\right)$

(b) $\text{Arccos}\left(-\dfrac{1}{2}\right)$

(c) $\sin\left(\text{Arccos}\,\dfrac{2}{3}\right)$

Solution.

(a) The angle in $[0, \pi]$ whose cosine is $1/2$ is $\pi/3$.

(b) The angle in $[0, \pi]$ whose cosine is $-1/2$ is $2\pi/3$.

(c) If α is the angle is $[0, \pi]$ with $\cos\alpha = 2/3$, then $\sin\alpha$ is positive (why?) and so

$$\sin\alpha = \sqrt{1 - \cos^2\alpha} = \frac{\sqrt{5}}{3}.$$

Arctangent

One of the most important of the inverse trigonometric functions occurs when we invert the tangent function. On the open interval $(-\pi/2, \pi/2)$ the tangent function is increasing and one to one [Fig. 10-23(a)]. The restriction of tangent to that interval has an inverse, the **arctangent** function, denoted Arctan.

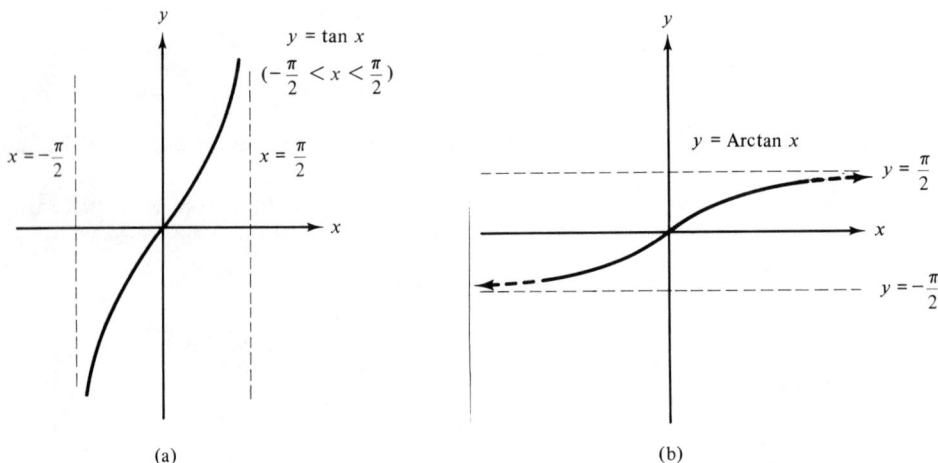

(a)   (b)

Figure 10-23

Properties of Arctan

1. The domain of Arctan is the set of all real numbers.
2. The range of Arctan is the open interval $(-\pi/2, \pi/2)$.
3. Arctan $(\tan x) = x$ for $-\pi/2 < x < \pi/2$.

---

4. tan (Arctan x) = x for all real numbers x.

5. y = Arctan x if and only if x = tan y and $-\pi/2 < y < \pi/2$.

We can describe the arctangent function in words as follows.

6. Arctan x is the number (or angle) in the interval $(-\pi/2, \pi/2)$ whose tangent is x.

The graph of y = Arctan x is obtained by reflecting the graph of y = tan x through the line $y = x$ [Fig. 10-23(b)]. In this reflection the vertical asymptotes $x = \pm\pi/2$ of y = tan x are carried over into horizontal asymptotes $y = \pm\pi/2$ of the graph of Arctan. In particular, we have the important relations

$$\text{Arctan } x \longrightarrow \frac{\pi}{2} \quad \text{ as } x \longrightarrow +\infty.$$

$$\text{Arctan } x \longrightarrow -\frac{\pi}{2} \quad \text{ as } x \longrightarrow -\infty.$$

Example 5

Evaluate.
 (a) Arctan 1
 (b) Arctan $\left(-\frac{\sqrt{3}}{3}\right)$
 (c) Arctan (-2.385)
 (d) cos (Arctan 3/4)
 (e) sin [Arctan $(-4/3)$]

Solution.
 (a) The angle between $-\pi/2$ and $\pi/2$ whose tangent is 1 is $\pi/4$.
 (b) The angle between $-\pi/2$ and $\pi/2$ whose tangent is $-\sqrt{3}/3$ is $-\pi/6$.
 (c) Using a table, we find tan $1.174 \approx 2.385$ and so Arctan $(-2.385) \approx -1.174$.
 (d) Arctan 3/4 is an acute angle, appearing on a 3-4-5 right triangle opposite the side 3; the cosine of this angle is 4/5.
 (e) The reference angle for Arctan $(-4/3)$ appears on the 3-4-5 right triangle opposite the side 4; the sine of the reference angle is 4/5. So

$$\sin\left[\text{Arctan}\left(-\frac{4}{3}\right)\right] = -\frac{4}{5}.$$

Example (Angle of Vision)

If a viewer observes a painting [see Fig. 10-24(a)] hanging on a vertical wall, the **angle of vision** is the angle subtended at the viewer's eye by the top and bottom of the painting. It is generally thought that the best view of the painting occurs when this angle of vision is greatest. Suppose that a painting 4 feet high is hung on a vertical wall 2 feet above the level of a viewer's eye. The viewer stands x feet from the wall. Express the angle of vision as a function of x.

Solution. In Figure 10-24(b) the viewer's eye is labeled A, the top and bottom of the painting B and C, respectively, and the point on the wall at the same height as the viewer's eye D. The length of AD is given as x; $BC = 4$ and $CD = 2$. We let β denote the angle DAB and γ the angle DAC. If α is the angle of vision CAB, then

$$\alpha = \beta - \gamma.$$

But $\tan \gamma = 2/x$ and $\tan \beta = 6/x$; and because γ and β are both acute, we have

$$\gamma = \text{Arctan}\, \frac{2}{x} \qquad \beta = \text{Arctan}\, \frac{6}{x}.$$

Then the angle of vision is given by

$$\alpha = \text{Arctan}\, \frac{6}{x} - \text{Arctan}\, \frac{2}{x} \qquad \text{for} \quad x > 0.$$

In studying calculus, you will learn how to maximize this function.

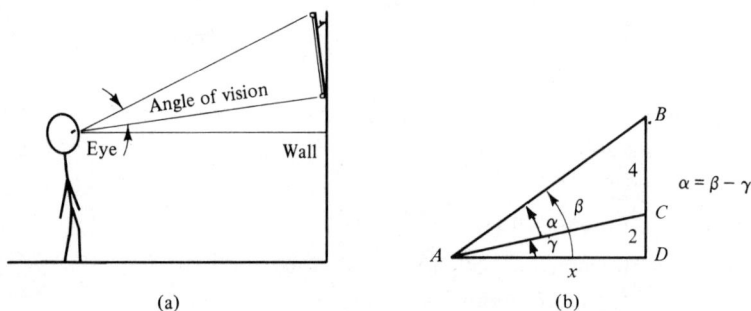

(a)   (b)

Figure 10-24

There are, of course, three other inverse trigonometric functions: Arccot, Arcsec, Arccsc. A definition of Arcsec is given in Problem 31. In general, it is quite unnecessary to use these other functions. If we want the number or angle whose cotangent is 2, for instance, we can refer to the number whose tangent is 1/2, Arctan 1/2.

EXERCISES 10-4

In Problems 1–24 evaluate the given expression.

1. $\text{Arcsin}\, \frac{\sqrt{3}}{2}$

2. $\text{Arcsin}\left(-\frac{1}{2}\right)$

3. $\text{Arccos}\, \frac{\sqrt{3}}{2}$

4. $\text{Arccos}\left(-\frac{\sqrt{3}}{2}\right)$

5. $\text{Arctan}\, \sqrt{3}$

6. $\text{Arctan}\, (-1)$

7. $\text{Arcsin}\, (-1)$

8. $\text{Arctan}\left(-\frac{1}{\sqrt{3}}\right)$

9. $\text{Arcsin}\, .7314$

10. $\text{Arctan}\, 1.327$

11. $\text{Arcsin}\, (-.4226)$

12. $\text{Arccos}\, (-.6991)$

13. $\cos\left(\text{Arccos}\, \frac{4}{5}\right)$

14. $\sin\left(\text{Arccos}\, \frac{4}{5}\right)$

15. $\tan\left(\text{Arccos}\, \frac{4}{5}\right)$

16. $\cos\left[\text{Arccos}\left(-\dfrac{4}{5}\right)\right]$ 17. $\sin\left[\text{Arccos}\left(-\dfrac{4}{5}\right)\right]$ 18. $\tan\left(\text{Arccos}\left(-\dfrac{4}{5}\right)\right)$

19. $\sin\left[\text{Arcsin}\left(-\dfrac{4}{5}\right)\right]$ 20. $\cos\left[\text{Arcsin}\left(-\dfrac{4}{5}\right)\right]$ 21. $\tan\left(\text{Arcsin}\left(-\dfrac{4}{5}\right)\right)$

22. $\sin\left(\text{Arccos}\,\dfrac{3}{5} + \text{Arcsin}\,\dfrac{1}{2}\right)$ 23. $\cos\left(\text{Arccos}\,\dfrac{3}{5} - \text{Arcsin}\,\dfrac{1}{2}\right)$

24. $\tan\left(\text{Arccos}\,\dfrac{3}{5} - \text{Arcsin}\,\dfrac{1}{2}\right)$

25. A girl flying a kite has let out 100 feet of cord. Express the angle of elevation of the kite as a function of
    (a) its altitude.
    (b) its horizontal distance from the girl.

26. A billboard is 8 feet high, its lower edge 10 feet above the ground. Express the angle of vision of the billboard as a function of the horizontal distance of the viewer from the billboard if the viewer's eye is 5 feet above the ground.

27. A searchlight 1 mile off a straight shore makes 2 revolutions every minute. How long does it take the beam of light to travel along the shore a distance of 2 miles from the point on the shore nearest the searchlight? 3 miles? 5 miles?

FOR THE ENTHUSIASTS

28. (a) Show that if an odd function has an inverse, then its inverse is also odd.
    (b) Show that Arcsin and Arctan are odd.

29. Verify the identity

$$\text{Arccos}\,x = \frac{\pi}{2} - \text{Arcsin}\,x.$$

30. Graph the function
    (a) $x\,\text{arcsin}\,x$
    (b) $\text{Arccos}\,x - \pi/2$
    (c) $\text{Arctan}\,(1/x)$

31. The **arcsecant** function is defined by the equation $\text{Arcsec}\,x = \text{Arccos}\,(1/x)$. Determine the domain and range of this function and sketch the graph.

10-5 POLAR COORDINATES

The study of angles and trigonometry allows us to draw some interesting graphs by means of a different system of coordinates, known as the **polar coordinate system**.

We choose a point O (the **pole**) and a directed half line (the **polar axis**) extending from O, as in Fig. 10-25. If θ is an angle (ordinarily measured in radians) with initial side along the polar axis, and r is any real number, then the point corresponding to the ordered pair (r, θ) is located as follows.

O **Figure 10-25**

1. If $r \geq 0$, then P is the point on the terminal side of the angle θ such that the distance between P and the pole is r.

2. If $r < 0$, we extend the terminal side of θ in the opposite direction; P is the point on this extension such that the distance between P and the pole is $|r|$.

We call r and θ **polar coordinates** of P. In polar coordinates r is a directed length; θ is also directed, the counterclockwise direction being positive.

Example 1

The points $P_1(2, \pi/4)$ and $P_2(-1, \pi/4)$ are shown in Fig. 10-26. Note that P_2 could be obtained also from the coordinates $\left(1, \dfrac{5\pi}{4}\right)$ or $\left(1, -\dfrac{3\pi}{4}\right)$.

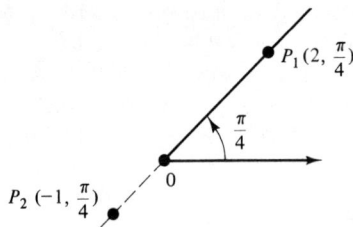

Figure 10-26

The pole itself has polar coordinates $(0, \theta)$, where θ is any angle; thus the pole has infinitely many sets of coordinates. In fact, if P is any point, say with coordinates (r, θ), then P also has the following coordinates:

$$(r, \theta + 2n\pi) \qquad \text{where } n \text{ is any integer}$$

$$(-r, \theta + (2n + 1)\pi) \qquad \text{where } n \text{ is any integer.}$$

So *every* point has infinitely many different sets of polar coordinates, in sharp contrast with the rectangular coordinate system, where each point has one and only one set of coordinates.

> **Definition.** The **polar graph** of an equation involving r and θ is the set of all points P in the plane such that some set of polar coordinates of P satisfies the equation.

Typically we will graph equations of the form $r = f(\theta)$, where f is a function.

A Circle

Example 2

Graph the equation $r = 5$ in polar coordinates.

Solution. This equation represents a constant function

$$r = f(\theta) = 5.$$

Its graph is a circle of radius 5 centered at the pole (Fig. 10-27).

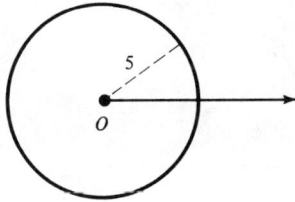

Figure 10-27

A Spiral

Example 3

Graph the equation $r = \sqrt{\theta}$ for $\theta \geq 0$ in polar coordinates.

Solution. Clearly r increases as θ increases through positive values. The result is a curve spiraling outward from the pole as θ turns counterclockwise (Fig. 10-28).

Approximate values
for $r = \sqrt{\theta}$

θ	$r = \sqrt{\theta}$
0	0
$\frac{\pi}{4}$	0.89
$\frac{\pi}{2}$	1.25
π	1.77
$\frac{3\pi}{2}$	2.17
2π	2.50

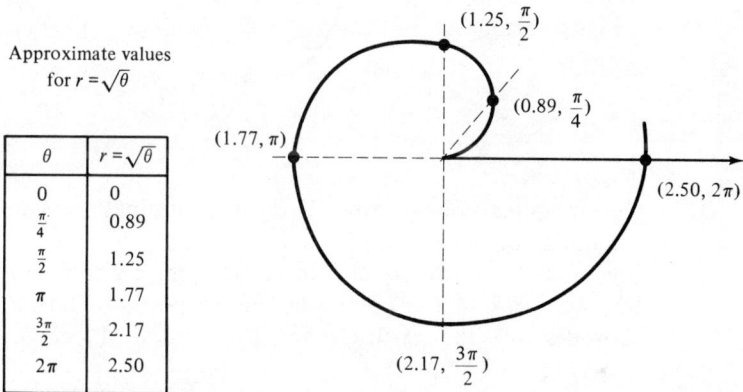

Figure 10-28

There are many other interesting spiral curves, several of which appear in the exercises.

Graphs with Loops

Example 4

Graph the function $f(\theta) = 2 \cos \theta$ in polar coordinates.

Solution. Because the function $f(\theta) = 2 \cos \theta$ has period 2π, it is sufficient to graph it for $0 \leq \theta \leq 2\pi$.

As θ increases from zero to $\pi/2$, $2 \cos \theta$ decreases from 2 to zero. The curve starts on the polar axis at $(2, 0)$ and comes in to the pole O. In Fig. 10-29(a) notice how the distance between P and O decreases as the corresponding value of θ increases from θ_1 to θ_2.

As θ increases from $\pi/2$ to π, $\cos \theta$ decreases from zero to -1 and r decreases from zero to -2. In this interval the values of r are negative. The

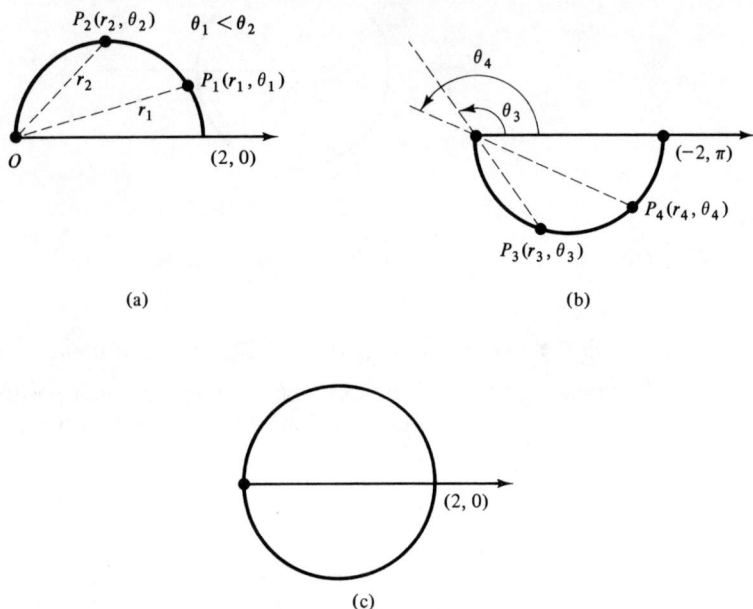

(a)

(b)

(c)

Figure 10-29

corresponding points do not lie on the terminal sides of θ but rather on the extensions of these terminal sides. Look, for instance, at points P_3 and P_4 in Fig. 10-29(b). Therefore as θ increases from $\pi/2$ to π, the curve continues from the pole back to the polar axis, at the point $(-2, \pi)$. This point, of course, coincides with the starting point $(2, 0)$ and so the curve has made a complete loop [Fig. 10-29(c)].

Letting θ increase now from π to 2π, we find that the graph retraces the same loop. It does so because the value of r at $\pi + \alpha$ is the negative of the value at α—a consequence of (9-6):

$$\cos(\pi + \alpha) = -\cos \alpha$$

The points (r, α) and $(-r, \pi + \alpha)$ coincide.

We see, then, that the graph retraces the loop in Fig. 10-29(c) each time θ increases through an interval of length π.

The resemblance of this graph to a circle is not accidental; we will show in Example 12 that it actually is a circle.

Example 5

Graph the function $f(\theta) = \sin 3\theta$ in polar coordinates.

Solution. First, we note that $0 \leq f(\theta) \leq 1$ so that all points on the graph of f are inside or on the circle $r = 1$. By means of the rays

$$\theta = 0, \frac{\pi}{6}, \frac{\pi}{3}, \ldots, \frac{11\pi}{6}$$

(Fig. 10-30), we divide the polar coordinate plane into 12 sectors. As θ increases from zero to $\pi/6$ in the first sector, $f(\theta)$ increases from zero to $f(3\pi/6) = f(\pi/2) = 1$. As θ increases from $\pi/6$ to $\pi/3$, $f(\theta)$ decreases from 1 to zero. Therefore in the first two sectors the graph of f has traced a loop from the pole out to $(1, \pi/6)$ on the unit circle and back again.

As θ increases from $\pi/3$ to $\pi/2$ and from $\pi/2$ to $2\pi/3$, the values of $f(\theta)$ are negative and the graph consists of points in the two sectors lying between $\theta = 4\pi/3$ to $5\pi/3$. Again a loop is formed, touching the unit circle at $(-1, \pi/2)$ or $(1, 3\pi/2)$. A third loop is formed over the interval $2\pi/3 \le \theta \le \pi$, touching the circle at $(1, 5\pi/6)$.

Values of θ from π to 2π do not produce new points on the graph; the earlier portions are retraced because $f(\theta + \pi) = -f(\theta)$.

The graph is called a **three-leaved rose**.

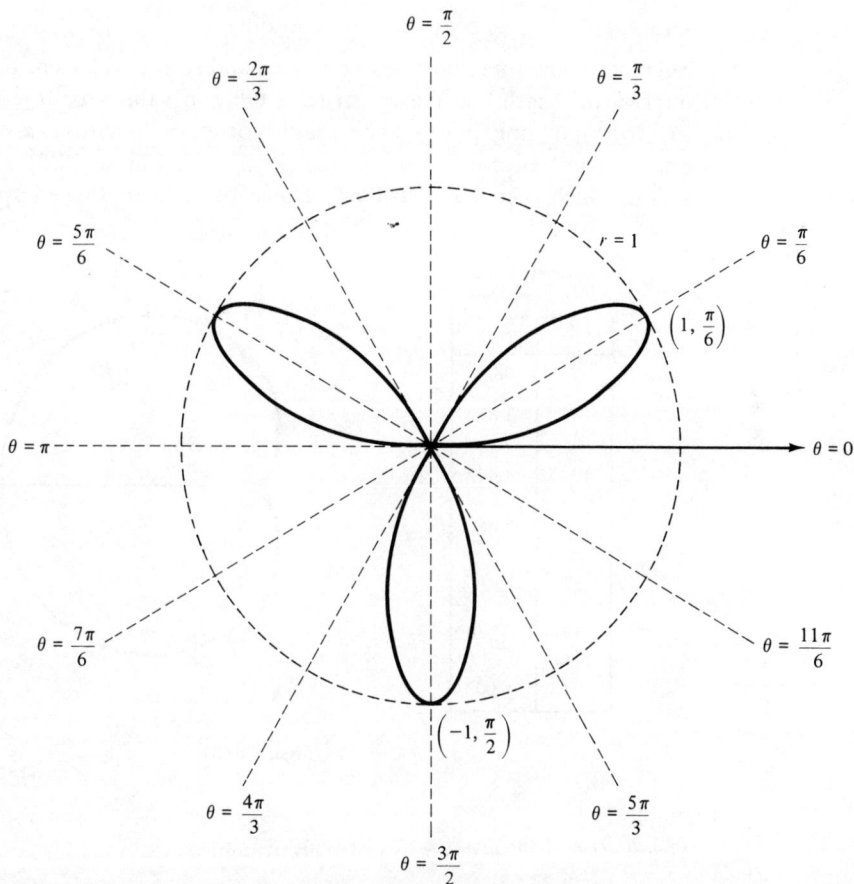

Figure 10-30

Symmetry

Symmetry of polar graphs can occur in many ways. The curve in Example 4, for instance, has symmetry with respect to the polar axis and the curve in Example 5 certainly appears to have symmetry with respect to the lines $\theta = \pi/6$, $\theta = \pi/2$, $\theta = 5\pi/6$. We will not try to formulate any general tests for symmetry, but it is useful in what follows to note that the polar graph of an even function $r = f(\theta)$ is symmetric with respect to the polar axis. [See Problem 51(a).]

Limaçons

The functions $f(\theta) = a + b\cos\theta$, where a and b are constants, have interesting graphs known as **limaçons**; their shapes depend on the relative sizes of a and b. Because f is even, all these graphs are symmetric with respect to the polar axis.

Example 6

Graph $f(\theta) = 1 + \cos\theta$ in polar coordinates.

Solution. The function f has period 2π and so it is sufficient to graph it for an interval of length 2π. We construct a table of values for $0 \le \theta \le \pi$ and plot the corresponding points. The graph corresponding to $-\pi \le \theta \le 0$ can be obtained by reflection through the polar axis. Figure 10-31 shows the entire graph, which is called a **cardioid** because of its heart shaped appearance.

θ	$f(\theta)$
0	2.00
$\dfrac{\pi}{6}$	1.87
$\dfrac{\pi}{3}$	1.50
$\dfrac{\pi}{2}$	1.00
$\dfrac{2\pi}{3}$	0.50
$\dfrac{5\pi}{6}$	0.13
π	0.00

Approximate values for $f(\theta) = 1 + \cos\theta$

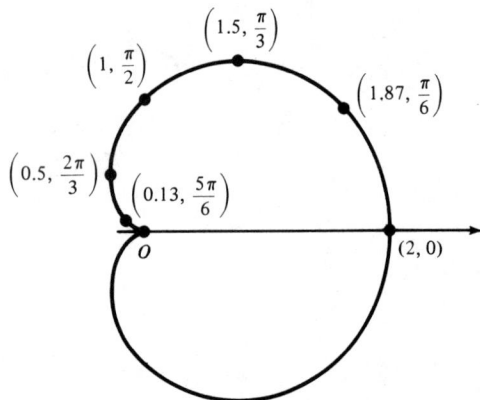

Figure 10-31

Example 7

Graph $f(\theta) = 2 + \cos\theta$ in polar coordinates.

Solution. Because $|\cos\theta| \le 1$ for all θ, the values of f satisfy $1 \le f(\theta) \le 3$. The graph does not pass through the pole. A table and graph are shown in Fig. 10-32.

Approximate values for
$f(\theta) = 2 + \cos \theta$

θ	$f(\theta)$
0	3
$\dfrac{\pi}{6}$	2.87
$\dfrac{\pi}{3}$	2.5
$\dfrac{\pi}{2}$	2
$\dfrac{2\pi}{3}$	1.5
$\dfrac{5\pi}{6}$	1.13
π	1

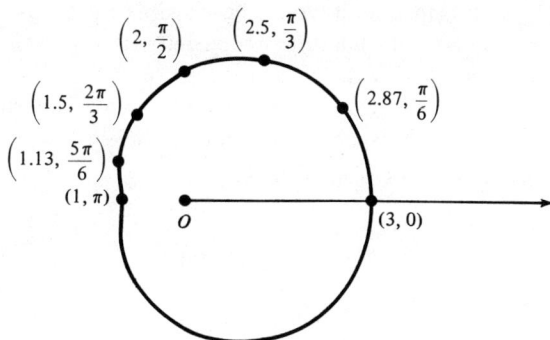

Figure 10-32

Example 8

Graph $f(\theta) = 1 + 2 \cos \theta$ in polar coordinates.

Solution. This limaçon is especially interesting because $f(\theta)$ changes sign as θ passes through $2\pi/3$. At this point the graph passes through the pole and $r = f(\theta)$ becomes negative.

The graph for $0 \le \theta \le \pi$ is shown in Fig. 10-33(a). Reflection through the polar axis produces a "loop within a loop" [Fig. 10-33(b)].

Approximate values for
$f(\theta) = 1 + 2 \cos \theta$

θ	$f(\theta)$
0	3
$\dfrac{\pi}{6}$	2.74
$\dfrac{\pi}{3}$	2
$\dfrac{\pi}{2}$	1
$\dfrac{2\pi}{3}$	0
$\dfrac{5\pi}{6}$	-0.73
π	-1

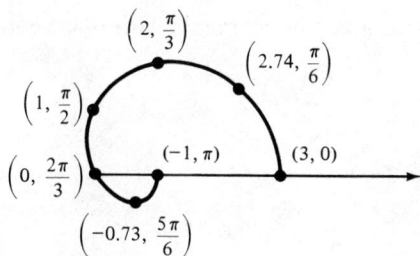

(a)

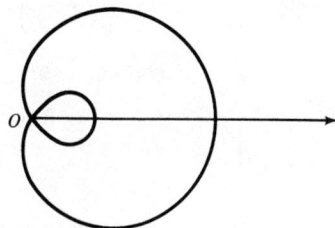

(b)

Figure 10-33

The graphs of $r = a + b \sin \theta$ are limaçons similar to Examples 6 to 8, although they are rotated from the positions shown above; these limaçons are symmetric with respect to the vertical line $\theta = \pi/2$ (Problem 52).

Relations Between Polar and Rectangular Coordinates

In principle, it would be possible to study polar graphs with no tools other than the polar coordinate system itself. It turns out, however, that some problems are best handled by taking advantage of two coordinate systems—rectangular and polar—on the same plane. For this purpose, we superimpose a rectangular system on the polar system in such a way that the origin is at the pole and the x axis coincides with the polar axis (Fig. 10-34).

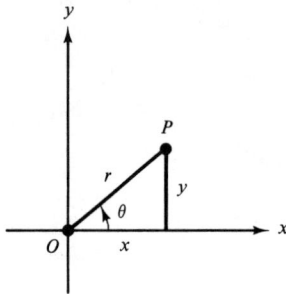

Figure 10-34

Now if (r, θ) is a set of polar coordinates for a point P, with $r > 0$, we see from Fig. 10-34 that rectangular coordinates for P are given by

$$\left. \begin{array}{l} x = r \cos \theta \\ y = r \sin \theta \end{array} \right\}. \tag{10-8}$$

In fact, these equations are simply the first two of Eqs. (10-4), which we derived in Section 10-1 in slightly different notation. The Equations (10-8) hold even if $r \leq 0$ (Problem 55) and give us a unique set of rectangular coordinates for any point whose polar coordinates are known.

Example 9
Find rectangular coordinates for the points with polar coordinates shown.

(a) $P_1\left(2, \dfrac{\pi}{4}\right)$ (b) $P_2\left(-1, \dfrac{2\pi}{3}\right)$

Solution.

(a) $x_1 = 2 \cos \dfrac{\pi}{4} = 2 \cdot \dfrac{\sqrt{2}}{2} = \sqrt{2}$

$y_1 = 2 \sin \dfrac{\pi}{4} = 2 \cdot \dfrac{\sqrt{2}}{2} = \sqrt{2}$

So P_1 has rectangular coordinates $(\sqrt{2}, \sqrt{2})$.

(b) $x_2 = -\cos \dfrac{2\pi}{3} = \dfrac{1}{2}$

$y_2 = -\sin \dfrac{2\pi}{3} = -\dfrac{\sqrt{3}}{2}$

So P_2 has rectangular coordinates $(1/2, -\sqrt{3}/2)$.

Because polar coordinates are generally not unique, we cannot determine a unique (r, θ) when given (x, y). Clearly, however, r and θ must satisfy relations

         Applications of Trigonometric Functions   Chap. 10

$$r^2 = x^2 + y^2$$
$$\tan \theta = \frac{y}{x} \Bigg\} . \qquad (10\text{-}9)$$

Example 10

Find two sets of polar coordinates for the rectangular point $P(1, 1)$.

Solution. By Eqs. (10-9), $r^2 = 2$ and $\tan \theta = 1$. One solution is to let $\theta = $ Arctan $1 = \pi/4$ and $r = \sqrt{2}$. A different solution is to take $\theta = 5\pi/4$ and $r = -\sqrt{2}$.

Example 11

Convert the rectangular equation $x^2 + y^2 = 25$ to polar coordinates.

Solution. By Eqs. (10-9), we obtain immediately $r^2 = 25$ or $r = \pm 5$.

Example 12

Convert the polar equation of Example 4 to rectangular coordinates.

Solution. Multiply both sides of the equation $r = 2 \cos \theta$ by r
$$r^2 = 2r \cos \theta$$
and substitute, using (10-9) and (10-8),
$$x^2 + y^2 = 2x.$$
This is the equation of a circle with center $(1, 0)$ and radius 1.

Example 13

Identify the polar graph $r = \sec \theta$.

Solution. Because $\sec \theta = 1/\cos \theta$, we can rewrite the equation as
$$r \cos \theta = 1$$
or in rectangular coordinates
$$x = 1.$$
So the graph is a vertical line (Fig. 10-35).

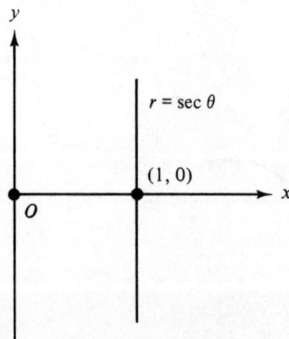

Figure 10-35

Example 14

Find the polar equation corresponding to $x^2 - y^2 = 4$.

Solution. Substituting Eqs. (10-8), we transform the given equation to

$$r^2 \cos^2 \theta - r^2 \sin^2 \theta = 4$$
$$r^2(\cos^2 \theta - \sin^2 \theta) = 4$$
$$r^2 \cos 2\theta = 4.$$

Example 15

Sketch the graph of the equation

$$(x^2 + y^2)^2 = 4(x^2 - y^2).$$

Solution. Converting to polar form, we obtain

$$(r^2)^2 = 4(r^2 \cos^2 \theta - r^2 \sin^2 \theta)$$
$$r^4 = 4r^2(\cos^2 \theta - \sin^2 \theta)$$
$$r^2 = 4(\cos^2 \theta - \sin^2 \theta)$$
$$r^2 = 4 \cos 2\theta.$$

Clearly r and θ are both restricted: $|r| \leq 2$ and $\cos 2\theta \geq 0$. Because $\cos 2\theta$ has period π, we can restrict our attention to the interval $[0, \pi]$. Moreover, because $\cos 2\theta \geq 0$, we exclude values of θ in $[\pi/4, 3\pi/4]$. We are left with θ in $[0, \pi/4]$ or in $[3\pi/4, \pi]$. For each value of θ (except $\pi/4$) there are two values of r:

$$r = \pm 2\sqrt{\cos 2\theta}.$$

As θ increases from zero to $\pi/4$, the positive values of r trace a curve from $(2, 0)$ that reaches the pole when $\theta = \pi/4$; the negative values of r trace a curve from $(-2, 0)$ that joins the first curve at the pole. The resulting curve is shown in Fig. 10-36(a). In the interval $[3\pi/4, \pi]$, as $\cos 2\theta$ increases from zero to 1, the values of r trace curves from the pole out to the points $(2, \pi)$ and $(-2, \pi)$ [Fig. 10-36(b)]. The entire curve, a **lemniscate**, appears in Fig. 10-36(c). It is clear from the rectangular equations that this curve has symmetry with respect to both the x and y axes and the origin.

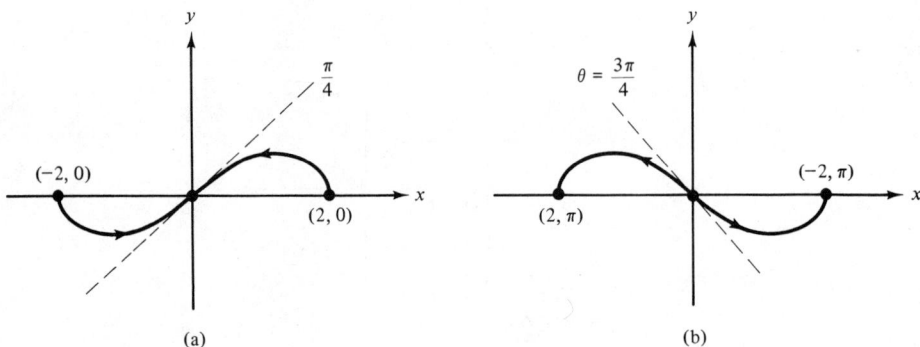

(a)   (b)

Figure 10-36 (a, b)

Applications of Trigonometric Functions   Chap. 10

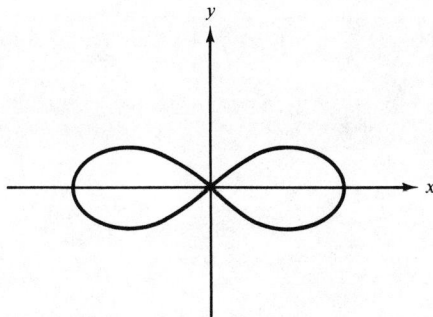

(c)

Figure 10-36 (c)

EXERCISES 10-5

In Problems 1–24 graph the function or equation in polar coordinates.

1. $r = 3$ **2.** $r = -3$ **3.** $r^2 = 9$

4. $r = \theta \; (\theta \geq 0)$ **5.** $r = \dfrac{1}{\theta} \; (\theta > 0)$ **6.** $f(\theta) = e^\theta \; (\theta > 0)$

7. $g(\theta) = \dfrac{\theta}{1 + \theta} \; (\theta > 0)$ **8.** $h(\theta) = e^{1/\theta} \; (\theta > 1)$ **9.** $r = 4 \cos \theta$

10. $g(\theta) = 6 \sin \theta$ **11.** $f(\theta) = \cos 2\theta$ **12.** $g(\theta) = \cos 3\theta$

13. $r = \sin 2\theta$ **14.** $r = \cos 4\theta$ **15.** $r = 2 \sin 4\theta$

16. $f(\theta) = 1 + \sin \theta$ **17.** $r = 2 - \sin \theta$ **18.** $r = 3 + 2 \cos \theta$

19. $r = 5 - 3 \sin \theta$ **20.** $r = 1 + 2 \sin \theta$ **21.** $r = \sqrt{3} - 2 \cos \theta$

22. $f(\theta) = 1 - 2 \cos \theta$ **23.** $r = \sqrt{3} + 2 \sin \theta$ **24.** $r = 4 + 5 \cos \theta$

In Problems 25–32 convert from polar to rectangular coordinates.

25. $\left(2, \dfrac{\pi}{6}\right)$ **26.** $\left(-3, \dfrac{\pi}{3}\right)$ **27.** $\left(-2, \dfrac{\pi}{3}\right)$ **28.** $\left(3, \dfrac{2\pi}{3}\right)$

29. $\left(1, -\dfrac{\pi}{4}\right)$ **30.** $\left(-1, -\dfrac{5\pi}{4}\right)$ **31.** $\left(0, \dfrac{7\pi}{6}\right)$ **32.** $(1, 3\pi)$

In Problems 33–38 you are given rectangular coordinates of P. Find polar coordinates (r, θ) for P such that $r \geq 0$ and $0 \leq \theta < 2\pi$.

33. $(2, 2)$ **34.** $(\sqrt{3}, 1)$ **35.** $(-1, \sqrt{3})$

36. $(-\sqrt{2}, -\sqrt{2})$ **37.** $(\sqrt{3}, -1)$ **38.** $(4, 0)$

In Problems 39–45 convert the equation from polar to rectangular form. Identify the curve.

39. $r = 3 \sin \theta$ **40.** $r = -6 \cos \theta$ **41.** $r = 4 \csc \theta$

42. $r = 2 + 2 \cos \theta$ **43.** $r = 1 - 2 \sin \theta$ **44.** $r = \dfrac{2}{1 + \cos \theta}$

45. $r = \dfrac{4}{1 - \sin \theta}$

In Problems 46–50 convert the equation from rectangular to polar form and sketch the graph.

46. $x^2 + y^2 = 36$ **47.** $x^2 + y^2 = 9$ **48.** $(x^2 + y^2)^2 = 16(x^2 - y^2)$

49. $(x^2 + y^2)^2 = 18xy$ **50.** $(x^2 + y^2)^3 = 8x^2y^2$

51. (a) Show that the polar graph of an even function $r = f(\theta)$ has symmetry with respect to the polar axis.
    (b) What kind of symmetry has the graph of an odd function?

52. (a) Show that the polar graph of $r = f(\theta)$ is symmetric with respect to the line $\theta = \pi/2$ if $f(\pi - \theta) = f(\theta)$ for every θ.
    (b) Show that the limaçons $f(\theta) = a + b \sin \theta$ are symmetric with respect to $\theta = \pi/2$.

53. (a) Show that the polar graph of $r = f(\theta)$ is symmetric with respect to the pole if $f(\theta + \pi) = f(\theta)$ for every θ.
    (b) Show that the graph of $r = \sin 2\theta$ is symmetric with respect to the pole.

54. Find the points of intersection of each pair of polar curves.
    (a) $r = 2 \sin \theta, r = 1$
    (b) $r^2 = 2 \cos 2\theta, r = 1$
    (c) $r = \cos \theta - 1, r = 1$

55. Show that Eqs. (10-8) hold even if $r \leq 0$.

†56. (a) With a calculator or a table of values of the sine function, examine the values of the function $(\sin \theta)/\theta$ as θ (radians) approaches zero.
    (b) Assuming that $(\sin \theta)/\theta \rightarrow 1$ as $\theta \rightarrow 0$, show that the spiral $r = 1/\theta$ approaches a horizontal asymptote, $y = 1$, as $\theta \rightarrow 0$.

SUMMARY

In Chapter 10 we studied the relationship of the trigonometric functions to angles and triangles.

1. You should know how to measure angles in degrees and in radians and be able to convert from one measure to the other.

2. You should be able to interpret the values of the trigonometric functions in terms of ratios of the numbers u, v, and r shown in Fig. 10-4. You should be able to interpret the values of the trigonometric functions of an acute angle as ratios of the sides of a right triangle.

3. You should be able to find the reference angle of a given angle and use the reference angle, together with Table 10-2, to find the trigonometric function values for any angle. Linear interpolation may be necessary.

4. You should be able to solve a right triangle, given one of its acute angles and one of its sides.

5. You should be able to state and prove the law of cosines and the law of sines.

6. You should be able to solve oblique triangles when given
    (i) two sides and the included angle.
    (ii) three sides.
    (iii) two angles and a side.

7. You should know the definitions and the basic properties of the functions Arcsin, Arccos, and Arctan.

8. You should know how to draw simple graphs in the polar coordinate system and be able to convert from a polar equation to a rectangular equation and vice versa.

REVIEW EXERCISES

1. For each angle, find its radian measure, reference angle, sine, and cosine.
   (a) 135°   (b) 210°   (c) 75°   (d) 285°

2. A wheel is spinning at 150π radians every minute. Express this rate
   (a) in revolutions per minute.
   (b) in degrees per minute.

3. Solve the right triangle.
   (a) $\alpha = 60°$, $a = 5$ (b) $\alpha = \dfrac{\pi}{6}$, $b = 12$
   (c) $\alpha = \dfrac{\pi}{8}$, $c = 3$ (d) $\alpha = 16°$, $a = 12$

4. Solve the oblique triangle.
   (a) $a = 7$, $b = 8$, $c = 9$
   (b) $\alpha = 24°$, $b = 2$, $c = 3$
   (c) $a = 7$, $\beta = \dfrac{\pi}{8}$, $c = 4$

5. The anchor of a fishing boat has a 50-foot rope. If the anchor is at the bottom of a lake with all the rope out (and no sag) and the rope makes an angle of 40° with the surface of the water, how deep is the lake?

6. The directions from the earth to each of two stars at a certain time make an angle of 70°. If the distances between the stars and the earth are 5 and 6 light-years, how far apart are the stars?

7. Without looking in the text, state the definition of each of the three functions Arcsin, Arccos, Arctan. Indicate the domain and range of each.

8. Draw the graph.
   (a) $f(x) = 3 \text{ Arcsin } (x - 1)$ for $0 \leq x \leq 2$
   (b) $f(x) = x \text{ Arctan } x$
   (c) $g(x) = \text{Arcsin } x + x$ for $-1 \leq x \leq 1$

9. Evaluate.
   (a) $\text{Arcsin}\left(-\dfrac{\sqrt{3}}{2}\right)$ (b) $\text{Arccos}\left(-\dfrac{\sqrt{2}}{2}\right)$
   (c) $\text{Arctan } (-1)$ (d) $\text{Arcsin } .7771$
   (e) $\text{Arctan } 3.078$ (f) $\tan\left(\text{Arcsin } \dfrac{5}{6}\right)$
   (g) $\sin\left[\text{Arctan}\left(-\dfrac{3}{4}\right)\right]$ (h) $\text{Arccos}\left(\sin \dfrac{2\pi}{3}\right)$

10. Draw the graph in polar coordinates. Convert the equation into rectangular coordinates.
    (a) $r = 3 \cos \theta$ (b) $r = 2 + \cos \theta$
    (c) $r = 1 - 2 \sin \theta$ (d) $r = 2 \sec \theta$

11. Under what circumstances involving a and b is it true that the limaçon $r = a + b \cos \theta$ has an inner loop?

11-1 PARABOLA, ELLIPSE, HYPERBOLA: DEFINITIONS AND SIMPLE EXAMPLES

Introduction

Earlier we studied graphs of various equations containing terms of second degree—for instance,

$$x^2 + y^2 = 25 \qquad \text{(a circle)}$$

$$y = x^2 + 1 \qquad \text{(a parabola)}$$

$$x^2 + 4y^2 = 4 \qquad \text{(an ellipse)}$$

$$xy = 1 \qquad \text{(a hyperbola)}.$$

In Section 2-3 we showed that a circle with center at (h, k) and radius r is the graph of an equation of the form

$$(x - h)^2 + (y - k)^2 = r^2$$

or equivalently

$$x^2 + y^2 - 2hx - 2ky + c = 0.$$

These equations are derived from the geometric definition of the circle as a set of points in a plane equidistant from a fixed point in that plane.

In the present chapter we give geometric definitions of parabola, ellipse, and hyperbola and derive equations for these curves. We will find that, like the circle, each of these curves has an equation of second degree:

$$Ax^2 + Bxy + Cy^2 + Dx + Ey + F = 0,$$

where A, B, C, D, E, and F are constants and not all three of A, B, C are zero. Because equations of this type occur frequently in important applications, we are also interested in the converse problem: Given an equation of second degree, to identify and sketch its graph.

The geometric definitions of parabola, ellipse, and hyperbola are presented in this section. The points, lines, and curves referred to in these definitions are all assumed to lie in a given plane. Because the distance concept is involved, we will always use coordinate systems with equal scales on the axes.

It will be convenient to abbreviate the distance between P and Q by the symbol $d(P, Q)$. If P is (x_1, y_1) and Q is (x_2, y_2), we recall from Section 2-2 that

$$d(P, Q) = \sqrt{(x_2 - x_1)^2 + (y_2 - y_1)^2}.$$

We will sometimes refer to the distance between a point P and a line L. By this we mean the distance between P and the foot of the perpendicular from P to L.

Parabola

Definition. Let D be a fixed line and F a fixed point not on D. The set of all points P equidistant from D and F is called a **parabola** (see Fig. 11-1). The line D is called the **directrix** and the point F the **focus** of this parabola.

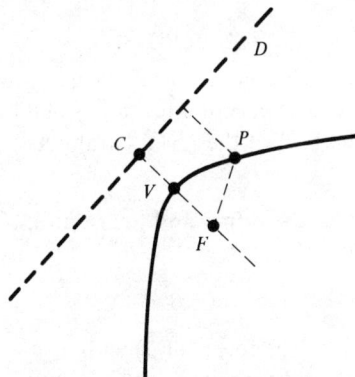

Figure 11-1

If the point C is the foot of the perpendicular from F to D, then the midpoint V of FC is clearly a point on the parabola, for V is equidistant from F and D. The point V is called the **vertex** of the parabola. A parabola is completely determined by its vertex and its focus. The line through F and C is called the **axis** of the parabola. It is easy to show that a parabola is symmetric with respect to its axis.

We will now derive an equation for the parabola relative to a coordinate system chosen as follows. The origin of this coordinate system is at the vertex and the focus is on the x axis at $F(a, 0)$ (see Fig. 11-2). It then follows that the x axis is the axis of this parabola and that the line $x = -a$ is the directrix.

If $P(x, y)$ is any point in the plane, the distance between P and the directrix is $|x + a|$ and the distance between P and the focus F is

$$d(P, F) = \sqrt{(x - a)^2 + y^2}.$$

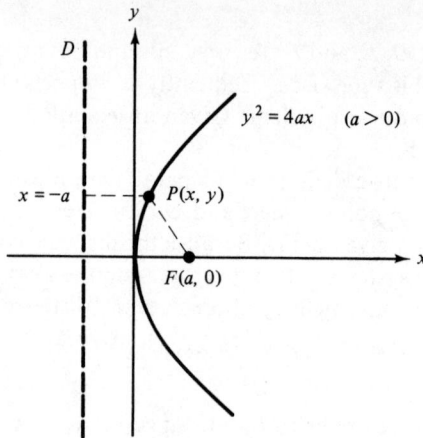

Figure 11-2

By definition, P lies on the parabola if and only if

$$(x - a)^2 + y^2 = |x + a|.$$

Squaring, we obtain the equivalent equation

$$x^2 - 2ax + a^2 + y^2 = x^2 + 2ax + a^2,$$

which can be simplified to

$$y^2 = 4ax. \tag{11-1}$$

This shows that every point on the given parabola has coordinates satisfying (11-1) and, conversely, every point whose coordinates satisfy (11-1) lies on the parabola. So, by definition, (11-1) is an equation of the given parabola.

Example 1

A parabola with vertex at the origin and focus at $(3, 0)$ has equation $y^2 = 4 \cdot 3x$ or $y^2 = 12x$ because $a = 3$.

Example 2

The equation $y^2 = 2x$ is of the form (11-1) with $a = 1/2$. Its graph is a parabola with vertex at the origin and focus $(1/2, 0)$.

Example 3

The graph of the equation $y^2 = -8x$ is of the form (11-1) with $a = -2$; it is a parabola with vertex at the origin and focus at $(-2, 0)$. Its directrix is the line $x = 2$.

The graph of Eq. (11-1) opens to the right when $a > 0$ and to the left when $a < 0$ (see Figs. 11-2 and 11-3).

If in (11-1) we interchange the two variables x and y, we obtain

$$x^2 = 4ay. \tag{11-2}$$

The graph of this equation is a parabola that can be obtained by reflecting the graph

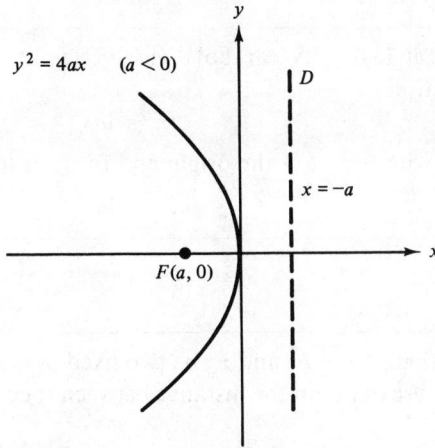

$y^2 = 4ax \quad (a < 0)$

D

$x = -a$

$F(a, 0)$

Figure 11-3

of (11-1) through the line $y = x$. Its vertex is at the origin, its focus at $(0, a)$, and its directrix the line $y = a$. The graph of (11-2) opens upward if $a > 0$, downward if $a < 0$ (see Fig. 11-4).

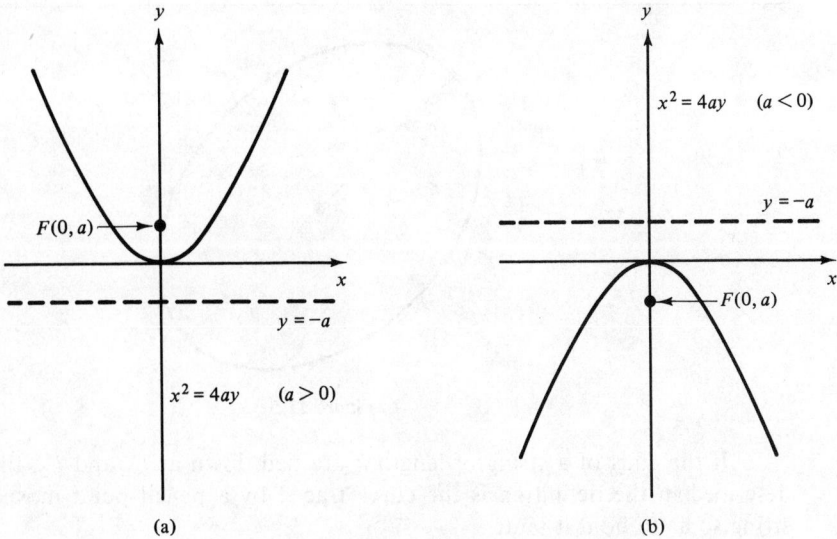

$F(0, a)$

$y = -a$

$x^2 = 4ay \quad (a > 0)$

(a)

$x^2 = 4ay \quad (a < 0)$

$y = -a$

$F(0, a)$

(b)

Figure 11-4

Example 4

The familiar equation $y = x^2$ is of the form (11-2) with $a = 1/4$; its graph is a parabola with vertex at the origin and focus at $(0, 1/4)$.

Let us summarize these results.

> **Theorem 11-1.** A parabola with vertex at the origin and focus at $(a, 0)$ has an equation
> $$y^2 = 4ax. \tag{11-1}$$
> A parabola with vertex at the origin and focus at $(0, a)$ has an equation
> $$x^2 = 4ay. \tag{11-2}$$

Ellipse

> **Definition.** Let F_1 and F_2 be two fixed points and let k be a fixed positive real number greater than the distance between them. The set of all points P such that
> $$d(P, F_1) + d(P, F_2) = k \tag{11-3}$$
> is called an **ellipse** (see Fig. 11-5). The points F_1 and F_2 are called the **foci** of the ellipse. The midpoint C of F_1F_2 is called its **center**.
>
> The line containing both foci and the center of an ellipse is called its **major axis**. The line perpendicular to the major axis at the center of the ellipse is called its **minor axis**. The points where the ellipse intersects its major axis are called its **vertices**.

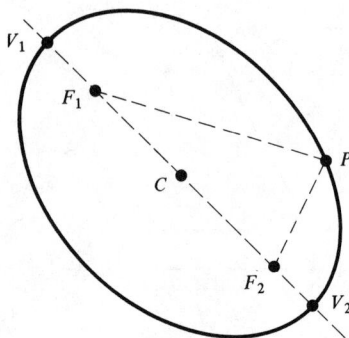

Figure 11-5

If the ends of a string of length k are tied down at F_1 and F_2, then the ellipse described in the definition is the curve traced by a pencil point moved against the string so as to hold it taut.

We will first derive an equation for an ellipse relative to a coordinate system in which the origin is its center and the x axis is its major axis (see Fig. 11-6). Because the foci lie on the x axis and are equidistant from the origin, we can assign them coordinates $(-c, 0)$ and $(c, 0)$, where c is a positive number. The geometric criterion for P to lie on this ellipse is expressed by Eq. (11-3). In this equation it will be convenient to set $a = k/2$ so that $2a = k$; we obtain

$$d(P, F_1) + d(P, F_2) = 2a. \tag{11-4}$$

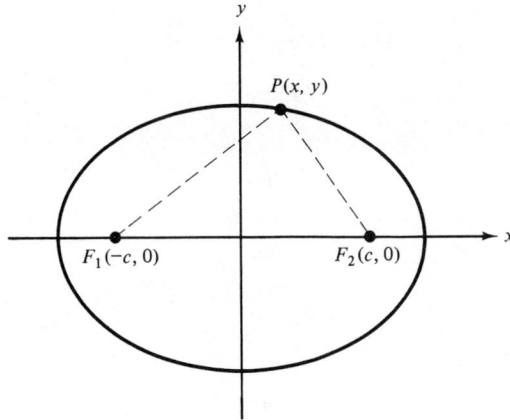

Figure 11-6

The definition requires that k be greater than the distance $2c$ between F_1 and F_2. Hence $2a > 2c$ and

$$c < a. \tag{11-5}$$

We will see later that a is an important geometric measurement of the ellipse.

Next, let $P(x, y)$ be any point on the given ellipse (see Fig. 11-6). Because

$$d(P, F_1) = \sqrt{(x + c)^2 + y^2} \quad \text{and} \quad d(P, F_2) = \sqrt{(x - c)^2 + y^2},$$

we can write Eq. (11-4) in the form

$$\sqrt{(x + c)^2 + y^2} + \sqrt{(x - c)^2 + y^2} = 2a.$$

In order to simplify this result, we subtract $\sqrt{(x - c)^2 + y^2}$ from both members and then square:

$$(x + c)^2 + y^2 = [2a - \sqrt{(x - c)^2 + y^2}]^2.$$

Simplifying,

$$a\sqrt{(x - c)^2 + y^2} = a^2 - cx.$$

Squaring once more, we obtain

$$a^2[(x - c)^2 + y^2] = (a^2 - cx)^2,$$

which can be written

$$(a^2 - c^2)x^2 + a^2y^2 = a^2(a^2 - c^2).$$

Because $0 < c < a$, $a^2 - c^2$ is positive and we can define $b = \sqrt{a^2 - c^2}$. Clearly $b < a$. Thus we have

$$b^2x^2 + a^2y^2 = a^2b^2, \quad (b < a). \tag{11-6}$$

To summarize, we have shown that if $P(x, y)$ lies on the ellipse, then (11-6) holds. To justify our claim that (11-6) is an equation of the ellipse, we must also show the converse: If condition (11-6) holds, then P satisfies the criterion (11-4) for being on the ellipse. (This is just a bit tricky.) Let $P(x, y)$ be a point whose coordinates satisfy (11-6), in which $b^2 = a^2 - c^2$. To prove (11-4), we begin by working out $d(P, F_2)$ and $d(P, F_1)$.

$$d(P, F_2) = \sqrt{(x - c)^2 + y^2}$$

$$= \sqrt{(x - c)^2 + \frac{a^2b^2 - b^2x^2}{a^2}}$$

$$= \sqrt{\frac{a^2(x - c)^2 + b^2(a^2 - x^2)}{a^2}}$$

$$= \sqrt{\frac{a^2(x - c)^2 + (a^2 - c^2)(a^2 - x^2)}{a^2}}$$

$$= \sqrt{\frac{(a^2 - cx)^2}{a^2}} = \frac{|a^2 - cx|}{a}$$

Similarly,

$$d(P, F_1) = \frac{|a^2 + cx|}{a}.$$

From (11-5) we have $c < a$ and from (11-6) it follows that $|x| \leq a$. Therefore $c|x| \leq ca < a^2$ and both $a^2 - cx$ and $a^2 + cx$ are positive. So

$$d(P, F_1) = \frac{a^2 + cx}{a} \quad \text{and} \quad d(P, F_2) = \frac{a^2 - cx}{a},$$

and (11-4) follows easily.

Equation (11-6) is frequently written in the form

$$\frac{x^2}{a^2} + \frac{y^2}{b^2} = 1 \qquad (b < a), \tag{11-7}$$

which is useful because the intercepts of the ellipse on the coordinate axes are easy to read from (11-7). They are the vertices $(-a, 0)$, $(a, 0)$ and intercepts $(0, -b)$, $(0, b)$ on the minor axis. It is easy to see from Eq. (11-7) that the graph has symmetry with respect to both coordinate axes and the origin. The variables x and y are restricted to certain intervals. We have already noted that $|x| \leq a$ and it follows from (11-7) that $|y| \leq b$.

Example 5

An ellipse with foci $F_1(-1, 0)$ and $F_2(1, 0)$ passes through $P(2, 0)$. Find its equation.

Solution. Because $d(P, F_1) = 3$ and $d(P, F_2) = 1$, we have $k = 4$ and so $a = 2$. Now $c = 1$; therefore $b^2 = a^2 - c^2 = 4 - 1 = 3$. So an equation of the given ellipse is

$$\frac{x^2}{4} + \frac{y^2}{3} = 1$$

or

$$3x^2 + 4x^2 = 12.$$

Example 6

The equation $9x^2 + 25y^2 = 225$ is of the form (11-6) with $b = 3$, $a = 5$. Because $c = \sqrt{a^2 - b^2} = 4$, we locate the foci at $(-4, 0)$ and $(4, 0)$ (see Fig. 11-7).

If we interchange the roles of x and y in the derivation of (11-7), we obtain

$$\frac{y^2}{a^2} + \frac{x^2}{b^2} = 1 \qquad (b < a). \tag{11-8}$$

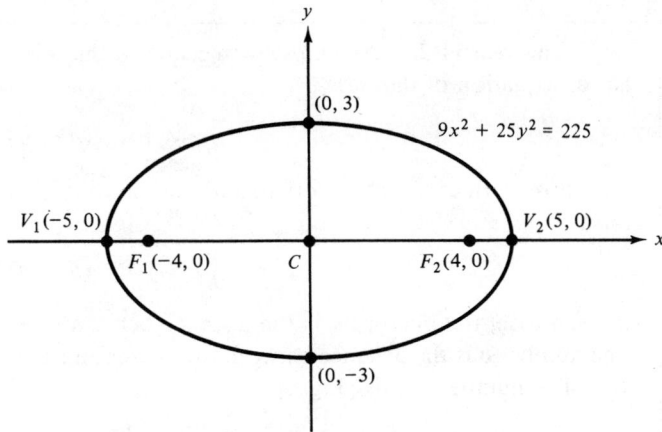

(0, 3)

$9x^2 + 25y^2 = 225$

$V_1(-5, 0)$

$V_2(5, 0)$

$F_1(-4, 0)$ C $F_2(4, 0)$

(0, −3)

Figure 11-7

This is an equation of an ellipse having the y axis as its major axis, with foci at $(0, -c)$ and $(0, c)$.

Example 7

Consider $2x^2 + y^2 = 4$, which may be written

$$\frac{x^2}{2} + \frac{y^2}{4} = 1.$$

Because b must be less than a, this is of the form (11-8), with $a = 2$, $b = \sqrt{2}$. Because $c = \sqrt{a^2 - b^2} = \sqrt{2}$, the foci of this ellipse are $(0, -\sqrt{2})$ and $(0, \sqrt{2})$; its vertices are $(0, -2)$ and $(0, 2)$; and its intercepts on the x axis are $(-\sqrt{2}, 0)$ and $(\sqrt{2}, 0)$ (Fig. 11-8).

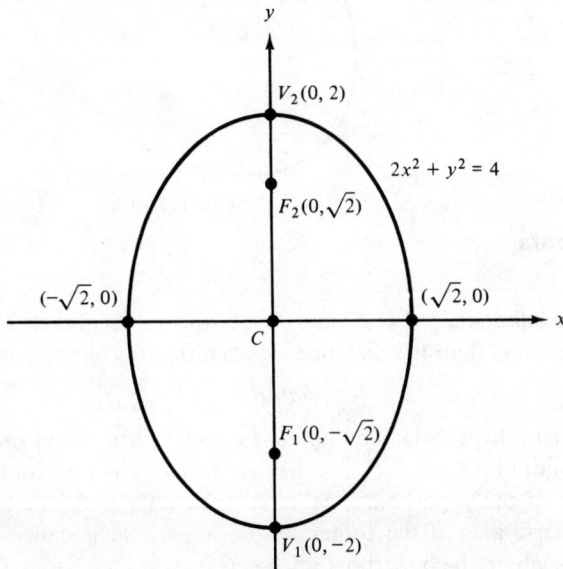

$V_2(0, 2)$

$2x^2 + y^2 = 4$

$F_2(0, \sqrt{2})$

$(-\sqrt{2}, 0)$ C $(\sqrt{2}, 0)$

$F_1(0, -\sqrt{2})$

$V_1(0, -2)$

Figure 11-8

Theorem 11-2. An ellipse with center at the origin and foci on the x axis has an equation of the form

$$\frac{x^2}{a^2} + \frac{y^2}{b^2} = 1 \qquad (b < a). \tag{11-7}$$

An ellipse with center at the origin and foci on the y axis has an equation of the form

$$\frac{y^2}{a^2} + \frac{x^2}{b^2} = 1 \qquad (b < a). \tag{11-8}$$

In each case, the number a is the distance between the center and the vertices. The number b is the distance between the center and the intercepts on the minor axis. The number c satisfying

$$c = \sqrt{a^2 - b^2}$$

is the distance between the center and the foci.

Note. The numbers a, b, c in Theorem 11-2 are geometrical measurements of the ellipse and do not depend on the position of the coordinate axes and origin. The number a is always greater than b. Figure 11-9 shows these measurements for an ellipse in a plane without a designated coordinate system.

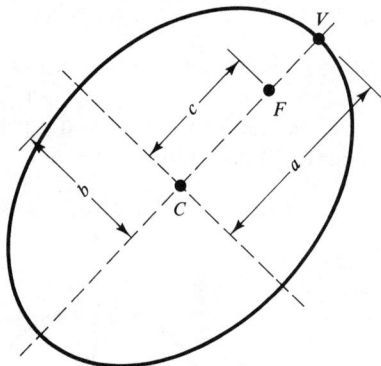

Figure 11-9

Hyperbola

Definition. Let F_1 and F_2 be two points and let k be a fixed positive real number less than the distance between them. The set of all points P such that

$$|d(P, F_1) - d(P, F_2)| = k \tag{11-9}$$

is called a **hyperbola**. The points F_1 and F_2 are called the **foci** of the hyperbola. The midpoint C of $F_1 F_2$ is called its **center**. The line containing the two foci and center of the hyperbola is called its **transverse axis**; the line perpendicular to the transverse axis at the center of the hyperbola is called its **conjugate axis**. The points where the hyperbola intersects its transverse axis are called its **vertices**.

We will first derive an equation for a hyperbola relative to a coordinate system in which the origin is its center and the x axis is its transverse axis (see Fig. 11-10). The procedure is similar to that used in obtaining an equation for an ellipse and we

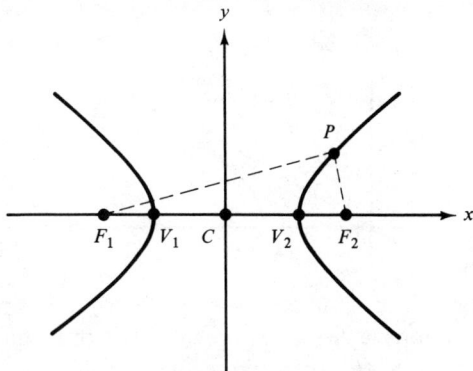

Figure 11-10

will present only an outline. The foci are located at $F_1(-c, 0)$ and $F_2(c, 0)$, where $c > 0$; the distance between them is $2c$. The criterion for a point P to lie on the hyperbola is expressed by Eq. (11-9), in which we set $k = 2a$:

$$|d(P, F_1) - d(P, F_2)| = 2a.$$

In accordance with the definition, $2a < 2c$ so that

$$a < c. \tag{11-10}$$

It can now be shown that $P(x, y)$ lies on the hyperbola if and only if

$$(c^2 - a^2)x^2 - a^2y^2 = a^2(c^2 - a^2).$$

Setting $b = \sqrt{c^2 - a^2}$ (which is permissible because $c^2 > a^2$), we obtain

$$b^2x^2 - a^2y^2 = a^2b^2$$

or

$$\frac{x^2}{a^2} - \frac{y^2}{b^2} = 1. \tag{11-11}$$

Equation (11-11) shows many features of this hyperbola. It is symmetric with respect to both coordinate axes and the origin. Intercepts on the x axis (the vertices) are at $(-a, 0)$ and $(a, 0)$; there are no intercepts on the y axis, for the equation $-y^2 = b^2$ has no solutions. Furthermore, for any point $P(x, y)$ on the hyperbola we have

$$\frac{x^2}{a^2} = 1 + \frac{y^2}{b^2}$$

so that $x^2/a^2 \geq 1$ and $|x| \geq a$.

Let us confine our attention to the first quadrant for the moment. If we solve Eq. (11-11) for y, choosing the positive square root, we obtain

$$y = \frac{b}{a}\sqrt{x^2 - a^2}.$$

As x increases, so does y. For large values of x, $\sqrt{x^2 - a^2}$ is approximately x. Hence, for large x, the point (x, y) on the hyperbola is close to the point $(x, (b/a)x)$ on the line $y = (b/a)x$. This observation suggests that the hyperbola approaches $y = (b/a)x$ as $x \rightarrow +\infty$ (see Fig. 11-11).

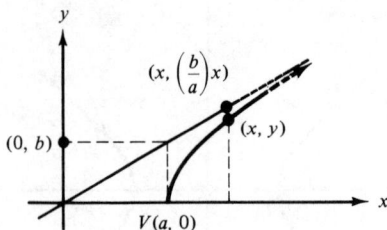

Figure 11-11

As a result of symmetry, similar descriptions apply in each quadrant: the hyperbola (11-11) approaches $y = (b/a)x$ in the third quadrant as well as in the first; it approaches $y = -(b/a)x$ in the second and fourth quadrants. Both the lines

$$y = \pm \frac{b}{a} x \tag{11-12}$$

are **asymptotes** of the hyperbola. They can be easily sketched by extending the diagonals of a rectangle of dimension $2a$ by $2b$ centered at the origin, as in Fig. 11-12. Notice that the hyperbola just touches the vertical sides of this rectangle at its vertices $(-a, 0)$ and $(a, 0)$. The region of the plane for which $-a < x < a$, separates the hyperbola into two branches that are symmetric to one another with respect to the y axis.

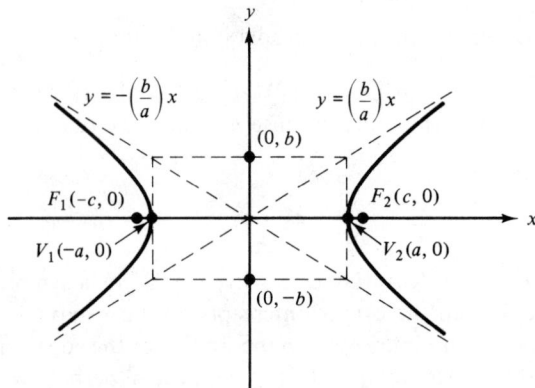

Figure 11-12

We can, of course, interchange the roles of x and y, thereby obtaining the general equation of a hyperbola with center at the origin and foci on the y axis. We summarize our results in the following theorem.

Theorem 11-3. A hyperbola with center at the origin and foci on the x axis has an equation of the form

$$\frac{x^2}{a^2} - \frac{y^2}{b^2} = 1. \tag{11-11}$$

The asymptotes of (11-11) are

$$y = \pm \frac{b}{a} x \qquad\qquad (11\text{-}12)$$

A hyperbola with center at the origin and foci on the y axis has an equation of the form

$$\frac{y^2}{a^2} - \frac{x^2}{b^2} = 1. \qquad\qquad (11\text{-}13)$$

The asymptotes of (11-13) are

$$y = \pm \frac{a}{b} x \cdot \qquad\qquad (11\text{-}14)$$

In both cases, the number a is the distance from the center to each vertex; the number c, satisfying

$$c = \sqrt{a^2 + b^2},$$

is the distance from the center to each focus.

Note. Just as in the case of the ellipse, the numbers a, b, and c are geometrical measurements of the hyperbola and do not depend on the position of the coordinate axes and origin. These measurements are shown in Fig. 11-13.

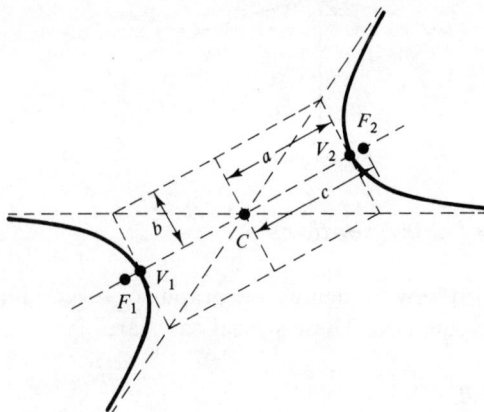

Figure 11-13

Example 8

Find an equation for the hyperbola with center at the origin, vertex at $(4, 0)$, and focus at $(5, 0)$.

Solution. Clearly the transverse axis is the x axis and so the equation is of the form (11-11). Because $c = 5$ and $a = 4$, we find that $b = \sqrt{c^2 - a^2} = 3$. Thus an equation of this hyperbola is

$$\frac{x^2}{16} - \frac{y^2}{9} = 1$$

or
$$9x^2 - 16y^2 = 144.$$

Its asymptotes are given by

$$y = \pm \frac{3}{4} x.$$

Example 9

The equation

$$\frac{y^2}{1} - \frac{x^2}{4} = 1$$

is that of a hyperbola with center at the origin and foci on the y axis. Because $a = 1, b = 2$, we find that $c = \sqrt{a^2 + b^2} = \sqrt{5}$. Asymptotes are lines

$$y = \pm\frac{1}{2}x.$$

The vertices are at $(0, -1)$ and $(0, 1)$; the foci at $(0, -\sqrt{5})$, $(0, \sqrt{5})$. This hyperbola is sketched in Fig. 11-14.

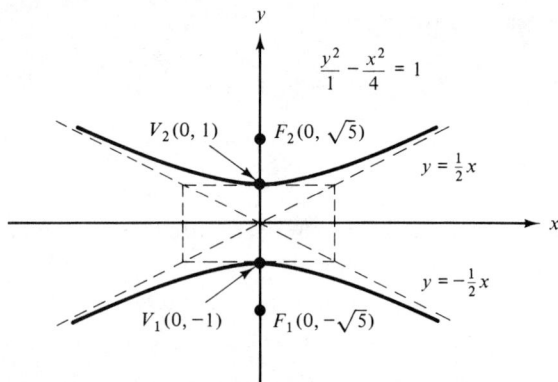

Figure 11-14

Identifying Graphs from Their Equations

In this section we have learned how to identify the graphs of several important cases of the general second-degree equation. These special cases are

1. $y = Ax^2$, where A is a nonzero constant.
2. $x = Cy^2$, where C is a nonzero constant.
3. $Ax^2 + Cy^2 = K$, where A, C, and K are nonzero constants.

Cases (1) and (2) can be identified immediately as parabolas, with vertex at the origin. In case (1) the axis is vertical; in case (2) it is horizontal. Case (3) gives either an ellipse or a hyperbola, depending on the signs of A, C, and K.

Example 10

Describe the graph.
  (a) $y + 3x^2 = 0$
  (b) $x - 3y^2 = 0$
  (c) $4x^2 + 9y^2 = 36$
  (d) $25x^2 - 4y^2 = 100$
  (e) $x^2 - y^2 = -10$

Solution.

(a) $y = -3x^2$ is a parabola with vertical axis, vertex at the origin, opening downward.

(b) $x = 3y^2$ is a parabola with horizontal axis, vertex at the origin, opening to the right.

(c) To write the given equation in a standard form, we divide by the constant on the right (36) so as to have a 1 there:

$$\frac{x^2}{9} + \frac{y^2}{4} = 1.$$

We see that the graph is an ellipse centered at the origin. Because a must be greater than b for an ellipse, we know that $a^2 = 9$; so the equation is of the form (11-7) and the major axis is horizontal.

(d) Again, we divide by the constant on the right:

$$\frac{x^2}{4} - \frac{y^2}{25} = 1$$

and see that the equation is of the form (11-11). This is a hyperbola with center at the origin; its transverse axis is horizontal.

(e) The equation is equivalent to

$$\frac{y^2}{10} - \frac{x^2}{10} = 1.$$

Because it is of the form (11-13), we know that the graph is a hyperbola with a vertical transverse axis.

Note that if one or more of constants A, C, K in case (3) are zero, a **degenerate** case may arise. The graph of $x^2 - 4y^2 = 0$, for instance, consists of two intersecting lines $y = \pm\frac{1}{2}x$; the graph of $x^2 = 4$ consists of two parallel lines $x = \pm 2$; the graph of $x^2 = 0$ is just the y axis.

EXERCISES 11-1

In Problems 1–7 sketch the parabola. Give the coordinates of the focus and the equation of the directrix.

1. $y^2 = 16x$ 2. $y^2 = -20x$ 3. $x = 4y^2$

4. $x^2 = 4y$ 5. $x^2 + 12y = 0$ 6. $3y^2 + x = 0$

7. $y = -x^2$

In Problems 8–13 use Theorem 11-1 to find an equation of the parabola with vertex at $(0, 0)$ having

8. focus $(0, 2)$.   9. focus $(\frac{1}{2}, 0)$.   10. focus $(-1, 0)$.

11. directrix $y = 2$.   12. directrix $x = -1$.   13. directrix $y = -3$.

In Problems 14–17 use the definition of parabola to derive an equation of the parabola having

14. focus $(1, 2)$, directrix $x = -1$.   15. focus $(2, -1)$, vertex $(2, -2)$.

16. vertex $(3, 2)$, directrix $x = 4$.   17. focus $(0, 0)$, directrix $y = -5$.

In Problems 18–23 sketch the given ellipse. Give the coordinates of the foci, vertices, and intercepts on the minor axis.

18. $4x^2 + 9y^2 = 36$ **19.** $x^2 + 2y^2 = 8$ **20.** $25x^2 + 16y^2 = 400$

21. $10x^2 + 25y^2 = 50$ **22.** $36x^2 + 4y^2 = 144$ **23.** $9x^2 + 8y^2 = 72$

In Problems 24–29 use Theorem 11-2 to find an equation of the ellipse having its center at the origin and one of the coordinate axes as its major axis, given that

24. a vertex is $(3, 0)$; an intercept on the minor axis is $(0, 2)$.

25. a vertex is $(0, 1)$; an intercept on the minor axis is $(-1/2, 0)$.

26. a vertex is $(3, 0)$; a focus is $(2, 0)$.

27. an intercept on the minor axis is $(0, 3)$; a focus is $(4, 0)$.

28. a vertex is $(3, 0)$; $(2, 3\sqrt{5}/5)$ is on the ellipse.

29. an intercept on the minor axis is $(0, 3)$; $(8, \sqrt{5})$ is on the ellipse.

In Problems 30–35 sketch the hyperbola. Find the coordinates of vertices and foci. Find the equations of the asymptotes.

30. $9x^2 - 16y^2 = 144$ **31.** $x^2 - y^2 = 1$ **32.** $y^2 - 2x^2 = 8$

33. $x^2 - y^2 = -1$ **34.** $y^2 - 4x^2 = 1$ **35.** $9y^2 - x^2 = 1$

In Problems 36–41 use Theorem 11-3 to find an equation of the hyperbola having its center at the origin and one of the coordinate axes as its transverse axis, given that

36. a vertex is $(0, 2)$; a focus is $(0, 3)$.

37. a focus is $(2, 0)$; $y = x$ is an asymptote.

38. a vertex is $(4, 0)$; $x = 2y$ is an asymptote.

39. a vertex is $(0, 3)$; a focus is $(0, -5)$.

40. a vertex is $(3\sqrt{2}, 0)$; asymptotes $y = \pm(4/3)x$.

41. a vertex is $(0, 3\sqrt{2})$; asymptotes $y = \pm(3/4)x$.

42. Show that the graph of the equation $(x^2/a^2) - (y^2/b^2) = 0$ is the union of the lines $y = (b/a)x$ and $y = -(b/a)x$.

In Problems 43–51 describe the graph.

43. $x^2 - y^2 = 10$ **44.** $x^2 + y^2 = 10$ **45.** $x^2 - y^2 = 0$

46. $3y^2 = 4x$ **47.** $3y^2 + 6x^2 = 24$ **48.** $y^2 - 4x^2 = -1$

49. $x^2 - 4y = 0$ **50.** $y^2 + 4x^2 = 0$

51. $(x^2 - y^2 - 1)(x^2 - y^2 + 1) = 0$

In Problems 52–55 find the points of intersection. Sketch.

52. $x^2 + 4y^2 = 16$, $x = 2y$ **53.** $x^2 + 4y^2 = 16$, $y^2 = -(3/2)x$

54. $y = x^2$, $y = 8 - x^2$ **55.** $x^2 - y^2 = 1$, $x^2 + y^2 = 9$

FOR THE ENTHUSIASTS

56. Show that
(a) any parabola is symmetric with respect to its axis.
(b) any ellipse is symmetric with respect to its major axis.
(c) any hyperbola is symmetric with respect to its center.

57. The line through the focus of a parabola and perpendicular to its axis intersects the parabola in two points P_1 and P_2. The segment $P_1 P_2$ is called the **latus rectum** of the parabola. Show that, for $y^2 = 4ax$, the length of the latus rectum is $4|a|$.

58. Using the definition of parabola, derive an equation of the parabola having focus $(1, 1)$ and directrix $x + y = -2$. [*Hint:* First show that the distance of a point (x_0, y_0) from the line $x + y = -2$ is $|x_0 + y_0 + 2|/\sqrt{2}$. This can be done by determining the point of intersection of the given line with the line perpendicular to it and passing through (x_0, y_0).]

59. The line through a focus of an ellipse and perpendicular to its major axis intersects the ellipse in two points D_2 and D_1. The segment $D_1 D_2$ is called a **latus rectum** of the ellipse. Find a formula for the length of a latus rectum in terms of a and b.

60. The **eccentricity** of an ellipse is defined to be the ratio c/a.
    (a) Show that the eccentricity of an ellipse is between zero and one.
    (b) Describe the shape of an ellipse whose eccentricity is near zero.
    (c) Describe the shape of an ellipse whose eccentricity is near one.

61. Show that if we allow the two foci in the definition of ellipse to coincide, then the resulting curve would be a circle. What relation between a and b would hold in this case?

62. A latus rectum for a hyperbola is defined in the same way as for an ellipse (see Problem 59). Find a formula for the length of a latus rectum of a hyperbola in terms of a and b.

63. A hyperbola for which $a = b$ is called **equilateral**.
    (a) Show that if an equilateral hyperbola has its foci on one of the coordinate axes and center at the origin, then its asymptotes are $y = x$ and $y = -x$.
    (b) Show that a hyperbola is equilateral if and only if its asymptotes are perpendicular.

64. Consider the hyperbola with foci at $(-\sqrt{2}, -\sqrt{2})$ and $(\sqrt{2}, \sqrt{2})$ and vertices at $(-1, -1)$ and $(1, 1)$.
    (a) Show that if $P(x, y)$ lies on the hyperbola described, then $xy = 1$.
    (b) Show that if $xy = 1$, then $P(x, y)$ lies on the hyperbola described.
    (c) Show that this hyperbola is equilateral (see Problem 63).
    (d) What lines are asymptotes of this hyperbola?
    (e) Draw the graph.

11-2 USING TRANSLATION OF AXES ON SECOND-DEGREE EQUATIONS

In Section 11-1 we learned to identify graphs of certain equations of second degree. Here we see how to extend our knowledge of second-degree graphs by using translation of axes. We deal with second-degree equations in which the coefficient B of xy is zero:

$$Ax^2 + Cy^2 + Dx + Ey + F = 0;$$

in other words, the xy term is missing. Let us begin by taking an example of a type already familiar.

Example 1

Graph $x^2 + 2x + y^2 - 2y = 0$.

Solution. We recognize this equation as belonging to a circle. To determine the coordinates of the center, we complete the squares.

$$x^2 + 2x \qquad + y^2 - 2y \qquad = 0$$
$$x^2 + 2x + 1 + y^2 - 2y + 1 = 2$$
$$(x + 1)^2 \qquad + (y - 1)^2 \qquad = 2.$$

The substitution

$$u = x + 1$$
$$v = y - 1$$

amounts to a translation of axes and simplifies the equation to

$$u^2 + v^2 = 2.$$

The graph of this equation is a circle of radius $\sqrt{2}$ in a coordinate system whose origin is at the point $(-1, 1)$ in the original x, y system (see Fig. 11-15).

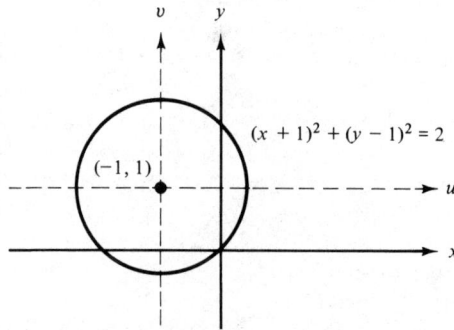

Figure 11-15

Example 2

Consider $4x^2 + 9y^2 - 8x + 36y + 4 = 0$. Completing the squares, we can rewrite this equation as follows:

$$4(x^2 - 2x + \quad) + 9(y^2 + 4y + \quad) = -4$$
$$4(x^2 - 2x + 1) + 9(y^2 + 4y + 4) = -4 + 4 + 36$$
$$4(x - 1)^2 + 9(y + 2)^2 = 36$$
$$\frac{(x - 1)^2}{9} + \frac{(y + 2)^2}{4} = 1.$$

Notice that if we substitute

$$u = x - 1$$
$$v = y + 2$$

we obtain

$$\frac{u^2}{9} + \frac{v^2}{4} = 1.$$

Graphs of Second-Degree Equations   Chap. 11

This translation of axes to the new origin $(1, -2)$ yields an equation belonging to an ellipse in the u, v coordinate system. It is easy to sketch this ellipse (see Fig. 11-16). Its major axis is the u axis. Because $a = 3$ and $b = 2$, we find $c = \sqrt{a^2 - b^2} = \sqrt{5}$. Thus the foci are at $(-\sqrt{5}, 0), (\sqrt{5}, 0)$, the vertices at $(-3, 0)$ and $(3, 0)$, and the intercepts on the v axis at $(0, -2)$ and $(0, 2)$.

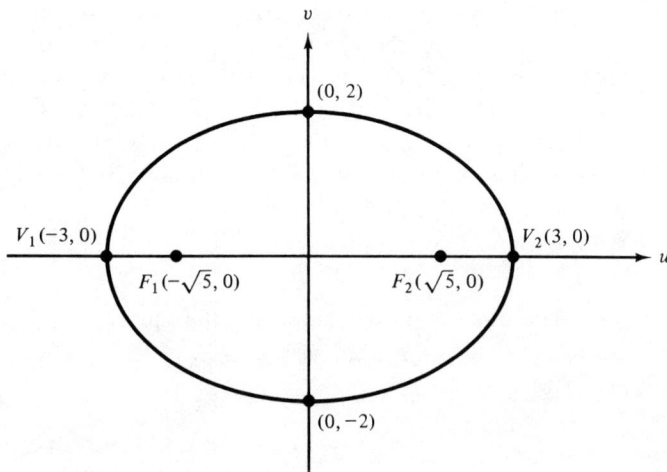

Figure 11-16

Example 3

Graph the equation of Example 2 in the x, y coordinate system.

Solution. We construct u, v axes at the point $(1, -2)$ in the x, y system (see Fig. 11-17) and copy the ellipse of Fig. 11-16. The major axis is the u axis, $v = 0$;

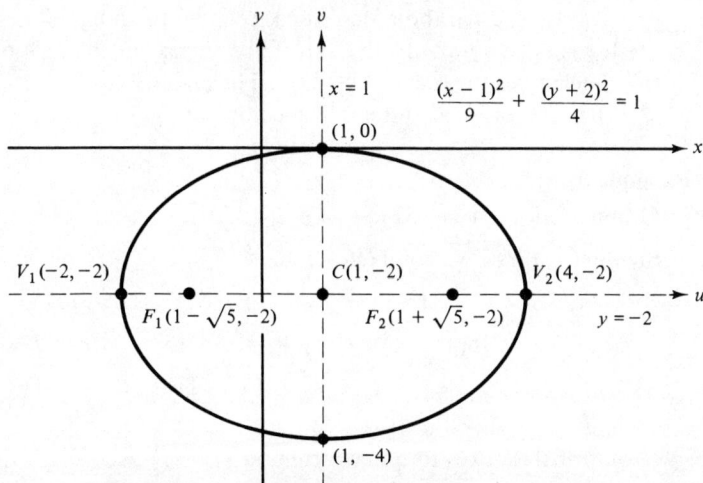

Figure 11-17

in the original coordinate system this line has equation $y + 2 = 0$ or $y = -2$. The minor axis is $u = 0$ or $x = 1$. By using the formulas

$$x = u + 1$$
$$y = v - 2$$

the original coordinates of the vertices, foci, and intercepts on the minor axis can be computed from the new coordinates. Another way is to locate these points relative to the center $(1, -2)$ by making use of the geometric interpretations of a, b, and c. The foci, for instance, are at a distance $\sqrt{5}$ from $(1, -2)$ and they lie on the line $y = -2$. So their coordinates are $(1 - \sqrt{5}, -2)$ and $(1 + \sqrt{5}, -2)$. Similarly, we find the vertices at $(-2, -2)$ and $(4, -2)$ and the intercepts on the line $x = 1$ at $(1, -4)$ and $(1, 0)$.

Example 4

Graph $y^2 + 8x - 6y + 17 = 0$.

Solution. We begin by completing the square.

$$y^2 - 6y + \quad = -8x - 17$$
$$y^2 - 6y + 9 = -8x - 17 + 9$$
$$(y - 3)^2 = -8(x + 1) = -4 \cdot 2(x + 1)$$

The resulting equation is of the form $v^2 = 4au$, where $a = -2$ and

$$v = y - 3$$
$$u = x + 1.$$

The graph in the u, v coordinate system is a parabola whose vertex is at the origin and whose axis of symmetry is the u axis; it opens to the left.

To obtain the graph of the given equation in the x, y coordinate system, we first construct the u and v axes, which intersect at $(-1, 3)$. Next, we can easily draw the parabola described, relative to the u, v axes. Because $a = -2$, the focus is two units to the left of the vertex on the axis of symmetry, $y = 3$; this determines the focus at $(-3, 3)$. The directrix $u = -a = 2$ has the equation $x = 1$ in the original system. The completed graph appears in Fig. 11-18.

Example 5

Graph $16x^2 - 9y^2 \quad 32x - 18y = 137$.

Solution. First, we complete squares.

$$16(x^2 - 2x + \quad) - 9(y^2 + 2y + \quad) = 137$$
$$16(x - 1)^2 - 9(y + 1)^2 = 137 + 16 - 9 = 144$$
$$\frac{(x - 1)^2}{9} - \frac{(y + 1)^2}{16} = 1$$

If we translate axes to a new origin at $(1, -1)$, we obtain the equation

$$\frac{u^2}{9} - \frac{v^2}{16} = 1$$

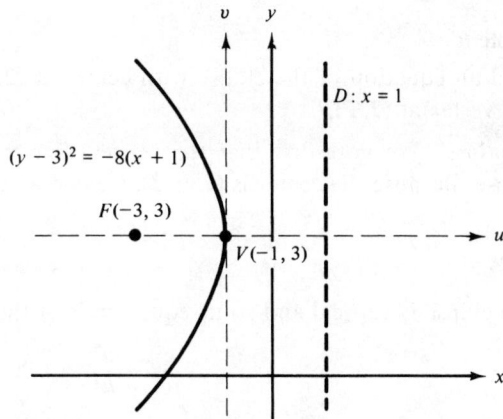

$(y - 3)^2 = -8(x + 1)$

$F(-3, 3)$

$V(-1, 3)$

$D: x = 1$

Figure 11-18

where
$$u = x - 1$$
$$v = y + 1.$$

The transverse axis of this hyperbola is the u axis—that is, $y + 1 = 0$. Because $a = 3$, the vertices are on this line at a distance 3 from $(1, -1)$; these are the points $(-2, -1)$ and $(4, -1)$. We find $c = \sqrt{a^2 + b^2} = 5$ so that the foci are $(-4, -1)$ and $(6, -1)$. Asymptotes are $v = \pm(4/3)u$, or

$$y + 1 = \pm\frac{4}{3}(x - 1).$$

The graph appears in Fig. 11-19.

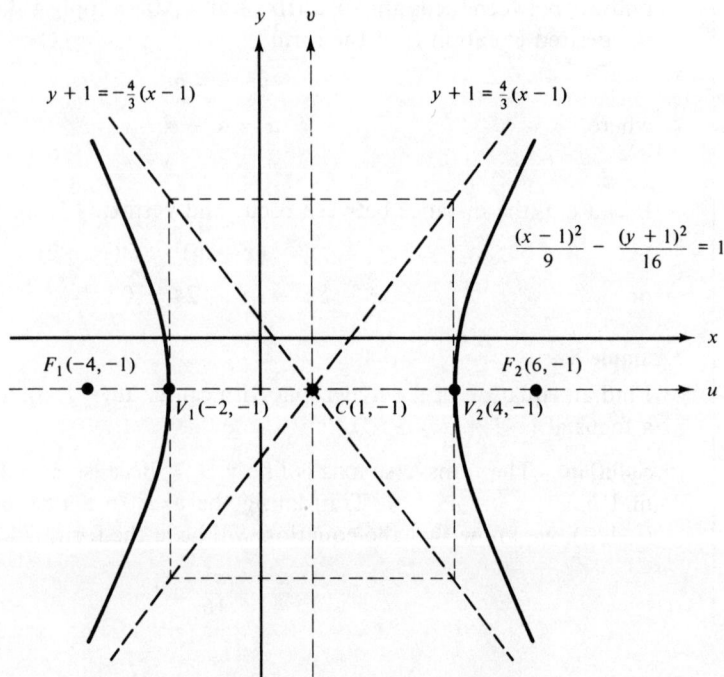

$y + 1 = -\frac{4}{3}(x - 1)$

$y + 1 = \frac{4}{3}(x - 1)$

$\dfrac{(x - 1)^2}{9} - \dfrac{(y + 1)^2}{16} = 1$

$F_1(-4, -1)$

$V_1(-2, -1)$

$C(1, -1)$

$V_2(4, -1)$

$F_2(6, -1)$

Figure 11-19

Example 6

Find an equation of the ellipse with center at $(2, -3)$, one focus at $(2, 1)$, and one vertex at $(2, 2)$.

Solution. We note first that the vertical line $x = 2$ is the major axis of this ellipse. Because the center is $(2, -3)$, we choose a u, v coordinate system where

$$u = x - 2$$
$$v = y + 3.$$

The ellipse is vertical and so its equation is of the form

$$\frac{v^2}{a^2} + \frac{u^2}{b^2} = 1.$$

Now c, the distance from the center to a focus, is easily found to be 4 and a, the distance from the center to a vertex, is 5. Therefore $b = \sqrt{a^2 - c^2} = 3$ and and the desired equation is

$$\frac{(y + 3)^2}{25} + \frac{(x - 2)^2}{9} = 1$$

or
$$25(x - 2)^2 + 9(y + 3)^2 = 225$$

or
$$25x^2 + 9y^2 - 100x + 54y - 44 = 0.$$

Example 7

Find an equation of the parabola with focus $(4, 3)$ and directrix $y = 1$.

Solution. Because the directrix is horizontal, the axis of the parabola is vertical. The focus is above the directrix and so the parabola opens upward. The vertex, midway between focus and directrix, is at $(4, 2)$. (Supply a sketch.) Consequently, the desired equation is of the form

$$u^2 = 4av,$$

where
$$u = x - 4$$
$$v = y - 2.$$

Because a, the distance between focus and vertex, is 1, we obtain finally

$$(x - 4)^2 = 4(y - 2)$$

or
$$x^2 - 8x - 4y + 24 = 0.$$

Example 8

Find an equation of the hyperbola with center at $(-1, 3)$, a vertex at $(1, 3)$, and a focus at $(-1 + 2\sqrt{5}, 3)$.

Solution. The transverse axis is line $y = 3$. Because $a = 2$ and $c = 2\sqrt{5}$, we find $b = \sqrt{c^2 - a^2} = 4$. Translating the axes to a new origin with center at $(-1, 3)$, we know that the equation will have the form

$$\frac{u^2}{4} - \frac{v^2}{16} = 1.$$

The translation equations are

$$u = x + 1$$
$$v = y - 3,$$

and so we obtain

$$\frac{(x+1)^2}{4} - \frac{(y-3)^2}{16} = 1$$

or

$$4x^2 - y^2 + 8x + 6y = 21.$$

Consider an equation of the form

$$Ax^2 + Cy^2 + Dx + Ey + F = 0,$$

where A and C are not both zero. By a translation of axes

$$u = x - h$$
$$v = y - k,$$

this equation can be written in terms of variables u and v in one of the forms

1. $v = Au^2$ $(A \neq 0)$
2. $u = Cv^2$ $(C \neq 0)$
3. $Au^2 + Cv^2 = K$, where A and C are both nonzero.

In either of cases (1) and (2), where $AC = 0$, we recognize the curve as a parabola. According to the analysis at the end of Section 11-1, we know that (barring degenerate cases) case (3) represents an ellipse or circle if A and C are of the same sign (AC positive) whereas it represents a hyperbola if they are of opposite signs (AC negative). The various possibilities can thus be described in terms of the sign of AC.

Barring degenerate cases, the graph of

$$Ax^2 + Cy^2 + Dx + Ey + F = 0 \quad (A, C \text{ not both zero})$$

is
1. an ellipse or circle if $AC > 0$.
2. a parabola if $AC = 0$.
3. a hyperbola if $AC < 0$.

EXERCISES 11-2

In Problems 1–7 determine a translation of axes that transforms the equation into one with no first-degree terms in the variables. Write the new equation. Identify the curve by name.

1. $x^2 + y^2 - 2x + 6y + 6 = 0$ **2.** $x^2 + y^2 + 4x - 4y + 7 = 0$

3. $9x^2 + 4y^2 - 16y - 20 = 0$ 4. $x^2 + 4y^2 - 4x + 8y + 4 = 0$

5. $x^2 + y^2 - 2x - 24 = 0$ 6. $x^2 - y^2 - 4x - 6y - 6 = 0$

7. $9x^2 - 4y^2 + 36x = 0$

In Problems 8–16 use a translation of axes to obtain the graph. Identify the curve and find whichever may be appropriate: center, foci, vertices, intercepts on minor axis, directrix, asymptotes.

8. $x^2 + 2y^2 - 2x - 4y + 1 = 0$ 9. $3x^2 + 4y^2 + 6x - 16y + 7 = 0$

10. $x^2 - 4x - 8y + 20 = 0$ 11. $x^2 - 4y^2 - 4x - 8y = 4$

12. $4x^2 + 9y^2 + 24x + 36y + 36 = 0$ 13. $y^2 + 6y + 4x - 3 = 0$

14. $4x^2 + 3y^2 - 24x - 6y + 27 = 0$ 15. $9x^2 - 16y^2 + 64y + 80 = 0$

16. $4y = x^2 + 8x + 16$

In Problems 17–25 determine an equation of the curve described.

17. An ellipse with center at $(2, 0)$, one focus at $(5, 0)$, and one vertex at $(7, 0)$

18. An ellipse with vertices at $(-1, \frac{1}{2})$, $(5, \frac{1}{2})$ and intercepts on the minor axis at $(2, -\frac{1}{2})$, $(2, \frac{3}{2})$

19. An ellipse with minor axis $y = -2$, one focus at $(1, 0)$, and one vertex at $(1, 1)$

20. An ellipse passing through the origin with major axis $x = -2$, center on the x axis, and one focus at $(-2, 2)$

21. A parabola with focus at $(3, 4)$ and directrix $y = -2$

22. A parabola with vertex at $(1, 1)$ and directrix $x = 2$

23. A parabola passing through the origin, with vertical axis, and having vertex at $(-2, 3)$

24. A hyperbola with center at $(3, 2)$, one vertex at $(3, 5)$, and one focus at $(3, 7)$

25. A hyperbola with one vertex at $(0, 1)$ having asymptotes $y = x + 3$ and $y = -(x + 1)$

11-3 ROTATION OF AXES

In this section we complete the discussions begun in Sections 11-1 and 11-2 on the second-degree equation

$$Ax^2 + Bxy + Cy^2 + Dx + Ey + F = 0.$$

So far our procedures for identifying the graph of this equation do not apply to the case in which the coefficient B of the xy term is different from zero.

Let us begin with a simple and familiar example: $xy = 1$. If you worked Problem 64 in Section 11-1, you showed from the definition that this graph is a hyperbola with foci $(-\sqrt{2}, -\sqrt{2})$, $(\sqrt{2}, \sqrt{2})$ and vertices $(-1, -1)$, $(1, 1)$. A glance at Fig. 11-20 suggests that this hyperbola would appear in a more familiar position if we were to change to a system of coordinates in which the "horizontal" axis coincides with the line $y = x$ and the "vertical" axis with $y = -x$. Such a coordinate system would result from a counterclockwise **rotation** of the axes about the origin through the angle $\pi/4$.

We investigate the effect of rotation of axes on the coordinates of points in much the same way as with translation of axes in Section 4-8. Let us assume that we have an "original" x, y coordinate system and that we construct a "new" u, v system (Fig. 11-21) having the same origin such that the positive u axis makes an acute angle α

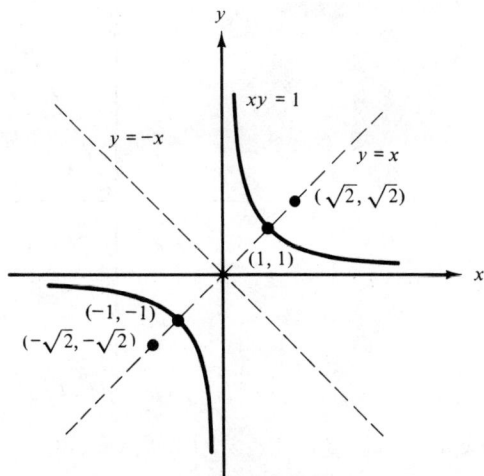

Figure 11-20

with the positive x axis. The scales on all four coordinate axes will be assumed equal. We want formulas that will enable us to convert from u, v coordinates to x, y coordinates and vice versa.

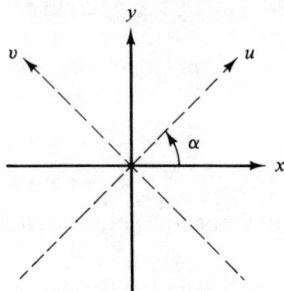

Figure 11-21

Suppose P is a point in the plane with original coordinates (x, y) and new coordinates (u, v). Denote by Q the foot of the perpendicular from P to the x axis and by R the foot of the perpendicular from P to the u axis. Let r denote the length of OP and let β denote the angle from the u axis to OP. Then (see Fig. 11-22) we have

$$\frac{u}{r} = \cos \beta \qquad \frac{v}{r} = \sin \beta.$$

Because the angle from the x axis to OP is $\alpha + \beta$, we also have

$$\frac{x}{r} = \cos (\alpha + \beta) \qquad \frac{y}{r} = \sin (\alpha + \beta).$$

Multiplying by r and using formulas (9-18) and (9-20) for $\cos (\alpha + \beta)$ and $\sin (\alpha + \beta)$, we get

$$x = r \cos (\alpha + \beta) = r(\cos \alpha \cos \beta - \sin \alpha \sin \beta),$$
$$y = r \sin (\alpha + \beta) = r(\sin \alpha \cos \beta + \cos \alpha \sin \beta).$$

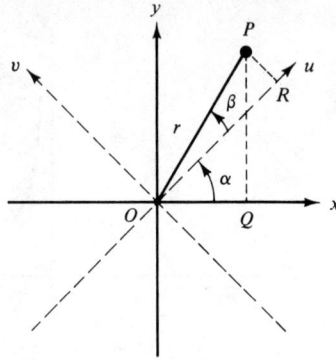

Figure 11-22

But because $u = r \cos \beta$ and $v = r \sin \beta$, we obtain

$$x = (r \cos \beta) \cos \alpha - (r \sin \beta) \sin \alpha = u \cos \alpha - v \sin \alpha,$$
$$y = (r \cos \beta) \sin \alpha + (r \sin \beta) \cos \alpha = u \sin \alpha + v \cos \alpha.$$

The formulas

$$\begin{aligned} x &= u \cos \alpha - v \sin \alpha \\ y &= u \sin \alpha + v \cos \alpha \end{aligned} \tag{11-15}$$

give us (x, y) in terms of (u, v). Solving these equations for (u, v) in terms of (x, y) gives

$$\begin{aligned} u &= x \cos \alpha + y \sin \alpha \\ v &= -x \sin \alpha + y \cos \alpha. \end{aligned} \tag{11-16}$$

You are asked to carry out this solution in Problem 44.

Example 1

(a) Write both sets of formulas (11-15) and (11-16) for a rotation through $\pi/3$.

(b) If P has x, y coordinates $(2, 4)$, what are its u, v coordinates?

(c) If Q has u, v coordinates $(-1, 2)$, what are its x, y coordinates?

Solution. (a) Because $\alpha = \pi/3$, we have $\cos \alpha = 1/2$, $\sin \alpha = \sqrt{3}/2$ and formulas (11-15) and (11-16) become

$$\left. \begin{aligned} x &= \frac{1}{2}u - \frac{\sqrt{3}}{2}v \\ y &= \frac{\sqrt{3}}{2}u + \frac{1}{2}v \end{aligned} \right\} \tag{11-17}$$

$$\left. \begin{aligned} u &= \frac{1}{2}x + \frac{\sqrt{3}}{2}y \\ v &= -\frac{\sqrt{3}}{2}x + \frac{1}{2}y \end{aligned} \right\} \tag{11-18}$$

Graphs of Second-Degree Equations   Chap. 11

(b) Applying (11-18) to the point (2, 4), we get

$$u = \frac{1}{2} \cdot 2 + \frac{\sqrt{3}}{2} \cdot 4 = 1 + 2\sqrt{3}$$

$$v = -\frac{\sqrt{3}}{2} \cdot 2 + \frac{1}{2} \cdot 4 = 2 - \sqrt{3}.$$

So the u, v coordinates are $(1 + 2\sqrt{3}, 2 - \sqrt{3})$.

(c) Applying (11-17) to the point $(-1, 2)$, we have

$$x = \frac{1}{2}(-1) - \frac{\sqrt{3}}{2} \cdot 2 = -\frac{1}{2} - \sqrt{3}$$

$$y = \frac{\sqrt{3}}{2}(-1) + \frac{1}{2} \cdot 2 = 1 - \frac{\sqrt{3}}{2}.$$

Example 2

Consider the equation $xy = 1$. We remarked earlier that a rotation of axes through $\pi/4$ should transform this equation into one that we recognize as a hyperbola. With $\alpha = \pi/4$ we have $\sin \alpha = \cos \alpha = \sqrt{2}/2$ and so (11-15) now becomes

$$x = \frac{\sqrt{2}}{2}u - \frac{\sqrt{2}}{2}v$$

$$y = \frac{\sqrt{2}}{2}u + \frac{\sqrt{2}}{2}v. \tag{11-19}$$

Substituting these expressions in $xy = 1$, we obtain

$$\frac{\sqrt{2}}{2}(u - v) \cdot \frac{\sqrt{2}}{2}(u + v) = 1$$

$$\frac{u^2}{2} - \frac{v^2}{2} = 1.$$

As expected, this is a hyperbola; the u axis is its transverse axis. Because $a = b = \sqrt{2}$, we have $c = \sqrt{a^2 + b^2} = 2$. The vertices are given in the u, v coordinate system by $(\pm\sqrt{2}, 0)$ and the foci by $(\pm 2, 0)$. The lines $u = \pm v$ are asymptotes.

We can use Eqs. (11-19) again to calculate the original coordinates of vertices and foci. The vertex with u, v coordinates $(\sqrt{2}, 0)$ has x, y coordinates $(1, 1)$, and the x, y coordinates of the focus $(2, 0)$ are $(\sqrt{2}, \sqrt{2})$. The equations

$$u = \frac{\sqrt{2}}{2}x + \frac{\sqrt{2}}{2}y$$

$$v = -\frac{\sqrt{2}}{2}x + \frac{\sqrt{2}}{2}y \tag{11-20}$$

can be used to show that the x, y equations of the asymptotes $u = \pm v$ are $x = 0, y = 0$; that is, the asymptotes are the x and y coordinate axes.

Example 3

Determine the effect of a rotation through $\pi/4$ on the equation

$$(x - \sqrt{2})^2 + (y - \sqrt{2})^2 = 4.$$

Solution. We can substitute the formulas (11-19) into the given equation. Doing so gives us the new equation

$$\left(\frac{\sqrt{2}}{2}u - \frac{\sqrt{2}}{2}v - \sqrt{2}\right)^2 + \left(\frac{\sqrt{2}}{2}u + \frac{\sqrt{2}}{2}v - \sqrt{2}\right)^2 = 4,$$

which reduces (after some calculation) to

$$u^2 + v^2 - 4u = 0.$$

But there is a better way do this. We recognize the original equation as belonging to a circle with radius 2 and center at $(\sqrt{2}, \sqrt{2})$ in the x, y system. The rotation assigns new u, v coordinates $(2, 0)$ to the center (why?). Because the radius is 2, we know that the u, v equation must be

$$(u - 2)^2 + v^2 = 4,$$

which, of course, is equivalent to our previous result.

Example 4

Find the u, v equation into which

$$x^2 - xy + y^2 = 6$$

is transformed by a rotation through $\pi/4$. Identify the graph and sketch it.

Solution. The formulas (11-19) and (11-20) apply. Substituting (11-19) in the given equation, we get

$$\frac{1}{2}(u^2 - 2uv + v^2) - \frac{1}{2}(u^2 - v^2) + \frac{1}{2}(u^2 + 2uv + v^2) = 6$$

$$\frac{1}{2}u^2 + \frac{3}{2}v^2 = 6$$

$$\frac{u^2}{12} + \frac{v^2}{4} = 1.$$

From the form of this equation we know that the graph is an ellipse with $a = 2\sqrt{3}$, $b = 2$; it follows that $c = 2\sqrt{2}$. In the u, v coordinate system, the vertices are $(\pm 2\sqrt{3}, 0)$ and the foci are $(\pm 2\sqrt{2}, 0)$; the intercepts on the v axis are $(0, \pm 2)$.

The x, y coordinates of these points are found from Eqs. (11-20). For the focus $(2\sqrt{2}, 0)$, for instance, we get

$$x = \frac{\sqrt{2}}{2} \cdot 2\sqrt{2} - 0 = 2,$$

$$y = \frac{\sqrt{2}}{2} \cdot 2\sqrt{2} + 0 = 2.$$

So the x, y coordinates are $(2, 2)$. Similarly, the x, y coordinates of the other focus are $(-2, -2)$ whereas those of the vertices are $(\sqrt{6}, \sqrt{6})$ and $(-\sqrt{6}, -\sqrt{6})$. Intercepts on the v axis have x, y coordinates $(-\sqrt{2}, \sqrt{2})$, $(\sqrt{2},$

$-\sqrt{2}$). Equations of the axes of the ellipse are $y = x$ (major axis) and $y = -x$ (minor axis).

Removal of Mixed Product Term

Examples 2 and 4 prompt us to ask whether the mixed product (xy) term in a second-degree equation

$$Ax^2 + Bxy + Cy^2 + Dx + Ey + F = 0 \qquad (11\text{-}21)$$

can always be removed by a rotation of axes. More precisely, is there an angle α such that the rotation through α given by (11-15) transforms the equation into the form

$$A'u^2 + B'uv + C'v^2 + D'u + E'v + F' = 0 \qquad (11\text{-}22)$$

ir which the coefficient B' of the mixed product term uv is zero? In order to answer this question, we begin by substituting Eqs. (11-15) into (11-21) and calculating the left side of the equation:

$$Ax^2 + Bxy + Cy^2 + Dx + Ey + F$$
$$= A(u \cos \alpha - v \sin \alpha)^2 + B(u \cos \alpha - v \sin \alpha)(u \sin \alpha + v \cos \alpha)$$
$$+ C(u \sin \alpha + v \cos \alpha)^2 + D(u \cos \alpha - v \sin \alpha)$$
$$+ E(u \sin \alpha + v \cos \alpha) + F$$

Squaring, multiplying, and collecting terms, we find that the coefficient A' of u^2 in the resulting equation is

$$A' = A \cos^2 \alpha + B \cos \alpha \sin \alpha + C \sin^2 \alpha.$$

Similarly, the coefficient B' of uv is

$$B' = -2A \cos \alpha \sin \alpha + B(\cos^2 \alpha - \sin^2 \alpha) + 2C \sin \alpha \cos \alpha$$
$$= B(\cos^2 \alpha - \sin^2 \alpha) - 2(A - C) \cos \alpha \sin \alpha.$$

The coefficients C', D', E', and F' are obtained in the same way; the results are summarized in the following theorem.

Theorem 11-4. The general equation of second degree

$$Ax^2 + Bxy + Cy^2 + Dx + Ey + F = 0 \qquad (11\text{-}21)$$

is transformed by rotation of axes through an acute angle α into

$$A'u^2 + B'uv + C'v^2 + D'u + E'v + F' = 0, \qquad (11\text{-}22)$$

where the new coefficients are given in terms of the original ones by the following equations:

$$A' = A \cos^2 \alpha + B \cos \alpha \sin \alpha + C \sin^2 \alpha$$
$$B' = B(\cos^2 \alpha - \sin^2 \alpha) - 2(A - C) \cos \alpha \sin \alpha$$
$$C' = A \sin^2 \alpha - B \sin \alpha \cos \alpha + C \cos^2 \alpha$$
$$D' = D \cos \alpha + E \sin \alpha$$
$$E' = -D \sin \alpha + E \cos \alpha$$
$$F' = F.$$

Let us return to the question raised earlier: can we *choose* α so that $B' = 0$? Doing so requires that

$$B(\cos^2 \alpha - \sin^2 \alpha) = 2(A - C)\sin \alpha \cos \alpha$$

$$B \cos 2\alpha = (A - C)\sin^2 \alpha$$

$$\cot 2\alpha = \frac{A - C}{B}.$$

Thus to remove the mixed product term $B'uv$, we must choose α so that

$$\cot 2\alpha = \frac{A - C}{B}. \tag{11-23}$$

The following examples indicate how we can do so in every case with an acute angle α.

Example 5

Let us consider the equations of Examples 2 and 4. In Example 2 $A = C = 0$, $B = 1$; in Example 4 $A = C = 1$, $B = -1$. Both times, the value of $(A - C)/B$ is zero and so we seek a value of α $(0 \le \alpha \le \pi/2)$ such that $\cot 2\alpha = 0$. Such an α is $\pi/4$. As we have seen, rotation through $\pi/4$ does indeed remove the mixed product term and enables us to determine the graph of each equation.

It is clear that **whenever $A = C$ and $B \ne 0$, the desired angle of rotation is $\pi/4$.**

Example 6

Identify the graph of the equation $x^2 - 4xy + y^2 + 12 = 0$.

Solution. Because $A = C = 1$, we rotate the axes through $\pi/4$. To get the new equation, we can either substitute Eqs. (11-19) directly or use the formulas from Theorem 11-4. We do it both ways. First, substituting (11-19) gives

$$x^2 - 4xy + y^2 + 12 = \left(\frac{\sqrt{2}}{2}u - \frac{\sqrt{2}}{2}v\right)^2 - 4\left(\frac{\sqrt{2}}{2}u - \frac{\sqrt{2}}{2}v\right)\left(\frac{\sqrt{2}}{2}u + \frac{\sqrt{2}}{2}v\right)$$

$$+ \left(\frac{\sqrt{2}}{2}u + \frac{\sqrt{2}}{2}v\right)^2 + 12$$

$$= -u^2 + 3v^2 + 12.$$

Using Theorem 11-4 to get the coefficients, we have

$$A' = A \cos^2 \alpha + B \cos \alpha \sin \alpha + C \sin^2 \alpha$$

$$= 1 \cdot \left(\frac{\sqrt{2}}{2}\right)^2 - 4\left(\frac{\sqrt{2}}{2}\right)\left(\frac{\sqrt{2}}{2}\right) + 1 \cdot \left(\frac{\sqrt{2}}{2}\right)^2 = -1.$$

Similarly, C' is found to be 3, $D' = E' = 0$, and $F' = 12$. We know that B' is zero because that is the way we chose α.[1]

Each method gives the new equation

$$-u^2 + 3v^2 + 12 = 0$$

[1] Nevertheless, calculating B' and verifying that it is zero provides a check on our work.

or
$$\frac{u^2}{12} - \frac{v^2}{4} = 1.$$

The graph is a hyperbola. Its transverse axis is the u axis whereas its conjugate axis is the v axis. Its asymptotes are $u = \pm \sqrt{3}\,v$; in the u, v system its vertices have coordinates $(\pm 2\sqrt{3}, 0)$, its foci $(\pm 4, 0)$.

Example 7

Find the x, y equations of the asymptotes and the x, y coordinates of the vertices and foci of the hyperbola in Example 6. Sketch the graph in the x, y system.

Solution. The asymptotes were found to be $u = \sqrt{3}\,v$ and $u = -\sqrt{3}\,v$. Substituting (11-20) into $u = \sqrt{3}\,v$, we get the x, y equations of one of these asymptotes

$$\frac{\sqrt{2}}{2}x + \frac{\sqrt{2}}{2}y = \sqrt{3}\left(-\frac{\sqrt{2}}{2}x + \frac{\sqrt{2}}{2}y\right).$$

To simplify, we first divide by $\sqrt{2}/2$:

$$x + y = \sqrt{3}(-x + y)$$
$$y - \sqrt{3}\,y = -x - \sqrt{3}\,x$$
$$(1 - \sqrt{3})y = -(1 + \sqrt{3})x$$
$$y = \frac{\sqrt{3} + 1}{\sqrt{3} - 1}x = (2 + \sqrt{3})x.$$

In the same way, we find that the asymptote $u = -\sqrt{3}\,v$ has x, y equation

$$y = (2 - \sqrt{3})x.$$

The x, y coordinates of vertices and foci are found from (11-19) to be $(\sqrt{6}, \sqrt{6}), (-\sqrt{6}, -\sqrt{6})$ and $(2\sqrt{2}, 2\sqrt{2}), (-2\sqrt{2}, -2\sqrt{2})$, respectively. The graph is shown in Fig. 11-23.

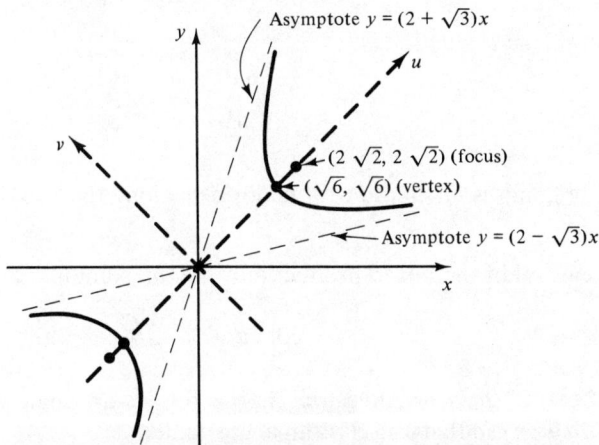

Figure 11-23

The next example is one for which $A \neq C$.

Example 8

Consider the equation $5x^2 + 4xy + 2y^2 = 12$. From $(A - C)/B = \frac{3}{4}$ we obtain $\cot 2\alpha = \frac{3}{4}$ or $\tan 2\alpha = \frac{4}{3}$ so that $\alpha = \frac{1}{2} \operatorname{Arctan} \frac{4}{3}$. (You should recognize α as half the larger acute angle in a 3-4-5 right triangle.) Rotation through α eliminates the mixed product term. In order to find the new equation, we can use either of the two methods illustrated in Example 6. But in each case we need to know $\sin \alpha$ and $\cos \alpha$. How do we find $\sin \alpha$ and $\cos \alpha$ when $\alpha = \frac{1}{2} \operatorname{Arctan} \frac{4}{3}$? We use the identities (9-23), (9-24) and take square roots, obtaining

$$\cos \alpha = \sqrt{\frac{1 + \cos 2\alpha}{2}}$$

$$\sin \alpha = \sqrt{\frac{1 - \cos 2\alpha}{2}}.$$

Because 2α is the larger angle in a 3-4-5 right triangle we know that $\cos 2\alpha$ is $3/5$. It follows, then, that

$$\sin \alpha = \sqrt{\frac{1 - 3/5}{2}} = \frac{1}{\sqrt{5}}$$

$$\cos \alpha = \sqrt{\frac{1 + 3/5}{2}} = \frac{2}{\sqrt{5}}.$$

Using Theorem 11-4 we calculate the new coefficients.

$$A' = 5 \cdot \frac{4}{5} + 4 \cdot \frac{2}{5} + 2 \cdot \frac{1}{5} = 6$$

$$B' = 0$$

$$C' = 5 \cdot \frac{1}{5} - 4 \cdot \frac{2}{5} + 2 \cdot \frac{4}{5} = 1$$

$$D' = 0 \qquad E' = 0 \qquad F' = 12$$

So the new equation is

$$6u^2 + v^2 = 12,$$

$$\frac{u^2}{2} + \frac{v^2}{12} = 1.$$

The graph is an ellipse with major axis along the v axis (Fig. 11-24).

Removal of the mixed product term requires finding α such that

$$\cot 2\alpha = \frac{A - C}{B}.$$

When $(A - C)/B$ is negative, we choose for 2α an angle in the second quadrant $(\pi/2 < 2\alpha < \pi)$ so that α itself is an acute angle.

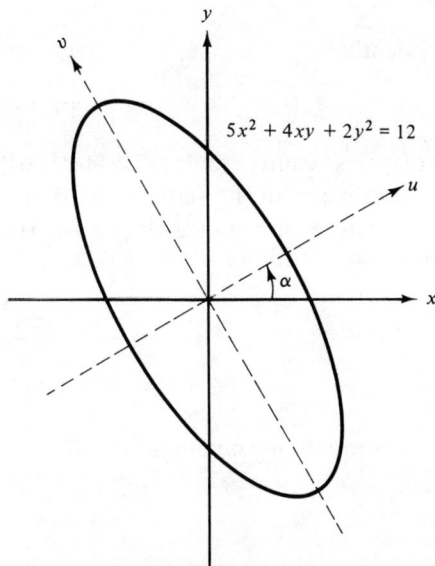

$5x^2 + 4xy + 2y^2 = 12$

Figure 11-24

Example 9

By a suitable rotation of axes, eliminate the mixed product term in equation $x^2 - \sqrt{3}\,xy = 6$. Identify and sketch the graph.

Solution. Setting

$$\cot 2\alpha = \frac{A - C}{B} = \frac{1 - 0}{-\sqrt{3}} = -\frac{1}{\sqrt{3}}$$

and looking for 2α in the second quadrant, we find

$$2\alpha = \frac{2\pi}{3}$$

so that

$$\alpha = \frac{\pi}{3}.$$

Formulas for this rotation were calculated in Example 1

$$x = \frac{1}{2}u - \frac{\sqrt{3}}{2}v$$

$$y = \frac{\sqrt{3}}{2}u + \frac{1}{2}v$$

(11-17)

and it is easy to obtain the new equation by substituting these values for x and y into the given equation. The result is

$$-\frac{1}{2}u^2 + \frac{3}{2}v^2 = 6$$

or equivalently

$$\frac{v^2}{4} - \frac{u^2}{12} = 1.$$

We see that this is a hyperbola (Fig. 11-25) with transverse axis along the v axis; the u, v coordinates of the vertices are $(0, \pm 2)$; the foci are $(0, \pm 4)$. Equations of the asymptotes are $v = \pm (1/\sqrt{3})u$. To find the x, y equations of these asymptotes, we substitute

$$u = \frac{1}{2}x + \frac{\sqrt{3}}{2}y$$

$$v = -\frac{\sqrt{3}}{2}x + \frac{1}{2}y$$

(11-18)

into the u, v equations, obtaining $x = 0$, $y = (\sqrt{3}/3)x$.

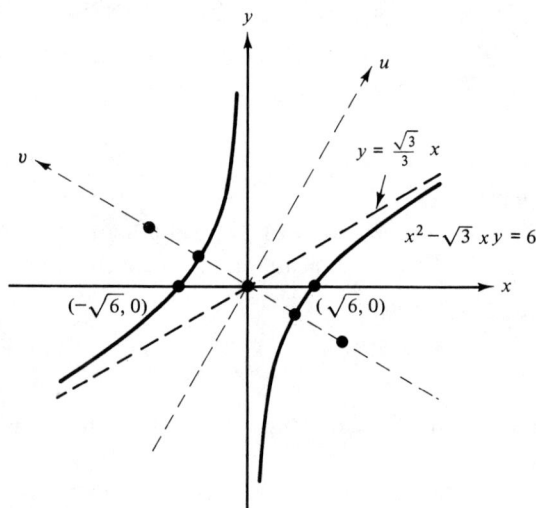

Figure 11-25

Example 10

Identify and sketch the graph of

$$8x^2 - 8xy + 2y^2 + \frac{\sqrt{5}}{5}x + \frac{2\sqrt{5}}{5}y = 0.$$

Solution. To eliminate the mixed product term, we choose α from the equation

$$\cot 2\alpha = \frac{A - C}{B} = -\frac{3}{4}.$$

Because $\cot 2\alpha < 0$, we choose 2α in the second quadrant. We have $\tan 2\alpha = -4/3$ and the reference angle β corresponding to 2α satisfies $\tan \beta = 4/3$. Thus β is the larger angle of a 3-4-5 right triangle. Because $\pi/2 < 2\alpha < \pi$,

$$\cos 2\alpha = -\cos \beta = -\frac{3}{5}.$$

Graphs of Second-Degree Equations   Chap. 11

Thus by (11-25),

$$\sin \alpha = \sqrt{\frac{1 - \cos 2\alpha}{2}} = \sqrt{\frac{1 + 3/5}{2}} = \frac{2}{\sqrt{5}}$$

$$\cos \alpha = \sqrt{\frac{1 + \cos 2\alpha}{2}} = \sqrt{\frac{1 - 3/5}{2}} = \frac{1}{\sqrt{5}}$$

and it follows (either by substitution or by Theorem 11-4) that the new equation is

$$10v^2 + u = 0.$$

We recognize this graph as a parabola, easily drawn in the u, v coordinate system (Fig. 11-26). The angle α of rotation is approximately 63.43°.

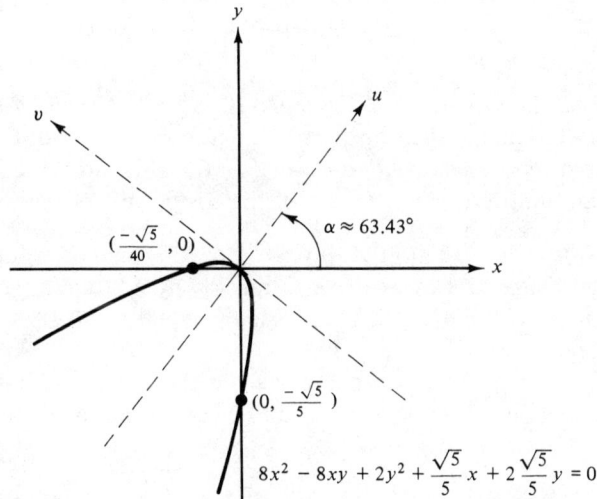

$$8x^2 - 8xy + 2y^2 + \frac{\sqrt{5}}{5}x + 2\frac{\sqrt{5}}{5}y = 0$$

Figure 11-26

Classifying Second-Degree Curves

In the following discussion we learn how to determine whether the graph of a second degree equation with a mixed product term ($B \neq 0$) is a parabola, ellipse, or hyperbola by examining the discriminant of the equation. Using the method described, it will be possible to make such a determination[2] *without* actually performing the rotation to eliminate the mixed product term. Our discussion is based on the concept of an invariant, of which the discriminant is an example.

Consider the second-degree equation

$$Ax^2 + Bxy + Cy^2 + Dx + Ey + F = 0 \qquad (B \neq 0)$$

and the transformed equation

$$A'u^2 + B'uv + C'v^2 + D'u + E'v + F' = 0$$

[2]Keep in mind (as was pointed out in Section 11-2) that there are degenerate cases in which the graph may degenerate to a line, a pair of lines, or a point; in some cases, there are no points at all satisfying the equation. These cases can be determined by performing the rotation and/or translation required to bring them to a simplified form.

after a rotation of axes through an angle α. We have already calculated expressions that give A', B', C', D', E', F' in terms of A, B, C, D, E, F. Explicit formulas are given in Theorem 11-4.

Notice that $F' = F$. The constant term is said to be an **invariant** under the rotation because it is unaffected by the rotation. The sum of the coefficients of the square terms is also an invariant:

$$
\begin{aligned}
A' + C' &= (A \cos^2 \alpha + B \cos \alpha \sin \alpha + C \sin^2 \alpha) \\
&\quad + (A \sin^2 \alpha - B \cos \alpha \sin \alpha + C \cos^2 \alpha) \\
&= A(\cos^2 \alpha + \sin^2 \alpha) + B(\cos \alpha \sin \alpha - \cos \alpha \sin \alpha) \\
&\quad + C(\sin^2 \alpha + \cos^2 \alpha) \\
&= A \cdot 1 + B \cdot 0 + C \cdot 1 = A + C.
\end{aligned}
$$

Example 11

For the equation of Example 10 $A = 8$, $C = 2$, $F = 0$. The corresponding coefficients in the transformed equation are $A' = 0$, $C' = 10$, $F' = 0$. The fact that $F = F'$ and $A + C = A' + C'$ illustrates the invariance of both the constant term and the sum of the coefficients of the square terms.

Note. The equation $A + C = A' + C'$ can be useful in either of two ways. Having calculated A', we can get C' quickly by subtracting A' from $A + C$. Or we can perform the calculation of C' and use the equation $A + C = A' + C'$ as a check on our work.

Less obvious but more important is the fact that the **discriminant** $B^2 - 4AC$ is also an invariant. This statement can be verified by calculating

$$
\begin{aligned}
(B')^2 - 4A'C' &= [B(\cos^2 \alpha - \sin^2 \alpha) - 2(A - C) \cos \alpha \sin \alpha]^2 \\
&\quad - 4[A \cos^2 \alpha + B \cos \alpha \sin \alpha + C \sin^2 \alpha][A \sin^2 \alpha \\
&\quad - B \sin \alpha \cos \alpha + C \cos^2 \alpha].
\end{aligned}
$$

Multiplying out and collecting terms give

$$
\begin{aligned}
(B')^2 - A'C' &= (B^2 - 4AC)(\cos^4 \alpha + 2 \sin^2 \alpha \cos^2 \alpha + \sin^4 \alpha) \\
&= (B^2 - 4AC)(\cos^2 \alpha + \sin^2 \alpha)^2 \\
&= B^2 - 4AC.
\end{aligned}
$$

The reason that $B^2 - 4AC$ is called the discriminant will soon be apparent.

The preceding discussion of invariants applies to any angle of rotation. Suppose, however, that the rotation angle α has been selected so as to make $B' = 0$. Then

$$
B^2 - 4AC = (B')^2 - 4A'C' = 0 - 4A'C'
$$

and so the sign of $A'C'$ is the opposite of the sign of the discriminant $B^2 - 4AC$.

Case 1. If $B^2 - 4AC < 0$, then $A'C'$ is positive. Because $B' = 0$ the new equation is of the form

$$
A'u^2 + C'v^2 + D'u + E'v + F' = 0
$$

where A' and C' have the same sign. The graph is therefore an ellipse. (It cannot be a circle because in the original equation $B \neq 0$.)

Case 2. If $B^2 - 4AC = 0$, then $A'C'$ is zero and so either $A' = 0$ or $C' = 0$ (but not both; see Problem 43). The new equation is either

$$A'u^2 + D'u + E'v + F = 0 \qquad (A' \neq 0)$$

or $\qquad\qquad C'v^2 + D'u + E'v + F = 0 \qquad (C' \neq 0)$

In each case the graph is a parabola.

Case 3. If $B^2 - 4AC > 0$, then $A'C'$ is negative so that A' and C' have opposite signs. In this case the graph of

$$A'u^2 + C'v^2 + D'u + E'v + F' = 0.$$

is a hyperbola.

Classification of Second-Degree Equation. Barring degenerate cases, the graph of

$$Ax^2 + Bxy + Cy^2 + Dx + Ey + F = 0 \qquad (B \neq 0)$$

is

1. an ellipse if $B^2 - 4AC < 0$.
2. a parabola if $B^2 - 4AC = 0$.
3. a hyperbola if $B^2 - 4AC > 0$.

Example 12

Classify the graphs of the following nondegenerate second-degree equations.
(a) $x^2 + xy + y^2 - x + y = 5$
(b) $\quad 3x^2 + 3xy - y^2 - 6x = 6$
(c) $\quad\quad x^2 + 2xy + y^2 + x = 3$
(d) $\quad 5x^2 + 5y^2 - 15x - 9 = 0$

Solution.
(a) $B^2 - 4AC = 1 - 4 < 0$ (ellipse)
(b) $B^2 - 4AC = 9 + 12 > 0$ (hyperbola)
(c) $B^2 - 4AC = 4 - 4 = 0$ (parabola)
(d) Because $B = 0$ and $A = C$, we have a circle.

Example 13

Identify the following degenerate cases.
(a) $x^2 + xy + y^2 + 1 = 0$
(b) $\quad\quad x^2 - 4xy + 4y^2 = 0$
(c) $\quad\quad x^2 + 2xy + y^2 = 4$
(d) $\quad\quad x^2 + 4xy + y^2 = 0$

Solution. (a) Although the discriminant is negative, suggesting an ellipse, we actually have a degenerate case. The mixed product term can be removed by a rotation through $\pi/4$. The transformed equation is

$$\frac{3}{2}u^2 + \frac{1}{2}v^2 = -1$$

and we now see that the graph is empty.

(b) The discriminant is zero, suggesting a parabola. However, the equation can be written

$$(x - 2y)^2 = 0,$$

which is equivalent to

$$x = 2y.$$

The graph is a line.

(c) Here the discriminant is zero again, but the equation can be written

$$(x + y)^2 = 4$$

$$x + y = \pm 2.$$

The graph is the union of two parallel lines, $x + y = 2$ and $x + y = -2$.

(d) Here the discriminant is positive, suggesting a hyperbola. Rotation through $\pi/4$ yields the new equation

$$3u^2 - v^2 = 0.$$

The graph is the union of a pair of intersecting lines, $v = \sqrt{3}\,u$ and $v = -\sqrt{3}\,u$.

EXERCISES 11-3

1. Consider a rotation of axes through $\alpha = \pi/6$.
   (a) Write both sets of formulas (11–15) and (11–16), expressing the x, y coordinates in terms of the u, v coordinates and vice versa.
   (b) If P has x, y coordinates (1, 3), find its u, v coordinates after this rotation.
   (c) If Q has u, v coordinates (2, -1), find its x, y coordinates.
   (d) Find the new u, v equation for the graph of $xy + y^2 = 2$.

2. Carry out the parts of Problem 1 for $\alpha = $ Arctan 3/4.

3. Carry out the parts of Problem 1 for $\alpha = $ Arctan 12/5.

In Problems 4–9 transform the given equation by a rotation of axes through $\alpha = \pi/4$. Find the new u, v equation and identify it. Sketch the curve giving vertices, foci, equations of axes and asymptotes (if any) in the x, y coordinate system.

4. $x^2 - 3xy + y^2 + 10 = 0$ 5. $2x^2 - xy + 2y^2 = 15$

6. $x^2 + xy + y^2 = 9$ 7. $x^2 + 3xy + y^2 = 10$

8. $x^2 - 2xy + y^2 - 4\sqrt{2}\,y = 0$ 9. $x^2 - 2xy + y^2 - 2\sqrt{2}\,x = 0$

In Problems 10–17 find an acute angle α such that rotation through α will eliminate the mixed product term in the given equation. Find the new u, v equation.

10. $x^2 + \sqrt{3}\,xy = 6$ 11. $\frac{1}{2}x^2 + \sqrt{3}\,xy - \frac{1}{2}y^2 = 4$

12. $x^2 - xy + y^2 = 1$ 13. $x^2 + 2xy + y^2 = 4$

14. $4x^2 - \sqrt{3}\,xy + y^2 = 10$ 15. $15x^2 - 24xy + 8y^2 + 3y = 11$

16. $4x^2 + 3xy + 2y = 5$ 17. $x^2 - 4xy + 4y^2 + 2x = 0$

In Problems 18–23 use the method of rotation of axes to identify and draw the graph of the given equation. Give the new u, v equation.

18. $x^2 + xy + y^2 = 2$ 19. $2x^2 + \sqrt{3}\,xy + y^2 = 5$

20. $x^2 + 4xy + 4y^2 = 10\sqrt{5}\,x - 5\sqrt{5}\,y$ 21. $11x^2 - 24xy + 4y^2 = 80$

22. $3x^2 + 3xy - y^2 = 21$ 23. $4x^2 + 4xy + y^2 = 10x - 20y$

Graphs of Second-Degree Equations   Chap. 11

In Problems 24–30 use the discriminant to identify the graph of the given equation as a parabola, circle, ellipse, or hyperbola. (You may assume that the graph is not degenerate.) Do not graph.

24. $x^2 - 3xy + y^2 - 2x + 4y = 0$ 25. $3x^2 - xy + y^2 - 6x + 5y + 12 = 0$

26. $x^2 + 2xy + 3y^2 - 5x + 4y - 9 = 0$ 27. $x^2 + 2xy + y^2 + 4x = 0$

28. $2x^2 + 4xy + y^2 + 4y = 0$ 29. $25x^2 - 10xy + y^2 - 10x + 4y = 10$

30. $9x^2 + 9y^2 + 54x + 36y + 81 = 0$

In Problems 31–36 identify the graph of the equation. (Do not assume that the graph is not degenerate.)

31. $x^2 + y^2 + 2x + 4y + 5 = 0$ 32. $x^2 + y^2 + 2x + 4y + 6 = 0$

33. $x^2 - y^2 + 2x + 4y + 5 = 0$ 34. $4x^2 - y^2 + 4x - 2y + 2 = 0$

35. $x^2 - 2xy + y^2 - 5 = 0$ 36. $4x^2 + 4xy + y^2 + 36 = 0$

37. $5x^2 - 6\sqrt{3}\,xy - y^2 = 0$

FOR THE ENTHUSIASTS

38. (a) Show that the graph of $\sqrt{x} + \sqrt{y} = 1$ is part of a parabola.
    (b) Find the x, y coordinates of the focus and the x, y equation of the directrix of the parabola in part (a).

39. Transform the equation
$$x^4 + 6x^2y^2 + y^4 = 8$$
by means of a rotation through $\pi/4$. Sketch the graph.

40. Transform the equation
$$|3y - 4x| + |3x + 4y| = 5$$
by means of a rotation through Arctan 4/3. Sketch the graph.

41. Show by a rotation of axes that the graph of the rational function $f(x) = x + (1/x)$ is a hyperbola. Find the x, y equation of its transverse axis and the x, y coordinates of its foci.

42. Show that $D^2 + E^2$ is invariant under any rotation of axes.

43. Show in the context of Case 2 that when $B^2 - 4AC = 0$, it is not possible for A' and C' both to be zero unless $A = B = C = 0$. (*Hint:* Use the invariance of $A + C$.)

44. Show that Eqs. (11-16) can be deduced from (11-15).

SUMMARY

In this chapter we concentrated on identifying and sketching graphs of second-degree equations. Except for degenerate cases, they can be classified into the categories (a) parabola, (b) circle or ellipse, and (c) hyperbola.

1. You should know the geometric definitions of each of these curves and their geometric features, such as center, vertices, foci, axes, and asymptotes.

2. You should be able to derive an equation for a parabola, ellipse, or hyperbola, given such data as center, vertices, foci, axes, asymptotes.

3. You should be able to classify simple equations of the form

$$Ax^2 + Cy^2 = K$$

by inspecting the coefficients A, C, K. You should be able to graph such equations.

4. By a translation of axes, you should be able to classify and sketch the graph of an equation of the form

$$Ax^2 + Cy^2 + Dx + Ey + F = 0$$

5. You should have learned the technique of rotation of axes. You should be able to convert a point or an equation to the new rotated system.

6. You should be able to choose an angle of rotation that removes the mixed product term from the equation

$$Ax^2 + Bxy + Cy^2 + Dx + Ey + F = 0.$$

7. You should be able to identify the graph of the equation in (6) by means of the discriminant and graph it in a rotated coordinate system.

REVIEW EXERCISES

1. Define.
   (a) parabola   (b) ellipse   (c) hyperbola

2. Derive an equation for the given figure:
   (a) parabola with vertex at $(0, 0)$ and focus at $(2, 0)$.
   (b) ellipse with foci $(-1, 0)$, $(1, 0)$ and vertices $(-2, 0)$, $(2, 0)$.
   (c) hyperbola with asymptotes $y = \pm 2x$ and vertex $(3, 0)$.

3. Identify the curve.
   (a) $x^2 + 3y^2 = 6$ (b) $x^2 - 3y^2 = 6$
   (c) $2x^2 + 3y = 6$ (d) $x^2 + 3y^2 - 8x + 6y = 6$
   (e) $x^2 - 3y^2 + 8x - 6y = 6$

4. Sketch.
   (a) $16x^2 + y^2 - 96x + 4y + 132 = 0$
   (b) $9x^2 - 4y^2 - 72x - 8y + 176 = 0$

5. (a) Give both sets of formulas for rotation of axes through the angle $\alpha = \pi/3$.
   (b) Find the equation into which $x^2 + \sqrt{3}\,xy + 2y^2 = 1$ is transformed by this rotation.
   (c) Graph the equation in part (b).
   (d) Find the x, y equation of the major axis and the x, y coordinates of the foci of the graph in part (c).

6. Identify the curve by means of the discriminant.
   (a) $x^2 - xy + y^2 = 3$ (b) $x^2 + \frac{1}{2}xy - y^2 = 3$
   (c) $x^2 - 4xy + 4y^2 = 3x + 4y$ (d) $5x^2 - 2\sqrt{2}\,xy = 36$

Answers to selected odd-numbered problems

ANSWERS TO EXERCISES 1-1

1. $\dfrac{3}{4}$　　**3.** 14　　**5.** 1584　　**7.** .001　　**9.** .000008　　**11.** .81　　**13.** .6561

15. $6a^2 - ab - b^2$　　**17.** $\dfrac{1}{2}u^2 + \dfrac{8}{3}uv - 2v^2$　　**19.** $1 - 6x + 11x^2 - 6x^3$

21. $2 - x^2 - x^4$　　**23.** $1 - 2x^2$　　**25.** $(x+4)(x-4)$　　**27.** $(u-v)(u-2v)$

29. $(u-v)(u+2v)$　　**31.** $(3x+1)(3x+2)$　　**33.** $(3x-4)(2x-5)$　　**35.** $(x+2)\left(x - \dfrac{1}{2}\right)$

37. $2(2t-5)\left(\dfrac{1}{3}t + 1\right)$　　**39.** $(t+3u)(t^2 - 3tu + 9u^2)$　　**41.** $\dfrac{1}{8}\left(x - \dfrac{y}{2}\right)\left(x^2 + \dfrac{xy}{2} + \dfrac{y^2}{4}\right)$

43. (a)

n	$\sqrt{n}$
4	2.000
5	2.236
6	2.449
8	2.828
9	3.000

45. $6\sqrt{2}$　　**47.** $2\sqrt{5}$　　**49.** $9\sqrt{3}$　　**51.** $4\sqrt{3}$　　**53.** $9 - 4\sqrt{2}$　　**55.** $8 + 5\sqrt{2}$

57. $\dfrac{7}{5}\sqrt{5}$　　**59.** $\dfrac{2\sqrt{5}}{5}$　　**61.** $\dfrac{3\sqrt{5}}{10}$　　**63.** $\dfrac{3\sqrt{5}}{5}$　　**65.** $\dfrac{6\sqrt{5}}{5}$　　**67.** $\dfrac{\sqrt{5}}{10}$

69. $3\sqrt[3]{3}$　　**71.** $-.5$　　**73.** $\dfrac{1}{3}\sqrt[3]{2}$　　**75.** $-\dfrac{2}{5}$　　**77.** $4\sqrt[3]{3}$　　**79.** equivalent

81. b implies a　　**83.** equivalent

ANSWERS TO EXERCISES 1-2

1. $\dfrac{3}{4}$　　**3.** 36　　**5.** $\dfrac{41}{9}$　　**7.** $80\dfrac{80}{81}$　　**9.** 1.64　　**11.** $39\dfrac{1}{16}$　　**13.** 25　　**15.** $a^2 - 1$

17. $\dfrac{a^3}{64}$ **19.** $2(a^2 + b^2)$ **21.** \$390   **23.** 9 years; 14 years; no   **25.** 6%   **27.** 81,000

29. (a) \$4550 (b) \$3185 (c) \$2229.50 (d) no   **31.** (a) 780, 3905, 19530 (b) 7 weeks

33. 6.65×10^{19} kilometers

ANSWERS TO EXERCISES 1-3

1. $a^2 + 4ab + 4b^2$ **3.** $x^3 + 3x + \dfrac{3}{x} + \dfrac{1}{x^3}$ **5.** $81t^4 + 216t^2 + 216 + \dfrac{96}{t^2} + \dfrac{16}{t^4}$

7. $t\sqrt{t} - 6t + 12\sqrt{t} - 8$ **9.** $\dfrac{x^3 - 6x^2y + 12xy^2 - 8y^3}{27}$

11. $x^2 - 2xy + y^2 + 4x - 4y + 4$ **13.** $9t^2 + 12st + 4s^2 - 6t - 4s + 1$

15. $x^{-4} - 4x^{-3} + 6x^{-2} - 4x^{-1} + 1$ **17.** $\dfrac{7}{4} - \sqrt{3}$ **19.** $37 - 30\sqrt{3}$

21. $17 + 12\sqrt{2}$ **23.** $5643 + 3258\sqrt{3}$

25. $(u + v)^7 = u^7 + 7u^6v + 21u^5v^2 + 35u^4v^3 + 35u^3v^4 + 21u^2v^5 + 7uv^6 + v^7$

$(u + v)^8 = u^8 + 8u^7v + 28u^6v^2 + 56u^5v^3 + 70u^4v^4 + 56u^3v^5 + 28u^2v^6 + 8uv^7 + v^8$

29. $(x - 2y)(x^3 + 2x^2y + 4xy^2 + 8y^3)$ **31.** $(\sqrt{3} - x)((\sqrt{3})^3 + 3x + \sqrt{3}x^2 + x^3)$

35. $\dfrac{1023}{512}$ **37.** $\dfrac{2059}{729}$

ANSWERS TO EXERCISES 1-4

5. -6 **7.** $8\dfrac{2}{3}$ **9.** $\dfrac{-1 + \sqrt{2}}{2}$ **11.** $5 > 1$ **13.** $-3 > -4$ **15.** $\sqrt{2} - 2 > -\dfrac{3}{5}$

17. $4x - 1 < 4x + 5$ **19.** not possible   **21.** not possible

23. (a) 3 (b) 4 (c) $\dfrac{3}{2}$ (d) $-\dfrac{9}{2}$ (e) $\dfrac{11}{15}$ (f) $\dfrac{7}{48}$

ANSWERS TO EXERCISES 1-5

1. $x = 2$ **3.** $x = \dfrac{3}{2}$ **5.** $x = \dfrac{3}{\sqrt{3} - 2}$ **7.** $x = \dfrac{1}{8}$ **9.** $y = \dfrac{5}{2}$

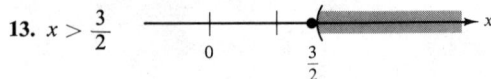

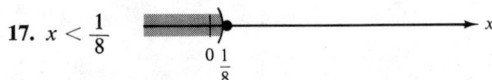

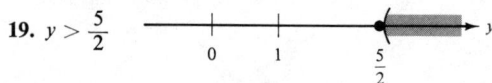

11. $x < 2$

13. $x > \dfrac{3}{2}$

15. $x < 1$

17. $x < \dfrac{1}{8}$

19. $y > \dfrac{5}{2}$

21. no solution   **23.** $x < 0$ **25.** Train cannot arrive on time.

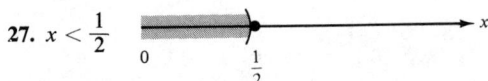

27. $x < \dfrac{1}{2}$

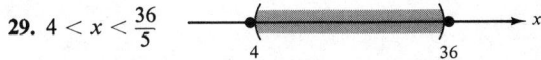

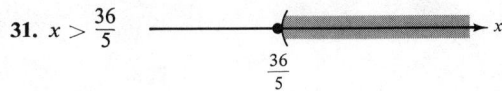

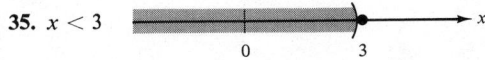

29. $4 < x < \dfrac{36}{5}$

31. $x > \dfrac{36}{5}$

33. $x < -5$ or $x > 5$

35. $x < 3$

39. (a) $19.9° < C < 20.1°$ (b) $67.82° < F < 68.18°$

ANSWERS TO EXERCISES 1-6

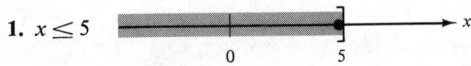

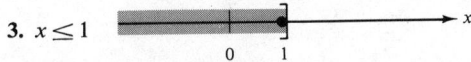

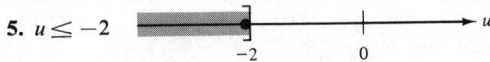

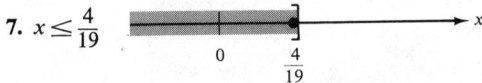

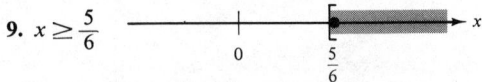

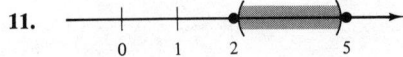

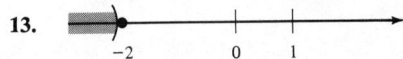

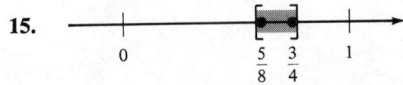

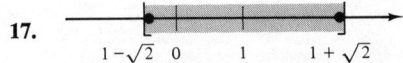

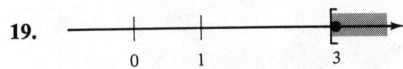

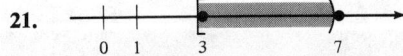

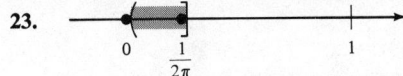

1. $x \le 5$

3. $x \le 1$

5. $u \le -2$

7. $x \le \dfrac{4}{19}$

9. $x \ge \dfrac{5}{6}$

11.

13.

15.

17.

19.

21.

23.

25.

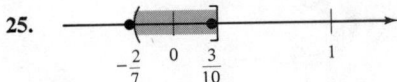

$$-\frac{2}{7} \quad 0 \quad \frac{3}{10} \qquad 1$$

27. $(-1, 0)$ **29.** $[0, \pi]$ **31.** $[1 - \sqrt{2}, 1 + \sqrt{2}]$ **33.** empty   **35.** $(-\infty, 5]$

37. $\left[\frac{13}{9}, +\infty\right)$ **39.** $\left[\frac{\sqrt{5}}{2}, +\infty\right)$ **41.** $(-\infty, -4)$ **43.** $\left(-\infty, \frac{\sqrt{3}-1}{3}\right)$ **45.** $(0, 4)$

47. $(3, 9)$ **49.** $(-\infty, 6]$ **51.** $(-\infty, \infty)$ **53.** $(-\infty, -1] \cup (3, +\infty)$ **55.** 12, 7, 7, 7

57. 4, 0, 4, 0   **59.** $\frac{-2 + 2\sqrt{2}}{3}, \frac{5}{3} - \sqrt{2}, \sqrt{2} - \frac{1}{3}, 2\sqrt{2} - \frac{23}{9}$ **61.** $(-2, 2)$

63. $\left(\frac{1}{2}, \frac{5}{2}\right)$ **65.** $\left[-\frac{3}{4}, \frac{7}{4}\right]$ **67.** $(-\infty, 0] \cup [2, +\infty)$

69. $\left(-\infty, \frac{3 - \sqrt{32}}{4}\right) \cup \left(\frac{3 + \sqrt{32}}{4}, +\infty\right)$ **71.** no solutions

ANSWERS TO EXERCISES 1-7

1. $x = -5$ or $x = 3$ **3.** $y = \frac{5}{3}$ or $y = \frac{1}{2}$ **5.** $x = -6$ or $x = 4$

7. $x = -3, -2, 2,$ or 3 **9.** $x = -2$ or $x = 4$ **11.** $x = -\frac{4}{5}$ or $x = \frac{1}{6}$

13. $x = \pm 1$ or ± 2 **15.** $t = 0, -\frac{5}{2},$ or 1

17. $(-\infty, -2) \cup (1, +\infty)$

19. $(-3, 1)$

21. $\left[-\frac{1}{3}, 0\right]$

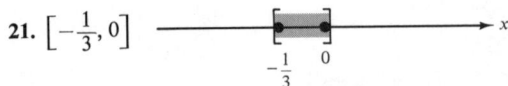

23. $(-\infty, +\infty)$ all real numbers

25. $\left[-\frac{2}{7}, \frac{3}{2}\right]$

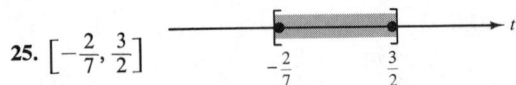

27. $(-\infty, 2] \cup [3, +\infty)$

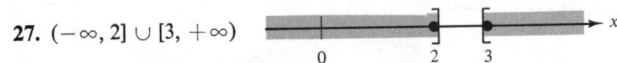

29. $[-6, 4]$

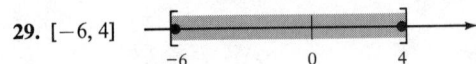

33. $(-3, 3)$ **35.** $(-4, 0)$ **37.** $(-\infty, -1 - \sqrt{5}) \cup (-1 + \sqrt{5}, +\infty)$

39. $[-3, 0] \cup [3, +\infty)$ **41.** $\left(-\infty, \frac{-\sqrt{3}}{2}\right) \cup \left(-\frac{1}{2}, \frac{1}{2}\right) \cup \left(\frac{-\sqrt{3}}{2}, +\infty\right)$

43. $(-\infty, -2] \cup [2, +\infty)$ **45.** $\left(\frac{1}{2}, +\infty\right)$ **47.** $[-1, 1]$ **49.** $\left(-\infty, -1\right) \cup (-\frac{2}{3} +\infty)$

51. $[0, +\infty)$

1. $a^2 - a - 6$ **3.** $2a^2 - 5ab - 12b^2$ **5.** $\frac{1}{2}x^2 - \frac{37}{12}x - 5$ **7.** $t^3 + 2t^2 - 9t - 18$

9. $a^6 - b^6$ **11.** $9y^2 + 12y + 4$ **13.** $125x^3 + 150x^2y + 60xy^2 + 8y^3$

15. $a^3 - 6a^2b^{-2} + 12ab^{-4} - 8b^{-6}$ **17.** $1 + 2x + \frac{3}{2}x^2 + \frac{1}{2}x^3 + \frac{x^4}{16}$

19. $8t^4 + 16t^3 + 8t^2$ **21.** $a^2x^2 + 2abxy + b^2y^2 + 2acx + 2bcy + c^2$

23. $t^8 - 18t^2 + 81t^{-4}$ **25.** $8p^2(q - 2p)$ **27.** $4(2x^{-1} + y^{-1})(2x^{-1} - y^{-1})$

29. $(v + 5u)(v + 2u)$ **31.** $(5x + 1)(2x + 1)$ **33.** $(2a - 3b)(4a^2 + 6ab + 9b^2)$

35. $(a - 2b)(a^3 + 2a^2b + 4ab^2 + 8b^3)$ **37.** 4680000000 **39.** 1.09512×10^{-1}

41. 2×10^{-2}

43. $(-\infty, -3)$
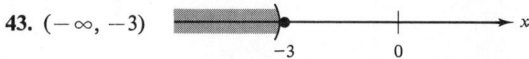

45. $(-\infty, -2) \cup (-1, +\infty)$

47. $(-3, 7)$
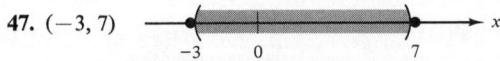

49. $(-\infty, 6]$

51. $\left[-\frac{1}{3}, \frac{1}{2}\right]$
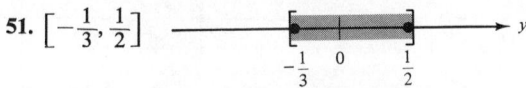

53. $[-1, 7]$
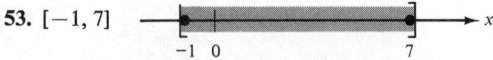

55. $\left(\frac{7}{3}, \frac{9}{2}\right)$
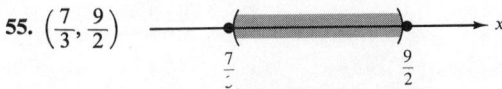

57. $(-10, -6]$
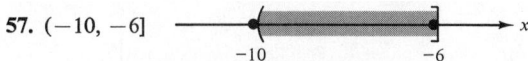

59. open
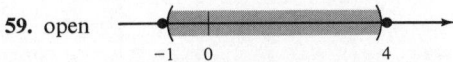

61. open
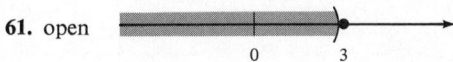

63. neither open nor closed
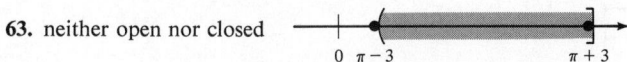

65. false   **67.** false   **69.** true   **71.** $45 <$ speed ≤ 55

ANSWERS TO EXERCISES 2-1

3.

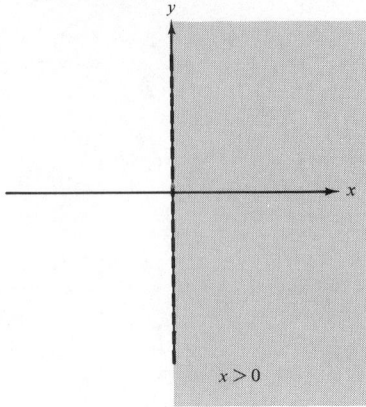

$x > 0$

5.

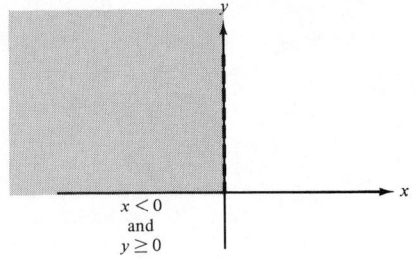

$x < 0$
and
$y \geq 0$

7.

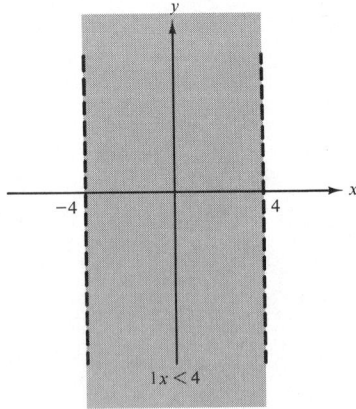

$1x < 4$

9.

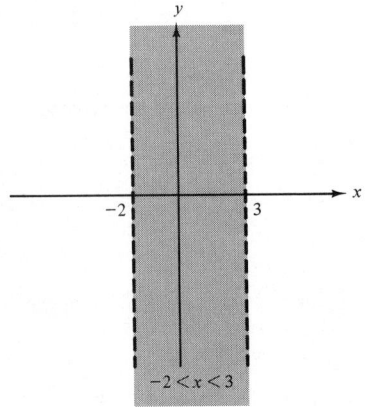

$-2 < x < 3$

11.

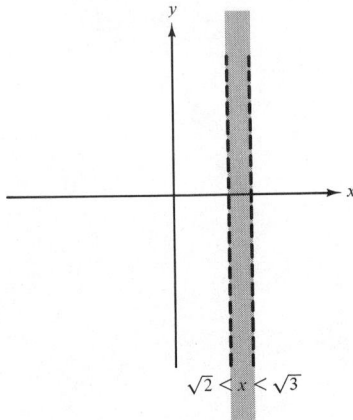

$\sqrt{2} < x < \sqrt{3}$

13.

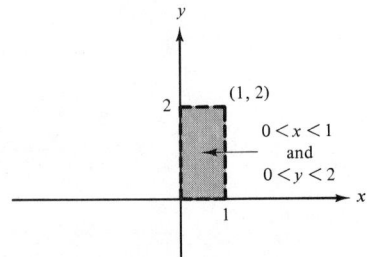

$(1, 2)$

$0 < x < 1$
and
$0 < y < 2$

15. $\dfrac{5}{2}$ **17.** -2 **19.** $\dfrac{11}{42}$ **21.** $-2\sqrt{3}$ **23.** $-\dfrac{1 - \sqrt{2}}{1 + \sqrt{2}}$ **25.** $\dfrac{3\sqrt{3} - 2\sqrt{5}}{\sqrt{3} - \sqrt{5}}$

27. (a) -1 (b) $2a + b$ (c) $-\dfrac{1}{a(a + b)}$ **29.** 6   **31.** 36   **33.** $\dfrac{\sqrt{7}}{8}$

35. (c)

t	T	Δt	ΔT	$\Delta T/\Delta t$
0	-29			
		.5	-3	-6
.5	-32			
		.5	1	2
1.0	-31			
		.5	0	0
1.5	-31			
		.5	2	4
2.0	-29			
		.5	3	6
2.5	-26			

(d) slope represents average rate of change of temperature with respect to time over each half hour interval.

(e) $-30.5°$; $-31°$; $-27°$. (f) $5{:}40$ $7{:}15$

ANSWERS TO EXERCISES 2-2

1. $\sqrt{5}$ **3.** 5 **5.** $\sqrt{22}$ **7.** $2\sqrt{5}$ **9.** $-2, 0$

11. $(x - 0)^2 + (y - 0)^2 = 9$; $x^2 + y^2 - 9 = 0$

13. $(x - 2)^2 + (y + 3)^2 = 16$; $x^2 + y^2 - 4x + 6y - 3 = 0$

15. $\left(x - \dfrac{1}{2}\right)^2 + \left(y - \dfrac{1}{3}\right)^2 = \dfrac{25}{36}$; $x^2 + y^2 - x - \dfrac{2y}{3} - \dfrac{1}{3} = 0$ **17.** $x = 2$ or -22

19. $x = \dfrac{2}{3}$ or $-\dfrac{5}{2}$ **21.** $x = \dfrac{5}{2} + \dfrac{\sqrt{17}}{2}$ or $\dfrac{5}{2} - \dfrac{\sqrt{17}}{2}$ **23.** $x = \dfrac{1}{2}$ or $-\dfrac{5}{2}$

25. $x = -\sqrt{2} + 3$ or $-2 - \sqrt{3}$

27. center $(3, 2)$ radius 2

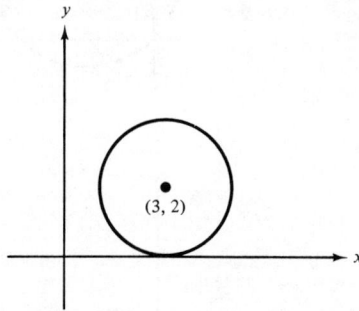

29. center $(-3, -2)$ radius 5

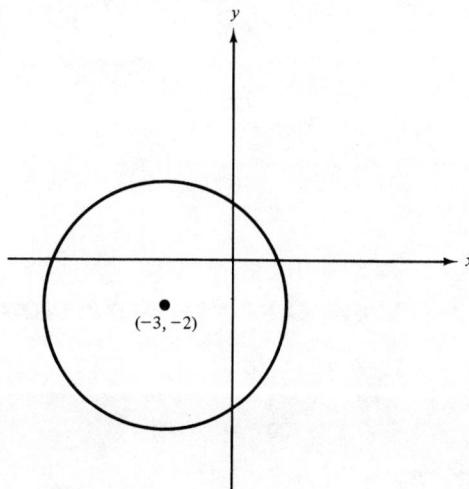

31. center $(1, -1)$ radius $2\sqrt{3}$

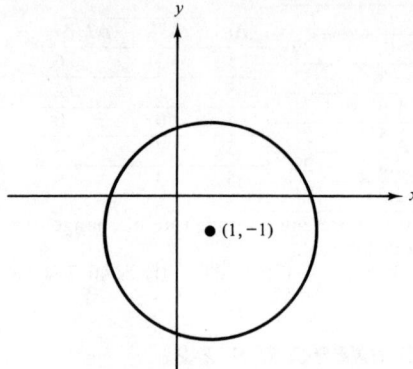

$\bullet\, (1, -1)$

33. center $\left(\dfrac{1}{2}, \dfrac{1}{2}\right)$ radius $\dfrac{\sqrt{2}}{2}$

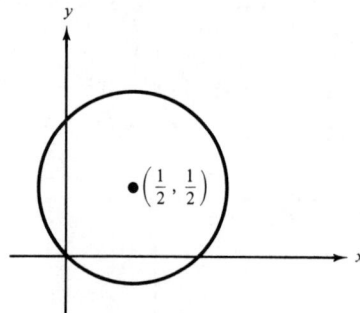

$\bullet\left(\dfrac{1}{2}, \dfrac{1}{2}\right)$

35. center $(\sqrt{2}, \sqrt{2})$ radius $\sqrt{2}$

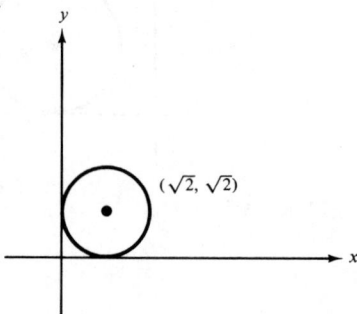

$(\sqrt{2}, \sqrt{2})$

37. outside   **39.** outside   **41.** on   **43.** inside

ANSWERS TO REVIEW EXERCISES FOR CHAPTER TWO

1. (b) $0, 5, 3, -14$ **3.** $\sqrt{5}$ **5.** $\sqrt{10}$ **7.** $x^2 + y^2 - 9 = 0$ **9.** $x^2 + y^2 - 8x + 6y = 0$

11. $x^2 + y^2 + 4x - 2y - 20 = 0$ **13.** $C(3, 2); r = 3$ **15.** $C\left(\dfrac{1}{3}, -\dfrac{2}{3}\right); r = 1$

17. $C(-\sqrt{2}, \sqrt{3}); r = \sqrt{5}$ **19.** on

35. (c)

t	T	Δt	ΔT	$\Delta T/\Delta t$
0	-29			
		.5	-3	-6
.5	-32			
		.5	1	2
1.0	-31			
		.5	0	0
1.5	-31			
		.5	2	4
2.0	-29			
		.5	3	6
2.5	-26			

(d) slope represents average rate of change of temperature with respect to time over each half hour interval.

(e) $-30.5°$; $-31°$; $-27°$. (f) 5:40 7:15

ANSWERS TO EXERCISES 2-2

1. $\sqrt{5}$ **3.** 5   **5.** $\sqrt{22}$ **7.** $2\sqrt{5}$ **9.** $-2, 0$

11. $(x-0)^2 + (y-0)^2 = 9$; $x^2 + y^2 - 9 = 0$

13. $(x-2)^2 + (y+3)^2 = 16$; $x^2 + y^2 - 4x + 6y - 3 = 0$

15. $\left(x - \frac{1}{2}\right)^2 + \left(y - \frac{1}{3}\right)^2 = \frac{25}{36}$; $x^2 + y^2 - x - \frac{2y}{3} - \frac{1}{3} = 0$ **17.** $x = 2$ or -22

19. $x = \frac{2}{3}$ or $-\frac{5}{2}$ **21.** $x = \frac{5}{2} + \frac{\sqrt{17}}{2}$ or $\frac{5}{2} - \frac{\sqrt{17}}{2}$ **23.** $x = \frac{1}{2}$ or $-\frac{5}{2}$

25. $x = -\sqrt{2} + 3$ or $-2 - \sqrt{3}$

27. center (3, 2) radius 2

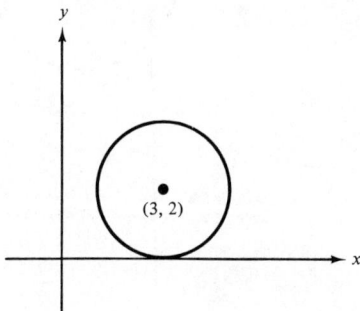

29. center $(-3, -2)$ radius 5

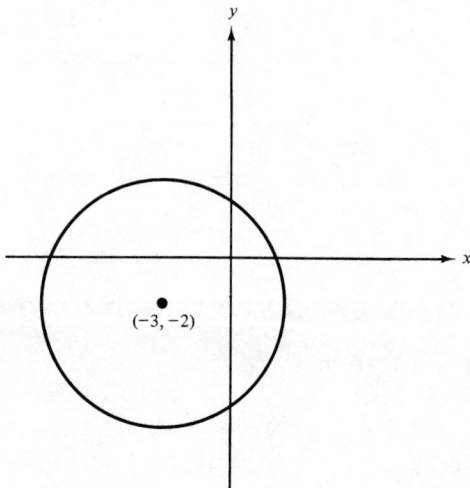

31. center $(1, -1)$ radius $2\sqrt{3}$

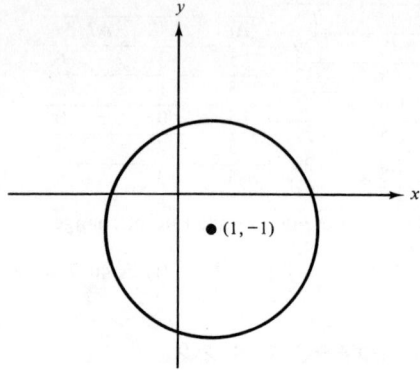

33. center $\left(\frac{1}{2}, \frac{1}{2}\right)$ radius $\frac{\sqrt{2}}{2}$

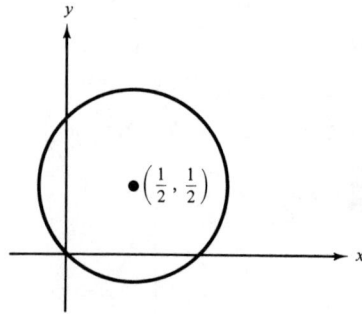

35. center $(\sqrt{2}, \sqrt{2})$ radius $\sqrt{2}$

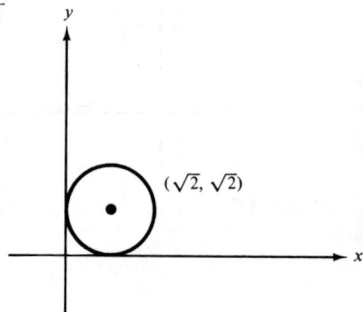

37. outside   **39.** outside   **41.** on   **43.** inside

ANSWERS TO REVIEW EXERCISES FOR CHAPTER TWO

1. (b) $0, 5, 3, -14$ **3.** $\sqrt{5}$ **5.** $\sqrt{10}$ **7.** $x^2 + y^2 - 9 = 0$ **9.** $x^2 + y^2 - 8x + 6y = 0$

11. $x^2 + y^2 + 4x - 2y - 20 = 0$ **13.** $C(3, 2); r = 3$ **15.** $C\left(\frac{1}{3}, -\frac{2}{3}\right); r = 1$

17. $C(-\sqrt{2}, \sqrt{3}); r = \sqrt{5}$ **19.** on

1. 3   **3.** $\frac{7}{18}$ **5.** $\sqrt{5}$ or $-\sqrt{5}$ **7.** no value   **9.**

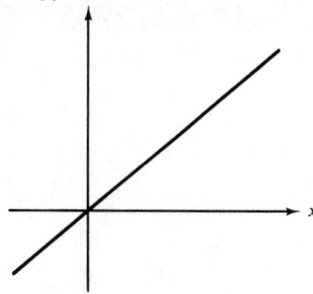

11.

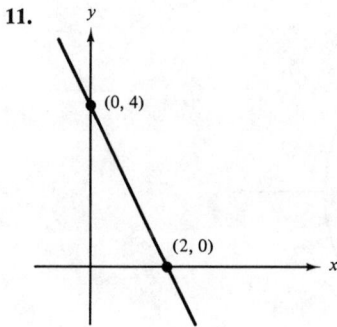

(0, 4)

(2, 0)

13. symmetric w.r.t. y axis; $y \geq 0$

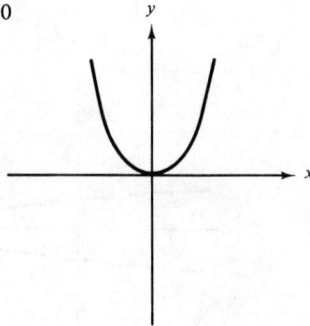

15. symmetric w.r.t. x axis; $x \geq 0$

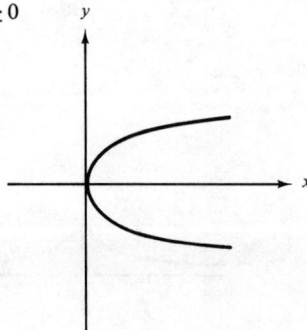

17. symmetric w.r.t. origin; $x \neq 0$ and $y \neq 0$

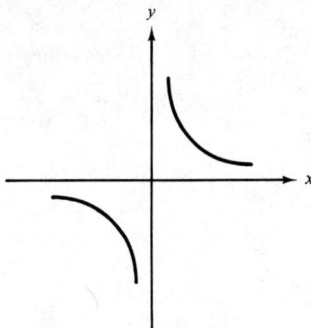

19. symmetry w.r.t. origin, x axis, and y axis $-2 \leq x \leq 2$, $-6 \leq y \leq 6$.

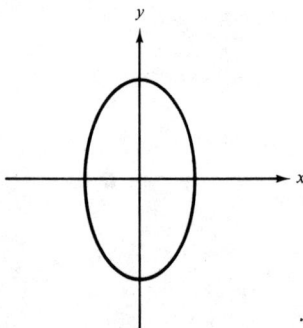

21. symmetry w.r.t. x axis; $x \leq 8$

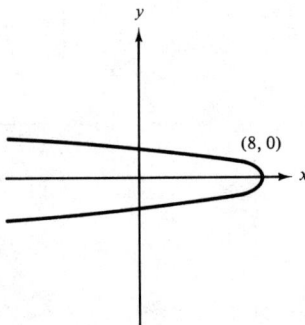

(8, 0)

23. symmetry w.r.t. y axis $0 < y \leq 1$

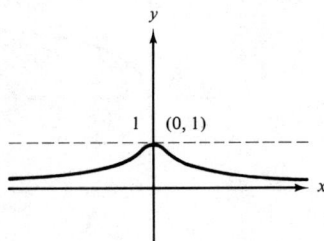

1 | (0, 1)

25. -1 and 1

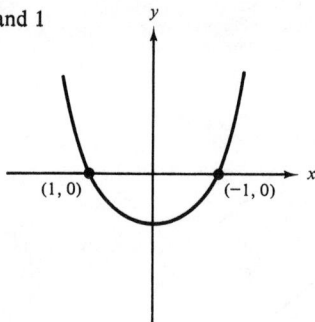

27. $\dfrac{10}{7}$

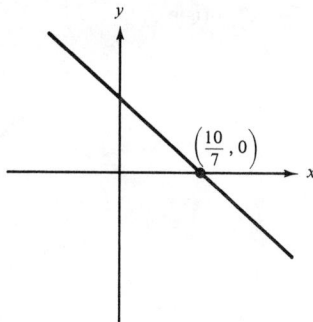

29. $-\dfrac{5\sqrt{2}}{2}$ and $\dfrac{5\sqrt{2}}{2}$

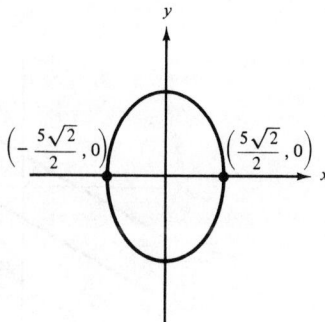

35. $y > 0$ for $x > 0$ or $x < -3$
$y < 0$ for $-3 < x < 0$

37. $y > 0$ for $x > 0$
$y < 0$ for $x < 0$

39. x axis y axis, origin; $-1 \leq x \leq 1$, $-1 \leq y \leq 1$

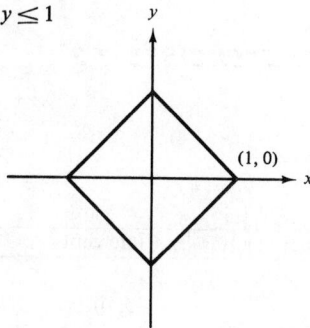

41. x axis, y axis, origin; $x \geq 1$ or $x \leq -1$

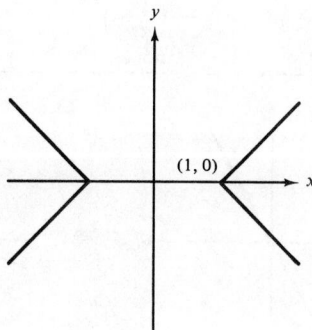

43. origin

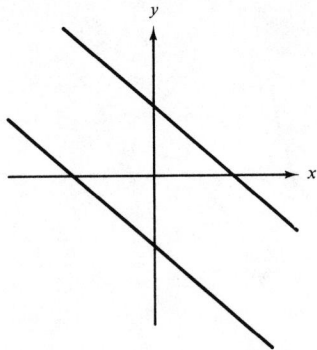

45. origin

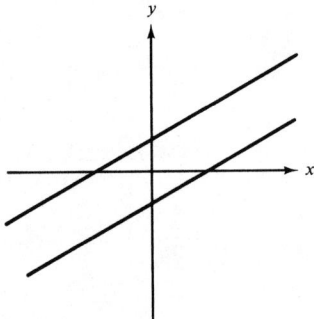

47. replace y by $4 - y$

ANSWERS TO EXERCISES 3-2

1. $y - 1 = -1(x - 4)$ **3.** $y - 1 = \frac{1}{3}(x + 2)$ **5.** $y = -(3 + 2\sqrt{2})x + 4 + 2\sqrt{2}$

7. $x = \frac{4}{7}$ **9.** $y = c$

	slope	x intercept	y intercept
11.	-1	$(5, 0)$	$(0, 5)$
13.	$\frac{1}{3}$	$(-5, 0)$	$\left(0, \frac{5}{3}\right)$
15.	$-(3 + 2\sqrt{2})$	$(4 - 2\sqrt{2}, 0)$	$(0, 4 + 2\sqrt{2})$
17.	no slope	$\left(\frac{4}{7}, 0\right)$	no intercept
19.	0	none	$(0, c)$

21.

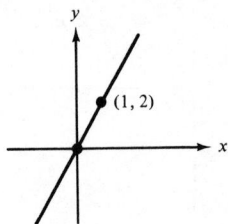

23.

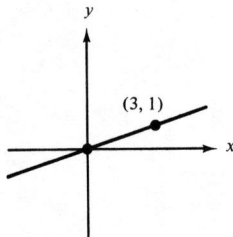

25.

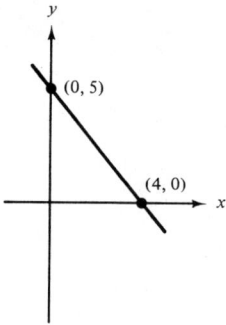

27.

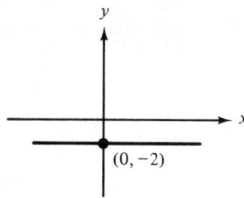

29.

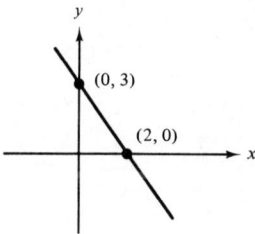

31.

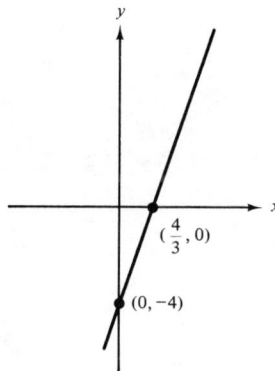

33. 14   **35.** $-\dfrac{1}{4}$ **37.** $\dfrac{5\sqrt{3}}{3} - 9$ **39.** 5

41. (b) $\dfrac{x}{-3} + \dfrac{y}{4} = 1$ (c) $\dfrac{x}{-13} + \dfrac{y}{13/3} = 1$ (c) $\dfrac{1}{2}|ab|$

ANSWERS TO EXERCISES 3-3

1. (1, 1) and (2, 4)

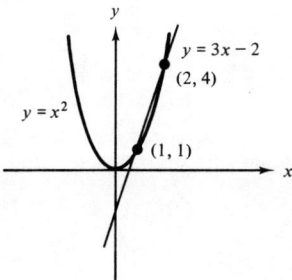

3. (5, -3) and (0, 2)

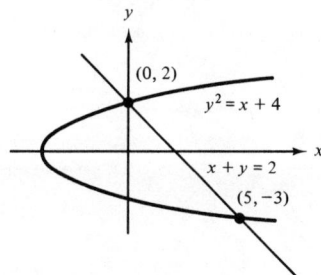

5. $\left(-\dfrac{3\sqrt{17}}{17}, -\dfrac{2\sqrt{17}}{17}\right)$ and $\left(\dfrac{3\sqrt{17}}{17}, \dfrac{2\sqrt{17}}{17}\right)$

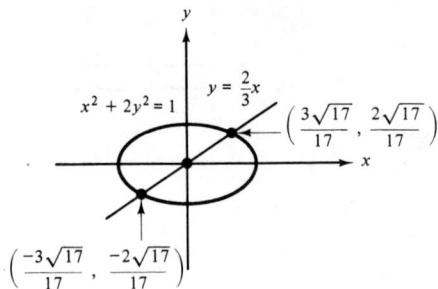

7. $(5, 1)$

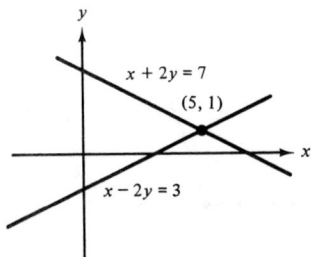

9. $\left(\dfrac{15}{4}, -\dfrac{1}{2}\right)$

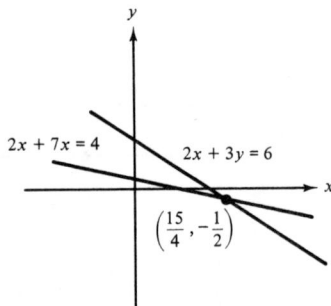

11. $(2, 4)$

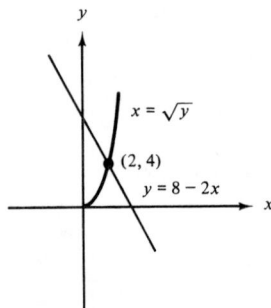

13. $(1, 0)$

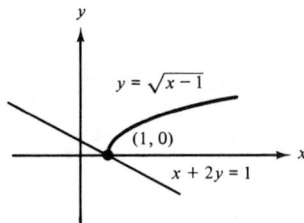

15. $(2, 1)$ and $\left(\dfrac{25}{18}, -\dfrac{5}{6}\right)$

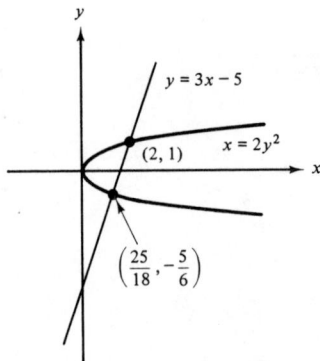

17. $(\pm 5, 0)$ and $(0, -25)$

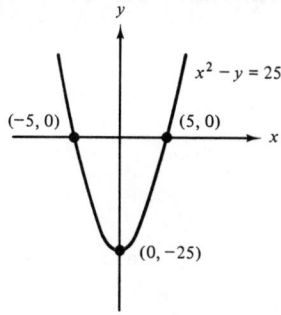

19. $(0, 0)$

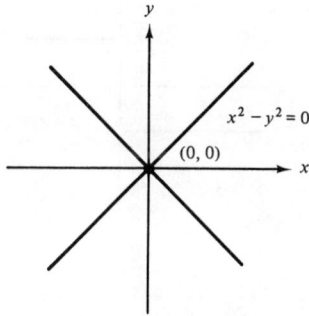

21. $(0, 0), (-\sqrt{15}, 0)$ and $(\sqrt{15}, 0)$

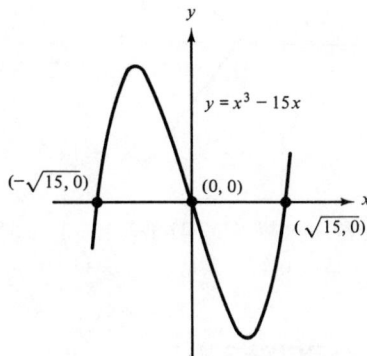

ANSWERS TO EXERCISES 3-4

1. $5x - 3y = -11$ **3.** $y = 4$ **5.** $x = 2$ **7.** $2x - 5y = 0$ **9.** $4x - 5y = -1$
11. $\dfrac{x}{3} - \dfrac{y}{2} = 0$ **13.** $x + 3y = 2$
21. $y - 3 = \dfrac{3}{8}(x - 1)$ **23.** $y + 2 = \dfrac{1}{13}(x + 2)$ **25.** $y - \dfrac{1}{2} = -\left(x + \dfrac{3}{2}\right)$
27. $x = -4$ **29.** (a) $x + y = 5$ (b) $y = 3x$ (c) $x + y = 4$ (d) $y = x$

ANSWERS TO REVIEW EXERCISES FOR CHAPTER THREE

	Symmetry with Respect to		
	x axis	y axis	origin
1.	no	yes	no
3.	yes	yes	yes
5.	no	no	no
7.	no	no	yes
9.	no	no	yes
11.	no	no	yes

13. $x \geq 1$; no restriction on y **15.** $x \geq 4$; no restriction on y

17. $-2 \leq x \leq 2$; $-3 \leq y \leq 3$

19.

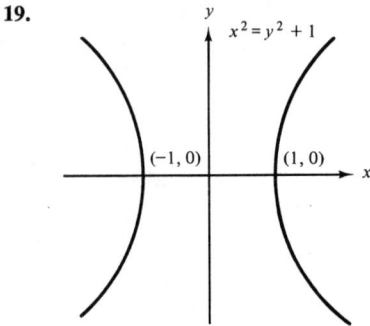

$x^2 = y^2 + 1$

$(-1, 0)$ $(1, 0)$

21.

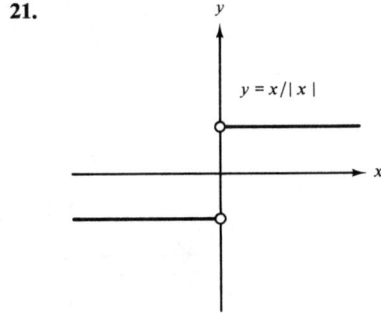

$y = x/|x|$

23.

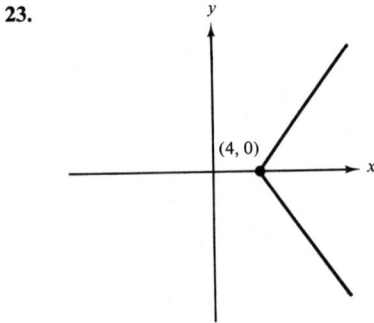

$(4, 0)$

25. $(-1, 0), (1, 2)$ **27.** $(2, -1), \left(\frac{12}{5}, -\frac{9}{5}\right)$ **29.** perpendicular   **31.** perpendicular

33. $4x + 3y - 25 = 0$ **35.** $x - 2y - 4 = 0$

ANSWERS TO EXERCISES 4-1

1. $C = 10.5x$ dollars; $C = \$31.50$ **3.** $1\frac{1}{4}$ seconds   **5.** $t \leq 17.5$ seconds

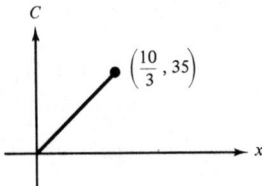

C

$\left(\frac{10}{3}, 35\right)$

7. (a) $86°$ (b) $23.9°$ (c) $-40°$ **9.** $I = \dfrac{2000}{d^2}$

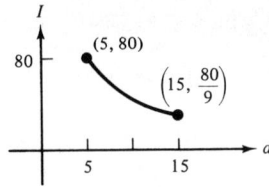

ANSWERS TO EXERCISES 4-2

1. (a) function; $y = 2x + 3$; $D = [-2, 1]$; $R = [-1, 5]$
  (b) function; $y = -\dfrac{1}{2}x + \dfrac{5}{2}$; $D = [0, 2]$; $R = \left[\dfrac{3}{2}, \dfrac{5}{2}\right]$
  (c) function; $y = -\dfrac{1}{2}x + \dfrac{5}{2}$; $D =$ all real numbers; $R =$ all real numbers
  (d) function not determined
  (e) function; $y = \sqrt{16 - x^2}$; $D = [0, 4]$; $R = [0, 4]$
  (f) function; $y = \dfrac{2}{x}$; $D = \{x \mid x \neq 0\}$; $R = \{y \mid y \neq 0\}$
  (g) function; $y = 16(x + 3)^2$; $D = [0, +\infty)$; $R = [144, +\infty)$
3. (a) $4, 9, 25, 36, \dfrac{4}{9}, a^4 + 2a^2 + 1, a^2 + 1$ (b) no (c) yes (d) yes
5. $\dfrac{1}{2}, \dfrac{1}{3}, \dfrac{1}{5}, \dfrac{3}{2}, \dfrac{1}{a^2 + 1}$ (b) no (c) yes (d) x **7.** all real numbers except 2
9. all real numbers except $1, -1$ **11.** all real numbers except $2, -2$
13. all real numbers except $-2, -4$ **15.** $\left[\dfrac{3}{4}, +\infty\right)$ **17.** $(-\infty, -2] \cup [2, +\infty)$
19. $(-\infty, 0] \cup [1, +\infty)$ **21.** all real numbers   **23.** $(-\infty, -5) \cup (0, +\infty)$
25. $(-\infty, -1) \cup (4, +\infty)$

ANSWERS TO EXERCISES 4-3

1. $v = e^3$ for $e \geq 0$

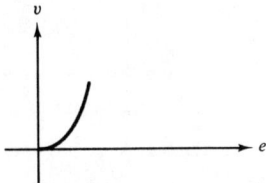

• **3.** (a) $S = 4\pi r^2$ for $r \geq 0$ (b) yes, $r = \sqrt{\dfrac{S}{4\pi}}$ for $S \geq 0$.

5. $A = w(200 - w)$ for $0 \leq w \leq 200$

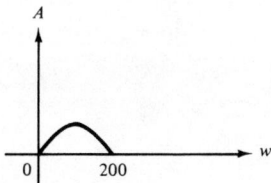

7. $S = \dfrac{11}{3}x^2$ for $2 \leq x \leq 4$

9. $V = \dfrac{3\pi h^3}{64}$ for $0 \le h \le 8$

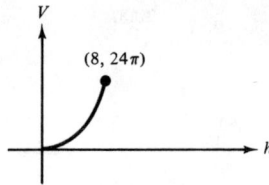

11. $S = \dfrac{50}{h} + 10\sqrt{\pi h}$ for $1 \le h \le 5$

13. $h = \sqrt{400 - x^2}$ for $0 \le x \le 20$

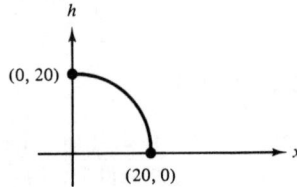

15. (a) y is a function of x
    (b) x is a function of y
    (c) y is a function of x and x is a function of y
    (d) y is a function of x and x is a function of y
    (e) y is a function of x and x is a function of y
    (f) x is a function of y

ANSWERS TO EXERCISES 4-4

1.

(a)

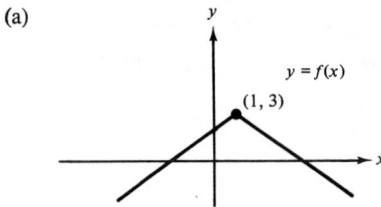

(b)

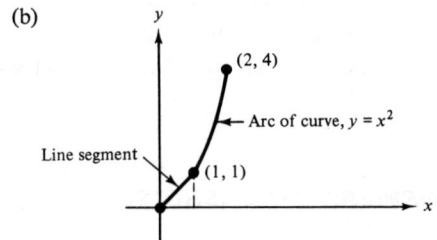

(c)

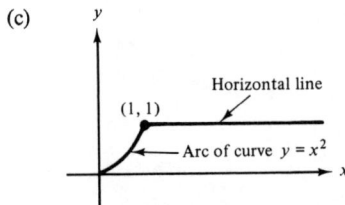

(d)

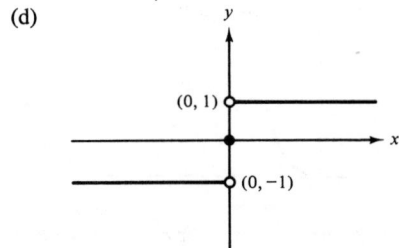

3.

$$T = \begin{cases} \$1.00 \text{ for } 0 < m \le \dfrac{1}{4} \\[4pt] 1.50 \text{ for } \dfrac{1}{4} < m \le \dfrac{1}{2} \\[4pt] 2.00 \text{ for } \dfrac{1}{2} < m \le \dfrac{3}{4} \\[4pt] 2.50 \text{ for } \dfrac{3}{4} < m \le 1 \\[4pt] 3.00 \text{ for } 1 < m \le \dfrac{5}{4} \\[4pt] 3.50 \text{ for } 54 < m \le \dfrac{3}{2} \end{cases}$$

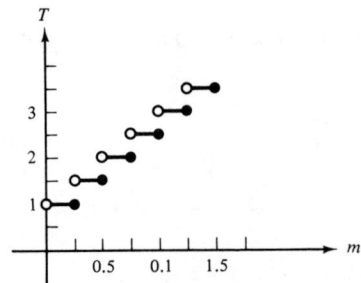

440

5. $W = f(d) = \begin{cases} \dfrac{w}{R} & 0 < d \le R \\ \dfrac{wR^2}{d^2} & d \ge R \end{cases}$

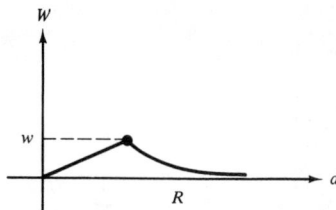

7.

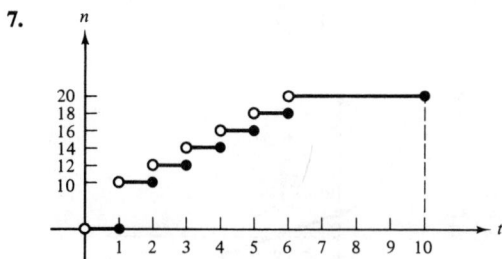

12 days; 20 days (the maximum has been earned after 7 years)

ANSWERS TO EXERCISES 4-5

1. (a) 6π (b) 4π (c) 5π **3.** $t = 20$ seconds; 80 feet per second; 560 feet per second
5. $4\pi(x_2 + x_1)$ **7.** (a) π (b) π (c) π **9.** (a) .05   (b) .06   (c) .07   (d) .062

ANSWERS TO EXERCISES 4-6

1. (a)

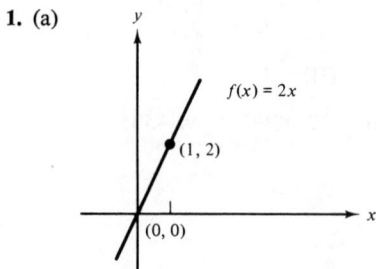

(b)

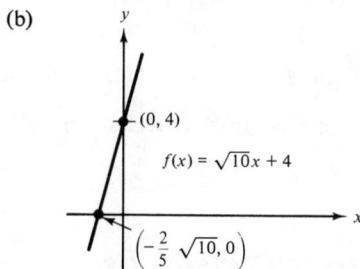

(c)

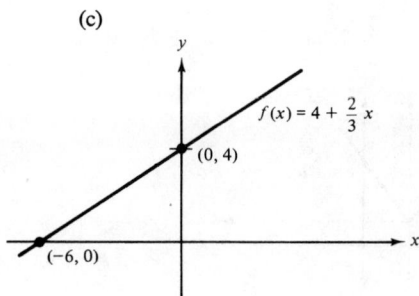

(d)

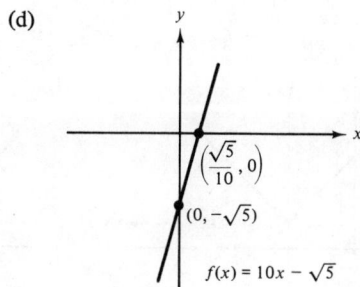

(e)

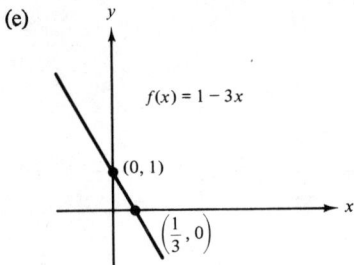

$f(x) = 1 - 3x$

$(0, 1)$

$\left(\frac{1}{3}, 0\right)$

(f)

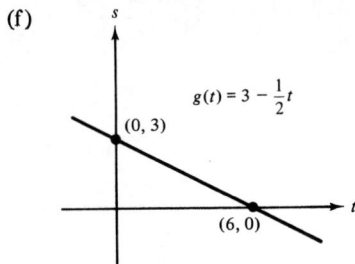

$g(t) = 3 - \frac{1}{2}t$

$(0, 3)$

$(6, 0)$

(g)

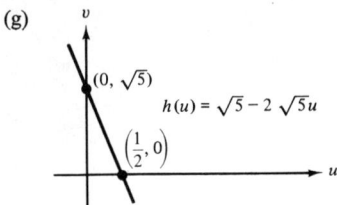

$(0, \sqrt{5})$

$h(u) = \sqrt{5} - 2\sqrt{5}u$

$\left(\frac{1}{2}, 0\right)$

(h)

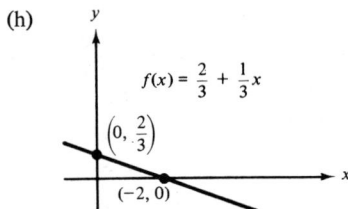

$f(x) = \frac{2}{3} + \frac{1}{3}x$

$\left(0, \frac{2}{3}\right)$

$(-2, 0)$

3. (a) $x = \frac{3}{4}y - 3$; $\frac{\Delta x}{\Delta y} = \frac{3}{4}$ (b) $x = -3y + 3$; $\frac{\Delta x}{\Delta y} = \frac{-3}{1}$

(c) $x = -\sqrt{\frac{3}{2}}y + \frac{1}{\sqrt{2}}$; $\frac{\Delta x}{\Delta y} = -\sqrt{\frac{3}{2}}$ (d) $y = \frac{3}{5}$ does not define unique value of x.

5. $v = 40 - 8t \ 0 \le t \le 5$ **7.** $3\frac{2}{3}$ centimeters   **9.** π

11. (a) .04   (b) 30 seconds after 5:25   (c) 4 hours, 10 minutes

13. $-\frac{16}{9}$; $\frac{17}{27}$ **15.** $-\frac{3}{8}$; $-\frac{9}{8}$

ANSWERS TO EXERCISES 4-7

1. (a) 0   (b) 8.7   (c) $-\frac{10}{7}$ **3.** (a) $f(1) \approx 4$ (b) $f(1) = 0$

5. (a) 3.5 meters   (b) $1\frac{3}{7}$ second   (c) $f(t) = 1.4t$ **7.** 8.397   **9.** 4.135

11. 16.804

ANSWERS TO EXERCISES 4-8

1. $u + v = 0$ **3.** $v = u + 1$ **5.** $u^2 + v^2 = 4$ **7.** $uv + 3u + 2v + 5 = 0$

9.

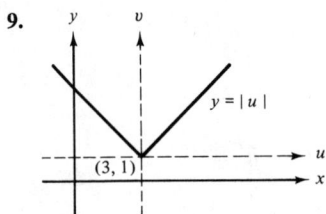

$y = |u|$

$(3, 1)$

11.

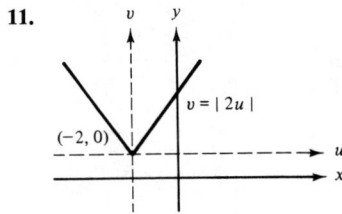

$v = |2u|$

$(-2, 0)$

13.

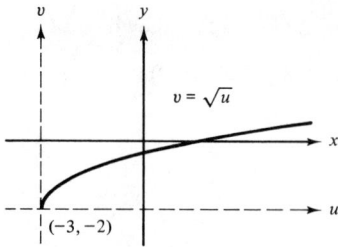

$v = \sqrt{u}$

$(-3, -2)$

15.

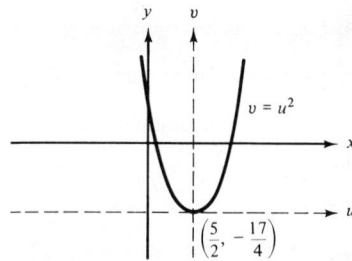

$v = u^2$

$\left(\frac{5}{2}, -\frac{17}{4}\right)$

17.

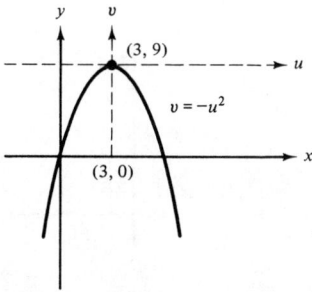

$(3, 9)$

$v = -u^2$

$(3, 0)$

19.

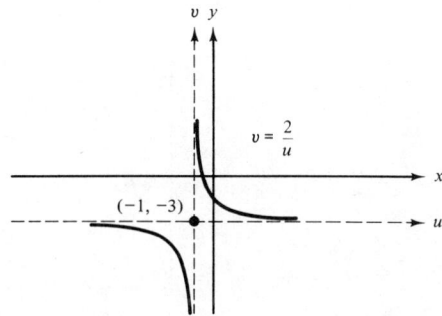

$v = \dfrac{2}{u}$

$(-1, -3)$

ANSWERS TO EXERCISES 4-9

1.

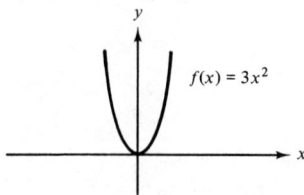

$f(x) = 3x^2$

3.

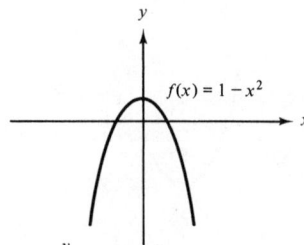

$f(x) = 1 - x^2$

5.

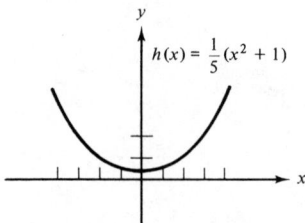

$h(x) = \frac{1}{5}(x^2 + 1)$

7.

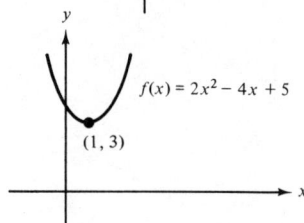

$f(x) = 2x^2 - 4x + 5$

$(1, 3)$

9.

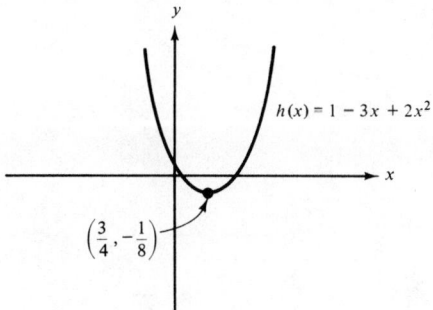

$h(x) = 1 - 3x + 2x^2$

$\left(\frac{3}{4}, -\frac{1}{8}\right)$

11.

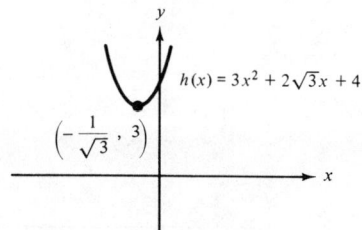

$h(x) = 3x^2 + 2\sqrt{3}x + 4$

$\left(-\dfrac{1}{\sqrt{3}}, 3\right)$

13. $(-3, 4)$ minimum   **15.** $\left(\frac{1}{2}, 2\right)$ minimum   **17.** $(0, 3)$ maximum

19. $\left(-\frac{5}{3}, -\frac{16}{3}\right)$ minimum   **21.** $\left(\frac{2}{5}, \frac{36}{5}\right)$ minimum   **23.** $(3, 9)$ maximum

25. $\left(-\frac{2}{3}, \frac{1}{3}\right)$ minimum

27. (a) $k = 160$ (b) 10 seconds   (c) $h = 160t - 16t^2$ for $0 < t < 10$ (d) 400 feet, at 5 seconds

ANSWERS TO EXERCISES 4-10

1. $-\frac{4}{3}$

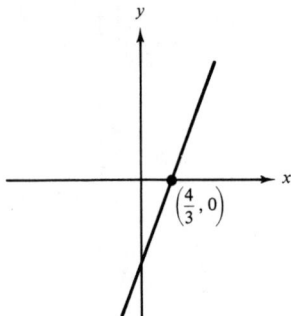

3. $-\frac{5}{2}$

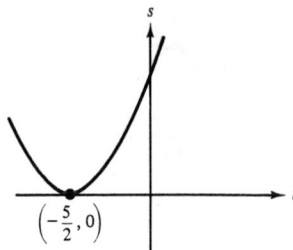

5. $-2, 2$

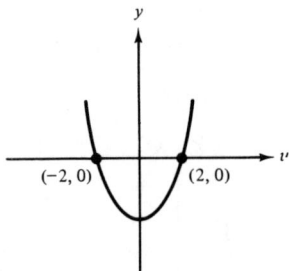

7. $1, 2$

9. $-\sqrt{5}, -1, 1, \sqrt{5}$

11. $-\frac{1}{2}, \frac{1}{3}$

13. no zeros

15. $-3, 3$

17. no zeros

19. no zeros

21. 0

23.

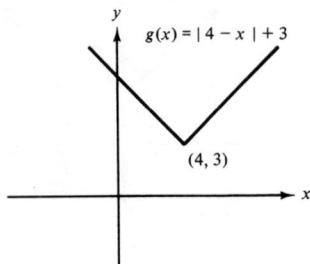

$g(x) = |4 - x| + 3$

$(4, 3)$

25. $-1, \frac{1}{2}$ **27.** no real zeros

29. $\frac{-5 - \sqrt{55}}{5}, \frac{-5 + \sqrt{55}}{5}$

31. no real zeros   **33.** no real zeros   **35.** no real zeros

37. $-1 - \sqrt{\frac{1}{2}}, 0, -1 + \sqrt{\frac{1}{2}}$

39.

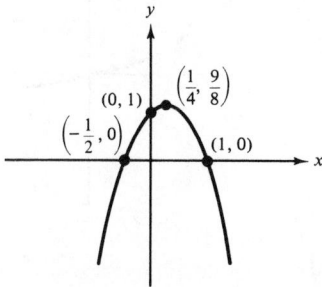

41.

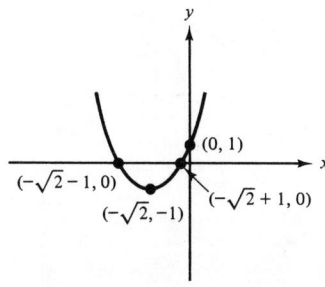

43. 6 seconds after it was thrown.   **45.** (a) $A < \frac{25}{12}$ (b) $A = \frac{25}{12}$ (c) $A > \frac{25}{12}$

ANSWERS TO EXERCISES 4-11

1. 100.75 feet × 198.5 feet   **3.** (a) 12.5, 12.5 (b) 12.5, 12.5 (c) 24.5, .5   **5.** 4

7. $V = 96x - 40x^2 + 4x^3$ for $0 \leq x \leq 4$; maximum volume 161.25 cubic inches, approximately

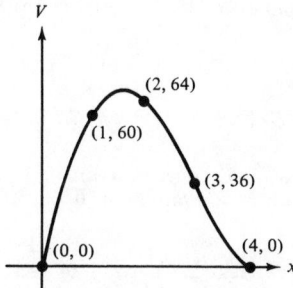

9. $C = 6r + \frac{12000}{r}$ for $40 \leq r \leq 55$; cost minimum when $r = 44.7$ miles per hour, approximately

ANSWERS TO REVIEW EXERCISES FOR CHAPTER FOUR

1. $v = f(s) = 8\sqrt{s}$ for $0 \leq s \leq 900$

3.

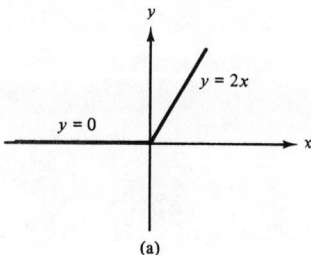

(a)

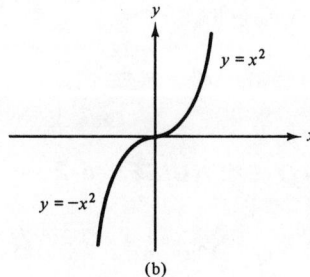

(b)

5. (a) a (b) $2\sqrt{2}$ (c) $2\sqrt{2}$ (d) $\dfrac{1}{a}\sqrt{1+4a^2}$ (e) $a + \dfrac{1}{a}$ (f) $-\left(a + \dfrac{1}{a}\right)$ (g) $\left|a + \dfrac{1}{a}\right|$

7. 43   **9.** (a) $V = f(T) = .6T + 332$ (b) 2429 meters (1.5 miles, approximately)

11.

$f(x) = 5 + |x - 1|$

$(1, 5)$

$g(x) = 2 - \sqrt{x+4}$

$(-4, 2)$

$h(x) = 4x^2 - 8x + 3$

$(1, -1)$

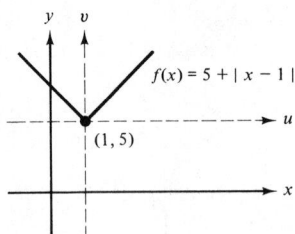

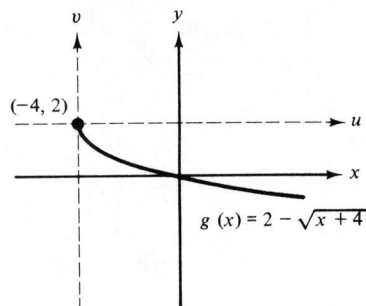

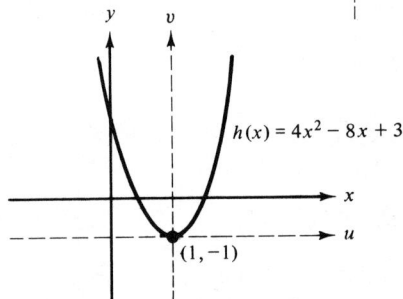

13. (a) $\dfrac{1}{2}, \dfrac{3}{2}$ (b) $0, 9$ (c) none   **15.** (a) 261 feet (b) 8.04 seconds, approximately

ANSWERS TO EXERCISES 5-1

1. degree: 2; leading coefficient: -1; constant term: 1
3. degree: 3; leading coefficient: 6; constant term: 0   **5.** not a polynomial
7. not a polynomial

	Sum	Product
9.	$1 + 3\sqrt{2}\,x^2 + 5x^3$	$3x + (2\sqrt{2} - 9)x^2 + (4 - 3\sqrt{2})x^3 - 5x^4 + 6\sqrt{2}\,x^5 + 4x^6$
11.	$2u + 2$	$-u^4 + u^2 + 2u + 1$

13. $-1, 1$ **15.** $-3, 3$ **17.** $\dfrac{-5 \pm \sqrt{57}}{4}$

	$Q(x)$	$R(x)$
19.	$3x^2 - 4x + 3$	$-3x + 4$
21.	$2x^2 - \dfrac{5}{2}x + 3$	$-7x - 6$
23.	$\dfrac{\sqrt{2}}{2}x + \dfrac{3}{2}$	$-\dfrac{7}{2}$
25.	2	-24
27.	0	$x + 2$

ANSWERS TO EXERCISES 5-2

1. $-1, 0, 1$ **3.** $-1, 1$ **5.** $-1, 1, 2$ **7.** $x(x + 1)(x - 4)$; $0, -1, 4$
9. $(u - 2)(u^2 - 10u + 4)$; $2, 5 \pm \sqrt{21}$ **11.** $x(x + 1)(x - 1)(x - 2)$; $-1, 0, 1, 2$

13. $x^2(x^2 + 2x + 6)$; 0 **15.** $(x + 2)(2x + 3)(2x - 3)$; $-2, -\dfrac{3}{2}, \dfrac{3}{2}$

17. $(t + 1)(t - 1)(t^2 - t + 1)(t^2 + t + 1)$; $-1, 1$

ANSWERS TO EXERCISES 5-3

1. even   **3.** odd   **5.** odd   **7.** neither

9. $P(x) = x^4 - 6x^2 + 8$ even; intercepts: $(0, 8)$, $(\pm 2, 0)$, $(\pm\sqrt{2}, 0)$ symmetry: y axis; positive on $(-\infty, -2) \cup (-\sqrt{2}, \sqrt{2}) \cup (2, +\infty)$

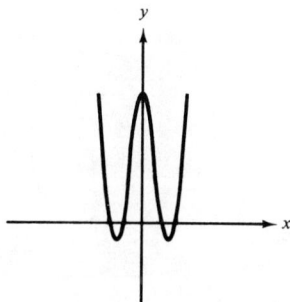

11. $P(x) = x^5 - 3x^3$ odd; intercepts: $(0, 0)$, $(\pm\sqrt{3}, 0)$ symmetry: origin; positive on $(-\sqrt{3}, 0) \cup (\sqrt{3}, +\infty)$

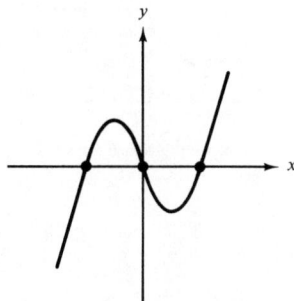

13. $P(x) = 5x - x^3$ odd; intercepts: $(0, 0)$, $(\pm\sqrt{5}, 0)$ symmetry: origin; positive on $(-\infty, -\sqrt{5}) \cup (0, \sqrt{5})$

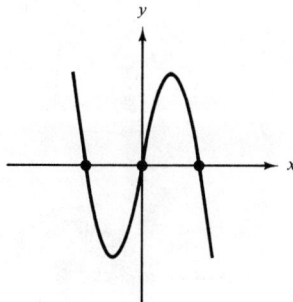

15. as $x \longrightarrow +\infty$ $f(x) \longrightarrow +\infty$
as $x \longrightarrow -\infty$ $f(x) \longrightarrow +\infty$

17. as $x \longrightarrow +\infty$ $g(x) \longrightarrow +\infty$
as $x \longrightarrow -\infty$ $g(x) \longrightarrow -\infty$

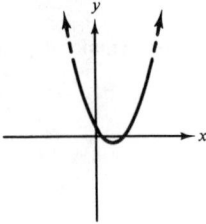

19. as $x \longrightarrow +\infty$ $g(x) \longrightarrow +\infty$
as $x \longrightarrow -\infty$ $g(x) \longrightarrow -\infty$

21. as $x \longrightarrow +\infty$ $g(x) \longrightarrow -\infty$
as $x \longrightarrow -\infty$ $g(x) \longrightarrow +\infty$

23. as $x \longrightarrow +\infty$ $g(x) \longrightarrow +\infty$
as $x \longrightarrow -\infty$ $g(x) \longrightarrow -\infty$

25. $f(x) = x^2 - 3x + 2$

27. $f(x) = x^2 + x + 10$

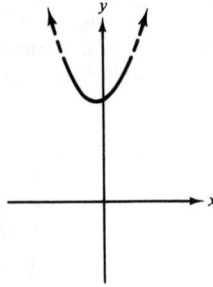

ANSWERS TO EXERCISES 5-4

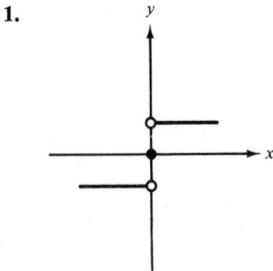

1.

Discontinuous at $x = 0$

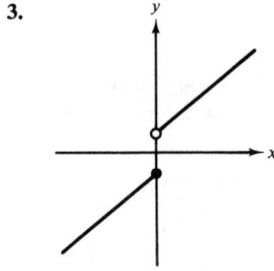

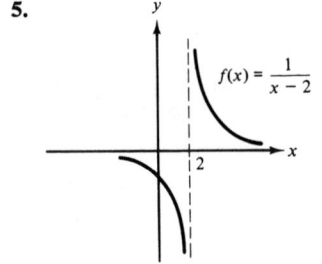

3.

Discontinuous at $x = 0$

5.

$f(x) = \dfrac{1}{x-2}$

Discontinuous at $x = 2$

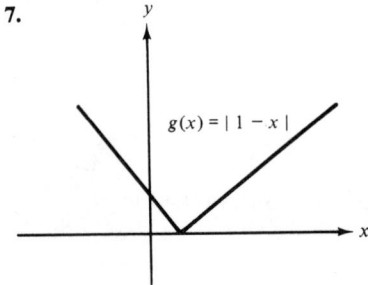

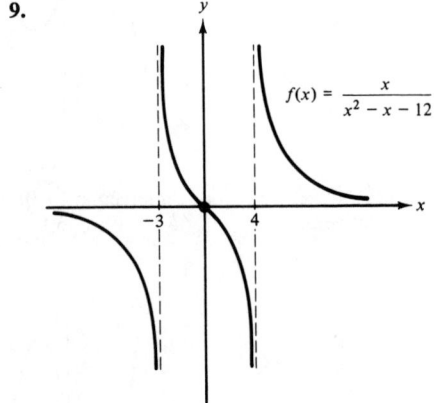

7.

$g(x) = |1-x|$

9.

$f(x) = \dfrac{x}{x^2 - x - 12}$

Discontinuous at $x = -3, 4$

11. (a) $\left.\begin{array}{l} g(3) = -2 \\ g(4) = 7 \end{array}\right\}$ So there is a zero between 3 and 4.

(b) $\left.\begin{array}{l} h(3) = 2 \\ h(4) = -28 \end{array}\right\}$ So there is a zero between 3 and 4.

13. $(0, +\infty)$ **15.** $(-3, 0) \cup (1, +\infty)$ **17.** $(-\infty, -1) \cup (2, +\infty)$
19. $(-2 - \sqrt{7}, 0) \cup (0, -2 + \sqrt{7})$ **21.** $(-\infty, -3) \cup (-1, 2)$ **23.** $(-\infty, -3) \cup (-2, 2)$

ANSWERS TO EXERCISES 5-5

1. $-5, -2$ **3.** 2   **5.** $-3, 1, 2$ **7.** $2, -15$ **9.** 3   **11.** $-\dfrac{3}{4}, \dfrac{1 \pm \sqrt{5}}{2}$ **13.** $0, \dfrac{4}{3}$
15. 2 is the only rational zero.   **17.** $-4/5$ is the only rational zero.   **19.** none

ANSWERS TO EXERCISES 5-6

1. (a) between 2.92 and 2.93   (b) between 1.40 and 1.41
(c) between 1.83 and 1.84   (d) between -1.36 and -1.35
3. One value is between .21 and .22, the other between 3.86 and 3.87.

ANSWERS TO EXERCISES 5-7

1. $\dfrac{17x + 1}{3x^2 - 5x - 2}$ **3.** $\dfrac{5x^3 - 4x^2 + 2x - 1}{x^4 - x^3 + x^2 - x}$ **5.** $\dfrac{203x^2 + 88x + 20}{-40x^2 + 64x + 32}$
7. $\dfrac{6u}{u^3 + 5u^2 - 4u - 20}$ **9.** $\dfrac{3w^2 + 15w}{2w^2 - 8}$ **11.** $\dfrac{4x^2 - 12x + 9}{9x^2 + 30x + 24}$ **13.** $-\dfrac{1}{x}$
15. $\dfrac{x - 1}{x + 3}$ Cancellation not valid at $x = -2$.   **17.** $\dfrac{2u}{2u^2 - 1}$
19. $\dfrac{x}{x - 3}$ Cancellation not valid at $x = -2$.   **21.** $w + 3$ **23.** 1, 2   **25.** 2   **27.** none
29. Discontinuities at $-\sqrt{3}, 0, \sqrt{3}$.   **31.** Discontinuities at $-2, 2$.   **33.** $(-1, 0) \cup (1, +\infty)$
35. $(-\infty, -1) \cup (1, +\infty)$ **37.** $(-2, -1) \cup (0, +\infty)$ **39.** $(-1, 1) \cup (-1, +\infty)$
41. $(-4, 4) \cup (8, +\infty)$
43. (a) No.
(b) No, because a variable can assume negative values.
(c) $[Q(x)]^2$ is positive.

ANSWERS TO EXERCISES 5-8

1. $x = 1$ **3.** $x = -\dfrac{5}{3}$ **5.** none   **7.** $x = 4$ **9.** $x = 0, x = 1$
11. $x = -2\sqrt{3}, x = 2\sqrt{3}$ **13.** $t = -2$ **15.** $x = 0$

17. $x = -2, x = 2$

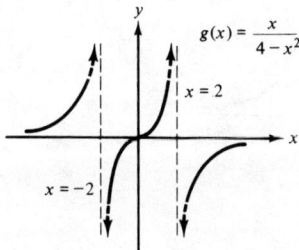

19. No vertical asymptotes.

21. $x = 0$

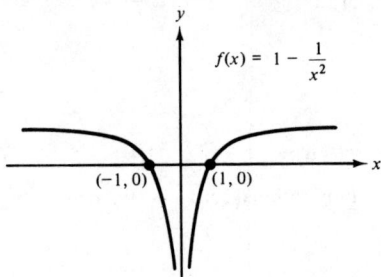

$f(x) = 1 - \dfrac{1}{x^2}$

$(-1, 0)$ $(1, 0)$

23. $\dfrac{x(x^2 - 4)}{x^2 - 1}$ **25.** $\dfrac{(x^2 - 4)^2}{x^2(x^2 - 1)}$ (Note: Answers to 23 and 25 are not unique.)

ANSWERS TO EXERCISES 5-9

1. $y = \dfrac{1}{3}$ **3.** $y = 4$ **5.** $y = 1$ **7.** $y = 0$ **9.** $y = \dfrac{1}{5}$ **11.** $y = 0$

13.

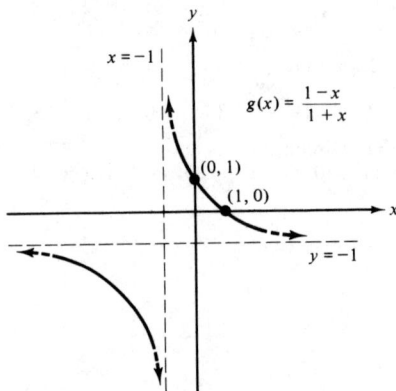

$x = -1$

$g(x) = \dfrac{1 - x}{1 + x}$

$(0, 1)$

$(1, 0)$

$y = -1$

15.

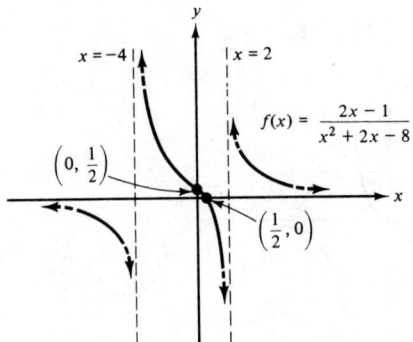

$x = -4$ $x = 2$

$f(x) = \dfrac{2x - 1}{x^2 + 2x - 8}$

$\left(0, \dfrac{1}{2}\right)$

$\left(\dfrac{1}{2}, 0\right)$

17.

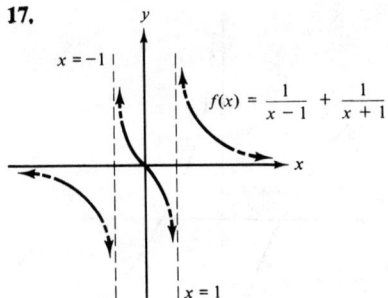

$$f(x) = \frac{1}{x-1} + \frac{1}{x+1}$$

with $x = -1$ and $x = 1$

19.

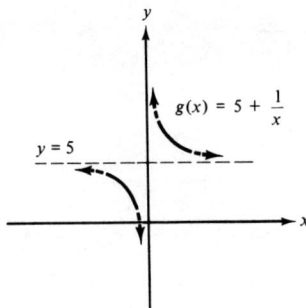

$$g(x) = 5 + \frac{1}{x}$$

with $y = 5$

21. $y = \dfrac{x^2 + 4}{x} = x + \dfrac{4}{x}$; $+\infty, -\infty$

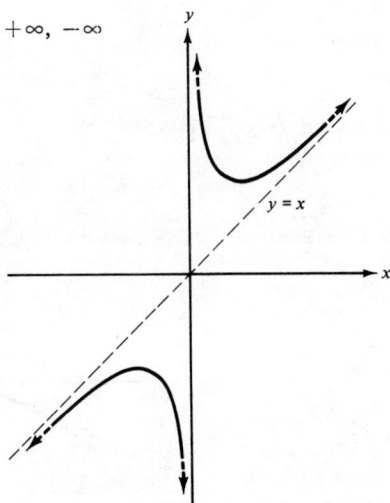

$y = x$

23. $y = \dfrac{x^2 + 3x + 1}{x - 2} = x + 5 + \dfrac{11}{x-2}$; $+\infty, -\infty$

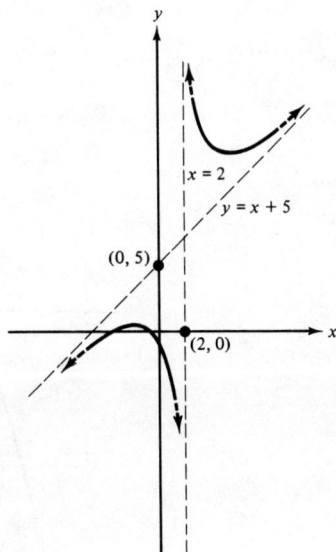

$x = 2$

$y = x + 5$

$(0, 5)$

$(2, 0)$

25. $y = \dfrac{9 - x^2}{x} = -x + \dfrac{9}{x}; \ -\infty, \ +\infty$

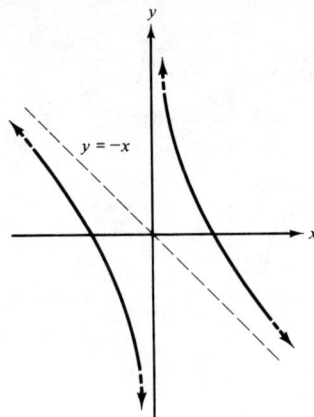

27. $y = \dfrac{3x^2 + 4x + 5}{x - 1} = 3x + 7 + \dfrac{12}{x - 1}; \ +\infty, \ -\infty$

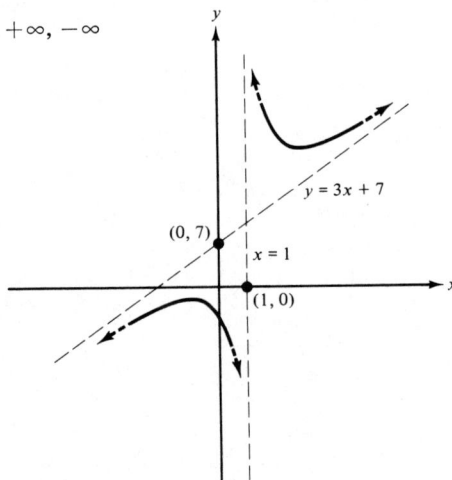

29. $y = \dfrac{2x^3 - 8x}{x^2 + 4} = 2x - \dfrac{16x}{x^2 + 4}; \ +\infty, \ -\infty$

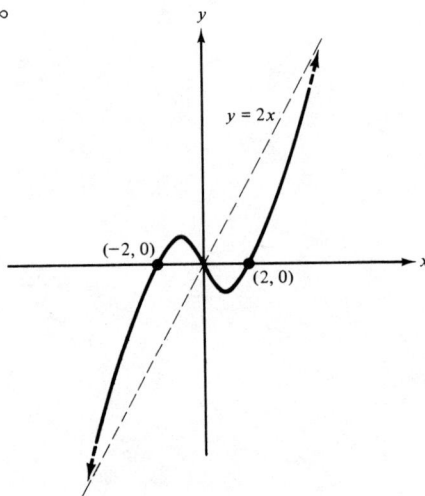

Answers to Selected Odd-Numbered Problems

33. $\dfrac{-3x^2}{2(x^2-1)}$ **35.** $\dfrac{5(x^2-4)x^2}{x^2-1}$ (Note: Answers to 33 and 35 are not unique.)

ANSWERS TO REVIEW EXERCISES FOR CHAPTER FIVE

1. (a) yes; degree 4; leading coefficient 1; constant term 8
   (b) yes; degree 3; leading coefficient 1; constant term -6
   (c) no
   (d) yes; degree 3; leading coefficient 1; constant term 0
   (e) yes; degree 2; leading coefficient 1; constant term -1
   (f) no

3. (a) 7 (b) 0 (c) -1

5. (a) $(0, 10)$ (b) $(-\infty, -3) \cup (0, 3)$
   (c) $(-\infty, -2) \cup (4, +\infty)$ (d) $(-\infty, 0) \cup (0, 2)$
   (e) $(-\infty, -2) \cup (-\sqrt{2}, \sqrt{2}) \cup (2, +\infty)$ (f) $(-\infty, -1) \cup (3, +\infty)$
   (g) $(-\infty, -3) \cup (-1, 1) \cup (1, +\infty)$ (h) $(-\infty, -3) \cup (-2, -1) \cup (-1, +\infty)$

7. (a) $-1, 1$ (b) none (c) none (d) 2, 3

9. (a) zero: 1; pole: -2

   (b) zero: 0; pole: 2

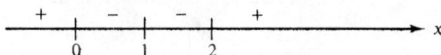

   (c) zeros: $-2, 2$; poles: $-1, 1$

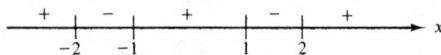

   (d) zero: 1; poles: $-3, -2$

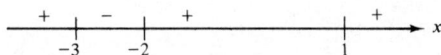

   (e) zeros: $-1, 1$; pole: 0

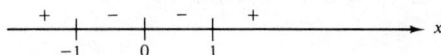

   (f) zeros: none; pole: $-1/2$

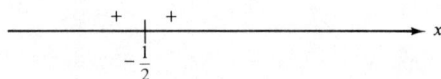

   (g) zeros: $-2/3, 1$; pole: 0

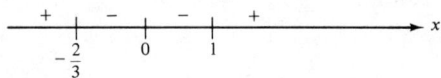

   (h) zero: $-1/3$; poles: $-3, 0$

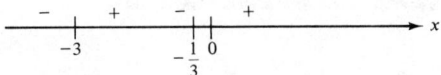

   (i) zeros: $-1/2, 1/2$; pole: 0

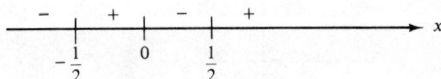

   (j) zeros: none; pole: 0

11. (a)

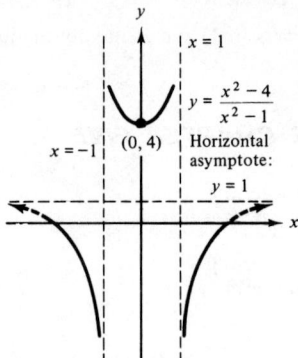

$x = 1$

$y = \dfrac{x^2 - 4}{x^2 - 1}$

Horizontal asymptote:
$y = 1$

$x = -1$ $(0, 4)$

(b)

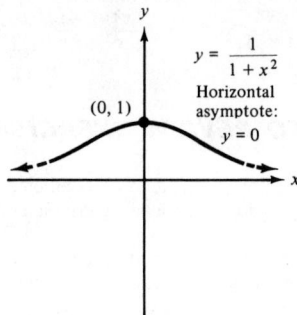

$y = \dfrac{1}{1 + x^2}$

Horizontal asymptote:
$y = 0$

$(0, 1)$

(c)

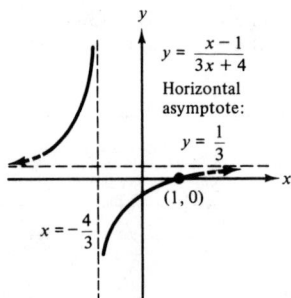

$y = \dfrac{x - 1}{3x + 4}$

Horizontal asymptote:
$y = \dfrac{1}{3}$

$(1, 0)$

$x = -\dfrac{4}{3}$

(d) no horizontal asymptotes
$F(x) \longrightarrow +\infty$ as $x \longrightarrow +\infty$
$F(x) \longrightarrow -\infty$ as $x \longrightarrow -\infty$

ANSWERS TO EXERCISES 6-1

1. i **3.** $-24i$ **5.** $\dfrac{3}{25} + \dfrac{4}{25}i$ **7.** $\dfrac{1}{5} - \dfrac{2}{5}i$ **9.** 1   **11.** 0   **13.** $1 - 4i$ **15.** -1

17. $\dfrac{3}{25} - \dfrac{4}{25}i$ **19.** $-\dfrac{1}{4}$ **21.** $3i$ **23.** $-9 + 6i$ **25.** $\dfrac{1}{5} + \dfrac{8}{5}i$ **27.** $\pm 10i$

29. $\pm 3\sqrt{2}\,i$ **31.** $\pm\dfrac{6}{5}i$ **33.** $\pm\dfrac{8\sqrt{5}}{5}i$ **35.** $\pm 5\sqrt{3}\,i$ **37.** $-\dfrac{1}{3} \pm \dfrac{1}{3}\sqrt{11}\,i$

39. $\dfrac{3}{4} \pm \dfrac{\sqrt{31}}{4}i$ **41.** $-2, 1 \pm \sqrt{3}\,i$ **43.** $\pm\sqrt{3}, \pm\sqrt{3}\,i$ **45.** $\pm\left(\dfrac{\sqrt{2}}{2} + \dfrac{\sqrt{2}}{2}i\right)$

ANSWERS TO EXERCISES 6-2

1.

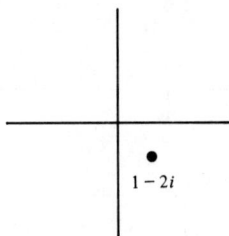

$1 - 2i$

3.

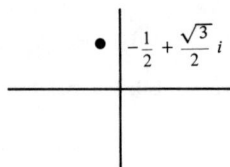

$-\dfrac{1}{2} + \dfrac{\sqrt{3}}{2}\,i$

Answers to Selected Odd-Numbered Problems

5. $2(1 + i) = 2 + 2i$

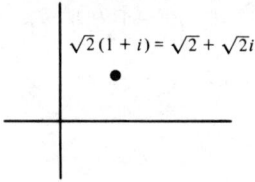

$$\sqrt{2}(1 + i) = \sqrt{2} + \sqrt{2}i$$

7. $\dfrac{2}{3 + 4i} = \dfrac{6}{25} - \dfrac{8}{25}i$

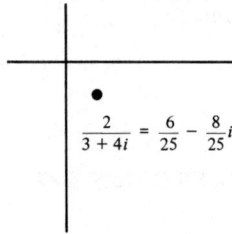

$$\frac{2}{3 + 4i} = \frac{6}{25} - \frac{8}{25}i$$

9. $\sqrt{5}, 1 + 2i$

11. $1, -\dfrac{1}{2} - \dfrac{\sqrt{3}}{2}i$

13. $2\pi, -2\pi i$

15. $\sqrt{26}, 5 - i$

17. $1, 1$

19. circle centered at $1 + 0i$, radius 2

21. region exterior to circle centered at -2, radius 1

23. region exterior to circle centered at -2, radius 1
   AND interior to circle centered at -2, radius 3

25. circle centered at $-2i$, radius 1

27. region exterior to circle centered at $(-1 + i)$, radius $\sqrt{2}$

29. $-\dfrac{3}{2} + \dfrac{\sqrt{7}}{2}i$ and $-\dfrac{3}{2} - \dfrac{\sqrt{7}}{2}i$ **31.** $0, -2 + 2i$ and $-2 - 2i$ **33.** $2, -2, 2i,$ and $-2i$

35. $2i, -2i, \sqrt{5}i,$ and $-\sqrt{5}i$

ANSWERS TO EXERCISES 6-3

1. degree 2; leading coefficient: $3 - 2i$

3. degree 2; leading coefficient: -1

5. degree 2; leading coefficient: $2i$

7. degree 1; leading coefficient: $\dfrac{3}{4}$

9. $Q(z) = 3z^2 - 6z + 14; R(z) = -23$ **11.** $Q(z) = \dfrac{2}{3}z^3 - \dfrac{2}{3}z^2 + z - 1; R(z) = \dfrac{7}{3}$

13. $Q(z) = 3iz + 1; R(z) = 3i$ **15.** $Q(z) = -12iz^2 - 19z + 35i; R(z) = 71$ **17.** $-1 + \dfrac{5}{3}i$

21. 3   **23.** 0   **25.** $(z - 2)(z + 1 - \sqrt{3}i)(z + 1 + \sqrt{3}i)$

27. $(z - i)(z + i)(z - \sqrt{2}i)(z + \sqrt{2}i)$ **29.** $(z - 2i)(z + 2i)(z - i)$

31. $(z + 1)(z - 1)\left[z - \dfrac{-1 + \sqrt{3}i}{2}\right]\left[z - \dfrac{-1 - \sqrt{3}i}{2}\right]$

33. $[z - (3 - 4i)][z - (3 + 4i)][z - 2i][z + 2i]$ **35.** $(z - 3i)^2(z + 3i)^2(z - i)(z + i)$

37. $2\left(z + \dfrac{3}{2}\right)\left(z - \dfrac{-1 + \sqrt{3}i}{2}\right)\left(z - \dfrac{-1 - \sqrt{3}i}{2}\right)$ **39.** $(z - 1)(z + 1)(z^2 + z + 1)$

41. $(z^2 + 4)(z^2 - 6z + 25)$ **43.** $(z^2 + 9)^2(z^2 + 1)$ **45.** $2\left(z - \dfrac{3}{2}\right)(z^2 + z + 1)$

47. $(z - i)(z - 2i)$ **49.** $(z - 1 - i)(z + 1 - i)$

51. $(z - 2i)^2(z - 3i)^3$ **53.** $(z^2 + 1)(z^2 + 4)$

55. $(z^2 - 2z + 2)(z^2 + 2z + 2)$ **57.** $(z^2 + 4)^2(z^2 + 9)^3$

ANSWERS TO REVIEW EXERCISES FOR CHAPTER SIX

1. (a) $-7 - 24i$ (b) $-\frac{1}{4} - \frac{1}{4}i$ **3.** (a) $-\frac{1}{13} + \frac{5}{13}i$ (b) $-\frac{i}{3}$ (c) $\frac{-1 \pm 2i}{5}$ (d) $\frac{i}{2}, -\frac{i}{3}$

7. (b) $\frac{i}{2}, -\frac{i}{2}, -\frac{2}{3}$ **9.** (a) $5(z^2 + 9)^2\left(z^2 + \frac{1}{5}z + \frac{1}{5}\right)$ (b) $25\left(z - \frac{4}{5}\right)\left(z^2 + \frac{1}{5}\right)$

ANSWERS TO EXERCISES 7-1

3. decreasing

ANSWERS TO EXERCISES 7-2

1. $g(h(x)) = (x + 2)^2$ for all real x; $h(g(x)) = x^2 + 2$ for all real x
3. $g(h(x)) = x^3 - 1$ for all real x; $h(g(x)) = (x - 1)^3$ for all real x
5. $g(h(x)) = \frac{1}{x^3} - 1$ for $x \neq 0$; $h(g(x)) = \frac{1}{x^3 - 1}$ for $x \neq 1$
(Note: Answers to 7–17 are not unique.)

7. $h(x) = x + 5$ for all real x; $g(x) = x^2$ for all real x

9. $h(x) = x^2 + 3$ for all real x; $g(x) = x^2$ for all real x

11. $h(x) = 1 + \frac{1}{x}$ for $x \neq 0, -1$; $g(x) = \frac{1}{x}$ for $x \neq 0$

13. $h(x) = 1 + \sqrt{x}$ for $x \geq 0$; $g(x) = \sqrt{x}$ for $x \geq 0$

15. $h(x) = 2x + 1$ for all real x; $g(x) = |x|$ for all real x

17. $h(x) = x^2 - 3x + 2$ for all real x; $g(x) = |x|$ for all real x

19.

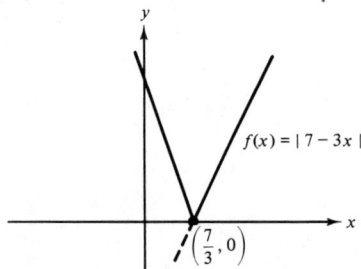

$f(x) = |7 - 3x|$, $\left(\frac{7}{3}, 0\right)$

21.

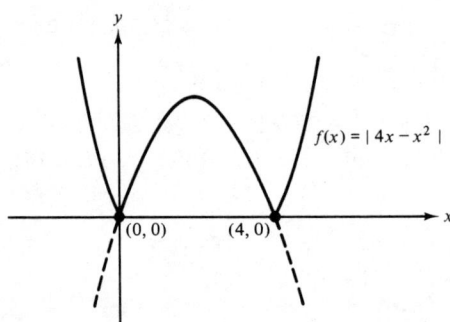

$f(x) = |4x - x^2|$, $(0, 0)$, $(4, 0)$

23.

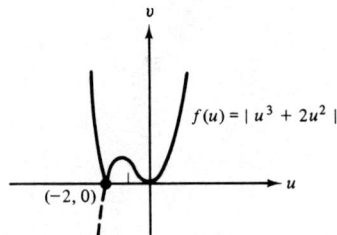

$f(u) = |u^3 + 2u^2|$, $(-2, 0)$

25.

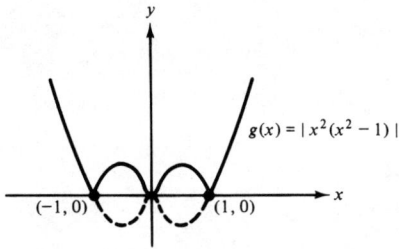

$g(x) = |x^2(x^2 - 1)|$

$(-1, 0)$ $(1, 0)$

27.

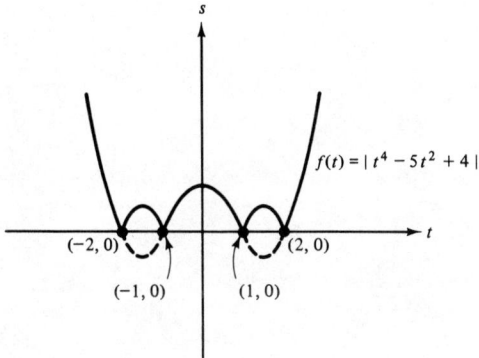

$f(t) = |t^4 - 5t^2 + 4|$

$(-2, 0)$ $(2, 0)$

$(-1, 0)$ $(1, 0)$

29. $f \circ f(x) = x$ for all real x

ANSWERS TO EXERCISES 7-3

1. $f^{-1}(u) = \dfrac{u}{3}$ for all real u **3.** $f^{-1}(u) = \dfrac{1}{3}(u - 4)$ for all real x

5. $f^{-1}(u) = \dfrac{1}{2}(u - 1)$ for $1 \leq u \leq 5$ **7.** $f^{-1}(u) = \sqrt{u - 1}$ for $u \geq 1$

9. $f^{-1}(u) = \sqrt[3]{u} - 2$ for all real all real u **11.** has an inverse (decreasing for $x \geq 0$)

13. no inverse (not one to one)   **15.** has an inverse (increasing for $x \geq 0$)

17. no inverse (not one to one)   **19.** has an inverse (increasing for $x \geq 1$)

21.

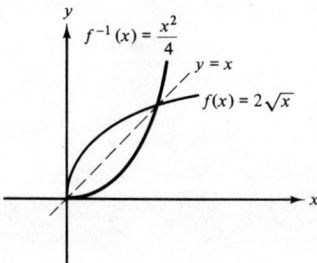

$f^{-1}(x) = \dfrac{x^2}{4}$

$y = x$

$f(x) = 2\sqrt{x}$

23.

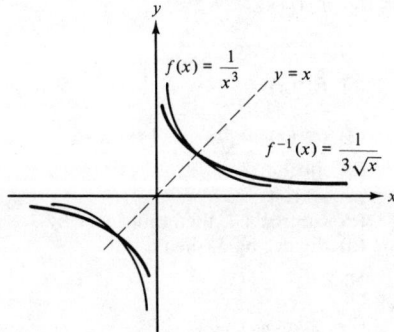

$f(x) = \dfrac{1}{x^3}$

$y = x$

$f^{-1}(x) = \dfrac{1}{3\sqrt{x}}$

25.

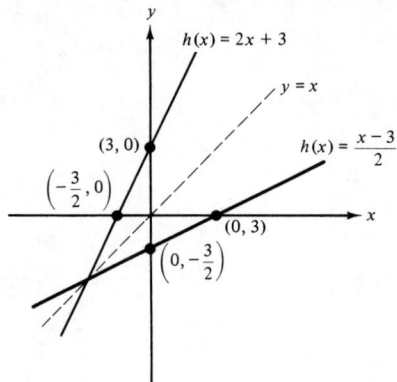

27.

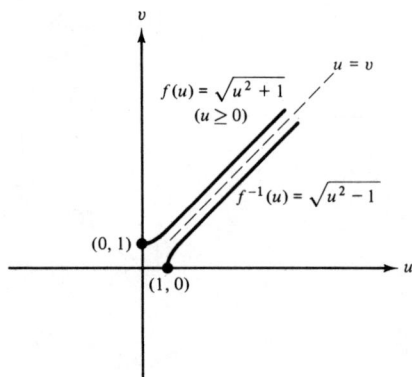

ANSWERS TO REVIEW EXERCISES FOR CHAPTER SEVEN

3. (a) $g(h(x)) = 1 - 6x^2$, $h(g(x)) = 2(1 - 3x)^2$

(b) $g(h(x)) = \sqrt{x^2} = |x|$, $h(g(x)) = (\sqrt{x})^2 = x$ for $x \geq 0$

(c) $g(h(x)) = \dfrac{1}{x+1}$ for $x \geq 0$, $h(g(x)) = \sqrt{\dfrac{1}{x^2+1}}$

5. (a) $f^{-1}(x) = \dfrac{1}{\sqrt{x}}$ for $x > 0$

(b) no inverse

(c) $f^{-1}(x) = \dfrac{-1 + \sqrt{1+4x}}{2}$ for $x > 0$

(d) $f^{-1}(x) = \sqrt{\dfrac{1}{x} + 1}$ for $x \leq -1$, $x > 0$

7. (a) divides by 3

(b) adds 4

(c) subtracts 4, then multiplies by 3

(d) divides by 3, then adds 4

$h(x) = 3x$, $g(x) = x - 4$, $h^{-1}(x) = \dfrac{1}{3}x$, $g^{-1}(x) = x + 4$, $(h \circ g)(x) = 3(x - 4)$,

$(g^{-1} \circ h^{-1})(x) = \dfrac{x}{3} + 4$

ANSWERS TO EXERCISES 8-1

1. $\frac{5\sqrt{2}}{2}$ 3. $\frac{5-\sqrt{5}}{5}$ 5. $7 + 5\sqrt{2}$ 7. $\frac{3\sqrt{5}-10}{-11}$ 9. $\frac{5-\sqrt{5}}{8}$

11. $\frac{6-5\sqrt{x}+x}{9-x}$ 13. $\frac{\sqrt{x}+\sqrt{y}}{x-y}$ 15. $\sqrt{x+4}+2$ 17. $(\sqrt{x}-\sqrt{y})(x+y)$

19. $\frac{\sqrt{2}}{4}$ 21. $\frac{7\sqrt{2}}{8}$ 23. $\frac{3}{2}$ 25. $\frac{9}{2}$ 27. 2 29. $19 + 6\sqrt{2}$ 31. $17 - 12\sqrt{2}$

33. $(a^2+1)^{1/2}(b^2+1)^{-1/2}$ 35. $8ab(a+4b)^{-3/5}$ 37. $(3x-2)^{3/2}$

39. $(x^2+y^2)^{1/2} - (x^3+8y^3)^{1/3}$ 41. $x^{4/3}y^{8/3}z^{20/3}$ 43. $\sqrt{\left(1+\frac{1}{a^2}\right)^3}$ 45. $\frac{1}{\sqrt[3]{(4-x)^2}}$

47. $\frac{6x}{\sqrt{1+3x^2}}$ 49. $\frac{5x^2+4x}{2\sqrt{1+x}}$ 51. $\frac{10t^2+6}{\sqrt[3]{t^2+1}}$ 53. $x = 3$ 55. $x = 0$

57. $x = 3$ 59. $t = -3, 3$ 61. none   69. $r = \frac{7}{2}$ 71. $r = 0$ 73. $r = \frac{4}{3}$

75. $r = 0, \frac{1}{2}$ 77. $r = -\frac{1}{2}, 0$ 79. 11.85 years

ANSWERS TO EXERCISES 8-2

1. $f(x) = 5^x$

x	0	1	2	3	$\frac{1}{2}$	$\frac{3}{2}$	$-\frac{1}{2}$	-1	-2
$f(x)$	1	5	25	125	$\sqrt{5}$	$5\sqrt{5}$	$\frac{1}{\sqrt{5}}$	$\frac{1}{5}$	$\frac{1}{25}$

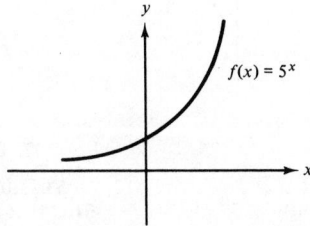

3. $x = 2$ 5. $t = 0$ 7. $y = -3$ 9. $x = 1, 2$ 11. $(-\infty, +\infty)$ 13. $(-\infty, +\infty)$
15. $(-3, +\infty)$ 17. $(2, \infty)$ 19. $(0, +\infty)$ 21. $(-\infty, -\sqrt{2}) \cup (\sqrt{2}, +\infty)$
23. $(-\sqrt{2}, \sqrt{2})$ 25. $(-\infty, 1) \cup \left(\frac{3}{2}, +\infty\right)$ 27. $(-\infty, -2) \cup (-1, 1) \cup (2, +\infty)$
29. $(-\infty, 2) \cup (3, +\infty)$ 31. $(-\infty, 4]$ 33. $(-\infty, 1)$ 35. $[-2, 2]$ 37. no solutions
39.

(a)

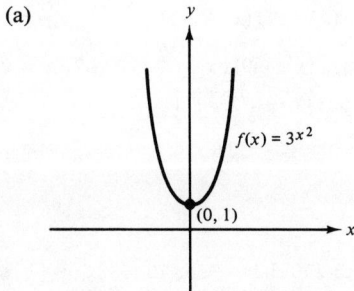

(b)

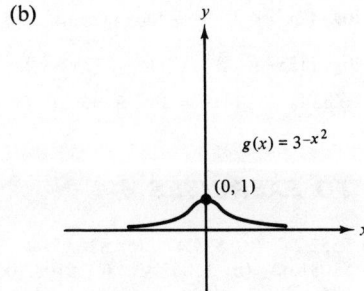

(c)

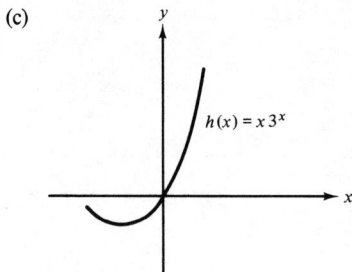

$h(x) = x\,3^x$

(d)

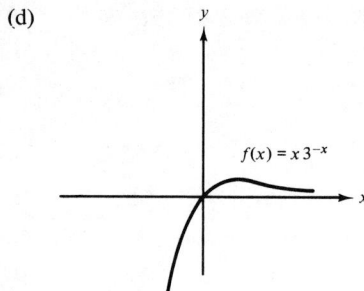

$f(x) = x\,3^{-x}$

ANSWERS TO EXERCISES 8-3

1. $\log_3 243 = 5$ **3.** $\log_3 (3\sqrt{3}) = \dfrac{3}{2}$ **5.** $\log_3 \left(\dfrac{1}{\sqrt[3]{9}}\right) = -\dfrac{2}{3}$ **7.** $3^4 = 81$

9. $4^{-2} = \dfrac{1}{16}$ **11.** $10^{3/2} = 10\sqrt{10}$ **13.** $\log_3 81 = 4$ **15.** $\log_2 \left(\dfrac{1}{1024}\right) = -10$

17. $\log_5 y = 2x$ **19.** **21.** -3 **23.** $-\dfrac{2}{3}$ **25.** 5

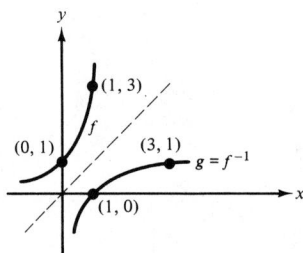

(1, 3)

f

(0, 1) (3, 1)

$g = f^{-1}$

(1, 0)

27. 5 **29.** 2 **31.** u **33.** 25 **35.** 11 **37.** 0 **39.** $-\dfrac{1}{3}$ **41.** 32 **43.** 1000
45. $2\sqrt{2}$ **47.** 3 **49.** $\sqrt{3}$ **51.** 512 **53.** 1.0792 **55.** 2.6021
57. .8194 (by interpolation) **59.** 1.8751 **61.** (a) .6171 (b) 10.3386 (c) 2.7248
63. $\log_{10} 5 + 3\log_{10} (x - 3)\ (x > 3)$ **65.** $\log_{10} (x^2 + 4) + 3\log_{10} (x^2 + 1)$ (all real x)
67. $-3\log_5 (3x - 1)\left(x > \dfrac{1}{3}\right)$ **69.** $\dfrac{1}{2}\log_2 (x^2 + 25)$ (all real x)
71. $\log_{10} (x^2 - 4) - \log_{10} (x^2 + 4)\ (x < -2,\ x > 2)$
73. $-\dfrac{1}{4}\log_2 (x^2 + 2x + 4)$ (all real x)

75. $\log_{10} x + 2\log_{10} (3x + 4) + \dfrac{1}{2}\log (3x - 2)\left(x > \dfrac{2}{3}\right)$

77. $2\log_5 (2x^2 + 3) - \dfrac{1}{2}\log_5 (x + 5) - 3\log_5 (x^2 + 1)\ (x > -5)$

79. $2\log_{10} (3x^2 + 4) + \dfrac{1}{2}\log_{10} (3x - 2) - \log_{10} (x^2 + 9)\left(x > \dfrac{2}{3}\right)$ **87.** $x = 2, 8$
89. $x = 1, 4$ **91.** $x = 16$ **93.** $h^{-1}(x) = \log_3 (x + \sqrt{x^2 + 1})$

ANSWERS TO EXERCISES 8-4

1. (a) \$552.24 (b) \$25.47 (c) \$26.77
3. (a) \$1081.60 (b) \$1082.43 (c) \$1083.00 (d) \$1083.28 **5.** 10.137%
7. $P = 25(1.2)^{t/10}$, where P is in millions, t in years; 36,000,000 **9.** 29.09 years
11. 16.43% remains after 75 years; 9% remains after 100 years

13.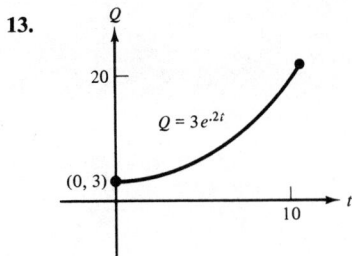

$Q = 3e^{.2t}$

$(0, 3)$

20

10

15.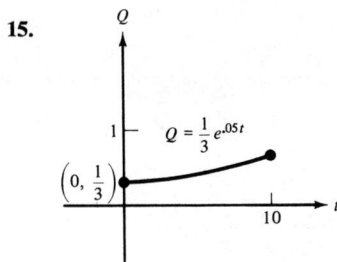

$Q = \frac{1}{3}e^{.05t}$

$\left(0, \frac{1}{3}\right)$

1

10

17.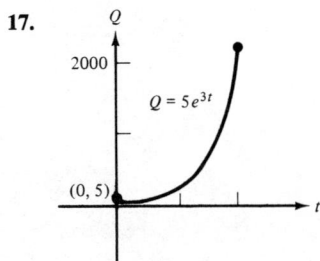

$Q = 5e^{3t}$

2000

$(0, 5)$

19.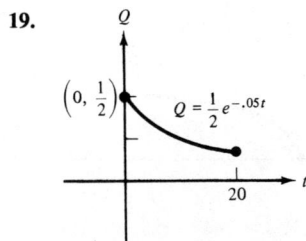

$\left(0, \frac{1}{2}\right)$

$Q = \frac{1}{2}e^{-.05t}$

20

21.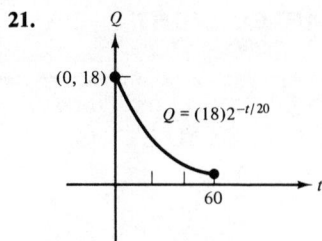

$(0, 18)$

$Q = (18)2^{-t/20}$

60

23.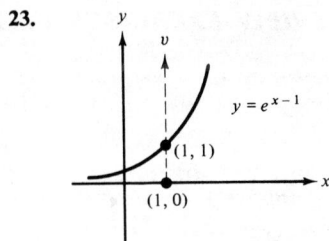

$y = e^{x-1}$

$(1, 1)$

$(1, 0)$

25.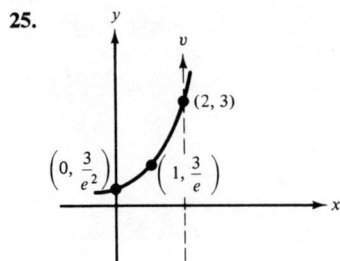

$(2, 3)$

$\left(0, \frac{3}{e^2}\right)$

$\left(1, \frac{3}{e}\right)$

27.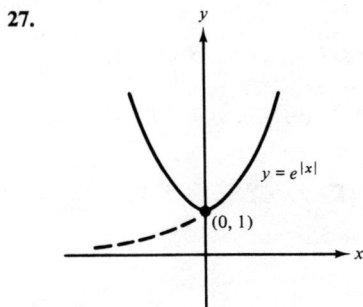

$y = e^{|x|}$

$(0, 1)$

29.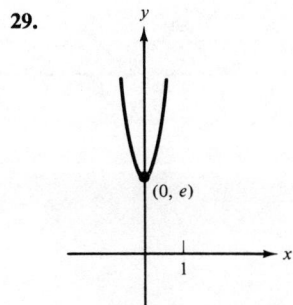

$(0, e)$

1

31.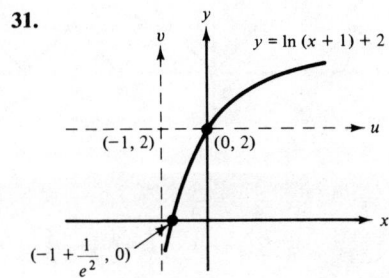

$y = \ln(x + 1) + 2$

$(-1, 2)$

$(0, 2)$

$\left(-1 + \frac{1}{e^2}, 0\right)$

33.

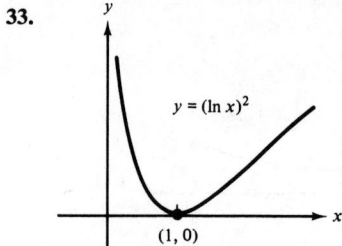

$y = (\ln x)^2$

$(1, 0)$

35.

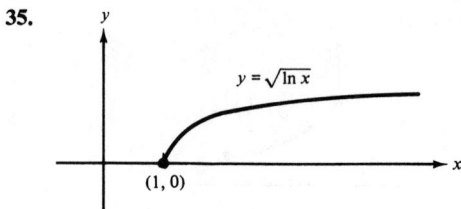

$y = \sqrt{\ln x}$

$(1, 0)$

ANSWERS TO REVIEW EXERCISES FOR CHAPTER EIGHT

1. (a) $a^{2/3}b^{1/3}c^{4/3}$ (b) $(x^2 + y^2)^{1/2}z^{-1/2}$ (c) $5x^{1/3} - 4x^{-2/3} + 7x^{-1}$ (d) $64x^3y^2$ (e) $a^{1/4}b^{1/6}$

3. (a) $-1, 1$ (b) $-2, 0, 1$ (c) -1 (d) 1 (e) $-1, 1$

5. (a) 3 (b) $-\dfrac{1}{2}, \dfrac{1}{2}$ (c) $-1, 1$ (d) $-1, -\dfrac{1}{2}$ (e) $(-\infty, +\infty)$ (f) $(-\infty, 0]$

 (g) $(-\infty, -2) \cup (2, +\infty)$ (h) $(-\infty, -\sqrt{5}] \cup [\sqrt{5}, +\infty)$

7. (a) $5^{1/x} = u$ (b) $5^{3x-1} = v$ (c) $5^{x^2+1} = w$

9. (a) $2 + \dfrac{5}{2}\log_5(x + 3)$

 (b) $\dfrac{2}{3}\log_5(x + 5) + 2\log_5(x + 3) - 4\log_5(x^2 + 1) - 1$

 (c) $\dfrac{1}{3}\ln(x^3 + 1) - \dfrac{1}{3}\ln(e^3 + 1)$

 (d) $2\ln(x + \sqrt{x^2 - 1})$

11. (a) $-2, 2$ (b) $-3, 3$ **13.** 71%

ANSWERS TO EXERCISES 9-1

1.

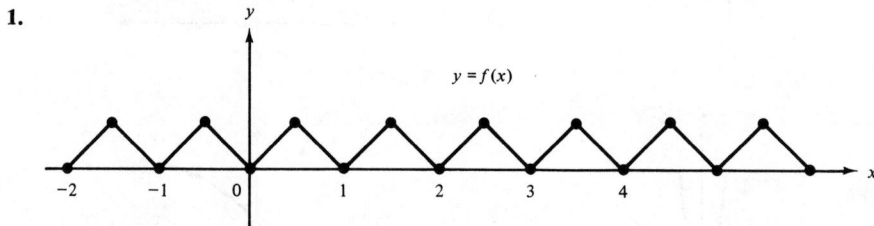

$y = f(x)$

$f(0) = 0 \quad f\left(\dfrac{3}{4}\right) = \dfrac{1}{4} \quad f(2) = 0 \quad f\left(-\dfrac{3}{4}\right) = \dfrac{1}{4}$

$f\left(\dfrac{1}{4}\right) = \dfrac{1}{4} \quad f(1) = 0 \quad f\left(-\dfrac{1}{4}\right) = \dfrac{1}{4} \quad f(-1) = 0$

$f\left(\dfrac{1}{2}\right) = \dfrac{1}{2} \quad f\left(\dfrac{3}{2}\right) = \dfrac{1}{2} \quad f\left(-\dfrac{1}{2}\right) = \dfrac{1}{2}$

Answers to Selected Odd-Numbered Problems

3.

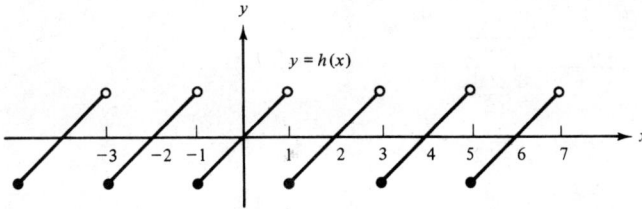

$y = h(x)$

$h(.1) = .1 \qquad h(1.1) = -.9 \qquad h(-.1) = -.1 \qquad h(-1.1) = .9$
$h(.5) = .5 \qquad h(1.5) = -.5 \qquad h(-.5) = -.5 \qquad h(-1.5) = .5$
$h(.9) = .9 \qquad h(1.9) = -.1 \qquad h(-.9) = -.9 \qquad h(-1.9) = .1$

5. $(-1, 0)$ **7.** $(-1, 0)$ **9.** $(-1, 0)$ **11.** $\left(-\dfrac{1}{\sqrt{2}}, -\dfrac{1}{\sqrt{2}}\right)$ **13.** $\left(-\dfrac{1}{\sqrt{2}}, -\dfrac{1}{\sqrt{2}}\right)$

15. $\left(\dfrac{1}{2}, \dfrac{\sqrt{3}}{2}\right)$ **17.** $\left(\dfrac{1}{2}, \dfrac{\sqrt{3}}{2}\right)$ **19.** 0   **21.** 0   **23.** 0   **25.** $\dfrac{1}{\sqrt{2}}$ **27.** $-\dfrac{1}{\sqrt{2}}$

29. $-\dfrac{1}{\sqrt{2}}$ **31.** $\dfrac{\sqrt{3}}{2}$ **33.** $\dfrac{\sqrt{3}}{2}$ **35.** $\dfrac{1}{2}$ **39.** $-\dfrac{1}{\sqrt{2}}$ **41.** $-\dfrac{1}{\sqrt{2}}$ **43.** $\dfrac{1}{\sqrt{2}}$

45. $-\dfrac{1}{\sqrt{2}}$ **47.** $-\dfrac{2}{3}\sqrt{2}$ or $\dfrac{2}{3}\sqrt{2}$; $-\dfrac{2}{3}\sqrt{2}$ **49.** $\beta = \pi - \alpha$

51. maximum at $2n\pi$, n any integer
minimum at $(2n + 1)\pi$, n any integer

ANSWERS TO EXERCISES 9-2

1. (a)

x	$\sin x$	$\cos x$
0	0	1
$\dfrac{\pi}{6}$	$\dfrac{1}{2}$	$\dfrac{\sqrt{3}}{2}$
$\dfrac{\pi}{4}$	$\dfrac{1}{\sqrt{2}}$	$\dfrac{1}{\sqrt{2}}$
$\dfrac{\pi}{3}$	$\dfrac{\sqrt{3}}{2}$	$\dfrac{1}{2}$
$\dfrac{\pi}{2}$	1	0

(b)

x	$\sin x$	$\cos x$
$\dfrac{2\pi}{3}$	$\dfrac{\sqrt{3}}{2}$	$-\dfrac{1}{2}$
$\dfrac{3}{4}\pi$	$\dfrac{1}{\sqrt{2}}$	$-\dfrac{1}{\sqrt{2}}$
$\dfrac{5}{6}\pi$	$\dfrac{1}{2}$	$-\dfrac{\sqrt{3}}{2}$
π	0	-1
$\dfrac{7\pi}{6}$	$-\dfrac{1}{2}$	$-\dfrac{\sqrt{3}}{2}$
$\dfrac{5\pi}{4}$	$-\dfrac{1}{\sqrt{2}}$	$-\dfrac{1}{\sqrt{2}}$
$\dfrac{4\pi}{3}$	$-\dfrac{\sqrt{3}}{2}$	$-\dfrac{1}{2}$
$\dfrac{3\pi}{2}$	-1	0
$\dfrac{5\pi}{3}$	$-\dfrac{\sqrt{3}}{2}$	$\dfrac{1}{2}$
$\dfrac{7\pi}{4}$	$-\dfrac{1}{\sqrt{2}}$	$\dfrac{1}{\sqrt{2}}$
$\dfrac{11\pi}{6}$	$-\dfrac{1}{2}$	$\dfrac{\sqrt{3}}{2}$
2π	0	1

(c)

x	$\sin x$	$\cos x$
$-\dfrac{\pi}{6}$	$-\dfrac{1}{2}$	$\dfrac{\sqrt{3}}{2}$
$-\dfrac{\pi}{4}$	$-\dfrac{1}{\sqrt{2}}$	$\dfrac{1}{\sqrt{2}}$
$-\dfrac{\pi}{3}$	$-\dfrac{\sqrt{3}}{2}$	$\dfrac{1}{2}$
$-\dfrac{\pi}{2}$	-1	0
$-\dfrac{2\pi}{3}$	$-\dfrac{\sqrt{3}}{2}$	$-\dfrac{1}{2}$
$-\dfrac{3\pi}{4}$	$-\dfrac{1}{\sqrt{2}}$	$-\dfrac{1}{\sqrt{2}}$
$-\dfrac{5\pi}{6}$	$-\dfrac{1}{2}$	$-\dfrac{\sqrt{3}}{2}$
$-\pi$	0	-1

3. $\dfrac{\pi}{6}$ or $\dfrac{5\pi}{6}$ **5.** $-\dfrac{5\pi}{3}, -\dfrac{\pi}{3}, \dfrac{\pi}{3},$ or $\dfrac{5\pi}{3}$ **7.** $\dfrac{\pi}{2}, \dfrac{3\pi}{2}, \dfrac{5\pi}{2}, \dfrac{7\pi}{2}$ **9.** $-\dfrac{3\pi}{4}, -\dfrac{\pi}{4}, \dfrac{\pi}{4}, \dfrac{3\pi}{4}$

11. $-\dfrac{3\pi}{4}, -\dfrac{\pi}{4}, \dfrac{\pi}{4}, \dfrac{3\pi}{4}, \dfrac{5\pi}{4}, \dfrac{7\pi}{4}, \dfrac{9\pi}{4}, \dfrac{11\pi}{4}$ **13.** $\dfrac{\pi}{6}, \dfrac{5\pi}{6}, \dfrac{13\pi}{6}, \dfrac{17\pi}{6}, \dfrac{\pi}{2}, \dfrac{5\pi}{2}$

15. $-2\pi, -\pi, 0, \pi, 2\pi, -\dfrac{5\pi}{3}, -\dfrac{\pi}{3}, \dfrac{\pi}{3}, \dfrac{5\pi}{3}$ **17.** $-\dfrac{3\pi}{2}, -\dfrac{5\pi}{6}, -\dfrac{\pi}{6}, \dfrac{\pi}{2}$ **19.** $0, 2\pi$

21. $-\dfrac{\pi}{3}, \dfrac{\pi}{3}$

23.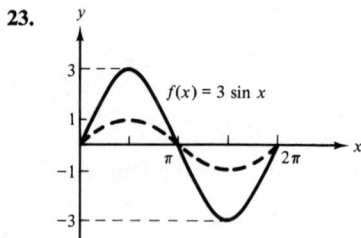

25.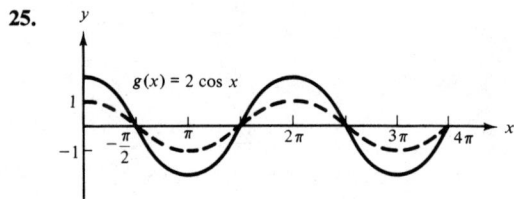

27.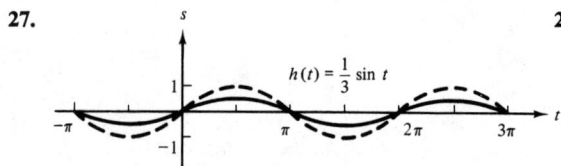

29.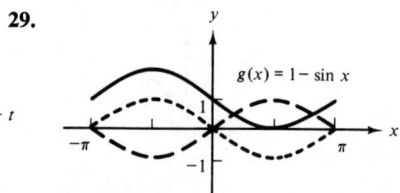

31.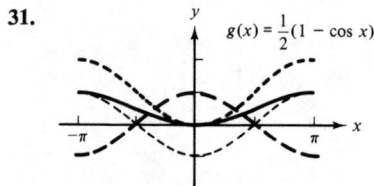

33.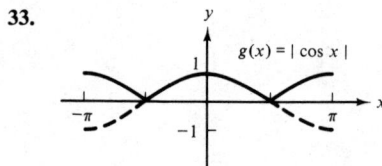

1.

x	$-\dfrac{\pi}{2}$	$-\dfrac{\pi}{3}$	$-\dfrac{\pi}{4}$	$-\dfrac{\pi}{6}$	0	$\dfrac{\pi}{6}$	$\dfrac{\pi}{4}$	$\dfrac{\pi}{3}$	$\dfrac{\pi}{2}$	$\dfrac{2\pi}{3}$	$\dfrac{3\pi}{4}$	$\dfrac{5\pi}{6}$	π
$\sin x$	-1	$-\dfrac{\sqrt{3}}{2}$	$-\dfrac{1}{\sqrt{2}}$	$-\dfrac{1}{2}$	0	$\dfrac{1}{2}$	$\dfrac{1}{\sqrt{2}}$	$\dfrac{\sqrt{3}}{2}$	1	$\dfrac{\sqrt{3}}{2}$	$\dfrac{1}{\sqrt{2}}$	$\dfrac{1}{2}$	0
$\cos x$	0	$\dfrac{1}{2}$	$\dfrac{1}{\sqrt{2}}$	$\dfrac{\sqrt{3}}{2}$	1	$\dfrac{\sqrt{3}}{2}$	$\dfrac{1}{\sqrt{2}}$	$\dfrac{1}{2}$	0	$-\dfrac{1}{2}$	$-\dfrac{1}{\sqrt{2}}$	$-\dfrac{\sqrt{3}}{2}$	-1
$\tan x$	$*$	$-\sqrt{3}$	-1	$-\dfrac{1}{\sqrt{3}}$	0	$\dfrac{1}{\sqrt{3}}$	1	$\sqrt{3}$	$*$	$-\sqrt{3}$	-1	$-\dfrac{1}{\sqrt{3}}$	0
$\cot x$	0	$-\dfrac{1}{\sqrt{3}}$	-1	$-\sqrt{3}$	$*$	$\sqrt{3}$	1	$\dfrac{1}{\sqrt{3}}$	0	$-\dfrac{1}{\sqrt{3}}$	-1	$-\sqrt{3}$	$*$
$\sec x$	$*$	2	$\sqrt{2}$	$\dfrac{2}{\sqrt{3}}$	1	$\dfrac{2}{\sqrt{3}}$	$\sqrt{2}$	2	$*$	-2	$-\sqrt{2}$	$-\dfrac{2}{\sqrt{3}}$	-1
$\csc x$	-1	$-\dfrac{2}{\sqrt{3}}$	$-\sqrt{2}$	-2	$*$	2	$\sqrt{2}$	$\dfrac{2}{\sqrt{3}}$	1	$\dfrac{2}{\sqrt{3}}$	$\sqrt{2}$	2	$*$

x	$\dfrac{7\pi}{6}$	$\dfrac{5\pi}{4}$	$\dfrac{4\pi}{3}$	$\dfrac{3\pi}{2}$	$\dfrac{5\pi}{3}$	$\dfrac{7\pi}{4}$	$\dfrac{11\pi}{6}$	2π
$\sin x$	$-\dfrac{1}{2}$	$-\dfrac{1}{\sqrt{2}}$	$-\dfrac{\sqrt{3}}{2}$	-1	$-\dfrac{\sqrt{3}}{2}$	$-\dfrac{1}{\sqrt{2}}$	$-\dfrac{1}{2}$	0
$\cos x$	$-\dfrac{\sqrt{3}}{2}$	$-\dfrac{1}{\sqrt{2}}$	$-\dfrac{1}{2}$	0	$\dfrac{1}{2}$	$\dfrac{1}{\sqrt{2}}$	$\dfrac{\sqrt{3}}{2}$	1
$\tan x$	$\dfrac{1}{\sqrt{3}}$	1	$\sqrt{3}$	$*$	$-\sqrt{3}$	-1	$-\dfrac{1}{\sqrt{3}}$	0
$\cot x$	$\sqrt{3}$	1	$\dfrac{1}{\sqrt{3}}$	0	$-\dfrac{1}{\sqrt{3}}$	-1	$-\sqrt{3}$	$*$
$\sec x$	$-\dfrac{2}{\sqrt{3}}$	$-\sqrt{2}$	-2	$*$	2	$\sqrt{2}$	$\dfrac{2}{\sqrt{3}}$	1
$\csc x$	-2	$-\sqrt{2}$	$-\dfrac{2}{\sqrt{3}}$	-1	$-\dfrac{2}{\sqrt{3}}$	$-\sqrt{2}$	-2	$*$

*means undefined

3.

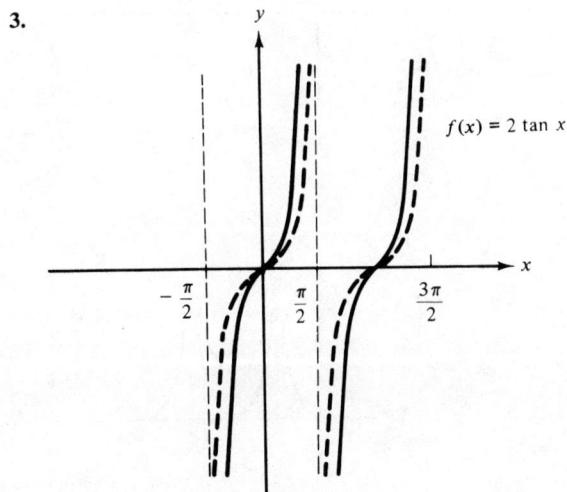

$f(x) = 2 \tan x$

5.

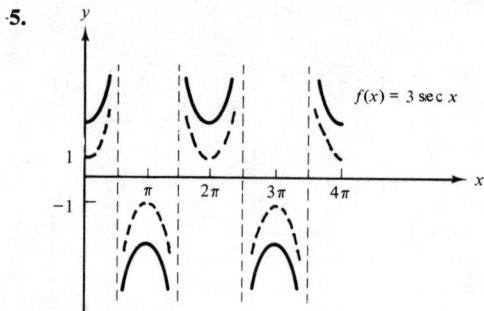

$f(x) = 3 \sec x$

7.

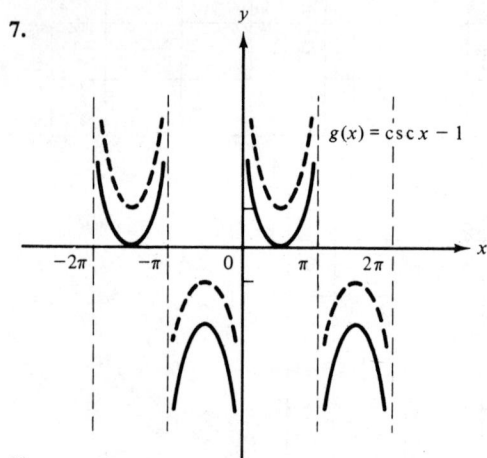

$g(x) = \csc x - 1$

9.

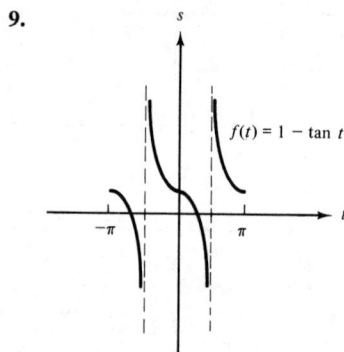

$f(t) = 1 - \tan t$

11.

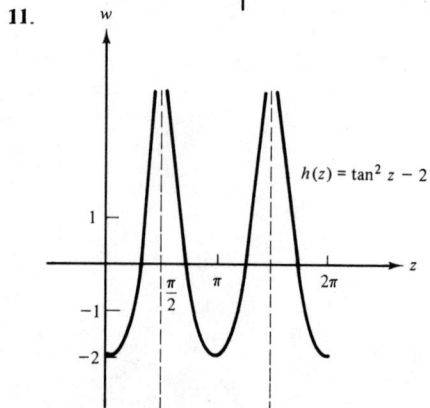

$h(z) = \tan^2 z - 2$

13.

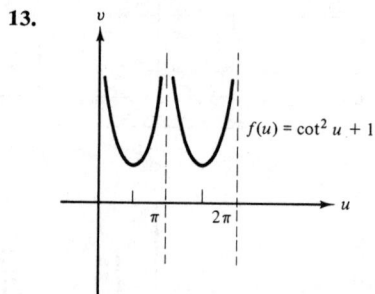

$f(u) = \cot^2 u + 1$

15.

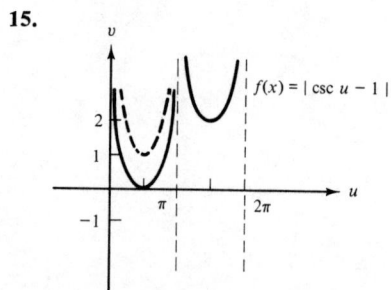

$f(x) = |\csc u - 1|$

17.

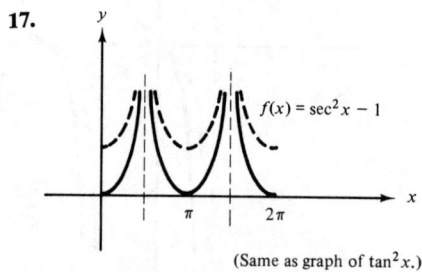

$f(x) = \sec^2 x - 1$

(Same as graph of $\tan^2 x$.)

ANSWERS TO EXERCISES 9-4

23. $\dfrac{\sqrt{2}}{4}(\sqrt{3}-1)$ **25.** $\dfrac{\sqrt{2}}{4}(\sqrt{3}-1)$ **27.** $2-\sqrt{3}$ **29.** $\sqrt{2}(\sqrt{3}-1)$

31. $\dfrac{4\sqrt{3}-3}{10}$ **33.** $\dfrac{7\sqrt{2}}{10}$ **35.** $\dfrac{-5-12\sqrt{3}}{26}$ **37.** $\dfrac{7}{25}$ **39.** $-\dfrac{119}{169}$ **41.** $-\dfrac{24}{7}$

ANSWERS TO EXERCISES 9-5

	Period	Maximum	Minimum
1.	$\dfrac{2\pi}{3}$	$\left(\dfrac{\pi}{6}, 4\right)$	$\left(\dfrac{\pi}{2}, -4\right)$
3.	$\dfrac{2\pi}{3}$	$\left(\dfrac{\pi}{6}, \dfrac{1}{2}\right)$	$\left(\dfrac{\pi}{2}, -\dfrac{1}{2}\right)$
5.	4π	$\left(3\pi, \dfrac{1}{3}\right)$	$\left(\pi, -\dfrac{1}{3}\right)$
7.	π	$(0, 3)$	$\left(\dfrac{\pi}{2}, -3\right)$
9.	$\dfrac{\pi}{3}$	$\left(\dfrac{\pi}{12}, 3\right)$	$\left(-\dfrac{\pi}{12}, -3\right)$
11.	4π	$\left(\dfrac{2\pi}{3}, \dfrac{2}{3}\right)$	$\left(\dfrac{8\pi}{3}, -\dfrac{2}{3}\right)$
13.	π	$\left(\dfrac{\pi}{4}, 1\right)$	$\left(\dfrac{3\pi}{4}, -5\right)$
15.	π	$(0, 1)$	$\left(\dfrac{\pi}{2}, -3\right)$

Note: In Problems 1–15 the first coordinates of the maximum and minimum points are not unique, but may differ by integral multiples of the period.

21.

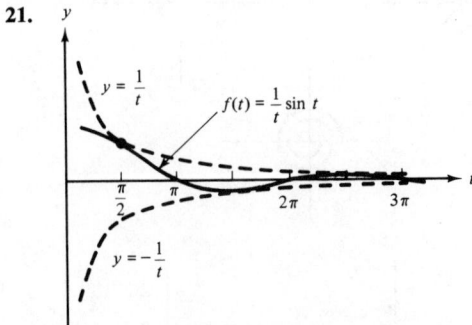

ANSWERS TO REVIEW EXERCISES FOR CHAPTER NINE

1.

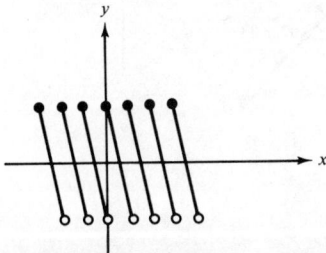

$f(x) = -4(x - n) + 2$ for $n \le x < n + 1$ (n is an integer)

3. $-\dfrac{\sqrt{5}}{3}; \ -\dfrac{2}{\sqrt{5}}; \ -\dfrac{3}{\sqrt{5}}$

5. (a) even (b) odd (c) odd (d) even (e) odd (f) even (g) even (h) odd (i) odd (j) even

9. (a)

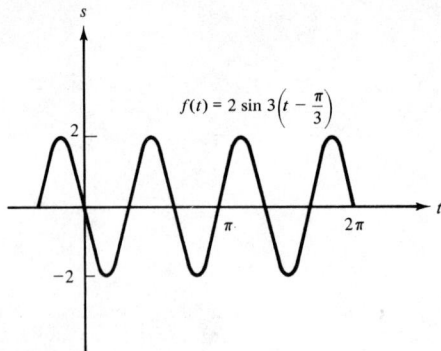

$$f(t) = 2 \sin 3\left(t - \frac{\pi}{3}\right)$$

amplitude 2; period $\frac{2\pi}{3}$; maximum $\left(\frac{\pi}{2}, 3\right)$, minimum $\left(\frac{5\pi}{6}, -3\right)$

(b) period $\frac{1}{60}$; maximum 110 occurs at $t = \frac{1}{240} + \frac{n}{60}$ were n is an integer

ANSWERS TO EXERCISES 10-1

1.

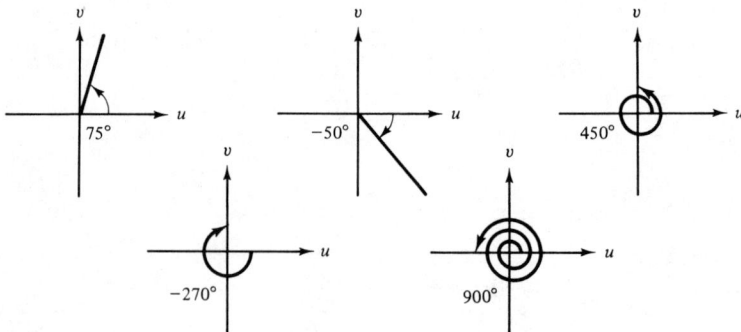

3. (a)

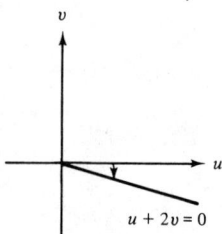

(b)

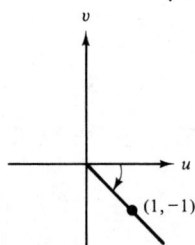

(c)

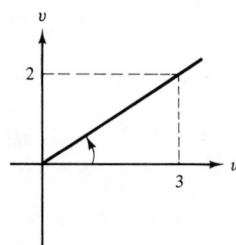

(d)

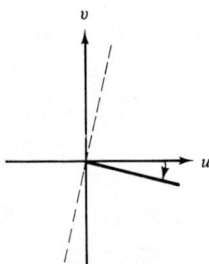

5. $\dfrac{5\pi}{6}$ **7.** $\dfrac{25\pi}{6}$ **9.** $-\dfrac{5\pi}{18}$ **11.** -5π **13.** $240°$ **15.** $1260°$ **17.** $\dfrac{270°}{\pi}$ **19.** $\dfrac{1440°}{\pi}$

21. 207π or $37260°$ **23.** $\dfrac{350\pi}{3}$ or $21000°$

25.

	(a)	(b)	(c)	(d)	(e)	(f)
sin	$\dfrac{2}{\sqrt{5}}$	$\dfrac{3}{\sqrt{10}}$	$-\dfrac{8}{\sqrt{73}}$	0	$\dfrac{1}{\sqrt{5}}$	$-\dfrac{2}{\sqrt{5}}$
cos	$\dfrac{1}{\sqrt{5}}$	$-\dfrac{1}{\sqrt{10}}$	$\dfrac{3}{\sqrt{73}}$	1	$\dfrac{2}{\sqrt{5}}$	$\dfrac{1}{\sqrt{5}}$
tan	2	-3	$-\dfrac{8}{3}$	0	$\dfrac{1}{2}$	-2
sec	$\sqrt{5}$	$-\sqrt{10}$	$\dfrac{\sqrt{73}}{3}$	1	$\dfrac{\sqrt{5}}{2}$	$\sqrt{5}$

27.

	$\dfrac{\pi}{6}$	$\dfrac{\pi}{3}$
sin	$\dfrac{1}{2}$	$\dfrac{\sqrt{3}}{2}$
cos	$\dfrac{\sqrt{3}}{2}$	$\dfrac{1}{2}$
tan	$\dfrac{1}{\sqrt{3}}$	$\sqrt{3}$
cot	$\sqrt{3}$	$\dfrac{1}{\sqrt{3}}$
sec	$\dfrac{2}{\sqrt{3}}$	2
csc	2	$\dfrac{2}{\sqrt{3}}$

ANSWERS TO EXERCISES 10-2 (Note: In many case we have written $=$ instead of $\approx$.)

1. .5878   **3.** .3090   **5.** .7002   **7.** 2.924   **9.** .4277   **11.** .7265   **13.** .8415
15. 1.008   **17.** $\sin 52° = .7880$ **19.** $\tan 63° = 1.963$ **21.** $-\cos 63° = -.4540$
23. $\tan 42° = .9004$ **25.** $-\csc 20° = -2.924$ **27.** $-\sin \dfrac{2\pi}{5} = -.9551$
29. $-\cos(4.132 - \pi) = -.5483$ **31.** $\dfrac{\pi}{6}$ **33.** $\dfrac{11\pi}{6}$ **35.** .5760   **37.** 3.805
39. 1.020   **41.** 2.149   **43.** 2.478   **45.** 2.234   **47.** $53.13°$ and $36.87°$

ANSWERS TO EXERCISES 10-3 (Note: In many cases we have written $=$ instead of $\approx$.)

1. $\beta = 60°, b = 8.66, c = 10$ **3.** $\beta = 40°, a = 2.30, b = 1.93$
5. $\alpha = 30°, a = \dfrac{4\sqrt{3}}{3}, c = \dfrac{8\sqrt{3}}{3}$ **7.** $\beta = \dfrac{\pi}{4}, b = 4, c = 4\sqrt{2}$
9. $\beta = \dfrac{\pi}{3}, a = \dfrac{4\sqrt{3}}{3}, c = \dfrac{8\sqrt{3}}{3}$ **11.** 14.43 meters   **13.** 83.91 yards   **15.** .47 meters
17. $a = 22.12, \beta = 53.44°, \gamma = 45.56°$ **19.** $b = 9.74, \alpha = 112.48°, \gamma = 27.52°$
21. $c = 5.42, \alpha = .6052, \beta = 1.654$ **23.** $\alpha = 21.78°, \beta = 38.21°, \gamma = 120°$
25. $\alpha = 36.87°, \beta = 53.13°, \gamma = 90°$ **27.** $\alpha = 97°, b = 7.11, c = 8.84$
29. $\beta = 119°, a = 7.39, c = 6.53$ **31.** $\gamma = 1.218, a = 17.88, b = 12.71$

ANSWERS TO EXERCISES 10-4

1. $\dfrac{\pi}{3}$ **3.** $\dfrac{\pi}{6}$ **5.** $\dfrac{\pi}{3}$ **7.** $-\dfrac{\pi}{2}$ **9.** .8203   **11.** $-.4363$ **13.** $\dfrac{4}{5}$ **15.** $\dfrac{3}{4}$
17. $\dfrac{3}{5}$ **19.** $-\dfrac{4}{5}$ **21.** $-\dfrac{4}{3}$ **23.** $\dfrac{3\sqrt{3}+4}{10}$ **25.** (a) $\text{Arcsin } \dfrac{h}{100}$ (b) $\text{Arccos } \dfrac{d}{100}$
27. .088 minute; .099 minute; .109 minute

ANSWERS TO EXERCISES 10-5

1.

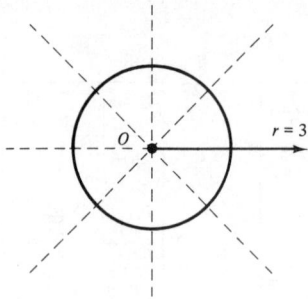

$r = 3$

3.

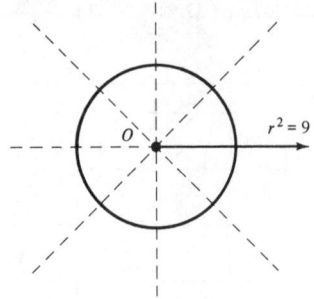

$r^2 = 9$

5.

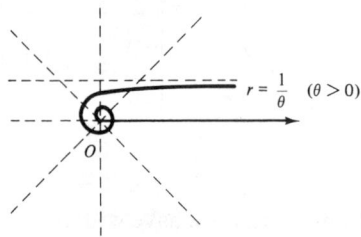

$r = \dfrac{1}{\theta}$ $(\theta > 0)$

7.

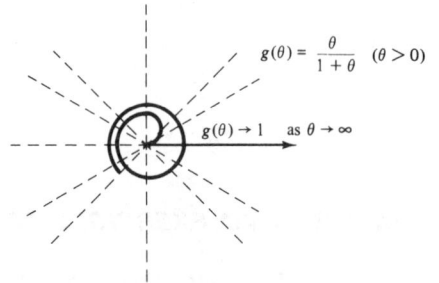

$g(\theta) = \dfrac{\theta}{1 + \theta}$ $(\theta > 0)$

$g(\theta) \to 1$ as $\theta \to \infty$

9.

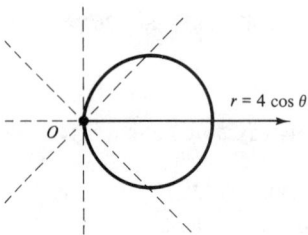

$r = 4 \cos \theta$

11.

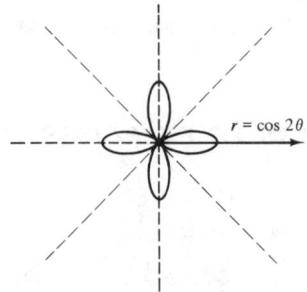

$r = \cos 2\theta$

13.

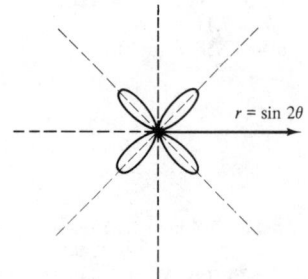

$r = \sin 2\theta$

15.

$r = 2 \sin 4\theta$

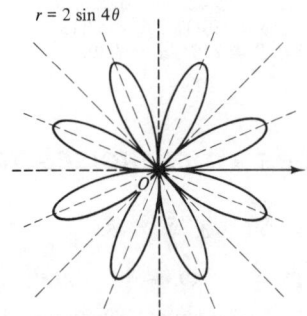

17.

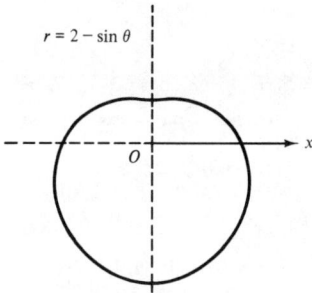

$r = 2 - \sin \theta$

19.

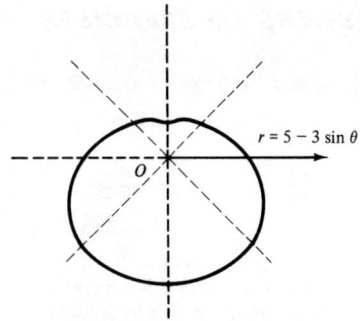

$r = 5 - 3 \sin \theta$

21.

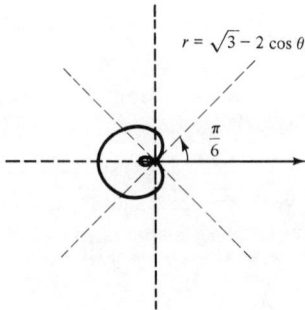

$r = \sqrt{3} - 2 \cos \theta$

$\dfrac{\pi}{6}$

23.

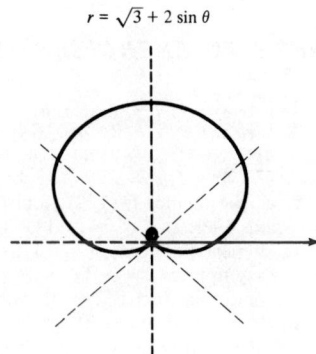

$r = \sqrt{3} + 2 \sin \theta$

25. $(\sqrt{3}, 1)$ **27.** $(-1, -\sqrt{3})$ **29.** $\left(\dfrac{\sqrt{2}}{2}, -\dfrac{\sqrt{2}}{2}\right)$ **31.** $(0, 0)$ **33.** $\left(2\sqrt{2}, \dfrac{\pi}{4}\right)$

35. $\left(2, \dfrac{2\pi}{3}\right)$ **37.** $\left(2, \dfrac{11\pi}{6}\right)$ **39.** $x^2 + y^2 = 3y$; circle **41.** $y = 4$; line

43. $(x^2 + y^2 + 2y)^2 = x^2 + y^2$; limaçon **45.** $x^2 = 8y + 16$; parabola **47.** $r^2 = 9$

49. $r^2 = 9 \sin 2\theta$

ANSWERS TO REVIEW EXERCISES FOR CHAPTER TEN

1. (a) $\dfrac{3\pi}{4}; \dfrac{\pi}{4}; \dfrac{\sqrt{2}}{2}; -\dfrac{\sqrt{2}}{2}$ (b) $\dfrac{7\pi}{6}; \dfrac{\pi}{6}; -\dfrac{1}{2}; -\dfrac{\sqrt{3}}{2}$

(c) $\dfrac{5\pi}{12}; 75°; .9659; .2588$ (d) $\dfrac{19\pi}{12}; 75°; -.9659; .2588$

3. (a) $b = \dfrac{5\sqrt{3}}{3}$ $c = \dfrac{10\sqrt{3}}{3}$ $\beta = 30°$

(b) $a = 4\sqrt{3}$ $c = 8\sqrt{3}$ $\beta = \dfrac{\pi}{3}$

(c) $a = 1.15$ $b = 2.77$ $\beta = \dfrac{3\pi}{8}$

(d) $b = 41.85$ $c = 43.54$ $\beta = 74°$

5. 32.14 feet

9. (a) $-\dfrac{\pi}{3}$ (b) $\dfrac{3\pi}{4}$ (c) $-\dfrac{\pi}{4}$ (d) .8901 (e) 1.257 (f) $\dfrac{5\sqrt{11}}{11}$ (g) $-\dfrac{3}{5}$ (h) $\dfrac{\pi}{6}$

11. $|a| < |b|$

ANSWERS TO EXERCISES 11-1

1. $F(4, 0)$; $x = -4$ **3.** $F\left(\frac{1}{16}, 0\right)$; $x = -\frac{1}{16}$ **5.** $F(0, -3)$; $y = 3$ **7.** $F\left(0, -\frac{1}{4}\right)$; $y = \frac{1}{4}$

9. $y^2 = 2x$ **11.** $x^2 = -8y$ **13.** $x^2 = 12y$ **15.** $x^2 - 4x - 4y - 4 = 0$

17. $x^2 - 10y - 25 = 0$ **19.** $(\pm 2, 0)$ $(\pm 2\sqrt{2}, 0)$ $(0, \pm 2)$

21. $(\pm\sqrt{3}, 0)$ $(\pm\sqrt{5}, 0)$ $(0, \pm\sqrt{2})$ **23.** $(0, \pm 1)$ $(0, \pm 3)$ $(\pm 2\sqrt{2}, 0)$ **25.** $4x^2 + y^2 = 1$

27. $9x^2 + 25y^2 = 225$ **29.** $x^2 + 16y^2 = 144$ **31.** $(\pm 1, 0)$ $(\pm\sqrt{2}, 0)$ $y = \pm x$

33. $(0, \pm 1)$ $(0, \pm\sqrt{2})$ $y = \pm x$ **35.** $\left(0, \pm\frac{1}{3}\right)$ $\left(0, \pm\frac{\sqrt{10}}{3}\right)$ $y = \pm\frac{1}{3}x$ **37.** $x^2 - y^2 = 2$

39. $16y^2 - 9x^2 = 144$ **41.** $16y^2 - 9x^2 = 288$ **43.** hyperbola, transverse axis vertical

45. two intersecting lines   **47.** ellipse, major axis vertical   **49.** parabola, opens upward

51. union of two hyperbolas   **53.** $(-2, \sqrt{3}), (-2, -\sqrt{3})$

55. $(\sqrt{5}, 2), (\sqrt{5}, -2), (-\sqrt{5}, 2), (-\sqrt{5}, -2)$ **59.** $\frac{2b^2}{a}$ **61.** $a = b$

ANSWERS TO EXERCISES 11-2

1. $u = x - 1$, $v = y + 3$; $u^2 + v^2 = 4$; circle

3. $u = x$, $v = y - 2$; $9u^2 + 4v^2 = 36$; ellipse

5. $u = x - 1$, $v = y$; $u^2 + v^2 = 25$; circle

7. $u = x + 2$, $v = y$; $9u^2 - 4v^2 = 36$; hyperbola

9. ellipse; center $(-1, 2)$; foci $(-2, 2)$ and $(0, 2)$; vertices $(-3, 2)$ and $(1, 2)$; intercepts on minor axis $(-1, 2 - \sqrt{3})$ and $(-1, 2 + \sqrt{3})$.

11. hyperbola; center $(2, -1)$; foci $(2 - \sqrt{5}, -1), (2 + \sqrt{5}, -1)$; vertices $(0, -1), (4, -1)$; asymptotes $2(y + 1) = \pm(x - 2)$.

13. parabola; focus $(2, -3)$; vertex $(3, -3)$; directrix $x = 4$.

15. hyperbola; center $(0, 2)$; foci $(0, -3), (0, 7)$; vertices $(0, -1), (0, 5)$; asymptotes $4(y - 2) = \pm 3x$.

17. $16(x - 2)^2 + 25y^2 = 400$ **19.** $9(x - 1)^2 + 5(y + 2)^2 = 45$ **21.** $(x - 3)^2 = 12(y - 1)$

23. $3(x + 2)^2 = -4(y - 3)$ **25.** $(x + 2)^2 - (y - 1)^2 = 4$

ANSWERS TO EXERCISES 11-3

1. (a) $x = \frac{1}{2}(\sqrt{3}\,u - v)$ $u = \frac{1}{2}(\sqrt{3}\,x + y)$

$y = \frac{1}{2}(u + \sqrt{3}\,v)$ $v = \frac{1}{2}(-x + \sqrt{3}\,y)$

(b) $\left(\frac{3 + \sqrt{3}}{2}, \frac{3\sqrt{3} - 1}{2}\right)$ (c) $\left(\frac{2\sqrt{3} + 1}{2}, \frac{2 - \sqrt{3}}{2}\right)$

(d) $\frac{\sqrt{3} + 1}{4}u^2 + \frac{\sqrt{3} + 1}{2}uv + \frac{3 - \sqrt{3}}{4}v^2 = 2$

3. (a) $x = \frac{1}{13}(5u - 12v)$ $u = \frac{1}{13}(5x + 12y)$

$y = \frac{1}{13}(12u + 5v)$ $v = \frac{1}{13}(-12x + 5y)$

(b) $\left(\frac{41}{13}, \frac{3}{13}\right)$ (c) $\left(\frac{22}{13}, \frac{19}{13}\right)$

(d) $204u^2 + uv - 35v^2 = 338$

5. $3u^2 + 5v^2 = 30$; ellipse; vertices $(-\sqrt{5}, -\sqrt{5}), (\sqrt{5}, \sqrt{5})$; foci $(-\sqrt{2}, -\sqrt{2}), (\sqrt{2}, \sqrt{2})$; major axis $y = x$; minor axis $y = -x$.

7. $5u^2 - v^2 = 20$; hyperbola; vertices $(-\sqrt{2}, -\sqrt{2}), (\sqrt{2}, \sqrt{2})$; foci $(-2\sqrt{3}, -2\sqrt{3}), (2\sqrt{3}, 2\sqrt{3})$; transverse axis $y = x$; asymptotes $y = \frac{1 + \sqrt{5}}{1 - \sqrt{5}}x, y = \frac{1 - \sqrt{5}}{1 + \sqrt{5}}x$

9. $\left(v + \frac{1}{2}\right)^2 = u + \frac{1}{4}$; parabola; vertex $\left(\frac{1}{8}\sqrt{2}, -\frac{3}{8}\sqrt{2}\right)$; focus $\left(\frac{1}{4}\sqrt{2}, -\frac{1}{4}\sqrt{2}\right)$; axis $x - y = \frac{1}{2}\sqrt{2}$

11. $\dfrac{\pi}{6}$; $u^2 - v^2 = 4$ **13.** $\dfrac{\pi}{4}$; $u^2 = 2$ **15.** $\text{Arcsin}\ \dfrac{4}{5}$; $120v^2 - 5u^2 + 12u + 9v = 55$

17. $\dfrac{1}{2}\ \text{Arctan}\ \dfrac{4}{3}$; $5v^2 + \dfrac{4\sqrt{5}}{5}u - \dfrac{2\sqrt{5}}{5}v = 0$

19. $\alpha = \dfrac{\pi}{6}$; $\dfrac{u^2}{2} + \dfrac{v^2}{10} = 1$; ellipse

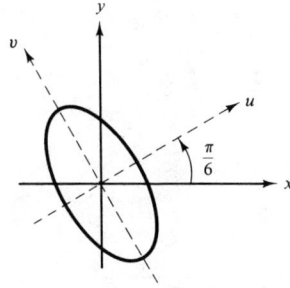

21. $\alpha = \text{Arcsin}\ \dfrac{4}{5}$; $\dfrac{v^2}{4} - \dfrac{u^2}{16} = 1$; hyperbola

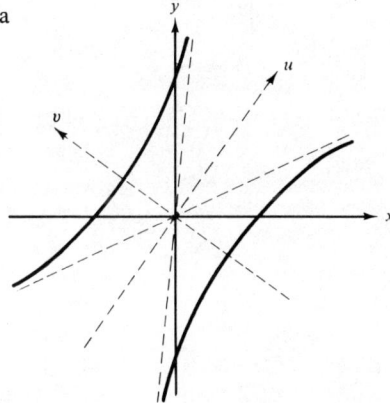

23. $\alpha = \text{Arcsin}\ \dfrac{\sqrt{5}}{5}$; $u^2 = -2\sqrt{5}\,v$; parabola

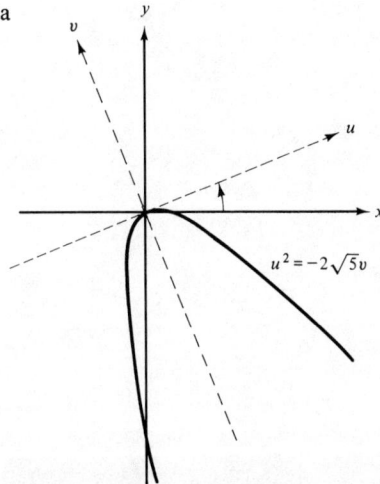

$u^2 = -2\sqrt{5}v$

25. ellipse   **27.** parabola   **29.** parabola   **31.** single point $(-1, -2)$ **33.** hyperbola
35. two parallel lines   **37.** two lines intersecting at origin

3. (a) ellipse (b) hyperbola (c) parabola (d) ellipse (e) hyperbola

5. (a) $x = \frac{1}{2}u - \frac{\sqrt{3}}{2}v$ $\qquad\qquad\qquad u = \frac{1}{2}x + \frac{\sqrt{3}}{2}y$

$\qquad y = \frac{\sqrt{3}}{2}u + \frac{1}{2}v$ $\qquad\qquad\qquad v = -\frac{\sqrt{3}}{2}x + \frac{1}{2}y$

(b) $5u^2 + v^2 = 2$

(c)

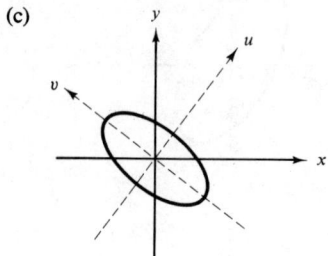

(d) $y = -\frac{\sqrt{3}}{3}x$; $\left(-\frac{\sqrt{30}}{5}, \frac{\sqrt{10}}{5}\right)$, $\left(\frac{\sqrt{30}}{5}, -\frac{\sqrt{10}}{5}\right)$

Index

SUMMARY OF TRIGONOMETRIC IDENTITIES

BASIC PROPERTIES

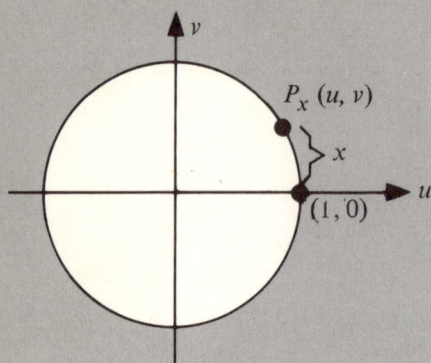

$P_x\,(u, v)$

x

$(1, 0)$

$u = \cos x$
$v = \sin x$

$$\tan x = \frac{\sin x}{\cos x} \qquad \cot x = \frac{\cos x}{\sin x}$$

$$\sec x = \frac{1}{\cos x} \qquad \csc x = \frac{1}{\sin x}$$

CIRCULAR PROPERTIES

$$\cos^2 x + \sin^2 x = 1$$
$$1 + \tan^2 x = \sec^2 x$$
$$\cot^2 x + 1 \quad = \csc^2 x$$

PERIODICITY PROPERTIES

$$\sin (x + 2\pi) = \sin x$$
$$\cos (x + 2\pi) = \cos x$$
$$\tan (x + \pi) \ = \tan x$$

SYMMETRY PROPERTIES

$$\sin (-x) = -\sin x$$
$$\cos (-x) = \cos x$$

$$\sin (\pi - x) = \sin x$$
$$\cos (\pi - x) = -\cos x$$

$$\sin \left(\frac{\pi}{2} - x\right) = \cos x$$
$$\cos \left(\frac{\pi}{2} - x\right) = \sin x$$

$$\sin (x + \pi) = -\sin x$$
$$\cos (x + \pi) = -\cos x$$

SUM AND DIFFERENCE FORMULAS

$$\cos (x - y) = \cos x \cos y + \sin x \sin y$$
$$\cos (x + y) = \cos x \cos y - \sin x \sin y \qquad \cos 2x = \cos^2 x - \sin^2 x$$
$$\sin (x - y) = \sin x \cos y - \cos x \sin y$$
$$\sin (x + y) = \sin x \cos y + \cos x \sin y \qquad \sin 2x = 2 \sin x \cos x$$

LAW OF COSINES

$$a^2 = b^2 + c^2 - 2bc \cos \alpha$$
$$b^2 = a^2 + c^2 - 2ac \cos \beta$$
$$c^2 = a^2 + b^2 - 2ab \cos \gamma$$

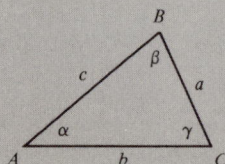

LAW OF SINES

$$\frac{a}{\sin \alpha} = \frac{b}{\sin \beta} = \frac{c}{\sin \gamma}$$